Lecture Notes in Computer Science **14366**

Founding Editors

Gerhard Goos
Juris Hartmanis

The series Lecture Notes in Computer Science (LNCS), including its subseries Lecture Notes in Artificial Intelligence (LNAI) and Lecture Notes in Bioinformatics (LNBI), has established itself as a medium for the publication of new developments in computer science and information technology research, teaching, and education.

LNCS enjoys close cooperation with the computer science R & D community, the series counts many renowned academics among its volume editors and paper authors, and collaborates with prestigious societies. Its mission is to serve this international community by providing an invaluable service, mainly focused on the publication of conference and workshop proceedings and postproceedings. LNCS commenced publication in 1973.

Gian Luca Foresti · Andrea Fusiello ·
Edwin Hancock
Editors

Image Analysis and Processing - ICIAP 2023 Workshops

Udine, Italy, September 11–15, 2023
Proceedings, Part II

Editors
Gian Luca Foresti
University of Udine
Udine, Italy

Andrea Fusiello
University of Udine
Udine, Italy

Edwin Hancock
University of York
York, UK

ISSN 0302-9743 ISSN 1611-3349 (electronic)
Lecture Notes in Computer Science
ISBN 978-3-031-51025-0 ISBN 978-3-031-51026-7 (eBook)
https://doi.org/10.1007/978-3-031-51026-7

This Springer imprint is published by the registered company Springer Nature Switzerland AG
The registered company address is: Gewerbestrasse 11, 6330 Cham, Switzerland

Paper in this product is recyclable.

Preface

This volume contains the papers accepted for presentation at the workshops hosted by the 22nd International Conference on Image Analysis and Processing (ICIAP 2023), held in Udine, Italy, from 11 to 15 September 2023. It was co-organised by the Department of Informatics, Mathematics and Physics (DMIF) and the Polytechnic Department of Engineering and Architecture (DPIA) of the University of Udine and sponsored by ST Microelectronics. The conference traditionally covers topics related to theoretical and experimental areas of Computer Vision, Image Processing, Pattern Recognition and Machine Learning, with emphasis on theoretical aspects and applications.

ICIAP 2023 also hosted 15 workshops (including 2 competitions), on topics of great relevance with respect to the state of the art. For the first time, workshops in the same field were collected in two hubs, the Medical Imaging Hub (MIH) and the Digital Humanities Hub (DHH). In addition, an industrial poster session was organised to bring together papers written by scientists working in industry and with a strong focus on application.

In total, 72 workshop papers and 10 industrial poster session papers were accepted for publication. As for the main conference, papers were selected through a double-blind peer review process, considering originality, significance, clarity, soundness, relevance and technical content. Each submission was assigned to at least three reviewers.

This volume contains 41 papers from the following workshops:

- Medical Imaging Hub:
  - Artificial Intelligence and Radiomics in Computer-Aided Diagnosis (AIRCAD)
  - Multi-Modal Medical Imaging Processing (M3IP)
  - Federated Learning in Medical Imaging and Vision (FedMed)
- Digital Humanities Hub:
  - Artificial Intelligence for Digital Humanities (AI4DH)
  - Fine Art Pattern Extraction and Recognition (FAPER)
  - Pattern Recognition for Cultural Heritage (PatReCH)
  - Visual Processing of Digital Manuscripts: Workflows, Pipelines, Best Practices (ViDiScript)

The 31 papers accepted for the other workshops, as well as 10 papers for the industrial poster session, are included in the companion volume (LNCS 14365).

AIRCAD: Artificial Intelligence and Radiomics in Computer-Aided Diagnosis, organized by Albert Comelli (Ri.MED Foundation, Italy), Cecilia Di Ruberto (University of Cagliari, Italy), Andrea Loddo (University of Cagliari, Italy), Lorenzo Putzu (University of Cagliari, Italy), and Alessandro Stefano (IBFM-CNR, Italy), provided an overview of recent advances in the field of biomedical image processing using machine learning, deep learning, artificial intelligence, and radiomics features with particular focus on practical applications.

M3IP: Multi-Modal Medical Imaging Processing, organized by Angel Garcia-Pedrero (Universidad Politecnica de Madrid, Spain), Michela Gravina (University of Naples Federico II, Italy), and Stefano Marrone (University of Naples Federico II, Italy), addressed the integration of heterogeneous sources of data for medical imaging applications with multimodal learning.

FedMed: Federated Learning in Medical Imaging and Vision, organized by Simone Palazzo (University of Catania, Italy), Federica Proietto Salanitri (University of Catania, Italy), Matteo Pennisi (University of Catania, Italy), Chen Chen (University of Central Florida, USA), Bruno Casella (University of Turin, Italy), and Gianluca Mittone (University of Turin, Italy), aimed at bringing together researchers and practitioners with common interest in federated learning for visual tasks, with a particular focus on medical imaging.

AI4DH: Artificial Intelligence for Digital Humanities, organized by Lorenzo Baraldi (University of Modena and Reggio Emilia, Italy), Silvia Cascianelli (University of Modena and Reggio Emilia, Italy), Marcella Cornia (University of Modena and Reggio Emilia, Italy), Francesca Matrone (Politecnico di Torino, Italy), Marina Paolanti (University of Macerata, Italy), and Roberto Pierdicca (Università Politecnica delle Marche, Italy), provided an overview on the usage of artificial intelligence in digital humanities, a new field at the intersection between social science and computer science.

FAPER: Fine Art Pattern Extraction and Recognition, organized by Gennaro Vessio (University of Bari, Italy), Giovanna Castellano (University of Bari, Italy), Fabio Bellavia (University of Palermo, Italy), Sinem Aslan (University of Venice, Italy), Eva Cetinic (University of Zurich, Switzerland), and Lucia Cipolina Kun (University of Bristol, UK), focused on cultural heritage with particular focus on understanding fine arts with automatic tools from computer vision.

PatReCH: Pattern Recognition for Cultural Heritage, organized by Dario Allegra (University of Catania, Italy), Mario Molinara (University of Cassino and Southern Lazio, Italy), Alessandra Scotto di Freca (University of Cassino and Southern Lazio, Italy), and Filippo Stanco (University of Catania, Italy), presented recent advances in pattern recognition techniques for data analysis and representation in the cultural heritage field.

ViDiScript: Visual Processing of Digital Manuscripts, Workflows, Pipelines, Best Practices, organized by Matteo Al Kalak (University of Modena and Reggio Emilia, Italy), Lorenzo Baraldi (University of Modena and Reggio Emilia, Italy), Andrea Brunello (University of Udine, Italy), Emanuela Colombi (University of Udine, Italy), Alessio Fagioli (Sapienza University of Rome, Italy), and Gian Luca Foresti (University of Udine, Italy), aimed at collecting and connecting the most recent applications of computer vision to historical handwritten document processing in its broader sense.

The success of ICIAP 2023 was due to the contribution of many people. Special thanks go to the Workshop Chairs (Federica Arrigoni and Lauro Snidaro) and we thank all the workshop organizers who made such an interesting program possible. We hope that you will find the papers in this volume interesting and informative, and that they will inspire you to further research in the field of image analysis and processing.

September 2023

Gian Luca Foresti
Andrea Fusiello
Edwin Hancock

Organization

General Chairs

Gian Luca Foresti	University of Udine, Italy
Andrea Fusiello	University of Udine, Italy
Edwin Hancock	University of York, UK

Program Chairs

Michael Bronstein	University of Oxford, UK
Barbara Caputo	Politecnico Torino, Italy
Giuseppe Serra	University of Udine, Italy

Steering Committee

Virginio Cantoni	University of Pavia, Italy
Luigi Pietro Cordella	University of Napoli Federico II, Italy
Rita Cucchiara	University of Modena-Reggio Emilia, Italy
Alberto Del Bimbo	University of Firenze, Italy
Marco Ferretti	University of Pavia, Italy
Gian Luca Foresti	University of Udine, Italy
Fabio Roli	University of Cagliari, Italy
Gabriella Sanniti di Baja	ICAR-CNR, Italy

Workshop Chairs

Federica Arrigoni	Politecnico Milano, Italy
Lauro Snidaro	University of Udine, Italy

Tutorial Chairs

Christian Micheloni	University of Udine, Italy
Francesca Odone	University of Genova, Italy

Publications Chairs

Claudio Piciarelli	University of Udine, Italy
Niki Martinel	University of Udine, Italy

Publicity/Social Chairs

Matteo Dunnhofer	University of Udine, Italy
Beatrice Portelli	University of Udine, Italy

Industrial Liaison Chair

Pasqualina Fragneto	STMicroelectronics, Italy

Local Organization Chairs

Eleonora Maset	University of Udine, Italy
Andrea Toma	University of Udine, Italy
Emanuela Colombi	University of Udine, Italy
Alex Falcon	University of Udine, Italy
Andrea Brunello	University of Udine, Italy

Area Chairs

Pattern Recognition

Raffaella Lanzarotti	University of Milano, Italy
Nicola Strisciuglio	University of Twente, The Netherlands

Machine Learning and Deep Learning

Tatiana Tommasi	Politecnico Torino, Italy
Timothy M. Hospedales	University of Edinburgh, UK

3D Computer Vision and Geometry

Luca Magri	Politecnico Milano, Italy
James Pritts	CTU Prague, Czech Republic

Image Analysis: Detection and Recognition

Giacomo Boracchi	Politecnico Milano, Italy
Mårten Sjöström	Mid Sweden University, Sweden

Video Analysis and Understanding

Elisa Ricci	University of Trento, Italy

Shape Representation, Recognition and Analysis

Efstratios Gavves	University of Amsterdam, The Netherlands

Biomedical and Assistive Technology

Marco Leo	CNR, Italy
Zhigang Zhu	City College of New York, USA

Digital Forensics and Biometrics

Alessandro Ortis	University of Catania, Italy
Christian Riess	Friedrich-Alexander University, Germany

Multimedia

Francesco Isgrò	University of Napoli Federico II, Italy
Oliver Schreer	Fraunhofer HHI, Germany

Cultural Heritage

Lorenzo Baraldi	University of Modena-Reggio Emilia, Italy
Christopher Kermorvant	Teklia, France

Robot Vision and Automotive

Alberto Pretto	University of Padova, Italy
Henrik Andreasson	Örebro University, Sweden
Emanuele Rodolà	Sapienza University of Rome, Italy
Zorah Laehner	University of Siegen, Germany

Augmented and Virtual Reality

Andrea Torsello	University of Venezia Ca' Foscari, Italy
Richard Wilson	University of York, UK

Geospatial Analysis

Enrico Magli	Politecnico Torino, Italy
Mozhdeh Shahbazi	University of Calgary, Canada

Computer Vision for UAVs

Danilo Avola	Sapienza University of Rome, Italy
Parameshachari B. D.	Nitte Meenakshi Institute of Technology, India

Brave New Ideas

Marco Cristani	University of Verona, Italy
Hichem Sahbi	Sorbonne University, France

Endorsing Institutions

International Association for Pattern Recognition (IAPR)
Italian Association for Computer Vision, Pattern Recognition and Machine Learning (CVPL)

Contents – Part II

Multi-modal Medical Imaging Processing (M3IP)

Federated Learning in Medical Imaging and Vision (FEDMED)

Fine Art Pattern Extraction and Recognition (FAPER)

Pattern Recognition for Cultural Heritage (PatReCH)

Contents – Part I

Artificial Intelligence and Radiomics in Computer-Aided Diagnosis (AIRCAD)

Leukocytes Classification Methods: Effectiveness and Robustness in a Real Application Scenario

Lorenzo Putzu[1(✉)] and Andrea Loddo[2]

[1] Department of Electrical and Electronic Engineering, University of Cagliari, Cagliari, Italy
lorenzo.putzu@unica.it

[2] Department of Mathematics and Computer Science, University of Cagliari, Cagliari, Italy
andrea.loddo@unica.it

Abstract. Classification and differentiation of leukocyte sub-types are important in peripheral blood smear analysis. Fully-automated systems for leukocyte analysis are grouped into segmentation- and detection-based methods. The accuracy of classification depends on the accuracy of segmentation and detection steps. Real-world applications often produce inaccurate ROIs due to image quality factors, e.g., colour and lighting conditions, absence of standards, or even density and presence of overlapping cells. To this end, we investigated the scenario in-depth with ROIs simulating segmentation and detection methods and evaluating different image descriptors on two tasks: differentiation of leukocyte sub-types and leukaemia detection. The obtained results show that even simpler approaches can lead to accurate and robust results in both tasks when exploiting appropriate images for model training. Traditional handcrafted features are more effective when extracted from tight bounding boxes or masks, while deep features are more effective when extracted from large bounding boxes or masks.

Keywords: Blood Smear Image · Leukocyte analysis · Cell Sub-types Classification · Leukaemia Detection

1 Introduction

Blood is a connective tissue containing plasma and cellular components produced in the bone marrow. These include platelets, which are responsible for homeostasis, red blood cells (RBCs), which transport oxygen to all parts of the body and remove carbon dioxide; and white blood cells (WBCs), which are an essential part of the immune system and help fight infections and diseases. WBCs are divided into granulocytes (neutrophils, eosinophils, and basophils) and agranulocytes (monocytes and lymphocytes). Leukaemia and other blood cancers are serious illnesses that have limited treatment options. To make an accurate diagnosis, it is important to examine the WBCs using a microscope, such as by performing a

G. L. Foresti et al. (Eds.): ICIAP 2023 Workshops, LNCS 14366, pp. 3–14, 2024.
https://doi.org/10.1007/978-3-031-51026-7_1

complete blood count (CBC) test to check their number in blood or bone marrow smears [16]. However, identifying different types of cells can be difficult due to their unique shapes, colours, and textures. These procedures require the expertise of trained professionals and are manual, subjective, time-consuming, repetitive, and error-prone. Therefore, an automated process is becoming increasingly important to prevent misdiagnosis and address procedural issues, e.g., in developing countries. Recent work has shown the potential for automation to handle this task effectively. Computer-aided diagnostics (CAD) systems for peripheral blood smear (PBS) analysis have evolved over time, transitioning from older methods that combined image processing and traditional machine learning (ML) techniques to newer approaches based on deep learning (DL). These CAD systems serve various purposes, ranging from basic cell counting to comprehensive cell analysis. In general, the main focus of research in this field is to differentiate and classify healthy and diseased WBCs [2,5,8], as well as differentiate the WBC subtypes [1,5,13,20]. On the one hand, Acevedo et al. [2] developed a classification system for hypo granulated neutrophils, which are a precursor to acute myeloid leukaemia (AML) known as myelodysplastic syndrome, while other works proposed novel architectures based on convolutional neural networks (CNNs) to perform binary classification of healthy and diseased WBCs [5,8]. On the other hand, several authors proposed CNN-based blood cells classification schemes to discover acute lymphoblastic leukaemia (ALL) [1,5] or AML [20] in PB, even with tailored attention mechanisms [13]. Obviously, the specific steps involved may vary depending on the task at hand and on the used approach, which could range from a complete pipeline with many intermediate steps, that act on full-size images [24] to single-step methods for direct image classification [3,8]. The latter approaches have attempted to detect diseases directly from full-size images [3,8]. However, recent research has shown that fine details like those in diseased white blood cells cannot be accurately extracted from full-size images [19], also leading to unexplainable results. These methods best suit monocentric cell images, which facilitate extracting important features. Nevertheless, obtaining precise single-cell images requires accurate detection or manual cropping of regions of interest, which can be challenging in real-world scenarios, except if the system still is not fully automated [18]. The most interesting approaches are definitely the ones exploiting a complete pipeline to analyse the PBS images, since they can be used without any manual intervention. Such pipelines can be divided into segmentation-based and detection-based. Segmentation-based systems are ideal for examining fine-grained details, such as identifying pathologies or parasites, while detection-based systems are better suited for quantitative or coarse-grained analysis, including cell counting and classification [24]. However, some drawbacks are associated with both segmentation- and detection-based approaches. In detection-based methods, excluding bounding boxes (BBs) with low confidence levels may lead to excluding relevant BBs. Also, BBs with high confidence levels could be inaccurately selected, causing misclassification in the subsequent step [18,19]. Segmentation-based methods are instead very sensitive to changes (in terms of colour, illumination, etc.) that may be present in the test images

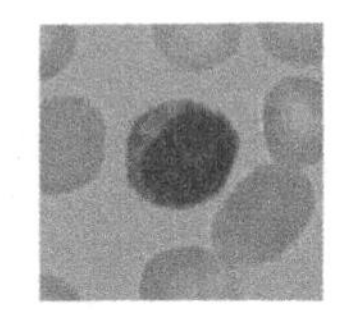 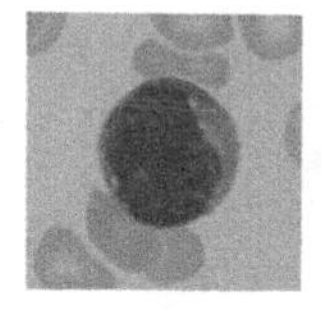

Fig. 1. WBC categories from ALL-IDB2: healthy lymphocyte (left) and lymphoblast (right).

compared to the images used for training, which could lead to under- or over-segmented binary masks (BMs). The progression of CAD systems development necessitates the resolution of these limitations to enhance the accuracy and reliability of PBS analysis. Our primary focus in this work is the impact of imperfect ROIs on the performance of a classification system. Existing limitations with both segmentation and detection methods must be addressed to improve the accuracy and reliability of CAD systems in PBS analysis. Indeed, imperfect ROIs can have a significant impact on the performance of a classification system. Therefore, in this work, the primary objective is to understand which limitations may be most impactful in such systems. Secondly, it is intended to experimentally test which descriptors best suit the different types of existing approaches. We thoroughly investigated automated systems for classifying WBC sub-types and detecting leukaemia. Our analysis focused on how ROIs' quality affects classification systems' performance. Specifically, we simulated the output of detection- and segmentation-based methods in a real-world application scenario. We then compared various image descriptors and classifiers to assess their effectiveness and robustness.

2 Materials and Methods

This section describes the materials and methods used in this work to perform the above evaluation. We first describe the data sets, methods employed, and experimental setup.

2.1 Data Sets

We used two public benchmark data sets: the Acute Lymphoblastic Leukaemia Image Database (ALL-IDB) version 2 (ALL-IDB2) [15], proposed for ALL detection, and Raabin-WBC (R-WBC) [14], a recently proposed data set for WBC sub-types classification. **ALL-IDB2** contains 260 images in JPG format with 24-bit colour depth, and each image presents a single-centred leukocyte, 50% of which are lymphoblasts, collected at the M. Tettamanti Research Centre for Childhood Leukaemia and Haematological Diseases, Monza, Italy. **R-WBC** is a large open-access data set collected from several laboratories. The images were labelled and processed to provide multiple subsets for different tasks; we exploited the subset containing WBC bounding boxes and the ground truth of

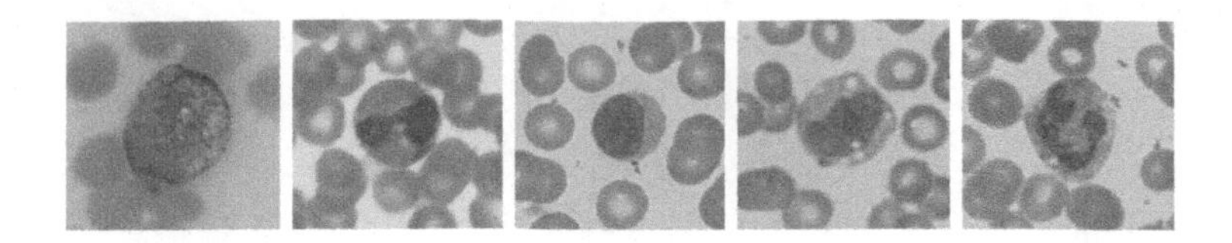

Fig. 2. WBC sub-type images from R-WBC. From left to right: Basophil, Eosinophil, Lymphocyte, Monocyte and Neutrophil.

the nucleus and cytoplasm. This subset includes 1145 images of selected WBCs, including 242 lymphocytes, 242 monocytes, 242 neutrophils, 201 eosinophils and 218 basophils. Figures 1 and 2 show a sample of healthy leukocytes and a lymphoblast, taken from ALL-IDB2 and a sample image for each WBC sub-type from R-WBC.

2.2 Data Pre-processing

As it can be observed from Figs. 1 and 2, each WBC is perfectly located in the image centre in both data set images. However, the image size is larger than the WBC size, resulting in the presence of many other cells (mostly RBCs) in the images. This aspect can significantly impact feature extraction and classification performance. As previously mentioned, the main goal of this work is to compare the effectiveness of different ROIs by evaluating if and to what extent the classification performance can be affected by inaccurate BBs (typically extracted with detection-based methods) or BMs (typically obtained with segmentation-based methods). To this aim, for each original data set, we created different alternative versions to simulate the output of detection- and segmentation-based methods in a controlled way. For this reason, we have not used accurate detection and segmentation methods precisely because we would not have control over the output. Instead, we aim at grouping in different sets of well-segmented, under-segmented and over-segmented images. To create these alternative versions, we exploited the pixel-wise ground truth in the form of a BM[1], where the foreground contains the WBCs only. **BBs** have been grouped into 4 different versions: *largeBB* version (the original data sets), *tightBB* version (BBs perfectly fit to the WBC size), *erodedBB* version (WBC is only partially enclosed) and *dilatedBB* version (WBC not centred with a consistent portion of background). For the latter 3 versions, we extracted the contour's extreme points (left, right, top and bottom) to re-crop the RGB images. The tight version has been re-cropped, adding just 3 pixels for each side of the box. The eroded/dilated version has been re-cropped, removing/adding 30 pixels for ALL-IDB and 60 pixels for R-WBC (whose resolution is double) from one side of the box randomly drawn. A sample image for each BB version on each class of ALL-IDB2 and R-WBC is shown in Fig. 3. **BMs** have been grouped into 6 different versions:*largeBM* version (largeBB masked with

[1] for ALL-IDB2, whose ground truth was not proposed by the authors, we provided a copy at ALL-IDB2 masks.

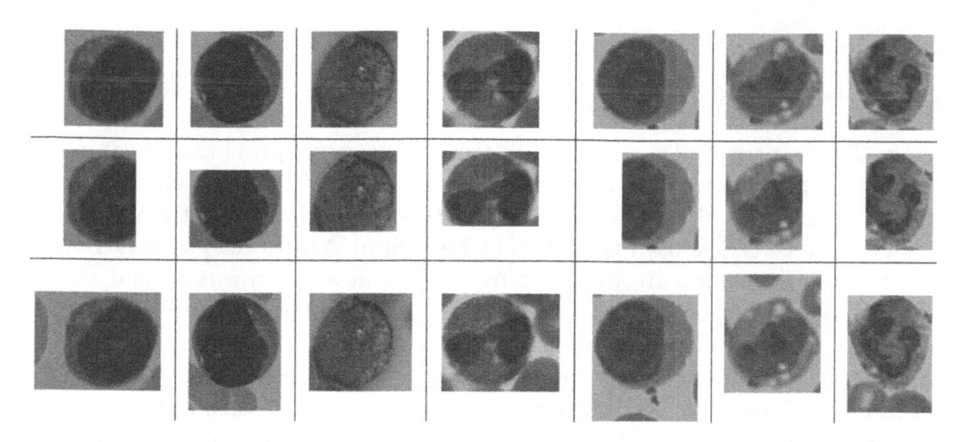

Fig. 3. *tightBB* (top), *erodedBB* (middle) and *dilatedBB*(bottom) versions of ALL-IDB2 (first two columns) and R-WBC (remaining columns) data sets.

the corresponding ground truth BMs), *large-erodedBM* version (largeBB with under-segmented WBC), *large-dilatedBM* version (largeBB with over-segmented WBC), *tightBM* version (tightBB masked with the corresponding ground truth BMs), *tight-erodedBM* version (tightBM with under-segmented WBC) and *tight-dilatedBM* version (tightBM with over-segmented WBC). To create the eroded and dilated versions, we used morphological erosion and dilation, respectively. Both operations have been performed with a line-shaped structuring element whose length is 15 pixels for ALL-IDB and 30 pixels for R-WBC (whose resolution is double), and the orientation is randomly drawn. The aim of such structuring element is to forge the cell shape in a non-homogeneous way, by eroding/dilating the BM only in a certain direction. A classic disk-shaped structuring element would have eroded/dilated the BM evenly, thus preserving the cell shape, which is an unrealistic error for segmentation-based methods. A sample image for each BM version on each class of ALL-IDB2 and R-WBC, is shown in Fig. 4.

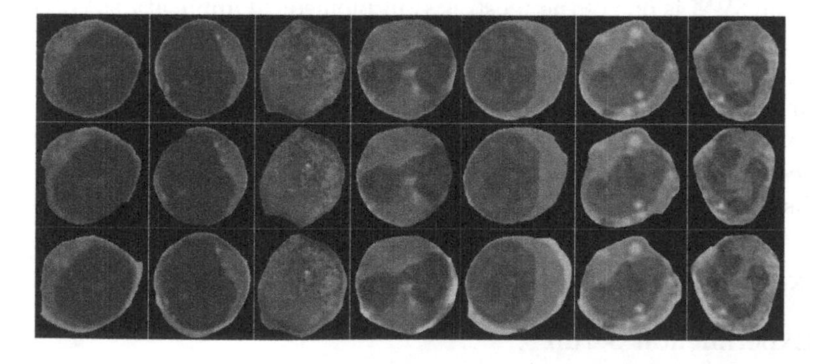

Fig. 4. *tightBM* (top), *tight-erodedBM* (middle) and *tight-dilatedBM*(bottom) versions of ALL-IDB2 (first two columns) and R-WBC (remaining columns) data sets.

2.3 Methods

Here we describe the image descriptors and classifiers used in our evaluation.

Image Descriptors. We evaluated different image descriptors that we grouped into four important classes: invariant moments, texture, colour and deep features. We excluded all shape- and contour-based descriptors that can only be extracted for BMs. We used two types of **Invariant Moments**, namely *Legendre* and *Zernike*, which are invariant to changes in scale, rotation, and reflection [28]. The order of the moments is set to 5 [10]. The **Texture Features** computed are the thirteen rotationally invariant features [11] Har_{ri} extracted from *Gray Level Co-occurrence Matrix* with $d = 1$ and $\theta = [0°, 45°, 90°, 135°]$ (for more details see [23]) and the rotation invariant *Local Binary Pattern (LBP)* features [22], with $r = 1$ and $n = 8$. For **Colour Features**, we extracted basic *colour histogram* including mean, standard deviation, smoothness, skewness, kurtosis, uniformity, and entropy, as well as *colour auto-correlogram* features [21] with four distance values: $d = 1, 2, 3, 4$. Lastly, we extracted **Deep Features** from three well-known CNN architectures namely VGG-19 [26], ResNet-50 [12] and Inceptionv3 [27]. As mentioned before, different ad-hoc architectures have been proposed so far for this task. However, we will focus on more versatile architectures that can be utilised for both pipeline types. Such architectures are all pre-trained on a well-known natural image data set (ImageNet [9]) and adapted to medical image tasks, following an established procedure for transfer learning and fine-tuning CNN models [25]. This process involves keeping all CNN layers except the last fully-connected one, which is substituted with a new layer to include new object categories. We extracted features from the penultimate fully connected layer (FC7) on VGG-19 and the last layer on ResNet-50 and Inceptionv3 architectures.

Classification. We used three different traditional classifiers: the k-Nearest Neighbour (**k-NN**) [7], Support Vector Machine (**SVM**) [17] and Random Forest (**RF**) [6]. The k-NN was used because it is one of the simplest and can document the effectiveness of the extracted features rather than assessing the classifier's performance. Here we used $k = 1$, computed using the Euclidean distance. Conversely, the SVM is one of the most used in biomedical applications [1,24]. Here we use a Gaussian radial basis function (RBF) trained using the *one VS rest* approach. To speed up the selection of the SVM hyperparameters, we employed an error-correcting output code mechanism [4] with 5-fold cross-validation to fit and automatically tune the hyperparameters. The RF was chosen because it combines many decision trees' results, thus reducing over-fitting and improving generalisation. Here we used a forest consisting of 100 trees.

3 Experimental Evaluation

3.1 Experimental Setup

We divided the data sets into training and testing sets, consisting of 70% and 30% of images, respectively. We further divided the original training set into a training and validation set to train the CNNs, with 80% and 20% of images, respectively. To further ease reproducibility, we ensured the splits were balanced using a stratified sampling procedure and organising each class's images in lexicographic order. All experiments were conducted on a single machine with an Intel(R) Core(TM) i9-8950HK @ 2.90GHz CPU, 32 GB RAM, and NVIDIA GTX1050 Ti 4GB GPU. The CNN architectures were fine-tuned using the Adam solver for 50 epochs, with a learning rate of 0.001, decreasing by a factor of 0.1 every 30 epochs. We also set the L_2 regularisation factor to avoid over-fitting during the training phase, as we did not use any image augmentation. We evaluated the classification performance using five standard metrics: Accuracy (A), Precision (P), Recall (R), and Specificity (S). For ALL-IDB, a two-class problem, we used the above metrics as is. However, for R-WBC, a multi-class problem, we calculated the performance for each class and then calculated the weighted average to obtain a single performance measure.

3.2 Experimental Results

Tables 1 and 2 report the performance of the different features and classifiers in a controlled (ideal) environment, as it happens in most benchmark data sets created specifically for model training, where the BBs and BMs are precise. The aim of this experiments is to evaluate if and to what extent the presence of a large portion of background in the ROIs can influence the extracted features and the consequent classification models.

Table 1. ALL-IDB2 performances obtained with different models created exploiting different data set version: (from the left) largeBB, tightBB, largeBM and tightBM.

Classifier	Descriptor	Target set															
		largeBB				tightBB				largeBM				tightBM			
		A	P	R	S	A	P	R	S	A	P	R	S	A	P	R	S
kNN	Legendre	56.4	55.6	64.1	48.7	69.2	68.3	71.8	66.7	67.9	65.9	74.4	61.5	73.1	70.5	79.5	66.7
	Zernike	35.9	33.3	28.2	43.6	47.4	47.1	41.0	53.8	38.5	38.5	38.5	38.5	65.4	70.0	53.8	76.9
	HARri	61.5	58.8	76.9	46.2	62.8	61.9	66.7	59.0	55.1	53.8	71.8	38.5	65.4	61.5	82.1	48.7
	LBP18	53.8	52.5	82.1	25.6	59.0	56.9	74.4	43.6	70.5	64.3	92.3	48.7	66.7	61.0	92.3	41.0
	Histogram	64.1	60.8	79.5	48.7	26.9	26.3	25.6	28.2	47.4	47.6	51.3	43.6	56.4	55.6	64.1	48.7
	Correlogram	50.0	50.0	10.3	**89.7**	56.4	55.3	66.7	46.2	48.7	48.7	48.7	48.7	48.7	48.7	48.7	48.7
	VGGNet-19	69.2	63.6	89.7	48.7	75.6	75.0	76.9	74.4	59.0	56.1	82.1	35.9	53.8	52.3	87.2	20.5
	ResNet-50	74.4	73.2	76.9	71.8	83.3	**79.5**	89.7	**76.9**	80.8	80.0	82.1	**79.5**	74.4	78.8	66.7	**82.1**
	Inceptionv3	**80.8**	**78.6**	84.6	76.9	80.8	75.0	92.3	69.2	79.5	79.5	79.5	**79.5**	75.6	76.3	74.4	76.9
	AVG	64.8	62.3	72.4	**57.2**	66.3	64.2	71.6	61.1	66.1	64.5	72.8	**59.5**	**64.3**	**63.4**	72.3	**56.3**
SVMRbf	Legendre	60.3	68.2	38.5	82.1	70.5	73.5	64.1	76.9	52.6	51.8	74.4	30.8	65.4	60.7	87.2	43.6
	Zernike	53.8	52.5	79.5	28.2	41.0	37.0	25.6	56.4	44.9	45.7	53.8	35.9	62.8	63.2	61.5	64.1
	HARri	57.7	54.8	87.2	28.2	62.8	63.9	59.0	66.7	51.3	51.1	61.5	41.0	65.4	61.5	82.1	48.7
	LBP18	61.5	58.2	82.1	41.0	66.7	61.8	87.2	46.2	69.2	64.2	87.2	51.3	67.9	62.1	92.3	43.6
	Histogram	39.7	36.7	28.2	51.3	44.9	43.3	33.3	56.4	60.3	61.1	56.4	64.1	66.7	63.3	79.5	53.8
	Correlogram	64.1	64.1	64.1	64.1	53.8	53.1	66.7	41.0	53.8	52.1	94.9	12.8	52.6	51.4	94.9	10.3
	VGGNet-19	**80.8**	75.0	92.3	69.2	70.5	71.1	69.2	71.8	64.1	58.5	**97.4**	30.8	50.0	50.0	**100.0**	0.0
	ResNet-50	75.6	68.5	94.9	56.4	**84.6**	76.5	**100.0**	69.2	79.5	79.5	79.5	**79.5**	71.8	73.0	69.2	74.4
	Inceptionv3	50.0	50.0	**100.0**	0.0	83.3	77.1	94.9	71.8	80.8	78.6	84.6	76.9	**78.2**	**80.6**	74.4	**82.1**
	AVG	**66.7**	**64.9**	**78.2**	55.3	**71.1**	**67.3**	77.6	**64.6**	64.3	65.1	**75.6**	52.9	63.1	62.0	**78.7**	47.4
RF	Legendre	65.4	63.6	71.8	59.0	74.4	77.1	69.2	79.5	50.0	50.0	74.4	25.6	62.8	60.0	76.9	48.7
	Zernike	23.1	20.0	17.9	28.2	33.3	33.3	33.3	33.3	50.0	50.0	61.5	38.5	48.7	48.9	59.0	38.5
	HARri	55.1	53.3	82.1	28.2	76.9	69.8	94.9	59.0	73.1	68.0	87.2	59.0	76.9	69.8	94.9	59.0
	LBP18	44.9	47.3	89.7	0.0	61.5	56.9	94.9	28.2	69.2	63.2	92.3	46.2	61.5	57.9	84.6	38.5
	Histogram	65.4	61.5	82.1	48.7	73.1	70.5	79.5	66.7	59.0	61.3	48.7	69.2	73.1	78.1	64.1	**82.1**
	Correlogram	71.8	71.8	71.8	71.8	61.5	60.0	69.2	53.8	56.4	54.0	87.2	25.6	55.1	53.8	71.8	38.5
	VGGNet-19	56.4	53.8	89.7	23.1	64.1	60.0	84.6	43.6	67.9	63.0	87.2	48.7	51.3	50.7	89.7	12.8
	ResNet-50	79.5	76.7	84.6	74.4	**84.6**	76.5	**100.0**	69.2	**84.6**	**81.4**	89.7	**79.5**	71.8	74.3	66.7	76.9
	Inceptionv3	79.5	75.6	87.2	71.8	83.3	76.0	97.4	69.2	78.2	77.5	79.5	76.9	70.5	68.2	76.9	64.1
	AVG	61.2	59.0	76.1	46.3	70.2	66.5	**80.8**	59.6	**66.2**	64.0	**79.2**	53.2	64.2	62.7	76.8	51.6

Concerning the use of BBs, as observed on both ALL-IDB2 and R-WBC data sets, by training with the tightBB versions, the performance (with few exceptions) is better than those obtained with largeBB. As for BMs, tightBM tends to be better than largeBM, except CNNs that behave worse with tightBM. This result can be attributed to the fact that CNNs require a larger area of the image to perform convolution operations than what tightBM can provide. In general, the performance is higher with the SVM, which, for this reason, we selected for the subsequent experiments devoted to evaluating the robustness of the different features in an uncontrolled (real) environment. In such a scenario, the features to train the classification models could still be extracted from controlled benchmark data sets composed of precise ROIs. Still, they are deployed in a real scenario where the ROIs often present errors, like the ones presented above. To simulate this scenario, in this experiment, we exploited the SVM models already trained in the previous ones and tested using the test sets belonging to the other versions of the data sets. The results are reported in the plots in Figs. 5 and 6, which present the accuracy trend for each descriptor when varying the data set version used for testing.

In this scenario, performance can be expected to worsen quickly; however, this trend has not occurred in all cases. In particular, by exploiting largeBB for training, a sharp degradation in performance is observed (especially in R-WBC) when testing with all the other data set versions. In contrast, by exploiting BBtight for training, the degradation is observed only when testing on largeBB or dilatedBB. For BMs, the behaviour is the opposite. Indeed, by using largeBM

Table 2. RWBC performances obtained with different models created exploiting different data set version: (from the left) largeBB, tightBB, largeBM and tightBM.

Classifier	Descriptor	Target set															
		largeBB				tightBB				largeBM				tightBM			
		A	P	R	S	A	P	R	S	A	P	R	S	A	P	R	S
kNN	Legendre	75.3	40.4	38.7	84.5	83.5	58.3	59.0	89.4	78.3	46.6	45.3	86.5	80.6	51.8	51.7	88.0
	Zernike	70.7	28.0	27.3	81.8	84.6	62.0	61.9	90.2	82.2	55.9	55.2	89.1	80.4	52.3	51.5	87.7
	HARri	69.8	24.9	24.7	81.1	78.3	48.3	45.9	86.5	75.4	38.6	38.7	84.5	80.8	53.1	52.6	88.0
	LBP18	77.3	44.5	43.9	85.8	87.6	69.7	68.9	92.3	88.4	70.7	70.6	92.7	89.5	74.3	73.5	93.4
	Histogram	70.8	27.4	27.6	81.6	76.8	43.3	42.4	85.3	85.6	65.0	63.7	91.0	79.3	52.1	49.1	86.8
	Correlogram	71.5	28.5	28.8	82.5	71.1	30.2	27.6	82.2	74.0	39.2	34.6	84.2	78.4	56.2	45.9	86.8
	VGGNet-19	81.1	52.7	53.2	88.4	92.2	80.6	80.8	95.2	97.9	95.0	94.8	98.6	88.7	73.2	72.1	92.8
	ResNet-50	82.6	58.6	56.7	89.3	94.8	87.3	86.9	96.8	97.7	94.6	94.5	98.5	94.4	86.7	86.3	96.4
	Inceptionv3	84.6	63.8	61.6	90.6	91.7	80.7	79.7	94.8	98.5	96.3	96.2	99.0	93.6	84.9	84.6	95.9
	AVG	78.7	47.8	47.1	86.8	87.0	68.2	67.6	91.8	89.2	73.7	73.1	93.3	87.0	69.5	67.9	91.8
SVMRbf	Legendre	75.4	35.0	38.7	84.8	84.8	60.2	62.2	90.2	83.3	57.9	58.1	89.6	83.7	58.5	59.9	89.7
	Zernike	75.1	37.7	38.4	84.3	87.8	70.1	69.8	92.2	85.1	62.1	62.8	90.6	84.7	63.4	62.2	90.1
	HARri	73.8	34.4	35.2	83.5	78.6	48.3	46.8	86.5	83.9	59.1	59.9	89.9	85.6	64.9	64.5	90.9
	LBP18	74.7	41.4	38.4	83.8	86.3	70.7	66.0	91.3	89.2	73.9	73.0	93.1	89.8	77.1	74.7	93.5
	Histogram	81.0	54.5	52.9	88.0	88.6	73.2	71.5	93.0	93.0	83.1	82.3	95.6	91.4	79.3	78.5	94.5
	Correlogram	79.8	51.6	50.3	87.3	83.0	58.6	57.6	89.3	90.2	76.4	75.9	93.8	89.8	75.5	74.7	93.5
	VGGNet-19	92.3	81.0	81.1	95.1	96.6	91.8	91.9	97.9	98.3	96.0	95.9	98.9	92.3	82.0	81.1	95.0
	ResNet-50	93.5	84.4	84.0	95.8	**98.5**	**96.2**	**96.2**	**99.0**	98.2	95.7	95.6	98.8	94.6	87.9	86.9	96.5
	Inceptionv3	**94.2**	**85.7**	**85.8**	**96.3**	96.7	92.0	91.9	98.0	98.1	95.4	95.3	98.8	95.0	88.3	87.8	96.8
	AVG	**85.1**	**63.3**	**63.2**	**90.6**	91.1	78.5	78.0	94.4	**92.1**	**80.0**	**80.3**	**94.9**	**90.6**	**77.7**	**76.9**	**94.0**
RF	Legendre	76.3	40.7	41.3	85.0	85.4	64.6	63.7	90.7	82.8	55.4	56.7	89.3	82.9	56.0	57.6	89.3
	Zernike	75.1	36.7	38.4	84.3	87.1	68.5	68.0	91.8	85.2	62.1	62.8	90.6	81.7	55.7	54.9	88.3
	HARri	82.6	58.0	56.7	89.2	89.1	74.8	73.3	93.1	91.2	78.0	77.9	94.4	92.7	82.8	82.0	95.3
	LBP18	78.3	48.6	45.9	86.7	89.3	72.4	73.0	93.2	90.8	76.5	77.0	94.1	91.6	78.9	79.1	94.6
	Histogram	80.7	53.2	52.3	87.9	88.2	71.1	70.3	92.7	87.0	68.1	67.2	91.8	88.5	72.0	71.2	92.6
	Correlogram	79.0	48.7	48.3	86.7	80.0	51.1	50.6	87.4	91.3	79.9	78.2	94.3	88.8	76.5	72.4	92.8
	VGGNet-19	87.6	69.2	69.5	92.1	95.9	90.0	89.8	97.3	98.2	95.8	95.6	98.8	89.3	75.7	73.5	93.1
	ResNet-50	88.3	71.9	71.2	92.4	96.7	92.3	91.9	97.9	98.1	95.5	95.3	98.7	**95.4**	**90.0**	**89.0**	**97.0**
	Inceptionv3	89.3	73.8	73.5	93.1	95.8	89.6	89.5	97.3	**98.6**	**96.5**	**96.5**	**99.1**	94.2	86.2	86.0	96.3
	AVG	84.3	61.4	61.2	90.1	**91.3**	**78.6**	**78.3**	**94.5**	92.0	79.4	80.1	**94.9**	90.5	77.4	76.7	93.9

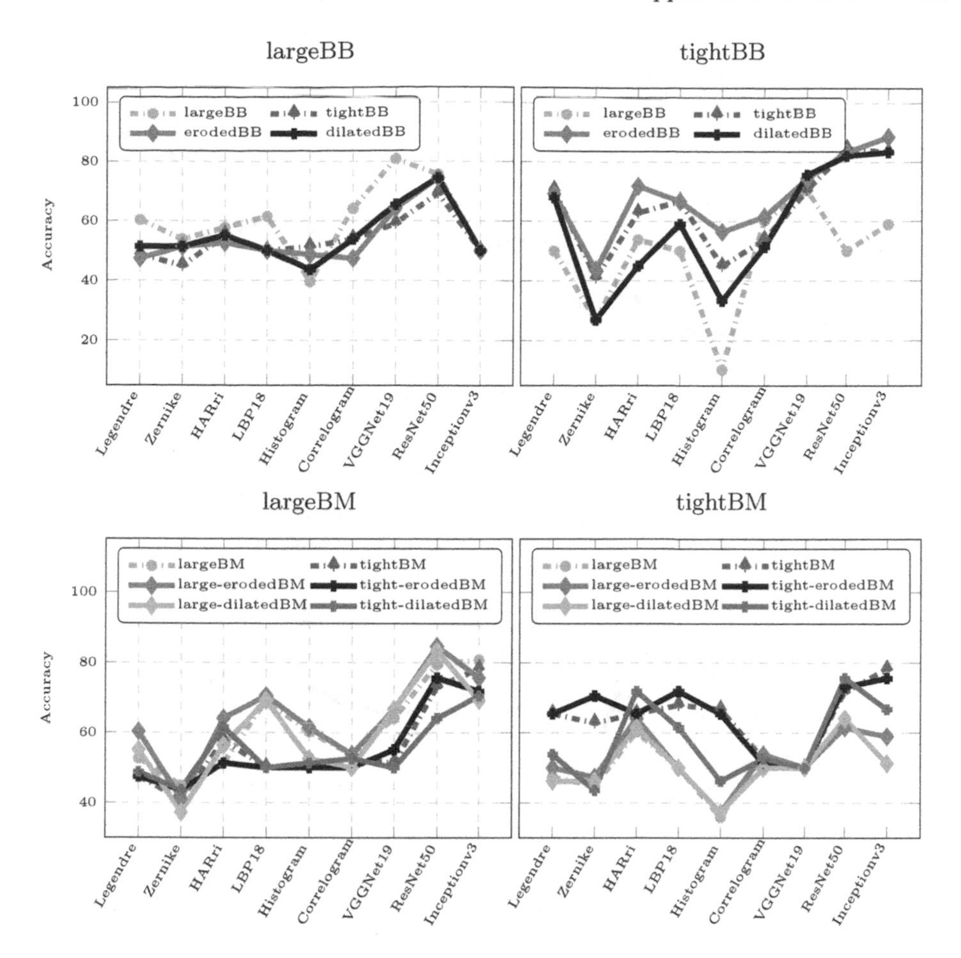

Fig. 5. Robustness of features extracted from different ROIs of ALL-IDB2. Top row, SVM models trained on largeBB and tightBB and tested on all BBs versions. Bottom row, SVM models trained on largeBM and tightBBM and tested on all BMs versions.

for training, the degradation is observed for the tight versions (tightBM, tight-erodedBM, tight-dilatedBM) only, while by exploiting tightBM for training, the degradation is observed for almost all (except tight-erodedBM) the other versions. In general, it was observed that the best and most stable performance for both data sets was achieved through training with largeBM. On ALL-IDB2, a relatively consistent trend was observed across all tested masks, especially with ResNet-50 and Inceptionv3. Indeed the accuracy with such descriptors remained acceptable, with results above 80%, meaning that they are highly representative of the imperfect and real scenarios assumed when training models with largeBM.

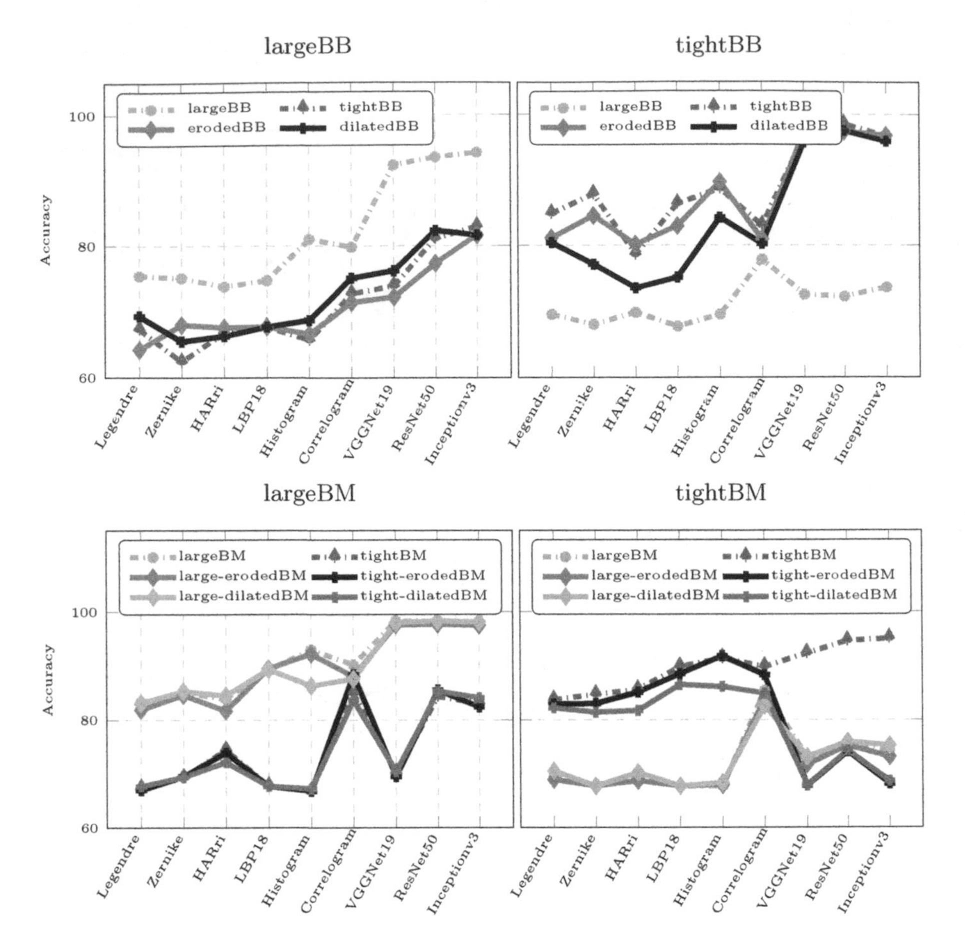

Fig. 6. Robustness of features extracted from different ROIs of R-WBC. Top row, SVM models trained on largeBB and tightBB and tested on all BBs versions. Bottom row, SVM models trained on largeBM and tightBM and tested on all BMs versions.

4 Conclusions

In this work, we investigated current leukocyte analysis methods in a real application scenario, where the segmentation and detection methods used to extract the ROIs automatically can be inaccurate or imprecise. We evaluated different image descriptors and classifiers to assess if and to what extent such factors can affect the performance of CAD systems for WBC sub-types classification and leukaemia detection. Experimental results confirmed that the performance in such a scenario could drop significantly if not addressed adequately. In particular, traditional handcrafted features are more effective when extracted from tight BBs or BMs. In contrast, deep features, primarily ResNet-50 and Inceptionv3 (generally the most robust), are more effective when extracted from large

BBs and (mostly) BMs. In the future, we plan to improve this work by verifying additional features, such as those extracted from vision transformers, and different classifiers. We will also explore this task from an explainable AI perspective. Finally, we aim to examine ad-hoc features for peripheral blood image analysis and suggest more robust ones.

References

1. Acevedo, A., Alférez, S., Merino, A., Puigví, L., Rodellar, J.: Recognition of peripheral blood cell images using convolutional neural networks. Comput. Methods Programs Biomed. 180 (2019). https://doi.org/10.1016/j.cmpb.2019.105020
2. Acevedo, A., Merino, A., Boldú, L., Molina, A., Alférez, S., Rodellar, J.: A new convolutional neural network predictive model for the automatic recognition of hypogranulated neutrophils in myelodysplastic syndromes. Comput. Biol. Medicine **134**, 104479 (2021). https://doi.org/10.1016/j.compbiomed.2021.104479
3. Anilkumar, K., Manoj, V., Sagi, T.: Automated detection of leukemia by pretrained deep neural networks and transfer learning: A comparison. Med. Eng. Phys. **98**, 8–19 (2021). https://doi.org/10.1016/j.medengphy.2021.10.006
4. a. Bagheri, M., Montazer, G.A., Escalera, S.: Error correcting output codes for multiclass classification: application to two image vision problems. In: The 16th CSI International Symposium on Artificial Intelligence and Signal Processing (AISP 2012), pp. 508–513 (2012)
5. Boldú, L., Merino, A., Acevedo, A., Molina, A., Rodellar, J.: A deep learning model (alnet) for the diagnosis of acute leukaemia lineage using peripheral blood cell images. Comput. Methods Programs Biomed. **202**, 105999 (2021). https://doi.org/10.1016/j.cmpb.2021.105999
6. Breiman, L.: Random forests. Mach. Learn. **4**, 5–32 (2001)
7. Cover, T.M., Hart, P.E.: Nearest neighbor pattern classification. IEEE Trans. Inf. Theory **13**(1), 21–27 (1967)
8. Das, P.K., Meher, S.: An efficient deep convolutional neural network based detection and classification of acute lymphoblastic leukemia. Expert Syst. Appl. **183**, 115311 (2021). https://doi.org/10.1016/j.eswa.2021.115311
9. Deng, J., Dong, W., Socher, R., Li, L.J., Li, K., Fei-Fei, L.: Imagenet: a large-scale hierarchical image database. In: Computer Vision and Pattern Recognition (CVPR), pp. 248–255 (2009)
10. Di Ruberto, C., Putzu, L., Rodriguez, G.: Fast and accurate computation of orthogonal moments for texture analysis. Pattern Recogn. **83**, 498–510 (2018)
11. Haralick, R.M., Shanmugam, K., Dinstein, I.H.: Textural features for image classification. IEEE Trans. Syst. Man Cybern. **SMC-3**(6), 610–621 (1973)
12. He, K., Zhang, X., Ren, S., Sun, J.: Deep residual learning for image recognition. In: 2016 IEEE Conference on Computer Vision and Pattern Recognition (CVPR), pp. 770–778 (2016). https://doi.org/10.1109/CVPR.2016.90
13. Huang, P.: Attention-aware residual network based manifold learning for white blood cells classification. IEEE J. Biomed. Health Informatics **25**(4), 1206–1214 (2021). https://doi.org/10.1109/JBHI.2020.3012711
14. Kouzehkanan, S.Z.M., Saghari, S., Tavakoli, E., Rostami, P., Abaszadeh, M., Satlsar, E.S., Mirzadeh, F., Gheidishahran, M., Gorgi, F., Mohammadi, S., Hosseini, R.: Raabin-wbc: a large free access dataset of white blood cells from normal peripheral blood. bioRxiv (2021). https://doi.org/10.1101/2021.05.02.442287

15. Labati, R.D., Piuri, V., Scotti, F.: All-IDB: the acute lymphoblastic leukemia image database for image processing. In: IEEE ICIP International Conference on Image Processing, pp. 2045–2048 (2011)
16. of Leeds, U.: The histology guide (2021). https://www.histology.leeds.ac.uk/blood/blood_wbc.php. Accessed 23 Jun 2023
17. Lin, Y., et al.: Large-scale image classification: Fast feature extraction and SVM training. In: The 24th IEEE Conference on Computer Vision and Pattern Recognition, CVPR 2011, Colorado Springs, CO, USA, 20–25 June 2011, pp. 1689–1696. IEEE Computer Society (2011)
18. Loddo, A., Putzu, L.: On the effectiveness of leukocytes classification methods in a real application scenario. AI **2**(3), 394–412 (2021). https://doi.org/10.3390/ai2030025
19. Loddo, A., Putzu, L.: On the reliability of cnns in clinical practice: A computer-aided diagnosis system case study. Appl. Sci. **12**(7) (2022). https://doi.org/10.3390/app12073269
20. Matek, C., Schwarz, S., Spiekermann, K., Marr, C.: Human-level recognition of blast cells in acute myeloid leukaemia with convolutional neural networks. Nat. Mach. Intell. **1**(11), 538–544 (2019). https://doi.org/10.1038/s42256-019-0101-9
21. Mitro, J.: Content-based image retrieval tutorial. ArXiv e-prints (2016)
22. Ojala, T., Pietikäinen, M., Mäenpää, T.: Multiresolution gray-scale and rotation invariant texture classification with local binary patterns. IEEE Trans. Pattern Anal. Mach. Intell. **24**(7), 971–987 (2002). https://doi.org/10.1109/TPAMI.2002.1017623
23. Putzu, L., Di Ruberto, C.: Rotation invariant co-occurrence matrix features. In: Battiato, S., Gallo, G., Schettini, R., Stanco, F. (eds.) ICIAP 2017. LNCS, vol. 10484, pp. 391–401. Springer, Cham (2017). https://doi.org/10.1007/978-3-319-68560-1_35
24. Ruberto, C.D., Loddo, A., Putzu, L.: A leucocytes count system from blood smear images segmentation and counting of white blood cells based on learning by sampling. Mach. Vis. Appl. **27**(8), 1151–1160 (2016)
25. Shin, H., et al.: Deep convolutional neural networks for computer-aided detection: CNN architectures, dataset characteristics and transfer learning. IEEE Trans. Med. Imaging **35**(5), 1285–1298 (2016). https://doi.org/10.1109/TMI.2016.2528162
26. Simonyan, K., Zisserman, A.: Very deep convolutional networks for large-scale image recognition. In: Bengio, Y., LeCun, Y. (eds.) 3rd International Conference on Learning Representations, ICLR 2015, San Diego, CA, USA, May 7–9, 2015, Conference Track Proceedings (2015)
27. Szegedy, C., Vanhoucke, V., Ioffe, S., Shlens, J., Wojna, Z.: Rethinking the inception architecture for computer vision. In: 2016 IEEE Conference on Computer Vision and Pattern Recognition (CVPR), pp. 2818–2826 (2016). https://doi.org/10.1109/CVPR.2016.308
28. Teague, M.R.: Image analysis via the general theory of moments. J. Opt. Soc. Am. **70**(8), 920–930 (1980). https://doi.org/10.1364/JOSA.70.000920

Vision Transformers for Breast Cancer Histology Image Classification

Giulia L. Baroni, Laura Rasotto, Kevin Roitero, Ameer Hamza Siraj, and Vincenzo Della Mea(✉)

Medical Informatics, Telemedicine & eHealth Lab (MITEL), Department of Mathematics, Computer Science and Physics, University of Udine, Udine, Italy
{giulia.baroni,kevin.roitero,ameerhamza.siraj, vincenzo.dellamea}@uniud.it, rasotto.laura@spes.uniud.it

Abstract. We propose a self-attention Vision Transformer (ViT) model tailored for breast cancer histology image classification. The proposed architecture uses a stack of transformer layers, with each layer consisting of a multi-head self-attention mechanism and a position-wise feed-forward network, and it is trained with different strategies and configurations, including pretraining, resize dimension, data augmentation, patch overlap, and patch size, to investigate their impact on performance on the histology image classification task. Experimental results show that pretraining on ImageNet and using geometric and color data augmentation techniques significantly improve the model's accuracy on the task. Additionally, a patch size of 16 × 16 and no patch overlap were found to be optimal for this task. These findings provide valuable insights for the design of future ViT-based models for similar image classification tasks.

Keywords: Breast Cancer · Deep Learning · Transformers

1 Introduction

Breast cancer is the most commonly diagnosed cancer in women worldwide, accounting for about 13% of all cancer diagnoses[1]. Early detection and accurate diagnosis are critical for effective treatment and improved patient outcomes. Medical imaging, particularly breast histology images, plays a crucial role in diagnosis. However, manually analyzing these images is time-consuming, prone to errors, and requires extensive training and expertise. Therefore, the development of automated methods has been an active area of research in recent years [4].

In this study, we propose a self-attention Vision Transformer (ViT) model [3] tailored for breast cancer histology image classification, which effectively captures both local and global contextual information. We evaluate the performance of the proposed model using different training strategies and configurations, including pretraining, resize dimension, data augmentation, patch overlap, and

The original version of the chapter has been revised. A correction to this chapter can be found at https://doi.org/10.1007/978-3-031-51026-7_42

[1] See https://www.iarc.who.int/cancer-type/breast-cancer/.

G. L. Foresti et al. (Eds.): ICIAP 2023 Workshops, LNCS 14366, pp. 15–26, 2024.
https://doi.org/10.1007/978-3-031-51026-7_2

patch size, to investigate their impact on performance. The experiments have been carried out on an image set that was part of the ICIAR 2018 Grand Challenge on BreAst Cancer Histology images (BACH) challenge [2], for which baselines and state-of-the-art are available. The experimental results demonstrate the effectiveness of the proposed ViT model on the breast cancer histology image classification task, and the findings can help inform the design of future ViT-based models for similar image classification tasks.

2 Background and Related Work

2.1 Deep Learning in Histopathology Images of Breast Cancer

Below, we briefly present some examples of deep learning applications on various datasets of breast cancer histopathology images. We start by showcasing the results obtained from the application of several convolutional neural networks and then proceed to demonstrate the outcomes achieved using vision transformers.

Vesal, et al. [15] present a transfer learning-based approach for breast histology image classification using the BACH 2018 Grand Challenge dataset. The data set is partitioned into training, validation, and test (60, 20, and 20 samples respectively) sets, by performing random sampling. The data set includes 400 images, 100 samples in each class. The images are initially normalized to correct for color variations that occurred during slide preparation [10]. Subsequently, image patches are extracted and utilized to fine-tune Inception-V3 and ResNet50. Patch-wise classification accuracy of ResNet50 for the validation and test sets are 0.91 and 0.94, respectively. The Inception-V3 network on the other hand achieves patch-wise classification accuracy of 0.87 and 0.86. ResNet50 achieves whole-image classification accuracy of 0.89 and 0.97, for the validation and test sets, respectively. Meanwhile, the Inception-V3 network achieves classification accuracy of 0.87 and 0.91 for the validation and test sets, respectively.

Roy, et al. [11] propose a patch-based classifier using CNNs for the classification of images from the ICIAR 2018 breast histology dataset. The PBC works in two modalities: one patch in one decision (OPOD) and all patches in one decision (APOD). OPOD entails independent classification of each patch, considered accurate when all patches match the image's class label. APOD aggregates class labels from all patches similar to OPOD but employs a majority vote to assign the final image label. Experimental data show the OPOD model attaining patch-wise classification accuracy rates of 0.77 and 0.84 for multiclass and binary histopathological categorizations. Moreover, APOD demonstrates image-wise classification accuracy rates of 0.90 and 0.92 for multiclass and binary classifications. Authors then achieve an accuracy of 0.87 on the hidden test dataset of ICIAR 2018.

Over the past few years, particularly since the emergence of ViTs in 2020, the medical imaging community has experienced a significant surge in the adoption of Transformer-based techniques for various tasks [8,14]. However, it is worth noting that only a few of studies have explored the application of ViTs in breast cancer detection, segmentation, and classification [18]. Furthermore, the full potential of ViTs in this domain has yet to be fully explored, as emphasized in recent research articles.

Wang et al. [16] propose to use a custom consistency training strategy to enhance the robustness of the model by combining supervised and unsupervised learning with image augmentation. The method is evaluated on two datasets comprising ultrasound and histopathology images (Dataset of breast ultrasound images – BUSI and the Breast Cancer Histopathological Image Classification – BreakHis). To strategically sample the most significant tokens from the input image, an adaptive token sampling technique is utilized. The core model employed in the training process is called ATS-ViT. The training procedure consists of two parts: supervised training, which improves the model's predictive ability, and consistency training, which enhances its generalization. These two parts are unified through an end-to-end training procedure. The best approach achieves an average test accuracy of 0.98 on the BreakHis dataset.

Tummala et al. [13] introduce an architecture called SwinT, a variant of ViT. SwinT operates on the concept of non-overlapping shifted windows. The study focuses on evaluating the performance of an ensemble of SwinTs in two-class classification (benign/malignant) and eight-class classification (four benign and four malignant subtypes) using the BreaKHis dataset. The distinctive aspect of SwinT is its utilization of shifted windows, organized in a non-overlapping manner. This hierarchical improved version of ViT benefits from linear complexity and scalability in window-based self-attention computations. The shifted window scheme in SwinT enhances efficiency by confining self-attention calculations to local, non-overlapping windows while facilitating cross-window connections. The ensemble of SwinTs, which includes the tiny, small, base, and large variants, achieves an average test accuracy of 0.96 for the eight-class classification and 0.99 for the two-class classification.

Alotaibi et al. [1] introduce an ensemble model that combines two pretrained models: ViT and DeiT. The proposed ViT-DeiT model focuses on classifying breast cancer histopathology images into eight classes using the BreakHis public dataset, with four classes classified as benign and the remaining as malignant. While the ViT model operates similarly to previous studies, the novelty lies in the utilization of DeiT, which is trained on the ImageNet dataset and exhibits improvements over previous ViT models. Both the ViT and DeiT models are pretrained on ImageNet with an input size of 224×224 and a batch size of 16, serving as the base models' representations. The ensemble technique employs a soft voting approach, where the predicted label for each sample is assigned based on the highest average probability. The experimental results showcase an accuracy of 0.98.

He et al. [7] propose a network model DecT, which incorporates color deconvolution through convolution layers. The DecT model utilizes a method similar to the residual connection network to combine information from both RGB and HED color space images. This fusion helps compensate for information loss when transferring RGB images to HED images. The training process of the

DecT model consists of two stages, allowing the parameters of the deconvolution layer to better adapt to different types of histopathological images. To mitigate overfitting during model training, color data augmentation is employed. On the BreakHis dataset, the DecT model achieves an average accuracy of 0.93. For the BACH and UC datasets, the DecT model achieves an average accuracy of 0.79 and 0.81, respectively.

2.2 Vision Transformers

Image classification has been a critical task in computer vision, with numerous applications in fields such as autonomous vehicles, medical imaging, and surveillance systems. CNNs have been the dominant approach to tackle image classification tasks, due to their unparalleled performance in capturing local patterns and hierarchical structures [6]. However, recent advancements in deep learning have paved the way for the emergence of a new paradigm known as ViT, which have demonstrated remarkable success in image classification tasks [3].

The success of Vision Transformers in image classification can be attributed to several factors. The self-attention mechanism provides the model with the ability to dynamically assign weights to different parts (referred to as patches) of the image, thereby enhancing the representation of the overall context. This capability is especially advantageous when the important information necessary for discrimination is scattered across distant areas of the image. Secondly, Vision Transformers can be easily scaled up to process larger input resolutions and deeper architectures, enabling improved performance on complex datasets. Lastly, the pretraining and fine-tuning process employed by Vision Transformers allows for effective transfer learning, making them particularly suitable for scenarios with limited labeled data.

2.3 BACH: Grand Challenge on Breast Cancer Histology Images

The Breast Cancer Histology (BACH) challenge [2] was a significant event in the field of medical imaging, aiming to promote the development of automated breast cancer detection and diagnosis methods. The BACH challenge was divided into two parts, A and B. Part A consisted of automatically classifying H&E stained breast histology microscopy images in four classes: 1) Normal, 2) Benign, 3) In situ carcinoma, and 4) Invasive carcinoma. Part B consisted of performing pixel-wise labeling of whole-slide breast histology images in the same four classes. In this paper, we participate in a single part of the challenge and focus on part A.

The BACH challenge made available two labeled training datasets for the registered participants. The Part A dataset is composed of microscopy images annotated image-wise by two expert pathologists. The Part B dataset contains pixel-wise annotated and non-annotated WSI images. For the WSI, annotations were performed by a pathologist and revised by a second expert. The training data is publicly available from the BACH challenge website (upon registration). Testing datasets were made public but without annotations.

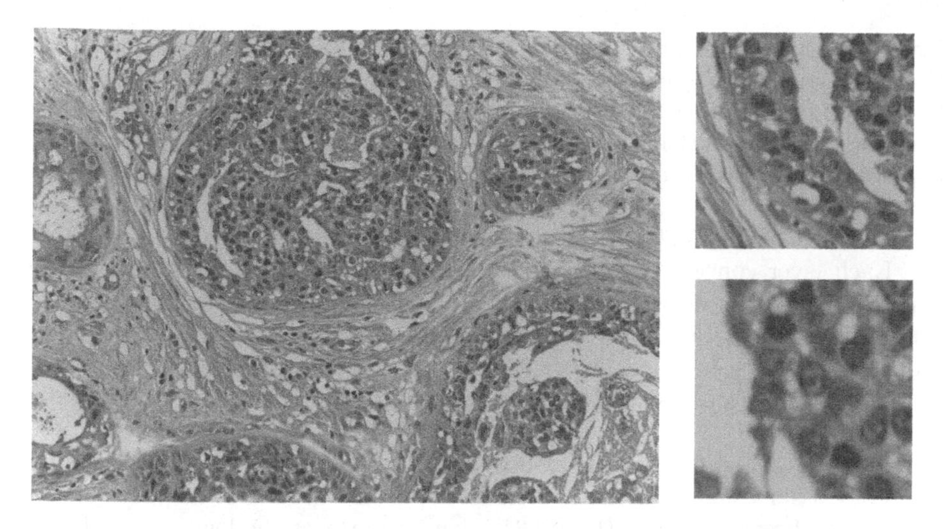

Fig. 1. On the left side, an in situ carcinoma image from the training set. On the right side, two sample patches in the two sizes used in our experiments.

In this paper, we focus on Part A, because the post-challenge submissions, which needed to have an evaluation of the results, were allowed only on Part A images and tasks. The Part A dataset is composed of 400 training images, with the four classes equally represented, and 100 test images. All images were acquired using a Leica DM 2000 LED microscope and a Leica ICC50 HD camera, and all patients are from the Porto and Castelo Branco regions in Portugal. Cases are from Ipatimup Diagnostics and come from three different hospitals. The annotation was performed by two medical experts, and images where there was disagreement between the Normal and Benign classes were discarded. The remaining doubtful cases were confirmed via immunohistochemical analysis. The provided images are in RGB .tiff format and have a size of 2048×1536 pixels with a pixel scale of $0.42\,\mu m \times 0.42\mu m$. Figure 1 shows an example image, with examples of the two patch sizes we tested in our experiments. Participants were given a partial patient-wise distribution of the images in the training set. The test data was collected from a completely different set of patients, ensuring a fairer evaluation of the methods.

The first two teams participating to the Part A challenge obtained an accuracy of 0.87 [2]. The submission was re-opened for post-challenge evaluations, and in this phase, some teams obtained slightly better results, sometimes including data from other datasets. For example, Yao et al. [17] reached an accuracy of 0.92 with a complex ensemble of CNN and RNN on the BACH test dataset.

3 Methodology

In this study, we propose a self-attention ViT tailored for breast cancer histology image classification. The architecture is designed to effectively capture both local and global contextual information, enabling accurate and robust predictions of breast cancer subtypes. The input to our ViT consists of a sequence of fixed-size image patches. Each patch is linearly embedded into a flat vector and combined with a positional embedding to retain spatial information. The resulting embeddings form the input sequence, which is then fed into the transformer layers.

The core of our ViT architecture is a stack of identical transformer layers. Each layer consists of a multi-head self-attention mechanism, followed by a position-wise feed-forward network. The self-attention mechanism computes attention scores for each patch in relation to all other patches in the input sequence, enabling the model to capture both local and global contextual information. The position-wise feed-forward network further refines the learned patch representations.

After processing the input sequence through the set of transformer layers, we apply a classification head in the form of a project applied to the final hidden state of the first token (i.e., the `[CLS]` token) to predict the logits of the breast cancer subtypes (i.e., the classes considered). The classification head consists of a linear layer that performs the desired projection followed by a softmax activation function, providing a probability distribution over the considered classes.

In this study, we consider different training strategies and configurations for our self-attention ViT model to explore their impact on the performance of breast cancer histology image classification. The parameters we investigate are as follows.

- *Base Model*: We use the ViT model as the base architecture for all our experiments. The ViT model has been proven effective in various computer vision tasks, including image classification. By using this model, we aim to leverage its self-attention mechanism and adapt it to the task of breast cancer histology image classification.
- *Pretraining Strategy*: We experiment with two strategies concerning the use of pretrained weights. The first strategy involves employing a ViT model that has been pretrained on a large-scale dataset. In more detail, we leverage the `google/vit-base-patch16-224` model[2] which was pretrained on ImageNet-21k (14M images and 22K classes) at resolution 224 × 224 and fine-tuned on ImageNet 2012 (1M images and 1K classes) at the same resolution. This approach is motivated by the transfer learning paradigm, which leverages the knowledge learned from the source domain to improve performance in the target domain. The second strategy involves using the same ViT architecture but initializing the weights randomly, effectively training the model from scratch. This allows us to investigate the performance of the ViT model when trained solely on the breast cancer histology image dataset, without any influence from pretrained weights.
- *Resize Dimension*: We investigate the impact of resizing the input images to a fixed dimension of 224 × 224 pixels, which is a common practice in image classification tasks. This standardization of input dimensions allows the model

[2] See https://huggingface.co/google/vit-base-patch16-224.

to process images more efficiently and ensures consistency across the dataset. In addition, we explore the effects of not resizing the input images, which enables the model to work with the original dimensions and potentially retain more information from the images.
- *Data Augmentation*: We delve into the utilization of geometric data augmentation methods, including rotation, scaling, and flipping, as a means to enhance the model's ability to generalize. In addition to geometric transformations, we also experiment with color data augmentation such as color jittering. By implementing these modifications to our input images, we can artificially expand and diversify our dataset, thereby enabling the model to acquire more robust and invariant characteristics. Furthermore, we contemplate a situation where no data augmentation is implemented to evaluate the model's performance devoid of the supplementary diversity introduced through these augmentation procedures. Lastly, we turn our attention to the Part B dataset. Finally, we incorporated tiles that have been cropped from Whole Slide Images (WSIs). A single whole-slide image can contain multiple regions classified as normal, benign, in situ carcinoma, and invasive carcinoma. The annotation of these whole-slide images was carried out by two medical professionals, and any images with conflicting classifications were excluded.
- *Tile Overlap*: The tile overlap is the number of pixels that two adjacent tiles share in common. Increasing the overlap can improve the model's performance by enabling better integration of information across tiles. However, this comes at the cost of increased computation, as the same region of the image is processed multiple times. Therefore, we aim to find the optimal overlap that balances model performance and computational efficiency.
- *Patch Size*: The patch size determines the amount of local information that the model can capture from each patch. Larger patches allow the model to capture more local details, while smaller patches enable the model to capture more global context. However, larger patches also increase the computational cost, as each patch contains more pixels. Therefore, we aim to find the optimal patch size that balances the trade-off between capturing local and global information while maintaining computational efficiency.

Let us make some remarks on the methodology used in this paper. In this study, our primary attention is directed toward Part A of the challenge dataset. The specific reasoning behind this decision is rooted in the fact that only post-challenge submissions were allowed on Part A images and their corresponding tasks. As such, to evaluate our results, we found it necessary to focus merely on Part A.

Moving forward, it is also important to clarify our decision to use exclusively the challenge dataset for this research. The choice to limit our dataset to the one provided by the challenge stems from our research objectives. By narrowing our focus, we intend to delve deeply into the nuances of the challenge dataset and focus solely on breast cancer histology images. By maintaining this strict focus, we ensure that our results are as pertinent as possible to the specific context of the challenge and of such specific research direction.

Lastly, in terms of the methodologies applied, we made a deliberate choice to employ only ViT-based models over the more commonly used convolutional networks. This choice is not arbitrary; instead, it is a decision based on the typical training strategy of ViT models. Traditional ViT training is based on "common images" such as cats and dogs, and not on medical images like histology images. Given the rising prominence of ViTs and transformers in general, which are becoming a de-facto standard in many applications, we aim to explore how ViTs behave when they are tasked with the analysis of histopathological images. This exploration will potentially help to uncover new insights into their applicability in fields beyond their typical usage.

4 Experimental Evaluation

To evaluate the performance of our self-attention ViT model for breast cancer histology image classification, we submit our predictions to the challenge submission site for evaluation. As we have no access to the labels for the test set, this is to be considered a full-blind submission, meaning that we do not have any information about the correctness of our predictions. The only information we receive in return from the submission system is a single accuracy score, which represents the percentage of correctly classified images in the test set.

Accuracy is a commonly used metric for classification tasks, as it provides a simple and easy-to-understand measure of how well our model performs on the given task. However, it is important to note that accuracy may not always be the best metric to evaluate model performance, as it can be biased towards the majority class in imbalanced datasets. In such cases, other metrics such as precision, recall, F1-score, and AUC-ROC may provide a better measure of the model's performance. Despite these potential limitations, in the specific context of the dataset used in this challenge, we are bound to rely exclusively on the accuracy metric to assess our model's effectiveness. This is mainly due to our lack of access to the actual labels of the images. Furthermore, adherence to the accuracy metric is necessitated by the fact that it is the official performance metric as used by the challenge. Consequently, our performance evaluation aligns with the challenge's evaluation standards, ensuring the comparability of our results with other submissions. Thus, in the following, we analyze the accuracy scores obtained on the test set to understand the performance of our model in different scenarios, such as when using different hyper-parameters, data augmentation techniques, or preprocessing strategies as defined above.

The experimental results showing the effectiveness scores for the breast cancer histology image classification task are presented in Table 1. We evaluated the performance of different ViT models (rows) using various strategies for pretraining, data augmentation, tile overlap, patch size, and image resizing (columns). The accuracy score is used as the evaluation metric (last column).

The first row of the table shows the results obtained with the base ViT model, using 512×512 tiles and an overlap of 256 with a patch size of 32×32,

Table 1. Experimental Results

Base Model	Pretrain Strategy	Resize Dimension	Data Augmentation	Tile Overlap	Patch Size	Accuracy Score
ViT	✗	✗	✗	256	32	0.53
ViT	✗	224 × 224	✗	256	16	0.53
ViT	✓	224 × 224	✗	256	16	0.84
ViT	✓	224 × 224	✗	✗	16	0.84
ViT	✓	224 × 224	Geometric, Color	✗	16	0.90
ViT	✓	224 × 224	Geometric, Color, Set B	✗	16	0.91

with no image resizing or pretraining. This resulted in an accuracy score of 0.53, suggesting that 32×32 patch size may not capture enough local information for the model to make accurate predictions.

In the second row, we used 512×512 tiles resized to 224×224. Only the breast cancer dataset was used to fine-tune the model. The model used a patch size of 16×16 and a tile overlap of 256. The accuracy score was 0.53.

The third row of the table shows the results obtained using the base ViT model which was pretrained on ImageNet and fine-tuned on the breast cancer dataset with 512×512 tiles resized to 224×224, the tile overlap to 256, maintaining the same patch size of 16×16. The accuracy score obtained by this model was 0.84, indicating the importance of pretraining on large-scale datasets like ImageNet.

In the fourth row, we turned off tile overlap. Images used have a size of 2048×1536 pixels and were resized to 224×224. The accuracy score obtained by this model was 0.84, indicating that a larger tile overlap does not necessarily improve the model's performance. In fact, it may lead to information redundancy and increased computation.

In the fifth row, we experimented using the same base ViT model which was pretrained on ImageNet and fine-tuned on the breast cancer dataset. The model used a patch size of 16×16, and there was no tile overlap. Additionally, geometric and color data augmentation techniques were used to improve the model's performance. The accuracy score obtained by this model was 0.90. This suggests that the geometric and color transformations were helpful in improving the model's ability to generalize to new images.

In the sixth row, we used the same ViT model that was pretrained on ImageNet, but this time we fine-tuned it on a breast cancer dataset consisting of two parts. The first part comprised original H&E stained breast histology microscopy images, which were labeled on an image-wise basis (referred to as Part A). The second part consisted of cropped tiles from WSI with pixel-wise labeling (referred to as Part B). The pixel scale for Part A was 0.42 µm, while for Part B it was 0.467 µm. Consequently, we deemed it necessary to re-scale the images in the second set. All the tiles generated and used in this experiment had a size of

2048×1536 pixels, with a pixel scale of $0.42\,\mu m \times 0.42\mu m$. In this final experiment, we normalized all the images to minimize stain variability. The method employed for normalizing the histology images was the Macenko color normalization method [9]. The accuracy score was 0.91.

In summary, our experimental results hint that by pretraining on ImageNet and employing geometric and color data augmentation techniques, we can enhance the performance of the ViT model when applied to the task of classifying breast cancer histology images. Furthermore, we found that a patch size of 16×16, with no tile overlap, yielded optimal results for this specific task. These findings provide valuable insights for the development of future ViT-based models designed for similar image classification tasks.

5 Discussion and Conclusion

Our results highlight the potential benefits of leveraging self-attention mechanisms via the ViT model in the challenging task of breast cancer histology image classification. We found that the pretraining task significantly improved the performance of the model, as evidenced by the accuracy drop when pretraining was removed. This result underscores the advantage of transferring knowledge from large-scale, general-purpose datasets to specific medical imaging tasks, a strategy that can potentially enhance the model's generalization capacity and robustness to dataset bias.

The obtained accuracy - 0.91 - improves the best results of the BACH challenge (accuracy: 0.87) [2], but is still slightly lower than the best post-challenge performance (accuracy: 0.92) [17]. However, the latter result has been obtained with an ensemble of five models (different CNNs and RNN), resulting in a computationally more intensive method than ours.

The effect of geometric and color data augmentation was another notable finding. With the augmentation turned off, the accuracy dropped from 0.90 to 0.84, suggesting that such augmentation techniques improve the model's ability to generalize to unseen images, likely by enhancing its robustness to minor variances in the data. It is also worth noting how the use of tiles derived from entire digital slides belonging to set part B and the application of Macenko's color normalization method has resulted in an accuracy of 0.91 on the test set.

We also explored the impact of patch size and overlap on model performance. The results indicated that a patch size of 16×16 without overlap yielded optimal results, reinforcing the importance of proper patch configuration in ViT models. It appears that larger tile overlap and large patch size may not be beneficial, possibly due to information redundancy and insufficient local information, respectively. However, the parameter that seemed to negatively impact more on accuracy is tiling. The worst results were obtained by splitting the images into tiles, followed or not by resizing. This in principle would have helped in preserving more information, yet somewhat unexpectedly it did not help. This is most likely due to the fact that while each original image was representative of a specific class, it did not contain only tissue of such class, in particular for pathological tissues. So, likely many tiles are unspecific.

In summary, our work presents a promising study on the application of self-attention ViT to breast cancer histology image classification. Our findings demonstrate the significance of pretraining, geometric and color data augmentation, and appropriate patch configuration in enhancing model performance. These insights could serve as valuable guidelines for designing future ViT-based models for similar image classification tasks in the medical imaging domain.

Moving forward, results from this paper open several potential research directions. One significant direction is the exploration of custom pretraining objectives [12] which can be tailored to better suit the task by incorporating objectives that are more closely aligned with the specifics of histology image analysis. In addition, we plan to examine the impact of custom pretraining datasets. While pretraining on large-scale, general-purpose datasets like ImageNet has proven beneficial, using pretraining datasets that are more domain-specific, such as large-scale histopathological image databases, could further enhance the model's performance. Finally, we plan to explore different attention mechanisms within the transformer architecture [5]. Although the standard global self-attention mechanism has shown promising results, attention variants such as local or hierarchical attention may offer advantages by restricting the attention scope to a local neighborhood and better capture the multi-scale structure of the data.

Acknowledgements. The present work is partially funded by the BosomShield Doctoral Network, HORIZON-MSCA-2021-DN-01-01 grant number 101073222.

References

1. Alotaibi, A., et al.: Vit-deit: an ensemble model for breast cancer histopathological images classification. In: 2023 1st International Conference on Advanced Innovations in Smart Cities (ICAISC), pp. 1–6. IEEE (2023)
2. Aresta, G., et al.: Bach: grand challenge on breast cancer histology images. Med. Image Anal. **56**, 122–139 (2019)
3. Dosovitskiy, A., et al.: An image is worth 16×16 words: transformers for image recognition at scale (2021). https://doi.org/10.48550/arXiv.2010.11929
4. Gardezi, S.J.S., Elazab, A., Lei, B., Wang, T.: Breast cancer detection and diagnosis using mammographic data: Systematic review. J. Med. Internet Res. **21**(7), e14464 (2019)
5. Guo, M.H., et al.: Attention mechanisms in computer vision: a survey. Comput. Vis. Media **8**(3), 331–368 (2022)
6. Han, K., et al.: A survey on vision transformer. IEEE Trans. Pattern Anal. Mach. Intell. **45**(1), 87–110 (2022)
7. He, Z., et al.: Deconv-transformer (DECT): a histopathological image classification model for breast cancer based on color deconvolution and transformer architecture. Inf. Sci. **608**, 1093–1112 (2022)
8. Lin, T., Wang, Y., Liu, X., Qiu, X.: A survey of transformers. AI (2022)
9. Macenko, M., et al.: A method for normalizing histology slides for quantitative analysis. IEEE International Symposium on Biomedical Imaging: From Nano to Macro, Boston, MA, USA, pp. 1107–1110 (2009)
10. Reinhard, E., Ashikhmin, M., Gooch, B., Shirley, P.: Color transfer between images. IEEE Comput. Graphics Appl. **21**, 34–41 (2001)

11. Roy, K., Banik, D., Bhattacharjee, D., Nasipuri, M.: Patch-based system for classification of breast histology images using deep learning. Comput. Med. Imaging Graph. **71**, 90–103 (2019)
12. Tay, Y., et al.: Scale efficiently: insights from pre-training and fine-tuning transformers. arXiv (2021)
13. Tummala, S., Kim, J., Kadry, S.: Breast-net: multi-class classification of breast cancer from histopathological images using ensemble of swin transformers. Mathematics **10**(21), 4109 (2022)
14. Vaswani, A., et al.: Attention is all you need. In: NIPS, vol. 30 (2017)
15. Vesal, S., Ravikumar, N., Davari, A.A., Ellmann, S., Maier, A.: Classification of breast cancer histology images using transfer learning. In: Campilho, A., Karray, F., ter Haar Romeny, B. (eds.) ICIAR 2018. LNCS, vol. 10882, pp. 812–819. Springer, Cham (2018). https://doi.org/10.1007/978-3-319-93000-8_92
16. Wang, W., Jiang, R., Cui, N., Li, Q., Yuan, F., Xiao, Z.: Semi-supervised vision transformer with adaptive token sampling for breast cancer classification. Front. Pharmacol. **13**, 929755 (2022)
17. Yao, H., Zhang, X., Zhou, X., Liu, S.: Parallel structure deep neural network using CNN and RNN with an attention mechanism for breast cancer histology image classification. Cancers **11**(12), 1901 (2019)
18. Zhao, Y., Zhang, J., Hu, D., Qu, H., Tian, Y., Cui, X.: Application of deep learning in histopathology images of breast cancer: a review. Micromachines **13**(12), 2197 (2022)

Editable Stain Transformation of Histological Images Using Unpaired GANs

Tibor Sloboda(✉), Lukáš Hudec, and Wanda Benešová

Vision and Graphics Group, Faculty of Informatics and Information Technology, Slovak University of Technology, Bratislava, Slovak Republic
xslobodat2@stuba.sk
https://vgg.fiit.stuba.sk/

Abstract. Double staining in histopathology is done to help identify tissue features and cell types differentiated between two tissue samples using two different dyes. In the case of metaplastic breast cancer, H&E and P63 are often used in conjunction for diagnosis. However, P63 tends to damage the tissue and is prohibitively expensive, motivating the development of virtual staining methods, or methods of using artificial intelligence in computer vision for diagnostic strain transformation. In this work, we present results of the new xAI-CycleGAN architecture's capability to transform from H&E pathology stain to the P63 pathology stain on samples of breast tissue with presence of metaplastic cancer. The architecture is based on Mask CycleGAN and explainability-enhanced training, and further enhanced by structure-preserving features, and the ability to edit the output to further bring generated samples to ground truth images. We demonstrate its ability to preserve structure well and produce superior quality images, and demonstrate the ability to use output editing to approach real images, and opening the doors for further tuning frameworks to perfect the model using the editing approach. Additionally, we present the results of a survey conducted with histopathologists, evaluating the realism of the generated images through a pairwise comparison task, where we demonstrate the approach produced high quality images that sometimes are indistinguishable from ground truth, and overall our model outputs get a high realism rating.

Keywords: CycleGAN · Explainability · Histopathology

1 Introduction

Histopathology images are often difficult to register properly, leading to issues in using paired approaches, which may yield better results. Often, there is also significant tissue damage caused by certain immunohistochemical stains used in particular to help pathologists identify the presence of cancerous cells in certain types of tissue.

G. L. Foresti et al. (Eds.): ICIAP 2023 Workshops, LNCS 14366, pp. 27–38, 2024.
https://doi.org/10.1007/978-3-031-51026-7_3

In our case, we're using a dataset consisting of H&E and P63 stained breast tissue images with the presence of metaplastic cancer. With cancer being among the top leading causes of disease-related death since 2016 [7,12], prevention by screening and early detection becomes paramount to combat this disease.

If an abnormal growth is found during inpatient screening for cancer, biopsies are a common next step in diagnosing the issue. This tissue is usually stained with Hematoxylin and Eosin (H&E) to aid screening for cancerous growths. While H&E stained tissue alone is sufficient and contains the necessary information to identify the presence of cancerous cells, this information is difficult to identify from the microscopy images. For this reason, immunohistochemical dyes are also used to help identify cancerous cells. Unfortunately, these dyes are often very expensive and also have the potential to damage the tissue, causing it to tear often, which is the case in our data as well. The effects of this are demonstrated in Fig. 1.

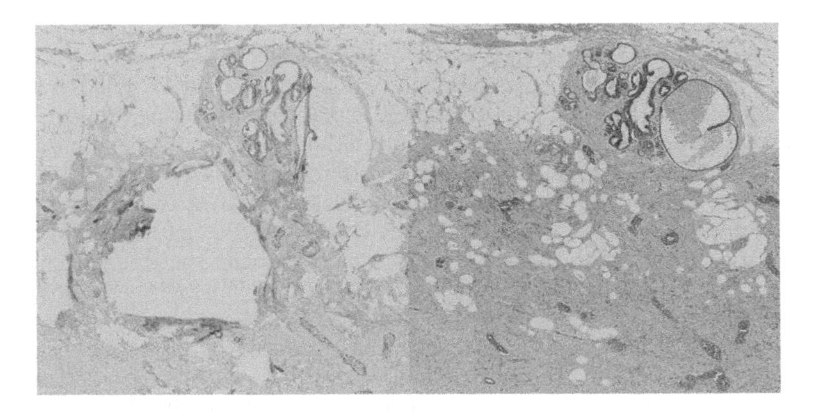

Fig. 1. Demonstration of tissue damage in paired and aligned P63 stained tissue (left) compared with its H&E counterpart (right). The issue is ripped in various places.

For this reason, CycleGAN [17] and its derivatives are popular in this field for unpaired unsupervised training, and we used a modified CycleGAN architecture in our case as well called xAI-CycleGAN [13]. We demonstrate its superior image quality in comparison to other existing methods.

Approach. We utilize xAI-CycleGAN with distilled features from the training data stored in the interpretable latent variable. We enhance this approach with an editable generation output on the basis of the interpretable latent variable using a modified semantic factorization algorithm, and introduce context loss to better preserve tissue structure and produce high quality images.

The output editing opens the doors to new possibilities in terms of the ability to control the generation of output using an easy-to-use tool that could serve to produce more data in cooperation with histopathologists.

We quantitatively evaluate our results by asking histopathologists to do a pair-wise comparison task in a survey to attempt to discern between a real and generated image, and then rate the realism of the image they believe not to be real. By this we show the approach has great potential but requires further work before it is usable in a medical setting. The code is available in our GitHub repository.

Our Contribution

- xAI-CycleGAN enhanced with an editable output using the interpretable latent variable
- Introduction of context loss for improved preservation of tissue structure
- Improved output quality and reduced counterfactuals compared to previous solution

2 Related Work

Various works have demonstrated relatively successful virtual staining of histopathological tissue [1] and transformation from one type of stained tissue to another [1,2,6]. Registration remains a consistent issue due to the nature of the problem, though it is possible. For paired transformation, various GAN architectures were used, from a simple GAN and DCGAN [10], as well as sequential cascaded GANs which show common use in the medical domain [3].

For the unpaired approach, CycleGAN [17] seems to be the dominant approach in nearly all cases. Therefore, we will look a bit closer at some related work that we predominantly use or relate to, below.

2.1 Overview of xAI-CycleGAN

xAI-CycleGAN is an enhanced version of the CycleGAN [17] architecture that aims to improve the convergence rate and image quality in unsupervised image-to-image transformation tasks [13]. It incorporates the concepts of explainability to provide a more powerful and versatile generative model.

One of the primary contributions of xAI-CycleGAN is the incorporation of explainability-driven training. Inspired by the work of Nagisetty et al. [8], xAI-CycleGAN utilizes saliency maps from the discriminator to mask the gradients of the generator during backpropagation.

In addition, xAI-CycleGAN leverages the insights from the work of Wang M.'s Mask CycleGAN [15], which introduces an interpretable latent variable using hard masks on the input. By combining these approaches, xAI-CycleGAN achieves enhanced explainability and convergence by taking advantage of information leakage into the interpretable latent variable.

One issue with this architecture is the production of counterfactuals by the generator and various artifacts and repeating patterns which may be worsened in the histopathological domain, and must be addressed.

We do this by introducing a new loss that attempts to preserve the structure and enforce context, by separating the context and style and forcing all style transformation to happen at the latent level of the generative model and preventing the encoder or decoder portion from doing any color transformation.

2.2 SeFa Algorithm for Editable Outputs

The Closed-Form Factorization of Latent Semantics in GANs (SeFa) algorithm enabled the means to control the generated output in semantically interpretable ways [11], such as the ability to modify the expression, color, posture, or other human-interpretable aspects.

The approach identifies semantically significant vectors in latent space with high variability, which are added to the activations from these layers by multiplying the eigenvector by an arbitrary factor.

Unfortunately, this approach only works with simpler generative networks that draw from noise distributions to generate images. In CycleGAN and other transformative generative models, the auto-encoder-like structure that starts and ends with an image does not have a 2-dimensional latent weight matrix, but instead a 4-dimensional one, making it impossible to apply this algorithm in its pure form.

We modify this approach to allow modification using the interpretable latent variable used in xAI-CycleGAN with some degree of success, but lose semantically interpretable aspects of this method since we only utilize one layer for this purpose due to difficulty of training and stability reasons.

Other methods also exist for image modification, such as the use of sequential GANs for editing [4,5,9] or inpainting, however all rely on a similar method as SeFa, with the difference of some using attention-based mechanisms [14] to achieve this. The common problem still remains that these methods cannot be easily adapted to an auto-encoder-like architecture for unpaired image transformation, such as CycleGAN.

2.3 cCGAN for Stain Transformation

cCGAN (conditional CycleGAN) [16] is a notable approach in the domain of stain transformation for histology images. It aims to translate images from one stain to another while preserving important structural information.

The cCGAN approach employs a conditional variant of the CycleGAN architecture, incorporating additional information such as staining information as input to guide the translation process.

While cCGAN has shown promise in transforming between stains, there are some limitations to consider. One challenge is the preservation of fine-grained structural details during the transformation process. Due to the absence of explicit constraints on structure preservation, cCGAN may struggle to accurately preserve important structural information, which is shown in their results.

We address the issue of structure preservation by incorporating context loss, which enforces consistency between encoded representations, and emphasizes the preservation of structural features during the transformation.

3 Methods

Here we will focus primarily on the context preserving approach by using context loss to improve carry-over of structural information and also fixing the artifacts and counterfactuals produced by xAI-CycleGAN. We also introduce a modified algorithm to edit the outputs of the generator to be able to adjust generation results to better match ground truth images of the matching domain.

3.1 Dataset

The dataset consists of 32 pairs of H&E and P63 stained tissue samples of varying resolutions, all around 3–5 gigapixels in total size. These were cut into 1024 by 1024 pixel slices, and then downscaled to 256 by 256 which was the final resolution used during training.

Each slice was converted to LAB color space before being passed to the model, and checked by a combined method of luminosity-based histogram and image entropy to determine whether the image contained a sufficient amount of tissue, or whether it only consisted of background. This resulted in a total of 110'564 training samples used to train the model.

The tissue samples were obtained internally, and thus are not available publicly.

3.2 Separating Structure from Style

In order to preserve the structure of tissue and enforce only the style alone to change, we introduce the context-preserving loss, or just context loss. Context loss takes the encoded features of the same original image from the encoder of both generators, in both domains, and compares them to each other using Huber loss, and also compares the encoded features of the fake/output image on input of both generators across both domains.

This ensures that the encoder focuses entirely on the structural aspects of the image rather than style aspects, in order to ensure that no style conversion is done by the encoder and that the structure is identical between both the transformed image and the original image after the encoding step or also before the decoding step of the cycled images. This confirms that structure is preserved while only style changes, with most structure information being stored in the weights of the encoder and decoder step (Fig. 2).

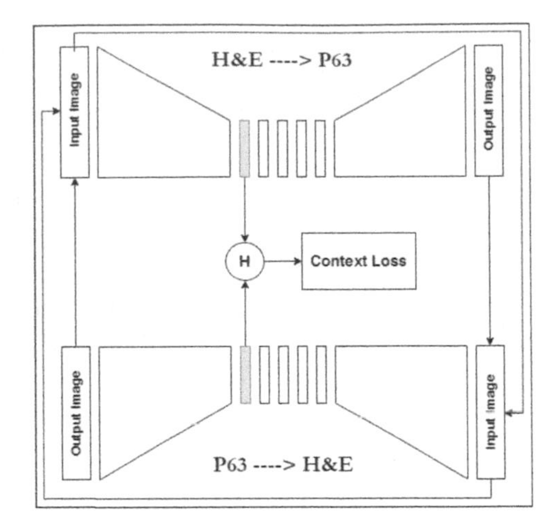

Fig. 2. Demonstration of the context loss computation. The encoded original image representation from both domains is compared to produce a partial loss, and added with the encoded fake/output representation of both domains in both generators to produce the final loss.

If we consider the H&E to P63 generator encoder as a function $X_e()$ and the P63 to H&E generator encoder as a function $Y_e()$, and then an original H&E image to be A and the generated counterpart A', then a P63 original image to be B and the generated counterpart B', and finally the Huber loss function as $H()$, then we can define the context loss as follows:

$$H(X_e(\alpha), Y_e(\beta)) = H_\gamma(\alpha, \beta) \tag{1}$$

$$\mathcal{L}_{context} = \frac{H_\gamma(A, A) + H_\gamma(B, B)}{2} + \frac{H_\gamma(A', B) + H_\gamma(B', A)}{2} \tag{2}$$

This loss is further adjusted by a parameter to weigh its importance in the total objective function for the generators, which is a hyperparameter. Our context loss enforces that the encoders should behave the same in both generators regardless of the input image, in order to preserve the structure. This technically means that after training only a single encoder and decoder is needed for both domains with only the latent transformation step being separate.

3.3 Editable Generation Results Using SeFa

We utilize the SeFa algorithm to extract important eigenvectors from the interpretable latent variable in xAI-CycleGAN, akin to the method in [11] and incorporate them into the generation process. The goal is to enable fine-grained editing of the generated images by manipulating specific attributes or semantics through the modification of the interpretable latent variable.

To apply SeFa in xAI-CycleGAN, we first extract the most significant eigenvectors from the interpretable latent variable. Once the important eigenvectors are obtained, we incorporate them back into the weights of the corresponding layer during the forward pass of the network. This is achieved by multiplying the most significant eigenvector with its transposed self to produce a matrix of the same size as the weight matrix of the interpretable latent variable.

$$W^* = W + \sum_{i=j}^{k} \eta_i^T \eta_i \cdot m_i \tag{3}$$

In Eq. 3, W^* represents the resulting weight matrix for the forward pass, with W being the base one, then η_i represents the i-th eigenvector of W, and m_i is the m-th multiplicative factor for the given eigenvector.

This is the only modification to the algorithm and presents a unique approach which allows us to use only the single layer at a cost of losing the semantically interpretable directions, but allowing us to utilize this algorithm in CycleGAN, an entirely new context for the algorithm, only made possible by the intentional leakage of semantically important information into the interpretable latent variable.

We allow a choice of any range of top 16 most important eigenvectors, where all are applied to the matrix with the same multiplicative factor in order.

We developed an easy-to-use tool using Streamlit shown in Fig. 3 to allow modifying generated outputs on the fly and explore the effects of this output editing, and to allow to interactively edit the images to more closely resemble the ground truth images.

4 Results

Using the already well-performing xAI-CycleGAN further enhanced with context loss, we vastly increase the visual fidelity and level of structure preservation in the images, while also preventing any counterfactuals or repeating patterns from appearing, which was an issue shown in xAI-CycleGAN.

We observe consistently high quality outputs for a variety of inputs seen in Fig. 4, however upon closer examination with the help of a histopathologist, there are inconsistencies or impossible arrangements in the produced tissue based on the coloring of the cells with the P63 dye, which shows the model was not able to fully and correctly learn the complexities of how the tissue should look, but it has shown relatively good attempts to correctly identify myoepithelial cells which are colored brown in the P63 stained images.

In some cases, we were able to use our new editing approach to modify the outputs of the generator in order to replicate the ground truth appearance relatively well, for instance in Fig. 5.

As visible in Fig. 5, the method can be successfully applied to get relatively close to the real counterpart. It is not perfect due to the differences between the two samples used to prepare the H&E stained tissue and P63 stained tissue.

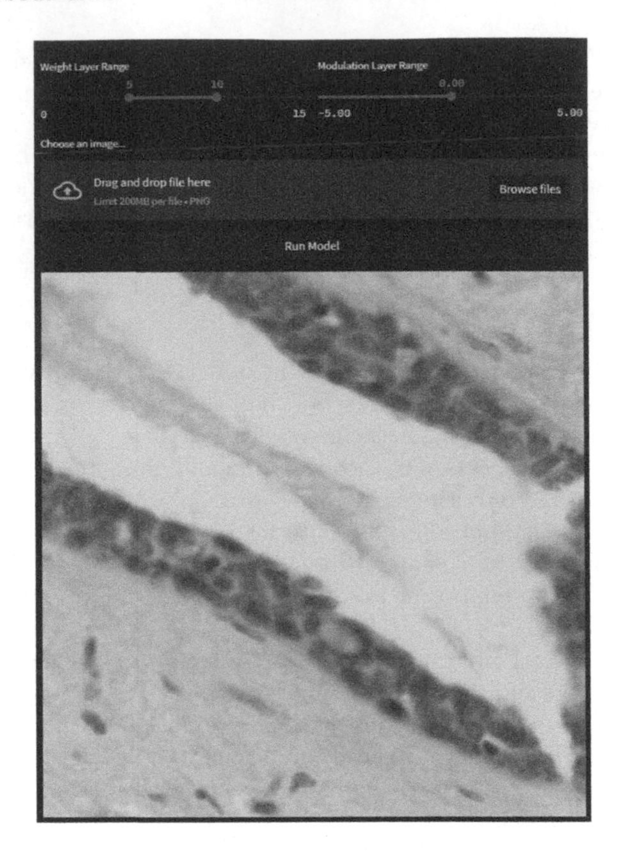

Fig. 3. The user interface to manipulate the image in real time before the image is transformed. Currently, a base H&E image is shown, not yet transformed. For simplicity, we use a single multiplicative factor for every range of eigenvectors due to only using one layer, as such there would be little benefit to manually manipulate each factor, due to lack of semantically interpretable directions with the lack of interpretable latent variable layers.

In addition to qualitative assessment, we also asked 4 histopathologists to take a survey with a pair-wise comparison task to identify the real image from two images, the other image being generated and edited using our method to best match the real counterpart. The histopathologist was also asked to rate the image they believed was generated/fake on a scale from 1 (not realistic) to 6 (very realistic).

In the case where an image was identified incorrectly, we considered the score to be 6 in order to quantitatively assess the performance of the model. These results are demonstrated in Table 1.

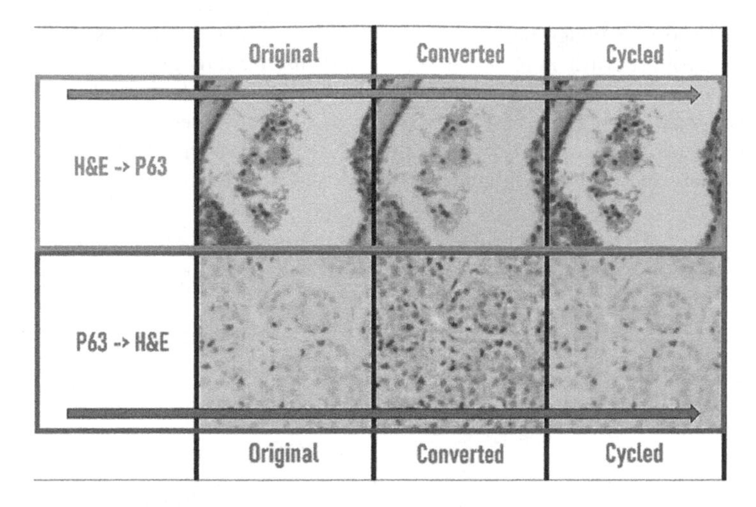

Fig. 4. Example of results generated from the model in both transformation directions, showing very high quality of outputs with preserved tissue structure.

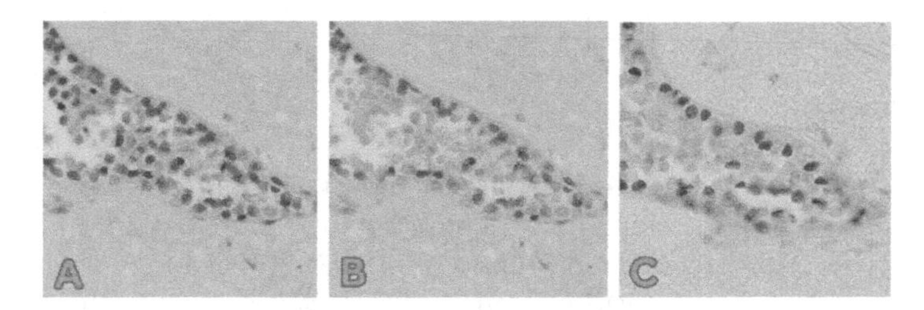

Fig. 5. A demonstration of successful editing capabilities of the tool. **A** contains image converted from H&E to P63 using the tool, without any modifications applied. **B** contains an image with modifications applied using the tool, to best match the real image. **C** contains the unmodified original P63 image of the same region.

We presented a total of 8 image pairs to 4 histopathologists. The results show a **total averaged realism score of 4.94** which is well above the midpoint, where the histopathologists were able to **correctly identify the real image 65.62% of the time**. The histopathologists commented that the visual quality is very high, though without a broader investigation of the whole converted tissue sample it is impossible to truly judge the performance of the model, thus more research is needed, and we see potential for further improving the method.

Table 1. A total of 8 image pairs were presented to 4 histopathologists; A generated image and a ground truth matched image. Each histopathologist was asked to guess the real image, and rate how realistic the generated image looks from 1 to 6. Incorrect guesses were assigned a score of 6 for realism of the generated image. 65.62% of the images were determined correctly, which is relatively close to the ideal 50% which would mean a complete inability to tell generated and real images apart.

	Correctly Identified	Incorrectly Identified	Average Realism Rating
Pair 1	2	2	5.5
Pair 2	2	2	5.25
Pair 3	2	2	5.25
Pair 4	3	1	5
Pair 5	3	1	4.75
Pair 6	3	1	5
Pair 7	2	2	4.75
Pair 8	4	0	4
Total/Average	**21**	**11**	**4.94** (max 6)

5 Discussion

The results obtained from xAI-CycleGAN demonstrate its potential for advancing the field of stain transformation. The structure preservation and editable outputs offer exciting prospects for improving the accuracy and realism of unsupervised image-to-image transformation tasks. However, despite the promising results, there is still considerable room for improvement before xAI-CycleGAN can be effectively deployed in a medical setting.

One area of further research lies in optimizing the editing capabilities of xAI-CycleGAN. While the current approach allows for fine-grained editing by manipulating the interpretable latent variable, there is still a need to identify the optimal editing settings for different images. A potential avenue is training a mini-network on top of xAI-CycleGAN using the data produced from the editing approach. This mini-network could learn to identify the correct editing settings based on the input image, leading to consistently good results. By automating the selection of editing parameters, the system becomes more user-friendly and efficient, reducing the reliance on manual adjustments.

Moreover, expanding the editing approach to include more interpretable layers could provide better control over semantically meaningful transformations. Currently, the editing process primarily focuses on modifying the interpretable latent variable, which may lack direct semantic interpretability. By incorporating additional interpretable layers into the editing framework, it becomes possible to manipulate specific attributes or semantics directly, leading to more intuitive and meaningful editing capabilities. This expansion would offer greater control and customization in generating histology images that closely resemble the ground truth.

While xAI-CycleGAN shows promising potential, several considerations need to be addressed before its practical adoption in medical settings. Robust evaluation on larger and diverse datasets is necessary to validate its performance across various staining protocols and histology image types. The integration of expert knowledge and feedback from histopathologists is crucial for refining the model and ensuring that the transformed images maintain diagnostic relevance and accuracy. Additionally, efforts to further optimize and fine-tune the architecture, loss functions, and training procedures are essential to maximize the model's performance and generalizability.

With further research and development, xAI-CycleGAN holds promise for enhancing the accuracy, efficiency, and clinical utility of histology image analysis in medical practice, as evidenced by the positive feedback from histopathologists that participated in our survey.

6 Future Work

Our present endeavor has laid the foundation for advancements in the field of virtual staining in histopathology, but there are aspects we seek to enhance in future work.

We plan to evaluate our xAI-CycleGAN architecture using a separate, diverse dataset. The intent is not only to affirm the robustness and adaptability of our model, but also to identify specific areas of potential improvements. An additional dataset will provide a more extensive ground for testing the limits and capabilities of our approach. With the extra information, we could better assess our approach's generality and reliability, and adjust our algorithm accordingly.

Additionally, as our model possesses the unique capability of editable outputs, we will explore how we can optimize this feature to improve the generated images. We aim to implement an iterative feedback loop where the edits are used to fine-tune the model, subsequently enhancing the generated outputs.

By moving ahead in these directions, we anticipate refining our approach to create a more precise and widely applicable tool for histopathological stain transformations. The ultimate goal is to establish a system that offers superior performance in generating virtual stains, thus offering the possibility of substantial savings in time, resources, and cost while maintaining diagnostic accuracy.

References

1. Bai, B., Yang, X., Li, Y., Zhang, Y., Pillar, N., Ozcan, A.: Deep learning-enabled virtual histological staining of biological samples. Light: Sci. Appl. **12**(1), 57 (2023)
2. de Bel, T., Hermsen, M., Kers, J., van der Laak, J., Litjens, G.: Stain-transforming cycle-consistent generative adversarial networks for improved segmentation of renal histopathology (2018)
3. Chen, H., Yan, S., Xie, M., Huang, J.: Application of cascaded GAN based on CT scan in the diagnosis of aortic dissection. Comput. Methods Programs Biomed. **226**, 107130 (2022). https://doi.org/10.1016/j.cmpb.2022.107130, https://www.sciencedirect.com/science/article/pii/S0169260722005119

4. Cheng, Y., Gan, Z., Li, Y., Liu, J., Gao, J.: Sequential attention GAN for interactive image editing. In: Proceedings of the 28th ACM International Conference on Multimedia, pp. 4383–4391 (2020)
5. Collins, E., Bala, R., Price, B., Susstrunk, S.: Editing in style: uncovering the local semantics of GANs. In: Proceedings of the IEEE/CVF Conference on Computer Vision and Pattern Recognition, pp. 5771–5780 (2020)
6. de Haan, K., et al.: Deep learning-based transformation of H&E stained tissues into special stains. Nat. Commun. **12**(1), 4884 (2021)
7. Harding, M.C., Sloan, C.D., Merrill, R.M., Harding, T.M., Thacker, B.J., Thacker, E.L.: Peer reviewed: transitions from heart disease to cancer as the leading cause of death in US States, 1999–2016. Prevent. Chronic Dis. **15**(12) (2018). https://doi.org/10.5888/PCD15.180151, https://www.ncbi.nlm.nih.gov/pmc/articles/PMC6307835/
8. Nagisetty, V., Graves, L., Scott, J., Ganesh, V.: xAI-GAN: enhancing generative adversarial networks via explainable AI systems (2020). https://doi.org/10.48550/arxiv.2002.10438, https://arxiv.org/abs/2002.10438v3
9. Pajouheshgar, E., Zhang, T., Süsstrunk, S.: Optimizing latent space directions for GAN-based local image editing. In: ICASSP 2022–2022 IEEE International Conference on Acoustics, Speech and Signal Processing (ICASSP), pp. 1740–1744. IEEE (2022)
10. Salimans, T., Goodfellow, I., Zaremba, W., Cheung, V., Radford, A., Chen, X.: Improved techniques for training GANs. In: Lee, D., Sugiyama, M., Luxburg, U., Guyon, I., Garnett, R. (eds.) Advances in Neural Information Processing Systems, pp. 2234–2242. Curran Associates, Inc. (2016)
11. Shen, Y., Zhou, B.: Closed-form factorization of latent semantics in GANs. In: Proceedings of the IEEE/CVF Conference on Computer Vision and Pattern Recognition (CVPR), pp. 1532–1540 (2021)
12. Siegel Mph, R.L., Miller, K.D., Sandeep, N., Mbbs, W., Ahmedin, Dvm, J., Siegel, R.L.: Cancer statistics, 2023. CA: Cancer J. Clinic. **73**(1), 17–48 (1 2023). https://doi.org/10.3322/CAAC.21763
13. Sloboda, T., Hudec, L., Benešová, W.: xai-cyclegan, a cycle-consistent generative assistive network. arXiv preprint: arXiv:2306.15760 (2023)
14. Vaswani, A., et al.: Attention is all you need. In: Advances in Neural Information Processing Systems, vol. 30 (2017)
15. Wang, M.: Mask CycleGAN: unpaired multi-modal domain translation with interpretable latent variable (2022). https://doi.org/10.48550/arxiv.2205.06969, https://arxiv.org/abs/2205.06969v1
16. Xu, Z., Huang, X., Moro, C.F., Bozóky, B., Zhang, Q.: GAN-based virtual re-staining: a promising solution for whole slide image analysis (2022)
17. Zhu, J.Y., Park, T., Isola, P., Efros, A.A., Research, B.A.: Unpaired image-to-image translation using cycle-consistent adversarial networks (2017). https://github.com/junyanz/CycleGAN

Assessing the Robustness and Reproducibility of CT Radiomics Features in Non-small-cell Lung Carcinoma

Giovanni Pasini[1,2](✉)

[1] Department of Mechanical and Aerospace Engineering, Sapienza University of Rome, Eudossiana 18, 00184 Rome, Italy
giovanni.pasini@uniroma1.it
[2] Institute of Molecular Bioimaging and Physiology, National Research Council (IBFM-CNR), Contrada, Pietrapollastra-Pisciotto, 90015 Cefalù, Italy

Abstract. The aim of this study was to investigate the robustness of radiomics features extracted from computed tomography (CT) images of patients affected by non-small-cell lung carcinoma (NSCLC). Specifically, the impact of manual segmentation on radiomics feature values and their variability were assessed. Therefore, 63 patients affected by squamous cell carcinoma (SCC) and adenocarcinoma (ADC) were retrospectively collected from a public dataset. Original segmentations (automated plus manual refinement approach) were provided together with CT images. Through the matRadiomics tool, manual segmentation of the volume of interest (VOI) was repeated by two training physicians and 107 features were extracted. Feature extraction was also performed using the original segmentations. Therefore, three datasets of extracted features were obtained and compared computing the difference percentage coefficient (DP) and the intraclass correlation coefficient (ICC). Moreover, feature reduction and selection on each dataset were performed using a hybrid descriptive inferential method and the differences among the three feature subsets were evaluated. Successively, three classification models were obtained using the Linear Discriminant Analysis (LDA) classifier. Validation was performed through 10 times repeated 5-fold stratified cross validation. As result, even if 87% features obtained an ICC > 0.8, showing robustness, an AVGDP (averaged DP) equal to 16.2% was observed between the datasets based on manual segmentation. Moreover, manual segmentation had an impact on the subsets of selected features, thus influencing study reproducibility and model explainability.

Keywords: Radiomics · computed tomography · non-small-cell lung carcinoma · segmentation · reproducibility · robustness · machine learning

1 Introduction

Lung cancer was the leading cause of cancer deaths in 2022 [1, 2] and it is estimated to be the leading cause of cancer deaths during 2023 [3]. Several types of lung cancer exist, which are grouped in two major categories: non-small cell lung carcinoma

G. L. Foresti et al. (Eds.): ICIAP 2023 Workshops, LNCS 14366, pp. 39–48, 2024.
https://doi.org/10.1007/978-3-031-51026-7_4

(NSCLC), which accounts for the 85% of lung cancer occurrences, and small-cell lung carcinoma (SCLC). Moreover, NSCLC is further classified into three main groups: adenocarcinoma (ADC, ~ 40%), squamous cell carcinoma (SCC, 25–30%) and large cell carcinoma (LCC, 5–10%). Difficulties in the diagnosis of NSCLC led to the creation of a fourth class, namely "not otherwise specified (NOS)", which includes NSCLC types that do not share common characteristics with the three main subtypes [4, 5]. Typically, medical imaging followed by biopsy is the gold standard to diagnose NSCLC, whose results are used by physicians to indicate the best treatment among surgery, chemotherapy, immunotherapy, and radiotherapy, depending on the grade and NSCLC subtypes [4, 6]. Therefore, since the differential diagnosis between NSCLC subtypes strongly relies on radiologists' experience, and biopsy is an invasive procedure, radiomics has emerged as a promising and non-invasive tool to support the clinical decision process and address these issues [7]. It involves the use of artificial intelligence to analyze medical images, either through classical machine learning pipelines or through advanced deep learning methods [8], to extract quantitative metrics from medical images. Several studies showed its potential in identifying the most predictive biomarkers for the differential diagnosis between cancer subtypes and lung diseases, to predict overall survival and responses to therapy [9, 10], to detect neurodegenerative diseases such as Alzheimer's [11] and Parkinson's [12] and to predict the expanded disability status scale (EDSS) [13] of patients affected by multiple sclerosis. Its main advantage over biopsy is that it is non-invasive, less time consuming and can be integrated in automated pipelines. Although, the Imaging Biomarker Standardization Initiative (IBSI) [14] provided the guidelines for computing handcrafted radiomics features, radiomics is still influenced by major challenges that could have a negative impact on study reproducibility and robustness [15], thus limiting its applications [16, 17]. For example, in the differential diagnosis of NSCLC subtypes, factors such as different segmentation methods (manual, semiautomatic, automatic), lack of standardization in the parameters used during the feature extraction process, and dissimilarities in the algorithms used for feature selection and machine learning, have an impact on radiomics results, which can vary greatly between studies [18]. Although, some studies proposed deep learning methods and whole automated workflows [19] for the segmentation of NSCLCs, manual segmentation is still the most used, either to obtain the final segmentations or to refine segmentations obtained automatically. Its main disadvantage is that, compared to automated methods [20–26], it is operator-dependent and time consuming. Therefore, the aim of this study is i) to verify if manual segmentation of NSCLCs performed by two different physicians could limit study reproducibility, thus influencing radiomics features values and the subsets of selected features, ii) to evaluate the differences in radiomics results when a manual approach (physicians' segmentations) and an automated plus manual refinement approach (dataset segmentations) were used, and iii) to assess feature robustness.

2 Materials and Methods

2.1 Dataset

The dataset used in this study was derived from the public NSCLC-Radiogenomics [27] dataset that contains CT images of NSCLC patients and their associated segmentations, both in DICOM (Digital Imaging and Communication in Medicine) format. Specifically, sixty-three patients (13 SCC, 50 ADC), which were visually inspected through the matRadiomics tool [28], were included in this study and constituted the final dataset (namely, the original dataset).

2.2 Segmentation

NSCLC VOIs was delineated by two training physicians (readers A and B), through the manual segmentation tool offered by matRadiomics. Both readers, inspected each slice of the DICOM volumes, and delineated the contours slice by slice. On the other hand, the segmentations provided in the datasets (namely, the gold standard) were initially obtained with an unpublished automatic segmentation algorithm, and refined by two experienced radiologists who reached consensus [29].

2.3 Image Pre-processing and Feature Extraction

Using matRadiomics, that integrates the Pyradiomics [30] extractor, the extraction of 107 features was repeated for each type of segmentation (reader A, reader B, and gold standard), thus obtaining three datasets of features (the reader A dataset, the reader B dataset and the gold standard dataset). The extracted features belonged to three major classes: (i) shape/morphological features, (ii) first order statistics features, (iii) texture features, further divided in 5 classes, which are the gray level co-occurrence matrix (GLCM), gray level run length matrix (GLRLM), gray level size zone matrix (GLSZM), neighboring gray tone difference matrix (NGTDM), and the gray level dependence matrix (GLDM) [31–35]. Before the extraction, the images were discretized setting the Pyradiomics "Bin Count" option to 64. The other feature extraction parameters were left to Pyradiomics' default. Finally, for each dataset, 14 Shape features, 18 First Order Statistics features, 24 GLCM features, 14 GLDM features, 16 GLRLM features, 16 GLSZM features and 5 NGTDM features were extracted.

2.4 Statistical Analysis

Firstly, the feature values were compared pairwise in terms of the difference percentage coefficient (DP):

$$DP = 100 \times ABS\left(\frac{FeatureValue_{dataset1} - FeatureValue_{dataset2}}{FeatureValue_{dataset1}}\right) \tag{1}$$

Therefore, three pairwise comparisons were performed: i) reader A – reader B, ii) reader A – gold standard dataset, iii) reader B – gold standard dataset. Moreover, for each feature, the averaged DP (avgDP) was computed.

$$avgDP_j = \frac{1}{N}\sum_{i=1}^{N} DP_i \tag{2}$$

where ***i*** indicates the i-th observation, ***j*** indicates the j-th feature, **N** the total number of observations (63), and $\boldsymbol{avgDP_j}$ indicates the j-th averaged DP.

Furthermore, the avgDPs were averaged over all the features and for each class of features to obtain the AVGDP and the $\mathrm{AVGDP}_{\mathrm{Shape}}$, $\mathrm{AVGDP}_{\mathrm{Statistics}}$, $\mathrm{AVGDP}_{\mathrm{GLCM}}$, $\mathrm{AVGDP}_{\mathrm{GLRLM}}$, $\mathrm{AVGDP}_{\mathrm{GLDM}}$, $\mathrm{AVGDP}_{\mathrm{GLSZM}}$, $\mathrm{AVGDP}_{\mathrm{NGTDM}}$:

$$AVGDP = \frac{1}{K}\sum_{j=1}^{K} avgDP_j \tag{3}$$

$$AVGDP_{Shape} = \frac{1}{K_{Shape}}\sum_{j=1}^{K_{Shape}} avgDP_{j,Shape} \tag{4}$$

$$AVGDP_{Statistics} = \frac{1}{K_{Statistics}}\sum_{j=1}^{K_{Statistics}} avgDP_{j,Statistics} \tag{5}$$

$$AVGDP_{GLCM} = \frac{1}{K_{GLCM}}\sum_{j=1}^{K_{GLCM}} avgDP_{j,GLCM} \tag{6}$$

$$AVGDP_{GLDM} = \frac{1}{K_{GLDM}}\sum_{j=1}^{K_{GLDM}} avgDP_{j,GLDM} \tag{7}$$

$$AVGDP_{GLRLM} = \frac{1}{K_{GLRLM}}\sum_{j=1}^{K_{GLRLM}} avgDP_{j,GLRLM} \tag{8}$$

$$AVGDP_{GLSZM} = \frac{1}{K_{GLSZM}}\sum_{j=1}^{K_{GLSZM}} avgDP_{j,GLSZM} \tag{9}$$

$$AVGDP_{NGTDM} = \frac{1}{K_{NGTDM}}\sum_{j=1}^{K_{NGTDM}} avgDP_{j,NGTDM} \tag{10}$$

where ***j*** indicates the j-th feature, ***K*** indicates the total number of features, $\boldsymbol{K_{Shape}}$ indicates the total number of features that belongs to the Shape class, $\boldsymbol{K_{Statistics}}$ indicates the total number of features that belongs to the First Order Statistics class, $\boldsymbol{K_{GLCM}}$ indicates the total number of features that belongs to the GLCM Class, $\boldsymbol{K_{GLRLM}}$ indicates the total number of features that belongs to the GLRLM Class, $\boldsymbol{K_{GLSZM}}$ indicates the total number of features that belongs to the GLSZM Class, $\boldsymbol{K_{GLDM}}$ indicates the total number of features that belongs to the GLDM Class, $\boldsymbol{K_{NGTDM}}$ indicates the total number of features that belongs to the NGTDM Class.

Secondly, the intraclass correlation coefficient (ICC) was computed for each feature to quantify inter-observer feature reproducibility and consequently feature robustness. It ranges between 0 and 1, indicating null and perfect reproducibility, respectively. It was computed using the formula proposed by McGraw and Wong in case 3A (A,1) [36] to measure absolute agreement as,

$$ICC = \frac{MS_R - MS_E}{MS_R + (k-1)MS_E + \frac{k}{n}(MS_C - MS_E)} \tag{11}$$

The ICCs were consequently averaged grouping features by feature class, thus obtaining the ICC_{Shape}, $ICC_{Statistics}$, ICC_{GLCM}, ICC_{GLDM}, ICC_{GLRLM}, ICC_{GLSZM}, ICC_{NGTDM}.

$$ICC_{Class} = \frac{1}{K_{Class}} \sum\nolimits_{j=1}^{K_{Class}} ICC_j \tag{12}$$

where $\mathbf{MS_R}$ = mean square for rows, $\mathbf{MS_E}$ = mean square error, $\mathbf{MS_C}$ = mean square for columns, $\mathbf{k}$ = number of observers involved and $\mathbf{n}$ = number of subjects, $\mathbf{j}$ indicates the j-th feature and $\mathbf{K_{Class}}$ the total number of features for each class.

Finally, boxplots that summarize the distributions of ICC values were obtained.

2.5 Feature Reduction, Selection, and Machine Learning

Then, feature selection was performed through matRadiomics, using the hybrid descriptive inferential method, extensively reported in [37], and repeated for each dataset of extracted features (reader A, reader B and gold standard dataset). Briefly, it uses a Point Biserial Correlation (PBC) to assign scores to features, order them by score, and then iteratively build a logistic regression model. At each cycle, the p-value of the model is compared with the p-value of the previous cycle. If the p-value does not decrease at the current cycle, the procedure stops, and the logistic regression model is obtained. The three datasets of selected features (reader A, reader B, gold standard dataset), were used to build three classifiers using the Linear Discriminant Analysis (LDA) [38]. 10 times repeated 5-fold stratified cross validation was used to validate models. The receiver operating characteristic (ROC) curves were computed, together with the accuracies, and results were averaged over the 10 repetitions.

3 Results

3.1 Statistical Analysis

To assess the differences between the three datasets (reader A, reader B and gold standard) the AVGDP, $AVGDP_{Shape}$, $AVGDP_{Statistics}$, $AVGDP_{GLCM}$, $AVGDP_{GLRLM}$, $AVGDP_{GLDM}$, $AVGDP_{GLSZM}$, and the $AVGDP_{NGTDM}$ were computed pairwise (reader A – reader B, reader A – gold standard dataset, reader B - gold standard dataset). Results are summarized in Table 1. The feature class that showed the lowest AVGDP was the Shape Feature class in all the pairwise comparisons, while the highest AVGDP was obtained for the features belonging to the First Order Statistics Class. Moreover, Table1 shows also that the differences in features values were lower (AVGDP = 13,2%) in Reader A – Reader B comparison. To assess feature robustness and reproducibility, the ICCs were computed for each feature and averaged by feature class. Table 2 shows the average ICCs by feature class while Fig. 1 shows the distribution of ICCs through box plots. The feature that obtained the highest ICC (0.99) was the "original_shape_LeastAxisLength" feature that belonged to the Shape feature class, while the feature that obtained the lowest ICC (0.43) was the "original_glszm_LargeAreaLowGrayLevelEmphasis" feature that belonged to the GLSZM feature class. Moreover, the feature class that obtained the highest ICC was the Shape feature class ($ICC_{Shape} = 0.99$), while the lowest ICC was obtained for the GLSZM class ($ICC_{GLSZM} = 0.83$). Overall, 87% of the features obtained an ICC value greater than 0.8.

Table 1. Pairwise differences between the three datasets (Reader A – Reader B, Reader B – Original datasets, Reader A – Original Dataset) measured with the following metrics: AVGDP, $AVGDP_{Shape}$, $AVGDP_{Statistics}$, $AVGDP_{GLCM}$, $AVGDP_{GLRLM}$, $AVGDP_{GLDM}$, $AVGDP_{GLSZM}$, and the $AVGDP_{NGTDM}$.

Metrics	Reader A – Reader B	Reader B – gold standard dataset	Reader A – gold standard dataset
AVGDP	13.2%	16%	17%
$AVGDP_{Shape}$	7.9%	7.8%	10%
$AVGDP_{Statistics}$	16.2%	22.3%	23.4%
$AVGDP_{GLCM}$	12.5%	14.7%	15.7%
$AVGDP_{GLDM}$	15.1%	18.1%	18.6%
$AVGDP_{GLRLM}$	12%	15.1%	15.3%
$AVGDP_{GLSZM}$	14.5%	16.8%	17.4%
$AVGDP_{NGTDM}$	15.2%	17.1%	19.5%

Table 2. Averaged ICCs by feature class.

ICCShape	ICCStatistics	ICCGLCM	ICCGLDM	ICCGLRLM	ICCGLSZM	ICCNGTDM
0.96	0.89	0.9	0.87	0.89	0.83	0.91

3.2 Feature Reduction, Selection, and Machine Learning

The hybrid descriptive inferential method was applied for feature selection. Three different subsets of selected features were obtained, and they are reported in Table 3. Finally, three models based on the three different subsets of selected features were built using a LDA classifier [38], and results (averaged area under curve (auc) and accuracy) are reported in Table 4. The highest accuracy (80%) was reached for the Reader B subsets of selected features.

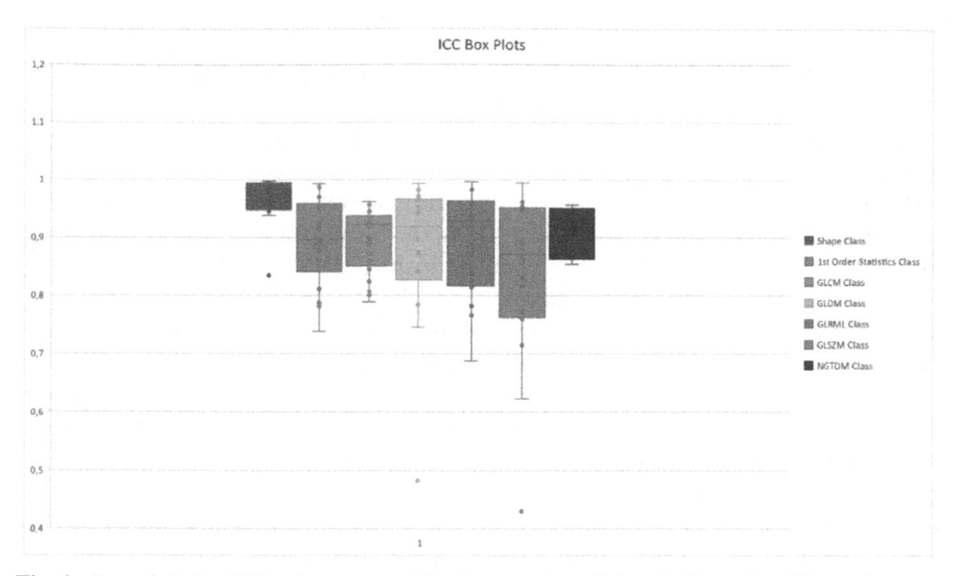

Fig. 1. Box plots for ICC values grouped by feature class. Colors indicate the different features classes; the legend is reported on the right.

Table 3. The three subsets of selected features

Reader A	Reader B	Original
–	original_glszm_LargeAreaHighGrayLevelEmphasis	–
original_firstorder_Mean	original_firstorder_Mean	–
original_firstorder_90Percentile	original_firstorder_90Percentile	–
–	–	original_glcm_ClusterShade

Table 4. Machine learning results for the three subsets of selected features using linear discriminant analysis.

Metrics	Reader A	Reader B	Original
area under curve	0.68	0.68	0.68
accuracy	0.79	0.8	0.79

4 Discussion and Conclusions

Segmentation methods influence radiomics analysis as reported in several studies, and have an impact on study reproducibility, that is still one of the major issues of radiomics itself [39]. Moreover, differences in radiomics results are higher when manual segmentation is adopted if compared with semi-automatic and automatic methods. Therefore, the impact of manual segmentation on radiomics analysis was assessed. As shown in Sect. 3.1, manual segmentation performed by different physicians (Reader A and B) lead to different feature values with a maximum AVGDP equal to 16,2% for the First Order Statistics feature class. Moreover, the differences between feature values were lower when comparing together the datasets of the two training physicians, rather than when comparing the Reader A dataset with the gold standard dataset and the Reader B dataset with the gold standard dataset. Indeed, the $AVGDP_{Statistics}$ increased to 22,3% and 23,4% when the readers datasets were compared with the gold standard dataset. This behavior was observed for all feature classes, except for the Shape feature class that obtained an $AVGDP_{Shape}$ value slightly lower when comparing the Reader B dataset with the gold standard dataset, rather than when comparing the two readers dataset together. This could be since the Shape feature class was the most robust one (averaged ICC = 0.96). Moreover, since manual segmentation was adopted for both the readers' datasets, the differences in feature values were lower when compared together, rather than when compared to the gold standard dataset, based, indeed, on an automated segmentation method plus manual refinement. Furthermore, even if 87% features obtained an ICC value greater than 0.8, thus showing robustness, the difference in feature values led to different subset of features. In this case, the subset of selected features was similar for Reader A and Reader B datasets. The Reader B subset of selected features presented only one more feature, namely "original_glszm_LargeAreaHighGrayLevelEmphasis", in addition to the "original_firstorder_Mean" and "original_firstorder_90Percentile" features present in both readers' datasets. This shows how manual segmentation, if performed by different readers, lead to different subsets of selected features, thus influencing study reproducibility. Although model performance was not impacted by the differences in the subsets of selected features, model explainability was affected. However, this is a preliminary study, whose main limitation is the small sample size. Finally, it is essential that the scientific community focuses on developing more robust and automated methods for lung cancer segmentation.

Acknowledgements. The author thanks the two training physicians, namely Accursio Scaduto and Francesco Cutrì, for having inspected the DICOM volume, manually segmented the NSCLCs and for feature extraction.

References

1. Siegel, R.L., Miller, K.D., Fuchs, H.E., Jemal, A.: Cancer statistics, 2022. Cancer J. Clin. **72**, 7–33 (2022). https://doi.org/10.3322/caac.21708
2. Dalmartello, M., et al.: European cancer mortality predictions for the year 2022 with focus on ovarian cancer. Ann. Oncol. **33**, 330–339 (2022). https://doi.org/10.1016/j.annonc.2021.12.007
3. Siegel, R.L., Miller, K.D., Wagle, N.S., Jemal, A.: Cancer statistics, 2023. Cancer J. Clin. **73**, 17–48 (2023). https://doi.org/10.3322/caac.21763
4. Duma, N., Santana-Davila, R., Molina, J.R.: Non-small cell lung cancer: epidemiology, screening, diagnosis, and treatment. Mayo Clin. Proc. **94**, 1623–1640 (2019). https://doi.org/10.1016/j.mayocp.2019.01.013
5. Travis, W.D., et al.: The 2015 world health organization classification of lung tumors: impact of genetic, clinical and radiologic advances since the 2004 classification. J. Thoracic Oncol. **10**, 1243–1260 (2015). https://doi.org/10.1097/JTO.0000000000000630

6. Xing, P.-Y., et al.: What are the clinical symptoms and physical signs for non-small cell lung cancer before diagnosis is made? A nation-wide multicenter 10-year retrospective study in China. Cancer Med. **8**, 4055–4069 (2019). https://doi.org/10.1002/cam4.2256
7. Vernuccio, F., Cannella, R., Comelli, A., Salvaggio, G., Lagalla, R., Midiri, M.: [Radiomics and artificial intelligence: new frontiers in medicine.]. Recenti Prog. Med. **111**, 130–135 (2020). https://doi.org/10.1701/3315.32853
8. Mayerhoefer, M.E., et al.: Introduction to radiomics. J. Nucl. Med. **61**, 488–495 (2020). https://doi.org/10.2967/jnumed.118.222893
9. Cuocolo, R., et al.: Machine learning applications in prostate cancer magnetic resonance imaging. Eur. Radiol. Exp. **3**, 35 (2019). https://doi.org/10.1186/s41747-019-0109-2
10. Comelli, A., et al.: Radiomics: a new biomedical workflow to create a predictive model. In: Papież, B.W., Namburete, A.I.L., Yaqub, M., Noble, J.A. (eds.) Medical Image Understanding and Analysis. Communications in Computer and Information Science, vol. 1248, pp. 280–293. Springer, Cham (2020). https://doi.org/10.1007/978-3-030-52791-4_22
11. Alongi, P., et al.: 18F-Florbetaben PET/CT to assess Alzheimer's disease: a new analysis method for regional amyloid quantification. J. Neuroimaging **29**, 383–393 (2019). https://doi.org/10.1111/jon.12601
12. Shu, Z.-Y., et al.: Predicting the progression of Parkinson's disease using conventional MRI and machine learning: an application of Radiomic biomarkers in whole-brain white matter. Magn. Reson. Med. **85**, 1611–1624 (2021). https://doi.org/10.1002/mrm.28522
13. Nepi, V., Pasini, G., Bini, F., Marinozzi, F., Russo, G., Stefano, A.: MRI-Based Radiomics analysis for identification of features correlated with the expanded disability status scale of multiple sclerosis patients. In: Mazzeo, P.L., Frontoni, E., Sclaroff, S., and Distante, C. (eds.) Image Analysis and Processing. ICIAP 2022 Workshops, pp. 362–373. Springer International Publishing, Cham (2022). https://doi.org/10.1007/978-3-031-13321-3_32
14. Zwanenburg, A., et al.: The image biomarker standardization initiative: standardized quantitative Radiomics for high-throughput image-based phenotyping. Radiology. **295**, 328–338 (2020). https://doi.org/10.1148/radiol.2020191145
15. Stefano, A., et al.: Robustness of PET Radiomics features: impact of co-registration with MRI. Appl. Sci. **11**, 10170 (2021). https://doi.org/10.3390/app112110170
16. van Timmeren, J.E., Cester, D., Tanadini-Lang, S., Alkadhi, H., Baessler, B.: Radiomics in medical imaging—"how-to" guide and critical reflection. Insights Imag. **11**, 91 (2020). https://doi.org/10.1186/s13244-020-00887-2
17. Cutaia, G., et al.: Radiomics and prostate MRI: current role and future applications. J. Imag. **7**, 34 (2021). https://doi.org/10.3390/jimaging7020034
18. Pasini, G., Stefano, A., Russo, G., Comelli, A., Marinozzi, F., Bini, F.: Phenotyping the histopathological subtypes of non-small-cell lung carcinoma: how beneficial is Radiomics? Diagnostics. **13**, 1167 (2023). https://doi.org/10.3390/diagnostics13061167
19. Primakov, S.P., et al.: Automated detection and segmentation of non-small cell lung cancer computed tomography images. Nat. Commun. **13**, 3423 (2022). https://doi.org/10.1038/s41467-022-30841-3
20. Stefano, A., et al.: A preliminary PET radiomics study of brain metastases using a fully automatic segmentation method. BMC Bioinform. **21**, 325 (2020). https://doi.org/10.1186/s12859-020-03647-7
21. Comelli, A., et al.: Development of a new fully three-dimensional methodology for tumours delineation in functional images. Comput. Biol. Med. **120**, 103701 (2020). https://doi.org/10.1016/j.compbiomed.2020.103701
22. Comelli, A., et al.: Tissue classification to support local active delineation of brain tumors. In: Zheng, Y., Williams, B.M., Chen, K. (eds.) Medical Image Understanding and Analysis. Communications in Computer and Information Science, vol. 1065, pp. 3–14. Springer, Cham (2020). https://doi.org/10.1007/978-3-030-39343-4_1

23. Banna, G.L., et al.: Predictive and prognostic value of early disease progression by PET evaluation in advanced non-small cell lung cancer. Oncology **92**, 39–47 (2017). https://doi.org/10.1159/000448005
24. Stefano, A., et al.: A fully automatic method for biological target volume segmentation of brain metastases. Int. J. Imag. Syst. Technol. **26**, 29–37 (2016). https://doi.org/10.1002/ima.22154
25. Stefano, A., et al.: A graph-based method for PET image segmentation in radiotherapy planning: a pilot study. In: Petrosino, A. (ed.) Image Analysis and Processing – ICIAP 2013. Lecture Notes in Computer Science, vol. 8157, pp. 711–720. Springer, Heidelberg (2013). https://doi.org/10.1007/978-3-642-41184-7_72
26. Agnello, L., Comelli, A., Ardizzone, E., Vitabile, S.: Unsupervised tissue classification of brain MR images for voxel-based morphometry analysis. Int. J. Imaging Syst. Technol. **26**, 136–150 (2016). https://doi.org/10.1002/ima.22168
27. Bakr, S., et al.: Data for NSCLC Radiogenomics collection (2017). https://wiki.cancerimagingarchive.net/x/W4G1AQ, https://doi.org/10.7937/K9/TCIA.2017.7HS46ERV
28. Pasini, G., Bini, F., Russo, G., Comelli, A., Marinozzi, F., Stefano, A.: MatRadiomics: a novel and complete Radiomics framework, from image visualization to predictive model. J. Imag. **8**, 221 (2022). https://doi.org/10.3390/jimaging8080221
29. Bakr, S., et al.: A Radiogenomic dataset of non-small cell lung cancer. Sci Data. **5**, 180202 (2018). https://doi.org/10.1038/sdata.2018.202
30. van Griethuysen, J.J.M., et al.: Computational Radiomics system to decode the radiographic phenotype. Can. Res. **77**, e104–e107 (2017). https://doi.org/10.1158/0008-5472.CAN-17-0339
31. Haralick, R.M., Shanmugam, K., Dinstein, I.: Textural features for image classification. IEEE Trans. Syst., Man Cybern. **SMC-3**, 610–621 (1973). https://doi.org/10.1109/TSMC.1973.4309314
32. Galloway, M.M.: Texture analysis using gray level run lengths. Comput. Graph. Image Process. **4**, 172–179 (1975). https://doi.org/10.1016/S0146-664X(75)80008-6
33. Thibault, G., Angulo, J., Meyer, F.: Advanced statistical matrices for texture characterization: application to cell classification. IEEE Trans. Biomed. Eng. **61**, 630–637 (2014). https://doi.org/10.1109/TBME.2013.2284600
34. Amadasun, M., King, R.: Textural features corresponding to textural properties. IEEE Trans. Syst. Man Cybern. **19**, 1264–1274 (1989). https://doi.org/10.1109/21.44046
35. Sun, C., Wee, W.G.: Neighboring gray level dependence matrix for texture classification. Comput. Vis., Graph. Image Process. **23**, 341–352 (1983). https://doi.org/10.1016/0734-189X(83)90032-4
36. McGraw, K.O., Wong, S.P.: Forming inferences about some intraclass correlation coefficients. Psychol. Methods **1**, 30–46 (1996). https://doi.org/10.1037/1082-989X.1.1.30
37. Barone, S., et al.: Hybrid descriptive-inferential method for key feature selection in prostate cancer Radiomics. Appl. Stoch. Model. Bus. Ind. **37**, 961–972 (2021). https://doi.org/10.1002/asmb.2642
38. Comelli, A., et al.: Active contour algorithm with discriminant analysis for delineating Tumors in positron emission tomography. Artif. Intell. Med. **94**, 67–78 (2019). https://doi.org/10.1016/j.artmed.2019.01.002
39. Parmar, C., et al.: Robust Radiomics feature quantification using semiautomatic volumetric segmentation. PLoS ONE **9**, e102107 (2014). https://doi.org/10.1371/journal.pone.0102107

Prediction of High Pathological Grade in Prostate Cancer Patients Undergoing [^{18}F]-PSMA PET/CT: A Preliminary Radiomics Study

Alessandro Stefano[1], Cristina Mantarro[2], Selene Richiusa[1], Giovanni Pasini[1,3](✉), Maria Gabriella Sabini[2], Sebastiano Cosentino[2], and Massimo Ippolito[2]

[1] Institute of Molecular Bioimaging and Physiology, National Research Council (IBFM-CNR), 90015 Cefalù, Italy
giovanni.pasini@uniroma1.it
[2] Cannizzaro Hospital, Catania, Italy
[3] Department of Mechanical and Aerospace Engineering, Sapienza University of Rome, Eudossiana 18, 00184 Rome, Italy

Abstract. The aim of this study was to evaluate the effectiveness of [^{18}F]-prostate-specific membrane antigen (PSMA) positron emission tomography/computed tomography (PET/CT) imaging in discriminating high pathological grade (Gleason score > 7), and low pathological grade (Gleason score < 7) using machine learning techniques. The study involved 81 patients with diagnosed prostate cancer who underwent positive [^{18}F]-SPMA PET/CT scans. The PET images were used to identify the primary lesions, and then radiomics analyses were performed using an Imaging Biomarker Standardization Initiative (IBSI) compliant software, namely *matRadiomics*. Machine learning approaches were employed to identify relevant radiomics features for predicting high-risk malignant disease. The performance of the models was validated using 10 times repeated 5-fold cross validation scheme. The results showed a value of 0.75 for the area under the curve and an accuracy of 72% using the support vector machine (SVM). In conclusion, the study showcased the clinical potential of [^{18}F]-SPMA PET/CT radiomics in differentiating high-risk and low-risk tumors, without the need for biopsy sampling. In-vivo PET/CT imaging could therefore be considered a noninvasive tool for virtual biopsy, facilitating personalized treatment management.

Keywords: Radiomics · Prostate · Machine Learning · Image Analysis · PET · PSMA

1 Introduction

Radiomics is a rapidly growing research field that uses advanced computational techniques to extract quantitative information from medical images, such as Positron Emission Tomography (PET) [1], Magnetic Resonance Imaging (MRI) [2], Computed

G. L. Foresti et al. (Eds.): ICIAP 2023 Workshops, LNCS 14366, pp. 49–58, 2024.
https://doi.org/10.1007/978-3-031-51026-7_5

Tomography (CT) [3], or molecular hybrid imaging [4]. Quantitative features are then used to develop predictive models that can aid in diagnosis, treatment planning, and prognosis in a variety of diseases, from oncology [5] to neurodegenerative diseases [6]. Radiomics seems to have several advantages over traditional imaging analysis methods, including its ability to capture a wealth of information from medical images that may not be immediately visible to the naked eye [7, 8].

Nevertheless, the effective utility and the challenges associated with radiomics, including the need for standardized imaging protocols of radiomics in medical imaging, has been highlighted recently in a lung-based study [9]: although the Imaging Biomarker Standardization Initiative (IBSI) offers guidelines for computing radiomics features, the absence of standardized parameters in the feature extraction process, as well as in the algorithms used for feature selection and machine learning, remain significant challenges and pose limitations to radiomics [10]. Proper reporting of all parameters utilized in preprocessing techniques is crucial since they heavily influence the extracted features. Despite the existing challenges, radiomics possesses the remarkable potential to bring about a paradigm shift in medical imaging, ultimately leading to enhanced patient outcomes [11]. Radiomics can revolutionize medical imaging and positively impact patient care are undeniably compelling. The application of radiomics in personalized medicine holds the promise of unlocking novel diagnostic and prognostic markers [12], guiding more targeted therapies, and ultimately enabling more effective and efficient healthcare delivery [13].

In this preliminary study, we aim to explore the possibility of using radiomics to predict the high pathological grade in prostate cancer (PCa) patients undergoing [^{18}F]-prostate-specific membrane antigen (PSMA)-1007 positron emission tomography/computed tomography (PET/CT) imaging [14, 15].

2 Materials and Methods

2.1 PET/CT Imaging

Eighty-one patients underwent [^{18}F]-PSMA-1007 PET/CT imaging using two different scanners (General Electric, Discovery 690FX&MOT and Siemens Biograph Horizon 4R). For the first scanner, 16 row helical CT scan was used with the following conditions: tube voltage (140 kVp), tube current (800 mAmax). PET collection time of every bed was 90 s, the whole-body scanning needed 7–8 beds. PET matrix was 128 * 128, and CT matrix was 512 $\times$ 512. The voxel size was 3.75 $\times$ 3.75 $\times$ 3.75 mm^3.

For the second scanner, 16 row helical CT scan was used with the following conditions: tube voltage (130 kVp), tube current (345 mAmax). PET collection time of every bed was 90 s, the whole-body scanning needed 6–7 beds. PET matrix was 512 * 512, and CT matrix was 512 $\times$ 512. The voxel size was 3 $\times$ 3 $\times$ 3 mm^3.

The tracer produced was intravenously injected into patients at a standardized dose of 4 MBq/kg. After the tracer administration, the patients rested in a quiet room for about 120 min before scanning.

2.2 Inclusion Criteria

Inclusion patients' criteria were: (1) diagnosis of PCa through biopsy; (2) elevated serum PSA value; (3) no therapy before PET scan, neither surgery, chemotherapy, radiotherapy, endocrine therapy, or anything else. Subjects who met the above three criteria simultaneously were included in our study. The patients without exhibiting relevant radiotracer uptake above background in the prostate on [^{18}F]-PSMA PET/CT imaging were excluded. Consequently, at least in this phase, only patients with positive PET scans were considered.

2.3 The Gleason Score

PCa frequently exhibit varying degrees of aggressiveness within different regions of the tumor due to its heterogeneity [16]. To characterize this variability, a grade is assigned to the two predominant areas of cancer tissue being analyzed. The sum of these grades yields the overall Gleason scores (GS). The first number represents the most prevalent grade within the tumor. For instance, a GS of $3 + 4 = 7$ indicates that the tumor is predominantly grade 3, with a smaller portion being grade 4. The two grades are combined to derive the final GS, which in this case is 7.

In general, the level of aggressiveness of PCa is divided into:

- GS ≤ 6: The tumor typically exhibits slow growth and does not spread to distant organs from the prostate (non-metastatic).
- GS $= 7$: The tumor displays intermediate aggressiveness.
- GS between 8 and 10: The tumor is aggressive (metastatic).

In our study, tumors with GS $3 + 3$ or $3 + 4$ were classified as low-grade, while tumors with GS $4 + 3$, $4 + 4$, or $4 + 5$ were categorized as high-grade.

2.4 Radiomics Analysis

The image analysis was done by a board-certified physician (C.M. author). The PET/CT images were primarily evaluated by visual inspection to assess the possible presence of potential artifacts, such as metal artifacts, or respiratory mobility. In case of absence of artifacts and movement, prostate lesions were manually segmented as volumes of interest (VOIs) on PET images. The final VOIs were confirmed by a senior nuclear medicine specialist (S.C. author). Subsequently, through quantitative analysis, 1781 radiomics features (107 original, 744 wavelets, 930 LoG) were extracted for each patient.

The whole radiomics analyses, from visual inspection to model implementation, was performed through matRadiomics [17]. Pyradiomics [18] has been integrated in matRadiomics as feature extractor module. Pre-processing and extraction options can be set by interacting with the matRadiomics user interface. The Pyradiomics configuration used in our analysis is reported in Table 1. The other parameters were left to Pyradiomics' default.

Subsequently, a hybrid-inferential descriptive method was used for the reduction and selection process. This method uses the point biserial correlation to assign scores to features, order them by score, and then iteratively build a logistic regression model, as

Table 1. Configuration used for the feature extraction process.

Bin width	0.25
Isotropic Resampling	2 × 2 × 2
Interpolator	SitkBSpline
Wavelet Method	Coif1
Log Sigma	[0.5, 1, 1.5, 2, 2.5, 3, 3.5, 4, 4.5, 5]
Normalization	True; Scale = 1

extensively reported in [19]. In summary, during each cycle, the p-value of the model is compared to the p-value of the previous cycle. If the p-value does not decrease in the current cycle, the procedure terminates, and the logistic regression model is obtained. Once the selection process is finished, the features are sorted by score, and the selected features are shown in red.

Finally, linear discriminant analysis (LDA), K-nearest neighbors (KNN), and support vector machines (SVM) [20] were used to build the radiomics model. K-fold cross validation and K-fold stratified cross validation were used to perform model validation. Performance metrics were calculated as accuracy, true positive rate, and false positive rate for each model produced by the K-fold cross-validation. Then, these metrics were used to build receiver operating characteristic (ROC) curves.

Furthermore, matRadiomics integrates ComBat package [21], one of the best-known methods for feature harmonization. The analyses were then repeated using this tool and compared to the previous procedure, i.e., without the use of harmonization techniques.

3 Results

A total of eighty-one patients were included in the study, and their PET/CT examinations as well as GS are documented in Table 2.

Table 2. Characteristics of the 81 patients with primary prostate cancer involved in this study.

PET/CT scanner (n = 81)	Total	Low Grade GS = 3 + 3/3 + 4
GE	30	20
Siemens	51	26

The GS ranged from 6 to 9, and a total of 46 low-grade and 35 high-grade lesions were identified by considering tumors with GS 3 + 3 or 3 + 4 as low-grade and tumors with GS 4 + 3, 4 + 4, or 4 + 5 as high-grade, as described in the Sect. 2.3.

Figure 1 presents two illustrative examples of PET/CT images. The first image showcases the segmentation of a low-grade tumor, while the second image displays the segmentation of a high-grade tumor. The target is surrounded by a blue closed line, while the segmentation is shown in transparent purple color overlapping the images.

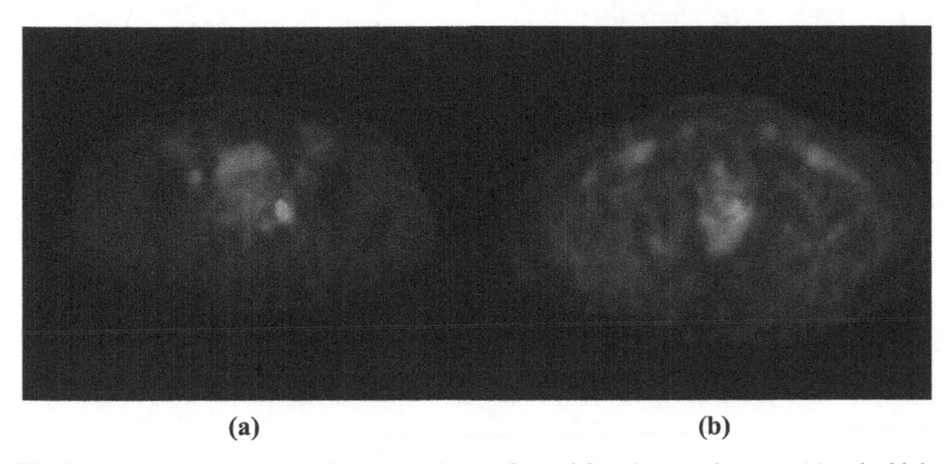

(a) **(b)**

Fig. 1. Two examples of manual segmentation performed for a low-grade tumor (a) and a high-grade tumor (b), respectively.

1781 radiomics features were extracted for each tumor, and feature selection results are shown in Table 3, with and without feature harmonization obtained through Combat package [21]. At this point, KNN, LDA, SVM were used to perform model implementation using 10 times repeated 5-fold cross validation. Performance metrics are shown in Tables 4 and 5, with and without feature harmonization. Examples of ROC curves using SVM, and accuracies are shown in the Figs. 2 and 3.

Table 3. Selected features with and without harmonization. Point Biserial Correlation (PBC) scores are reported in the last column.

Without harmonization	With harmonization	PBC
–	wavelet_LHH_glszm_ LowGrayLevelZoneEmphasis	0.424
–	original_firstorder_10Percentile	0.420
wavelet_LLL_firstorder_Minimum	–	0.454

Table 4. Machine learning performance metrics for the non-harmonized group. Averaged on 10 repetitions of K-fold cross validation.

Classifier	AUC	Accuracy	Sensitivity	Specificity
LDA	0.76	0.7	0.48	0.87
SVM	0.75	0.72	0.57	0.84
KNN	0.61	0.6	0.48	0.65

Table 5. Machine learning performance metrics for the harmonized group. Averaged on 10 repetitions of K-fold cross validation.

Classifier	AUC	Accuracy	Sensitivity	Specificity
LDA	0.77	0.69	0.56	0.78
SVM	0.78	0.69	0.55	0.78
KNN	0.67	0.67	0.58	0.72

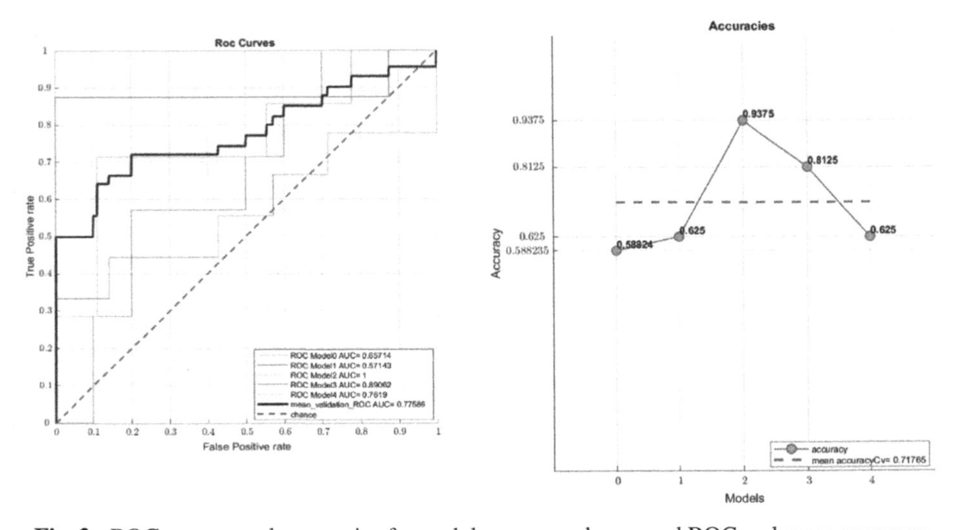

Fig. 2. ROC curves, and accuracies for each k group, and averaged ROC and mean accuracy.

4 Discussions and Conclusion

The GS is the most important tissue-based marker for PCa patients. The pathological grade of PCa, determined through biopsy or surgical resection, provides critical information about the aggressiveness and potential clinical outcomes of the disease. Nevertheless, the ability to avoid biopsies and the associated potential complications to determine the GS would have a strong clinical impact [22]. For this reason, in this preliminary study, we aimed to explore the possibility of using radiomics to predict the high pathological grade in PCa patients undergoing [^{18}F]-PSMA PET/CT.

Nuclear medicine is an area of growing interest and significance in the field of oncology [23]. Specifically, [^{18}F]-PSMA PET/CT has emerged as a valuable tool for the assessment of PCa, offering enhanced sensitivity and specificity in detecting tumor lesions. By leveraging the quantitative information obtained through radiomics from [^{18}F]-PSMA PET/CT scans, researchers and clinicians are striving to develop robust predictive models that can accurately identify patients at risk of having high pathological PCa grade. This holds tremendous potential for optimizing treatment strategies, facilitating early intervention, and ultimately improving patient management and outcomes. The integration of [^{18}F]-PSMA PET/CT imaging with predictive modeling approaches represents a promising avenue for advancing personalized medicine in PCa care. Radiomics features, which include intensity, shape, texture, and wavelet, can provide comprehensive information such as cancer phenotype to discover and predict clinical cancer pathological and biological characteristics.

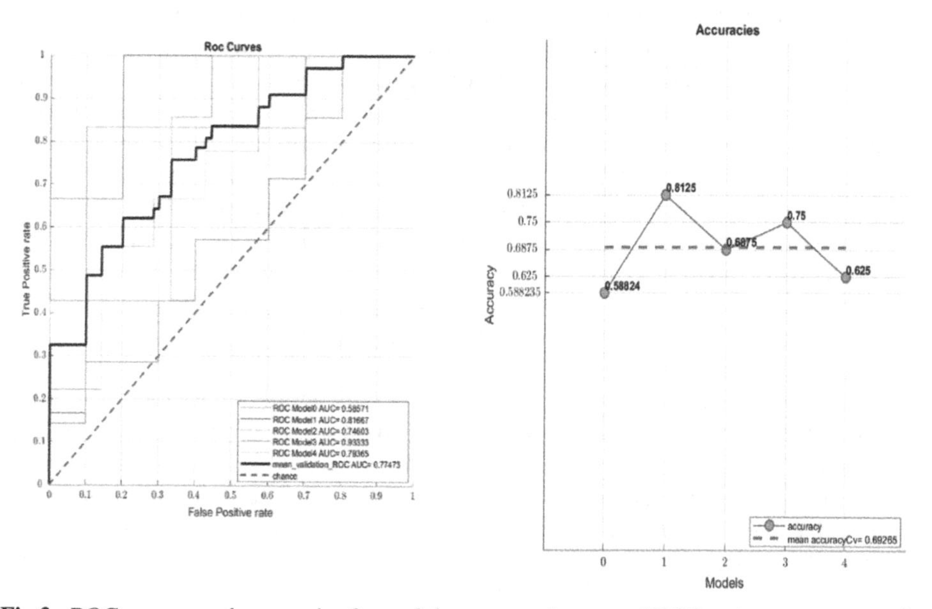

Fig.3. ROC curves, and accuracies for each k group, and averaged ROC and mean accuracy for the harmonized study.

In our study, to identify the target from which to extract the radiomics features, the segmentation was manually performed in PET images. This could be a serious problem for the reproducibility of the proposed radiomics model. As reported in a very recent study [24]: "*Radiologists can flexibly delineate targets manually resulting in highly accurate segmentations. Nevertheless, manual segmentation is labor-intensive, time-consuming, and not always feasible for radiomics analysis requiring huge datasets. Additionally, manual segmentation is subject to inter- and intra-observer variability.*

Hence, many semi-automatic delineation algorithms, such as region growing or thresholding, are used in the clinical environment although less precise than manual segmentation." In other words, radiomics studies are meaningful when the process of extracting data is reproducible, which relies on the approach used for segmentation. Although manual outlining appears to be the most intuitive and easily executable method for obtaining the target volume, it heavily relies on the operator, resulting in inconsistent outcomes. Different operators or the same operator at different times may yield varied segmentations, making reproducibility challenging [25]. To reduce the operator interaction in the segmentation process, improving the reproducibility of radiomics studies, automatic or semi-automatic approaches should be used, e.g., [26, 27]. In alternative, in the case of manual delineations performed by different radiologists, one solution is the STAPLE tool [28] that can be used to overcome the limitation by producing a consolidated reference between the different operators.

To avoid encountering further reproducibility issues, among all the tools used for radiomics analyses, we used matRadiomics [17], a IBSI compliant tool, that allows the user to fulfill the whole radiomics workflow, from target identification to model implementation, with the aim of simplifying the radiomics workflow. Moreover, this tool holds significant promise in enhancing the robustness and reliability of radiomics-based analyses, especially when applied to multi-center datasets to address the issue of data harmonization. Specifically, data harmonization refers to the process of standardizing and integrating data from different sources to ensure compatibility and comparability. In radiomics, data harmonization is crucial because it involves extracting features from medical images obtained using different imaging devices, and acquisition protocols such as in this study. These variations can lead to inconsistencies and hinder the reliability and reproducibility of radiomics analyses. Efforts to address the issue of data harmonization in radiomics are underway. Advancements in image processing techniques, such as normalization and resampling, can help mitigate the impact of technical differences in image acquisition and protocols.

What sets matRadiomics apart from other state-of-the-art radiomics tools is its integration of Combat (Comprehensive Analysis of RObustness for Imaging biomarker Translation) [21], an essential component for addressing these variations among different imaging scanners, such as in our study where two different PET scanners were considered. Combat tool that was originally developed to address batch effects in gene expression microarray data but has been adapted for use in radiomics to mitigate sources of variability that can arise from different imaging centers, scanners, protocols, and acquisition parameters. Combat aims to reduce variations by scanners by employing an empirical Bayes framework to estimate the influence of each source of variability on the radiomics features. It then applies an adjustment procedure to normalize the features across datasets, thereby harmonizing the data and allowing for more robust comparisons and combined analyses. By incorporating Combat, matRadiomics tackles an important challenge in radiomics, ensuring the harmonization of data from diverse sources and increasing the consistency and generalizability of the results. Consequently, it represents an asset for researchers and clinicians working with large-scale, multi-center radiomics studies, enabling more accurate and robust analyses. Furthermore, matRadiomics can compare the results obtained with and without feature harmonization.

In this study, although the selected features were different in the harmonized dataset and the non-harmonized one, we obtained similar results using the three machine learning classifiers. The best result was achieved by SVM in the non-harmonized group, with an accuracy of 0.72 (AUC = 0.75). The worst result (accuracy = 0.6) was obtained with KNN in the non-harmonized group, which improved to 0.67 after harmonization. In general, at least in this study, harmonization did not lead to a significant improvement in performance. However, although the proposed model showed promise in accurately grading prostate tumor, the results need to be validated on a larger sample size and in an external patient cohort to verify if the performance improves and if the obtained results remain consistent, i.e. harmonization has little impact and SVM is the best classifier.

Finally, as a future work, in addition to increasing the number of patients and including patients with negative [^{18}F]-SPMA PET scans, we plane to investigate whether radiomics outperforms better than a combined evaluation of radiology and nuclear medicine (MRI+PSMA-PET/CT).

References

1. Banna, G.L., et al.: Predictive and prognostic value of early disease progression by PET evaluation in advanced non-small cell lung cancer. Oncology (Switzerland) **92**, 39–47 (2017). https://doi.org/10.1159/000448005
2. Cutaia, G., et al.: Radiomics and prostate MRI: current role and future applications. J. Imaging. **7**, 34 (2021). https://doi.org/10.3390/jimaging7020034
3. Torrisi, S.E., et al.: Assessment of survival in patients with idiopathic pulmonary fibrosis using quantitative HRCT indexes. Multidiscip Respir Med. **13**, 1–8 (2018). https://doi.org/10.1186/s40248-018-0155-2
4. Liberini, V., et al.: Radiomics and artificial intelligence in prostate cancer: new tools for molecular hybrid imaging and theragnostics. Eur Radiol Exp. **6** (2022). https://doi.org/10.1186/S41747-022-00282-0
5. Vernuccio, F., et al.: Lo: diagnostic performance of qualitative and radiomics approach to parotid gland tumors: which is the added benefit of texture analysis? Br. J. Radiol. **94** (2021). https://doi.org/10.1259/bjr.20210340
6. Alongi, P., et al.: 18F-Florbetaben PET/CT to assess Alzheimer's disease: a new analysis method for regional amyloid quantification. J. Neuroimaging **29**, 383–393 (2019). https://doi.org/10.1111/jon.12601
7. Castiglioni, I., Gilardi, M.C.: Radiomics: is it time to compose the puzzle? Clin Transl Imaging. (2018). https://doi.org/10.1007/s40336-018-0302-y
8. Vernuccio, F., Cannella, R., Comelli, A., Salvaggio, G., Lagalla, R., Midiri, M.: Radiomics and artificial intelligence: new frontiers in medicine. Recent. Prog. Med. **111**, 130–135 (2020). https://doi.org/10.1701/3315.32853
9. Pasini, G., Stefano, A., Russo, G., Comelli, A., Marinozzi, F., Bini, F.: Phenotyping the histopathological subtypes of non-small-cell lung carcinoma: how beneficial is radiomics? Diagnostics **13** (2023). https://doi.org/10.3390/diagnostics13061167
10. Russo, G., et al.: Feasibility on the use of radiomics features of 11[C]-MET PET/CT in central nervous system tumours: Preliminary results on potential grading discrimination using a machine learning model. Curr. Oncol. **28**, 5318–5331 (2021). https://doi.org/10.3390/curroncol28060444
11. Laudicella, R., et al.: Artificial neural networks in cardiovascular diseases and its potential for clinical application in molecular imaging. Curr. Radiopharm. **14**, 209–219 (2020). https://doi.org/10.2174/1874471013666200621191259

12. Benfante, V., et al.: A new preclinical decision support system based on PET radiomics: a preliminary study on the evaluation of an innovative 64Cu-labeled chelator in mouse models. J. Imaging. **8**, 92 (2022). https://doi.org/10.3390/jimaging8040092
13. Cuocolo, R., et al.: Clinically significant prostate cancer detection on MRI: a radiomic shape features study. Eur. J. Radiol. **116**, 144–149 (2019). https://doi.org/10.1016/j.ejrad.2019.05.006
14. Alongi, P., et al.: PSMA and choline PET for the assessment of response to therapy and survival outcomes in prostate cancer patients: a systematic review from the literature. Cancers (Basel) **14** (2022). https://doi.org/10.3390/CANCERS14071770
15. Evangelista, L., et al.: [68Ga]Ga-PSMA Versus [18F]PSMA positron emission tomography/computed tomography in the staging of primary and recurrent prostate cancer. A systematic review of the literature. Eur. Urol. Oncol. **5**, 273–282 (2022). https://doi.org/10.1016/J.EUO.2022.03.004
16. Laudicella, R., et al.: Preliminary findings of the role of FAPi in prostate cancer theranostics. Diagnostics (Basel) **13** (2023). https://doi.org/10.3390/DIAGNOSTICS13061175
17. Pasini, G., Bini, F., Russo, G., Marinozzi, F., Stefano, A.: matRadiomics: from biomedical image visualization to predictive model implementation. In: Mazzeo, P.L., Frontoni, E., Sclaroff, S., Distante, C. (eds.) ICIAP 2022. LNCS, vol. 13373, pp. 374–385. Springer, Cham (2022). https://doi.org/10.1007/978-3-031-13321-3_33
18. pyradiomics Documentation Release v3.0.post5+gf06ac1d pyradiomics community (2020)
19. Barone, S., et al.: Hybrid descriptive-inferential method for key feature selection in prostate cancer radiomics. Appl. Stoch Models Bus. Ind. **37**, 961–972 (2021). https://doi.org/10.1002/asmb.2642
20. Comelli, A., et al.: A kernel support vector machine based technique for Crohn's disease classification in human patients. In: Barolli, L., Terzo, O. (eds.) CISIS 2017. AISC, vol. 611, pp. 262–273. Springer, Cham (2018). https://doi.org/10.1007/978-3-319-61566-0_25
21. Horng, H., et al.: Generalized ComBat harmonization methods for radiomic features with multi-modal distributions and multiple batch effects. Sci. Rep. **12**, 1–12 (2022). https://doi.org/10.1038/s41598-022-08412-9
22. Ferraro, D.A., et al.: Hot needles can confirm accurate lesion sampling intraoperatively using [18F]PSMA-1007 PET/CT-guided biopsy in patients with suspected prostate cancer. Eur. J. Nucl. Med. Mol. Imaging **49**, 1721–1730 (2022). https://doi.org/10.1007/S00259-021-05599-3
23. Laudicella, R., et al.: [68 Ga]DOTATOC PET/CT Radiomics to Predict the Response in GEP-NETs Undergoing [177 Lu]DOTATOC PRRT: The "Theragnomics" Concept. Cancers (Basel) **14**, 984 (2022). https://doi.org/10.3390/cancers14040984
24. Stefano, A., et al.: Robustness of pet radiomics features: impact of co-registration with MRI. Appl. Sci. (Switzerland) **11**, 10170 (2021). https://doi.org/10.3390/app112110170
25. Stefano, A., et al.: A fully automatic method for biological target volume segmentation of brain metastases. Int. J. Imaging Syst. Technol. 26, 29–37 (2016). https://doi.org/10.1002/ima.22154
26. Stefano, A., et al.: A graph-based method for PET image segmentation in radiotherapy planning: a pilot study. In: Petrosino, A. (ed.) ICIAP 2013. LNCS, vol. 8157, pp. 711–720. Springer, Heidelberg (2013). https://doi.org/10.1007/978-3-642-41184-7_72
27. Comelli, A., et al.: Tissue classification to support local active delineation of brain tumors. In: Zheng, Y., Williams, B.M., Chen, K. (eds.) MIUA 2019. CCIS, vol. 1065, pp. 3–14. Springer, Cham (2020). https://doi.org/10.1007/978-3-030-39343-4_1
28. Warfield, S.K., Zou, K.H., Wells, W.M.: Simultaneous truth and performance level estimation (STAPLE): an algorithm for the validation of image segmentation. IEEE Trans. Med. Imaging **23**, 903–921 (2004). https://doi.org/10.1109/TMI.2004.828354

MTANet: Multi-Type Attention Ensemble for Malaria Parasite Detection

Luca Zedda, Andrea Loddo(✉), and Cecilia Di Ruberto

Department of Mathematics and Computer Science, University of Cagliari, Cagliari, Italy
{luca.zedda,andrea.loddo,cecilia.dir}@unica.it

Abstract. Malaria is a severe infectious disease caused by the Plasmodium parasite. Diagnosing and treating the disease is crucial to increase the chances of survival. However, detecting malaria parasites is still a manual process performed by experts examining blood smears, especially in less developed countries. This task is time-consuming and prone to errors. Fortunately, deep learning-based object detection methods have shown promising results in automating this task, allowing quick diagnosis and treatment. In this work, we proposed an object detection ensemble architecture, MTANet, that efficiently detects malaria parasite species using one tailored YOLOv5 version integrated with an attention-based approach. We compared its performance against several methods in the literature. The experimental results have shown that MTANet can efficiently and accurately address the detection of different species with a single model.

Keywords: Deep Learning · Computer Vision · Object Detection · Image Processing · Blood Smear Images · Malaria Parasite Detection

1 Introduction

Malaria is a serious disease caused by the Plasmodium parasite, which can be fatal if left untreated. The infection is transmitted through the bites of female Anopheles mosquitoes carrying the parasite. In 2021, there were 247 million malaria cases worldwide, with 619,000 deaths, mostly occurring in the African region and affecting children under five, responsible alone for 80% of deaths [24].

The parasites of the genus Plasmodium attack red blood cells in humans and can be of five species, P. ovale (Po), P. malariae (Pm), P. knowlesi (Pk), P. falciparum (Pf), and P. vivax (Pv). The latter two pose the greatest threat [24]. Understanding the human host's stages of the parasite's life cycle, i.e., ring, trophozoite, schizont, and gametocyte, is crucial for developing effective treatments and prevention strategies.

According to the WHO, malaria is preventable and treatable if diagnosed promptly, as it can lead to severe complications. Diagnosis can be made using various techniques, including microscopy, rapid diagnostic tests (RDT), or polymerase chain reaction (PCR). However, false negative results can occur, leading to delayed treatment and an increased risk of spreading the disease [21,24].

G. L. Foresti et al. (Eds.): ICIAP 2023 Workshops, LNCS 14366, pp. 59–70, 2024.
https://doi.org/10.1007/978-3-031-51026-7_6

Pathologists prefer using microscopy as a diagnostic method because it is sensitive, affordable, and can identify parasite species and density [12,13]. However, this method has some drawbacks, such as the need for skilled microscopists, limited availability in certain regions, and the possibility of misdiagnosis. It is crucial to control infectious diseases, especially in underdeveloped countries [21].

Pathologists can use computer-aided diagnosis (CAD) systems to diagnose diseases and monitor post-therapy progress. CAD systems provide more precise and faster results than manual analysis, while also reducing subjectivity [12,13].

To address the limitations presented above, this paper introduces a new CAD approach based on a tailored version of the one-stage detector YOLOv5 integrated with an attention-based approach. We proposed an ensemble architecture, MTANet, to simultaneously identify malaria parasites of different species within the same system efficiently and accurately. We compared its performance against several methods in the literature, intending to assist pathologists and address the difficulties encountered in gold-standard microscopy.

The main contributions of our work are listed as follows: 1. We have designed a new deep learning (DL)-based architecture for detecting various types of malaria parasites in real time. 2. Our detection system can identify four different malaria species and life stages, even in mixed or intra-species situations. 3. We have thoroughly evaluated our approach using a high-variation public dataset and compared it with three off-the-shelf object detectors and a previous proposal realized on the same dataset.

2 Related Work

In recent years, the computer vision (CV) community has developed CAD-based solutions to automatically detect malaria parasites. This reduces the problems of manual analysis and provides a more consistent interpretation of blood samples, making diagnosis less expensive [11,19].

Before DL techniques, classical image analysis methods were used to detect malaria parasites. This involved several steps: image preprocessing, object segmentation, feature extraction, and classification. The steps included using mathematical morphology techniques for preprocessing and segmentation [4,20], and handcrafted feature extraction [2] to train machine learning classification methods. In the past decade, numerous studies have proposed deep learning approaches as alternatives to classical methods for this task [1,13,19].

Existing works on malaria detection using DL approaches can be divided into two categories. The first involves classifying images containing single cells to identify the most appropriate classifier to differentiate between parasitized and healthy cells. These studies typically use NIH [15,18] as a reference. The second category involves full pipelines that detect parasites from whole images and rely on several existing datasets, such as MP-IDB [10,25], IML [1], or M5 [19].

IML, proposed by Arshad et al. [1], contains four life cycle stages of the Pv malaria species. The authors used a deep learning-based life cycle stage classification, choosing the ResNet-50v2 network for single-stage multi-class clas-

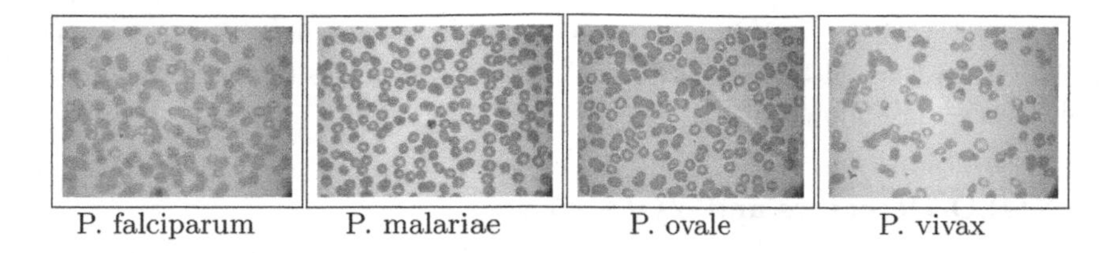

Fig. 1. Samples of the four classes of malaria parasites contained in MP-IDB.

sification. M5, collected by Sultani et al. [19], is a multi-microscope, multi-magnification malaria image dataset containing thin-blood smear slides. The authors obtained two versions of the dataset, low-cost microscopes (LCM) and high-cost microscopes (HCM), to address the real-world problem of image acquisition in resource-constrained settings. They employed object detectors to handle the malaria detection task and tested some off-the-shelf domain adaptation methods to address the microscope domain adaptation tasks. The authors found that ranking combined with triplet loss produced the best overall performance, with HCM as the source domain and LCM as the target domain.

Since malaria parasites infect erythrocytes, any automatic malaria detection must analyze them to determine whether the parasite infects them and further categorize the life stages. Most related work only addresses the classification problem without considering the detection problem. Additionally, researchers have put much effort into developing mobile devices that enable cheaper and faster malaria diagnosis in underdeveloped areas where more expensive laboratories do not exist [2].

Compared to the previous work, this study aims to realize a lightweight and effective pipeline for detecting malaria parasites of any species at once on the high-variation dataset MP-IDB. IML and M5 were not considered in this work as they contain only one malaria parasite species.

3 Materials and Methods

In this section, in Sect. 3.1, we provide information about the public dataset used in this study. Then, we summarize the YOLOv5 object detector used as the baseline and describe the CBAM attention module adopted in this study in Sects. 3.2 and 3.3 respectively. Our proposed approach is presented in Sect. 3.4, and finally, we describe the chosen evaluation metrics Sect. 3.5.

3.1 Dataset

MP-IDB [12] consists of 210 images representing four different species of malaria parasites. These include 104 images of the P. falciparum species, 37 of the P. malariae species, 29 of the P. ovale species, and 40 of the P. vivax species. Each species has four life stages: ring, trophozoite, schizont, and gametocyte. Each image in the dataset has a corresponding ground truth, indicating the parasites' position and life stages. All images were captured at a resolution of 2592×1944 and 24-bit color depth. To give a glimpse of the dataset, some sample images are shown in Fig. 1.

3.2 YOLO Detectors and YOLOv5

Convolutional neural networks (CNN)-based methods form the basis of modern detectors, which can be categorized into two types: one-stage and two-stage. Two-stage architectures, exemplified by Faster R-CNN (FRCNN) [17], follow a coarse-to-fine approach where regions of interest are extracted and subsequent processes involve classification and bounding box regression. In contrast, one-stage detectors like RetinaNet [7] and the YOLO family [16] directly generate bounding boxes and classes from predefined anchors on predicted feature maps. One-stage detectors, faster and more compact, are better suited for time-critical applications and computationally constrained edge devices [26].

Instead of following the traditional two-step approach involving region selection, the YOLO family of detectors adopts an end-to-end differentiable network that integrates bounding box estimation and object identification. YOLO divides the input image into a constant-size grid of $S \times S$, and a CNN predicts bounding boxes and classes for each grid. If the confidence of a bounding box exceeds a certain threshold, it is selected to locate the object in the image. The CNN performs a single pass to make predictions and, after non-maximum suppression (NMS), outputs recognized objects and their bounding boxes, ensuring that each object is detected only once.

YOLOv5 is a collection of architectures and models for object detection [5], pretrained on the Common Object in Context (COCO) dataset [8]. This family includes five models that share the same architecture but vary in size, depth, and the number of trainable parameters: YOLOv5n (nano), YOLOv5s (small), YOLOv5m (medium), YOLOv5l (large), and YOLOv5x (extra large). Each model can be pretrained on 640×640 or 1280×1280 resolution images. The parameters range from 1.9 million for YOLOv5n to 140.7 million for YOLOv5x for the respective resolutions. The YOLOv5 architecture comprises three components, similar to other single-stage object detectors: backbone, neck, and prediction head. The backbone is a pretrained network responsible for image feature extraction, which reduces spatial resolution while increasing feature resolution. The neck combines the extracted features and generates feature pyramids at three different scales to facilitate the detection of objects of varying sizes. The prediction head employs anchor boxes on the feature maps to detect objects. YOLOv5 utilizes the CSPDarknet53 architecture with a Spatial Pyramid Pooling layer [6] as the backbone, employs Path Aggregation Network [9] as the neck, and incorporates the YOLO detection head [16].

3.3 Convolutional Block Attention Module (CBAM)

The concept of attention involves assigning weights to input features to determine their importance [14]. In object detection, attention can guide the model's focus towards relevant regions, improving accuracy and reducing the computational burden associated with processing irrelevant information. Previous research has demonstrated that CBAM is a valuable addition that enhances classification and detection tasks [23].

A widely used method for adding attention to CV is by using CNNs with attention modules [22]. These modules contain adjustable weights that determine the significance of various image regions. During training, the model learns to prioritize the most significant areas for the task at hand by modifying these weights.

In this context, CBAM is an attention mechanism proposed for CNNs to enhance their performance in various image recognition tasks [23]. It consists of two sub-modules: the *channel* and *spatial* attention module. The channel attention module assigns weights to different channels in a feature map based on their relevance. In contrast, the spatial attention module selectively emphasizes important spatial regions within the feature map.

In CV tasks, attention mechanisms can be broadly categorized into two types: *spatial* and *channel* attention. The first directly focuses on specific positions, enabling neural networks to learn where to concentrate. Conversely, the second exploits the inter-channel relationship of features and generates a channel attention map.

More formally, CBAM can be expressed by Eq. (1):

$$W_c = \sigma(\text{MLP}(A_{\text{pool}}(X))) \tag{1}$$

where A_{pool} represents global average pooling, which aggregates the spatial dimensions of the feature map. MLP refers to a two-layer feedforward network with ReLU activations, and σ denotes the sigmoid activation function.

The spatial attention module operates on the feature map X, and the channel attention weights W_c. Firstly, it computes a set of spatial attention weights W_s by applying a sigmoid activation to the feature map passed through a convolutional layer, as shown in Eq. (2):

$$W_s = \sigma(\text{Conv}7x7(A\text{pool}(X) \otimes W_c)) \tag{2}$$

In this equation, $\otimes$ represents element-wise multiplication, Conv denotes a convolutional layer, and A_{pool} is as defined previously. The spatial attention weights are then utilized to modulate the feature map, resulting in the output feature map Y, as illustrated in Eq. (3):

$$Y = W_s \otimes X \tag{3}$$

where Y denotes the output feature map.

3.4 Our Proposed Method: MTANet

This study aims to create a malaria parasite detector that can simultaneously detect multiple types of malaria. To achieve this objective, we first created a

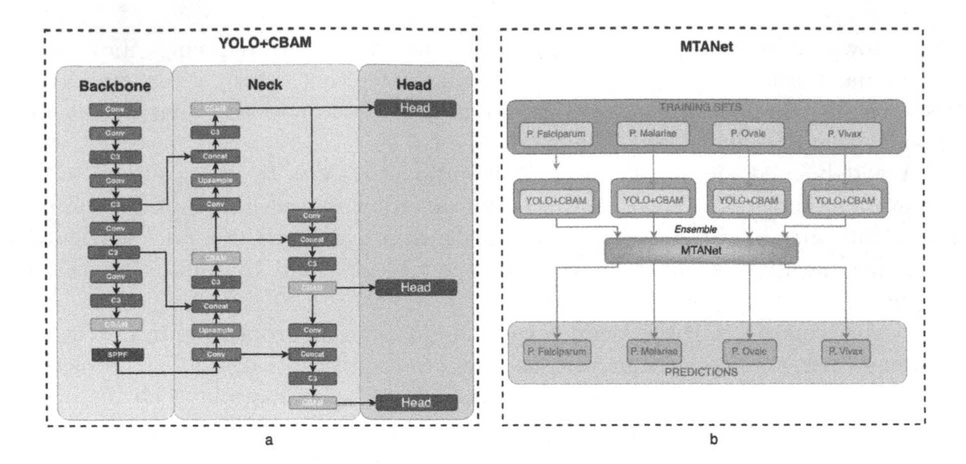

Fig. 2. Representation of our proposed approach. Figure a shows the architecture of YOLO with the addition of multiple CBAMs. Figure b shows the proposed MTANet, realized with an ensemble strategy of four *YOLO+CBAM* trained on each of the class.

baseline architecture by modifying the structure of YOLOv5. Specifically, we included multiple CBAMs in the backbone and neck to improve the intra-class detection by enhancing their characterization by attention. We will refer to this architecture as *YOLO+CBAM*. Then, we create MTANet, an ensemble of four distinct *YOLO+CBAM* models, each trained on a specific parasite species to enhance the out-of-class detection capability. Its predictions consist of the aggregation of the predictions of each single model through an ensembling strategy. The final prediction is refined by an NMS strategy to avoid bounding box duplicates. Figure 2 shows the architectures of *YOLO+CBAM* and MTANet.

3.5 Metrics

To evaluate the results obtained by the proposed method, we employ four metrics commonly used to assess object detection methods:

- **Precision (P)** refers to the model's ability to accurately identify relevant objects and defined in Eq. (4):

$$P = \frac{TP}{TP + FP} = \frac{TP}{alldetections} \tag{4}$$

- **Recall (R)** measures the model's ability to identify all relevant cases, among all ground-truth bounding boxes, as defined in Eq. (5):

$$R = \frac{TP}{TP + FN} = \frac{TP}{allgroundtruths} \tag{5}$$

- **Average Precision (AP)** is evaluated with 10 different IOUs varying in a range of 50% to 95% with steps of 5%;

Table 1. Image augmentation setup.

Augmentation	Parameters	Probability (%)
Rotation	range iterations: [0, 3]	100
Gaussian Noise	variance range: [50, 100]	30
HSV - Hue	shift limit: 20	30
HSV - Saturation	shift limit: 30	30
HSV - Value	shift limit: 20	30

- $\mathbf{AP}_{50}$ is evaluated with a single values of IOU corresponding to 50%.

With reference to Eqs. (4) and (5), it must be noted that the accuracy of detections is determined using the concept of Intersection over Union (IoU). It calculates the ratio of the overlapping area between the predicted bounding box and the actual object to the total area encompassed by both. If the IoU exceeds a specific threshold, the detection is considered accurate and labeled as a *true positive* (TP). However, if the IoU falls below the threshold, the detection is labeled a *false positive* (FP). In addition, when the model fails to detect an object in the ground truth, it is referred to as a *false negative* (FN).

4 Experimental Results and Discussion

4.1 Experimental Setup

The experiments were conducted on a workstation equipped with the following hardware specifications: an *Intel(R) Core(TM) i9-8950HK* CPU operating at *2.90GHz*, *32GB* RAM, and an *NVIDIA GTX1050 Ti GPU* with *4GB* memory. We utilized the PyTorch implementation of YOLOv5[1], developed by the Ultralytics LLC team [5].

Training Details: Two different backbones were used: ResNet-50 for FRCNN and RetinaNet and Darknet53 for YOLO. The pre-trained weights from the COCO2017 dataset [8] were used to initialize all the networks. We employed the *Adam* optimizer with a learning rate of 0.001 and momentum of 0.9. Each model was trained for 100 epochs, with a batch size of 4, on every class of MP-IDB. MP-IDB was split into three parts for each parasite class: training (60%), validation (20%), and testing (20%).

Data Augmentation: We employed a preliminary step of data augmentation to overcome the dataset size limitations, address data imbalance, strengthen models against object rotations, enhance diversity, and enable targeted generalization capabilities. We utilized the *Albumentation* library [3], which provides

[1] Available at: https://github.com/ultralytics/yolov5.

Table 2. Quantitative performance results obtained by MTANet on the four classes of MP-IDB [12]. The table reports a comparison with three off-the-shelf object detectors, viz. FRCNN, RetinaNet, and YOLOv5 (first three rows); with the work of Zedda et al. [25] only on Pf split; and with a custom *Ensemble* realized with the YOLOv5 trained on the four different splits.

Method	**P. falciparum**				**P. malariae**				**P. ovale**				**P. vivax**			
	AP(%)	AP_{50}	P	R	AP	AP_{50}	P	R	AP	AP_{50}	P	R	AP	AP_{50}	P	R
FRCNN	39.2	80.6	83.1	82.2	75.1	98.4	98.6	96.2	71.0	89.1	91.7	84.9	60.3	87.7	91.1	83.2
RetinaNet	34.0	78.5	81.1	74.6	76.0	95.0	96.1	91.3	44.2	81.8	85.0	77.6	62.8	85.5	89.3	79.9
YOLO	62.5	93.5	95.9	95.3	80.0	96.4	**99.9**	92.9	83.9	96.8	93.8	**99.9**	83.1	**93.2**	**91.8**	89.8
YOLO+CBAM	**74.7**	**98.7**	**96.9**	**98.3**	**94.1**	**99.5**	99.7	**99.9**	**93.8**	**99.5**	**99.9**	**99.9**	**83.6**	92.9	89.0	**92.1**
Zedda et al. [25]	-	95.2	92.8	95.2	-	-	-	-	-	-	-	-	-	-	-	-
YOLO-Ensemble	56.9	86.8	86.6	92.1	68.6	96.0	88.9	95.3	64.5	94.1	82.1	93.9	65.3	88.5	80.8	85.7
MTANet	**70.1**	**96.0**	**93.4**	**95.2**	**90.1**	**99.5**	**98.9**	**99.9**	**83.6**	**97.1**	**93.6**	**99.9**	**75.4**	**91.3**	**88.9**	**88.9**

a wide range of transformations. Hence, we created 35 augmented samples for each original sample for each original one, across all species. We adopted a careful approach due to the concerns about the excessive changes given by the augmentations applied to some parasites, especially the smallest [25]. The applied augmentations combined several transformations based on a specific probability, as shown in Table 1.

4.2 Experimental Results

Quantitative Results. To ensure statistical significance, we tested our method using a 5-fold cross-validation process. We divided the samples into training and validation sets, using 80% of the samples for this purpose. Of the 80%, we used 60% for training and 20% for validation. The remaining 20% was used as the test set. We obtained the final results by averaging all the folds.

Table 2 summarized the results of our experiments. Specifically, we present the performance obtained by MTANet on the four classes of MP-IDB [12]. These results are compared with five class-to-class methods (i.e., trained and tested on the same species): FRCNN, RetinaNet, YOLO, *YOLO+CBAM* and the proposal of Zedda et al. [25] (which only reported the Pf split), and with *YOLO-Ensemble*, i.e., an ensemble of the *YOLO* models trained on the four species separately, used to compare MTANet with a baseline approach that operates without the use of an attention mechanism.

We found that each class-to-class model behaves differently depending on the parasite class used. For instance, YOLO performs significantly better with the Pm, Po, and Pv classes, achieving 80.0%, 83.9%, and 83.1% AP, respectively. Even though it struggles with Pf, it is the third-best performing in this specific class in terms of AP. *YOLO+CBAM* produced a substantial improvement to YOLO, outperforming every class-to-class detector in every metric, except for precision in Pm split (99.7% vs. 99.9% of YOLO), and AP_{50} (92.9% vs. 93.2% of YOLO) and precision (89.0% vs. 91.8% of YOLO) in Pv one. These results

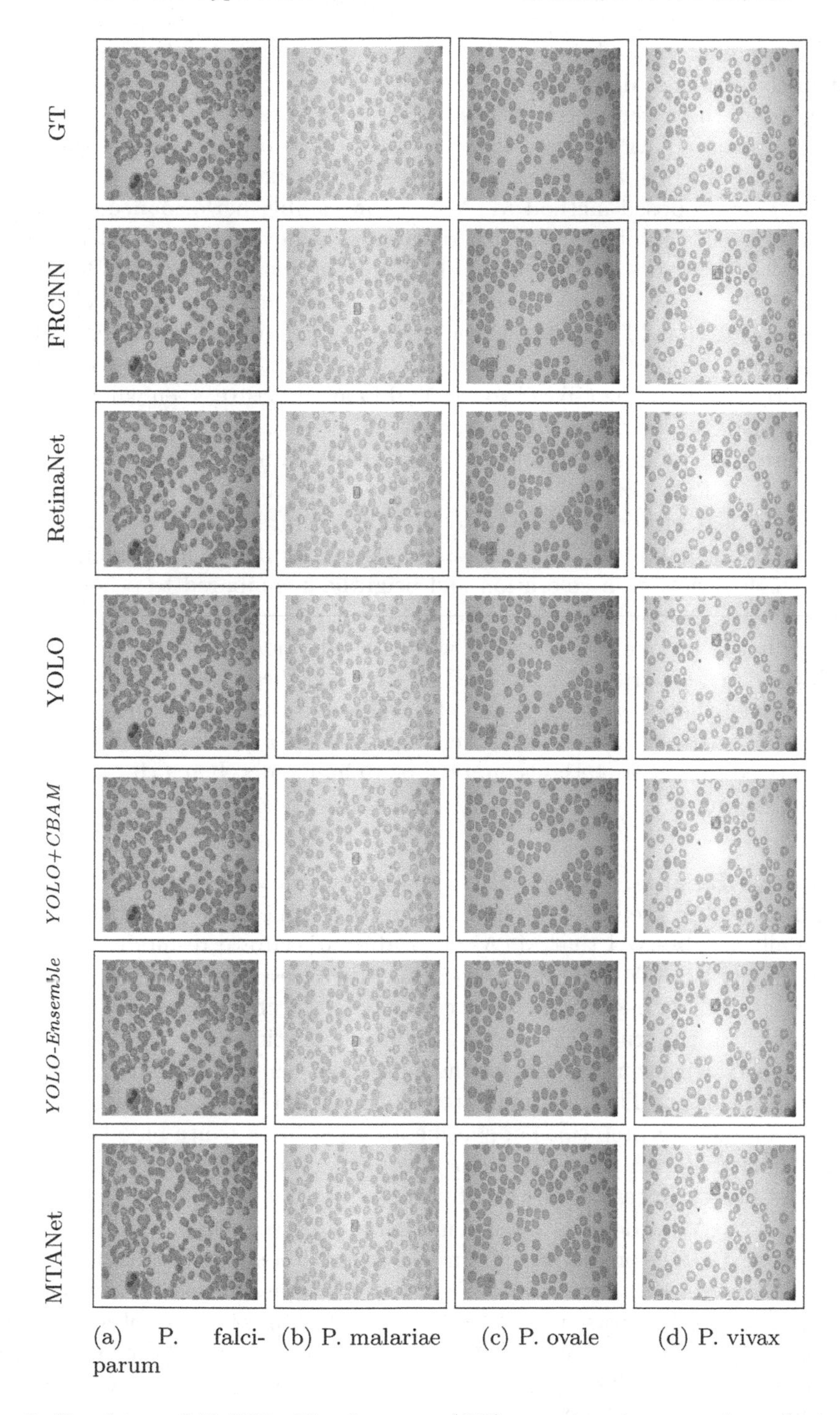

Fig. 3. Results on MP-IDB. The first row (GT) contains the ground truths in □, while every subsequent row shows the results of the object detectors in □ (Color figure online).

demonstrate how the addition of CBAMs was essential to produce an improvement in class-to-class detection.

Regarding the ensemble strategies, the *YOLO-Ensemble* model is the worst and never achieved AP above 68.6%. However, in some cases, it demonstrates superior performance with respect to class-to-class detectors, for instance, in Pv split. Our proposed method MTANet shows stable performance across the different classes. It exhibits the best performance, achieving the highest values for the ensemble strategies and the second-best absolute performance for Pf (70.1%) and Po (90.1%), as well as excellent precision and recall scores for all classes. It is worth noting that MTANet is in line with the second best-performing YOLO (83.6% vs. 83.9% of YOLO) concerning Po. These outstanding results confirmed the importance of adding CBAM to the YOLO architecture, even when multiple classes need to be predicted with a single architecture.

Despite the training set containing a limited number of samples in some cases (e.g., 18 Pf schizonts, 7 Pf gametocytes [25]) and the high variation of every single class, either intra-class or out-of-class, our method achieved excellent generalization capabilities. MP-IDB images are particularly challenging because of their high variation in coloration and brightness compared to other existing datasets [1,19]. Nevertheless, MTANet achieved remarkable performance, considering that it targets every species with a single model, not only concerning the object detectors trained and tested on each specific class but also compared to the state-of-the-art work of Zedda et al. [25] and the generalization strategy represented by the *Ensemble*. MTANet improved by 1.7% the AP_{50} on Pf and outperformed the other methods on 9 out of 16 indicators, being the second-best in the remaining 7.

Qualitative Results. The visual results of our study are presented in Fig. 3, which displays the predicted bounding boxes of the methods we considered.

Overall, we observed that MTANet produced accurate results compared to the ground truth, while other detectors generated excess predictions, particularly with Pf species. The struggle in detecting Pf parasites is mainly because it contains several tiny rings representing an early infection and could be found in large numbers, with some images having more than ten parasites. It is important to note that this condition is unique to the Pf split. This surplus of predictions led to incorrect interpretations of specific areas, such as white blood cell nuclei being classified as parasites. This error was particularly evident in FRCNN and RetinaNet detection results on the Pf class (see Fig. 3a).

Finally, MTANet demonstrated remarkable progress compared to the other detectors, accurately predicting all four classes with a single architecture. However, MTANet sometimes generated excess predictions, as seen in Fig. 3d. This was due to several factors, including the limited number of samples in the Pv split, its more substantial variations wrt. the other classes, the production of extra predictions, resulting in FPs (e.g., the border ones kept for fairness and exemplified in Fig. 3d). Addressing these issues is the key to improving performance in this direction.

5 Conclusions

Our study has made significant progress in improving malaria detection. Our main goal was to find a solution to the challenge of simultaneously detecting parasites of different species. We implemented MTANet, which has proved to be highly effective in the significant challenge of simultaneously detecting multiple species of parasites, surpassing the current state-of-the-art on MP-IDB. Our approach helps to address both multi-species infections in the same patient and diverse single-species infections from multiple patients.

We intend to continue researching to enhance our methodology to detect all types of malaria parasites with greater precision. Although we have achieved encouraging results with our current dataset, we aim to refine our system to operate on a cross-dataset model. This will allow it to effectively tackle environmental variations between different datasets. Our ultimate goal is to extend our approach to encompass a multi-magnification image representation of the same blood smear. This aspect will enable us to accurately identify malaria parasites across different magnifications.

References

1. Arshad, Q.A., et al.: A dataset and benchmark for malaria life-cycle classification in thin blood smear images. Neural Comput. Appl. **34**(6), 4473–4485 (2022)
2. Bias, S., Reni, S., Kale, P.I.: Mobile hardware based implementation of a novel, efficient, fuzzy logic inspired edge detection technique for analysis of malaria infected microscopic thin blood images. Proced. Comput. Sci. **141**, 374–381 (2018)
3. Buslaev, A., Iglovikov, V.I., Khvedchenya, E., Parinov, A., Druzhinin, M., Kalinin, A.A.: Albumentations: fast and flexible image augmentations. Information **11**(2), 125 (2020)
4. Di Ruberto, C., Dempster, A., Khan, S., Jarra, B.: Analysis of infected blood cell images using morphological operators. Image Vis. Comput. **20**(2), 133–146 (2002)
5. Jocher, G., et al.: ultralytics/yolov5: v6.0 - YOLOv5n 'Nano' models, Roboflow integration, TensorFlow export, OpenCV DNN support (2021). https://doi.org/10.5281/zenodo.5563715
6. He, K., Zhang, X., Ren, S., Sun, J.: Spatial pyramid pooling in deep convolutional networks for visual recognition. IEEE Trans. Pattern Anal. Mach. Intell. **37**(9), 1904–1916 (2015)
7. Lin, T., Goyal, P., Girshick, R.B., He, K., Dollár, P.: Focal loss for dense object detection. In: IEEE International Conference on Computer Vision, ICCV 2017, Venice, Italy, 22–29 October 2017, pp. 2999–3007. IEEE Computer Society (2017)
8. Lin, T.-Y., et al.: Microsoft COCO: common objects in context. In: Fleet, D., Pajdla, T., Schiele, B., Tuytelaars, T. (eds.) ECCV 2014. LNCS, vol. 8693, pp. 740–755. Springer, Cham (2014). https://doi.org/10.1007/978-3-319-10602-1_48
9. Liu, S., Qi, L., Qin, H., Shi, J., Jia, J.: Path aggregation network for instance segmentation. In: 2018 IEEE Conference on Computer Vision and Pattern Recognition, CVPR 2018, Salt Lake City, UT, USA, 18–22 June, 2018, pp. 8759–8768. Computer Vision Foundation / IEEE Computer Society (2018)
10. Loddo, A., Fadda, C., Ruberto, C.D.: An empirical evaluation of convolutional networks for malaria diagnosis. J. Imaging **8**(3), 66 (2022)

11. Loddo, A., Ruberto, C.D., Kocher, M.: Recent advances of malaria parasites detection systems based on mathematical morphology. Sensors **18**(2), 513 (2018)
12. Loddo, A., Di Ruberto, C., Kocher, M., Prod'Hom, G.: MP-IDB: the malaria parasite image database for image processing and analysis. In: Lepore, N., Brieva, J., Romero, E., Racoceanu, D., Joskowicz, L. (eds.) SaMBa 2018. LNCS, vol. 11379, pp. 57–65. Springer, Cham (2019). https://doi.org/10.1007/978-3-030-13835-6_7
13. Maity, M., Jaiswal, A., Gantait, K., Chatterjee, J., Mukherjee, A.: Quantification of malaria parasitaemia using trainable semantic segmentation and capsnet. Pattern Recogn. Lett. **138**, 88–94 (2020)
14. Niu, Z., Zhong, G., Yu, H.: A review on the attention mechanism of deep learning. Neurocomputing **452**, 48–62 (2021)
15. Rajaraman, S., et al.: Pre-trained convolutional neural networks as feature extractors toward improved malaria parasite detection in thin blood smear images. PeerJ **6**, e4568 (2018)
16. Redmon, J., Divvala, S.K., Girshick, R.B., Farhadi, A.: You only look once: unified, real-time object detection. In: 2016 IEEE Conference on Computer Vision and Pattern Recognition, CVPR 2016, Las Vegas, NV, USA, 27–30 June 2016, pp. 779–788. IEEE Computer Society (2016)
17. Ren, S., He, K., Girshick, R.B., Sun, J.: Faster R-CNN: towards real-time object detection with region proposal networks. In: Advances in Neural Information Processing Systems 28: Annual Conference on Neural Information Processing Systems 2015, 7–12 December 2015, Montreal, Quebec, Canada, pp. 91–99 (2015)
18. Sengar, N., Burget, R., Dutta, M.K.: A vision transformer based approach for analysis of plasmodium vivax life cycle for malaria prediction using thin blood smear microscopic images. Comput. Methods Programs Biomed. **224**, 106996 (2022)
19. Sultani, W., Nawaz, W., Javed, S., Danish, M.S., Saadia, A., Ali, M.: Towards low-cost and efficient malaria detection. In: IEEE/CVF Conference on Computer Vision and Pattern Recognition, CVPR 2022, New Orleans, LA, USA, 18–24 June 2022, pp. 20655–20664. IEEE (2022)
20. Tek, F.B., Dempster, A.G., Kale, I.: Malaria parasite detection in peripheral blood images. BMVA (2006)
21. United States' Centers for Disease Control and Prevention. https://www.cdc.gov/malaria/about/biology/index.html (2021). Accessed 03 July 2023
22. Wang, F., et al.: Residual attention network for image classification. In: 2017 IEEE Conference on Computer Vision and Pattern Recognition, CVPR 2017, Honolulu, HI, USA, 21–26 July 2017, pp. 6450–6458. IEEE Computer Society (2017)
23. Woo, S., Park, J., Lee, J.-Y., Kweon, I.S.: CBAM: convolutional block attention module. In: Ferrari, V., Hebert, M., Sminchisescu, C., Weiss, Y. (eds.) ECCV 2018. LNCS, vol. 11211, pp. 3–19. Springer, Cham (2018). https://doi.org/10.1007/978-3-030-01234-2_1
24. World Health Organization. https://www.who.int/news-room/fact-sheets/detail/malaria (2021). Accessed 03 July 2023
25. Zedda, L., Loddo, A., Di Ruberto, C.: A deep learning based framework for malaria diagnosis on high variation data set. In: Sclaroff, S., Distante, C., Leo, M., Farinella, G.M., Tombari, F. (eds.) Image Analysis and Processing – ICIAP 2022. ICIAP 2022. LNCS, vol. 13232. Springer, Cham (2022). https://doi.org/10.1007/978-3-031-06430-2_30
26. Zou, Z., Chen, K., Shi, Z., Guo, Y., Ye, J.: Object detection in 20 years: a survey. Proc. IEEE **111**(3), 257–276 (2023)

Breast Mass Detection and Classification Using Transfer Learning on OPTIMAM Dataset Through RadImageNet Weights

Ruth Kehali Kassahun[1](✉), Mario Molinara[1], Alessandro Bria[1], Claudio Marrocco[1], and Francesco Tortorella[2]

[1] Department of Electrical and Information Engineering, University of Cassino and Southern Lazio, Cassino, Italy
ruthkehali@gmail.com

[2] Department of Information and Electrical Engineering and Applied Mathematics, University of Salerno, Salerno, Italy

Abstract. A significant number of women are diagnosed with breast cancer each year. Early detection of breast masses is crucial in improving patient prognosis and survival rates. In recent years, deep learning techniques, particularly object detection models, have shown remarkable success in medical imaging, providing promising tools for the early detection of breast masses. This paper uses transfer learning methodologies to present an end-to-end breast mass detection and classification pipeline. Our approach involves a two-step process: initial detection of breast masses using variants of the YOLO object detection models, followed by classification of the detected masses into benign or malignant categories. We used a subset of OPTIMAM (OMI-DB) dataset for our study. We leveraged the weights of RadImageNet, a set of models specifically trained on medical images, to enhance our object detection models. Among the publicly available RadImageNet weights, DenseNet-121 coupled with the yolov5m model gives 0.718 mean average precision(mAP) at 0.5 IoU threshold and a True Positive Rate (TPR) of 0.97 at 0.85 False Positives Per Image (FPPI). For the classification task, we implement a transfer learning approach with fine-tuning, demonstrating the ability to effectively classify breast masses into benign and malignant categories. We used a combination of class weighting and weight decay methods to tackle the class imbalance problem for the classification task.

Keywords: Breast Cancer · Breast Mass Detection · Breast Mass Classification · RadImageNet · YOLO ObjectDetection · Transfer Learning · Computer Aided Diagnosis

1 Introduction

Breast cancer is a significant public health concern affecting women predominantly. It is the second leading cause of cancer death among women, following

G. L. Foresti et al. (Eds.): ICIAP 2023 Workshops, LNCS 14366, pp. 71–82, 2024.
https://doi.org/10.1007/978-3-031-51026-7_7

lung and bronchus cancer [1]. In 2022, it attributed to 31% of all women's cancers. Despite advances in screening and treatment, breast cancer remains challenging to diagnose, making early detection and prevention critical to reducing mortality rates [2].

One of the primary methods for detecting breast cancer is mammography screening, which involves taking X-ray images of the breast to visualize the internal structure of the breast. Early mammography screening for breast cancer is a widely used clinical procedure. By enabling the detection of breast cancer in its initial stages, when it is more manageable, and the chances for effective treatment are higher, mammography screenings significantly contribute to reducing breast cancer mortality rates [3].

Breast masses, also known as breast lumps, are swellings, bulges, or bumps in the breast that differ from the surrounding tissue. They vary significantly in size, shape, and texture. These masses, such as fibroadenomas or cysts, can be benign (non-cancerous), and some can be malignant (cancerous). Regarding their relationship with breast cancer, not all breast masses indicate malignancy. However, they are often the first noticeable symptom of breast cancer. Therefore, any new or unusual breast mass suggests medical evaluations.

However, the detection of breast masses are not always a straightforward task. At times, the anomalies may be subtly embedded within the breast tissue, making it challenging for even experienced radiologists to distinguish them [4]. Additionally, the variety in the shape, size, and density of breast masses further complicates their identification. Some masses may appear as tiny specks or calcifications, while others might present as larger, irregularly shaped structures. [5]. The characteristics of these masses can be highly variable. This heterogeneity in presentation often makes it difficult to consistently detect breast masses, emphasizing the need for advanced imaging techniques and tools to aid in accurate diagnosis.

This paper proposes an approach to breast cancer diagnosis by developing a computer-aided diagnosis(CAD) system. This system aims to detect and classify masses present in mammography images from the OPTIMAM (OMI-DB) dataset. To accomplish this, state-of-the-art object detection algorithms, particularly YOLO based algorithms are utilized to detect and localize breast masses accurately. In addition, the paper leverages the RadImageNet-trained weights. The research uses weights initialized by the MS COCO (Microsoft Common Objects in COntext) dataset and RadImageNet dataset weights. MS COCO is an object detection and captioning dataset known for its high-quality annotations and diversity of object categories. It is an ideal choice for weight initialization in object detection tasks [6]. The RadImageNet dataset comprises five million professionally annotated medical images for effective transfer learning [7]. Consequently, the detected masses are further classified into benign and malignant classes, which is crucial for making informed decisions regarding patient treatment plans.

Current clinical methods for breast mass detection are largely based on radiologist interpretation of mammography images. Radiologists use various indicating factors such as the shape, margin, and density of abnormal tissue to determine the presence of mass [8]. In some cases, CAD systems were used. The rise of deep learning, machine learning, and artificial neural networks has greatly contributed to the effective implementation of CAD systems.

For breast mass detection, various Convolutional Neural Network(CNN)-based object detection algorithms have been proposed. [9] proposes a Faster R-CNN model to detect breast masses by classifying the dataset into benign, malignant, and other categories. The model achieves an area under the curve (AUC) of 0.91 (95% CI: 0.89, 0.93), with a specificity of 77.3% (95% CI: 69.2%,85.4%) at a sensitivity of 87%.

YOLO region proposals have been used to effectively detect breast masses in INbreast and DDSM-CBIS (Digital Database for Screening Mammography) datasets using both patch level and dual view mammographs [10]. In this study, they integrated the mass-matching technique and achieved 94.78% as AUC score for detection and classification accuracy of 0.87.

This paper [11] applied the Faster-RCNN model in Full-Field Digital Mammograms (FFDM). It achieved a true positive rate of 0.87 at 0.84 false positives per image (FPPI). This study serves as one of the first research to benchmark on large-scale OPTIMAM (OMI-DB). [12] showcases the potential of transformer models to achieve remarkable results when combined with convolutional layers for prediction tasks. This study uses multi-scale swin transformers as a backbone model, Representative Points, and the Deformable Detection Transformer (DETR). This research notably achieved a high TPR of 0.903 at 0.8 FPPI.

The work presented by [13] proposes a two-stage methodology for detecting and classifying breast masses in the OPTIMAM (OMI-DB) dataset. The author used RetinaNet variation of ResNet backbones for breast mass detection alongside different weight initialization, mainly ImageNet, COCO weights, and model trained from scratch. Additionally, the study emphasized analyzing the whole mammograph and patches taken from the mammogram images. The results show a TPR of 0.959 at 0.84 FPPI when using the RetinaNet with the ResNet151 backbone and ImageNet weights.

CNNs have been extensively employed to classify breast masses particularly in this study [14] to classify breast masses in the DDSM dataset using different architectures, specifically shallow CNN, AlexNet, and GoogLeNet.

2 Methods

2.1 Dataset

The OPTIMAM Mammography Image Database (OMI-DB) is a collection of mammography images. Currently, there are 2.5 million images gathered from 173,319 women from three main UK breast screening centers [18]. For this particular study, we were provided with a total of 7,629 images from the OPTIMAM committee. 3,529 of the mammograms are identified as having breast masses, and the remaining 4,100 mammograms have no breast masses. The data was split into training, validation, and test subsets in a 70%, 20%, and 10% ratio, respectively.

2.2 Proposed Method

In this study, our objective is to develop an effective and robust algorithm for detecting breast masses, focusing on utilizing YOLO based object detection techniques. YOLO has emerged as a popular and powerful approach in the field of computer vision due to its real-time performance and accurate object localization capabilities. To achieve our goal, we explored and implemented various versions of the YOLO algorithm, including YOLOv5, YOLOv6, YOLOv7, and YOLOv8. Each version offers distinct architectural enhancements and optimization strategies, which we carefully evaluated and compared in our experiments. Our evaluation consists of various metrics, such as mAP, precision, recall, and TPR per FPPI, to ensure the assessment of each algorithm's performance.

Our proposed method involves a two-step procedure: breast mass detection followed by a classification task. The object detection stage is mainly used variants of YOLO models. As shown in Fig. 1, the first step involves feeding the mammogram images into the YOLO model.

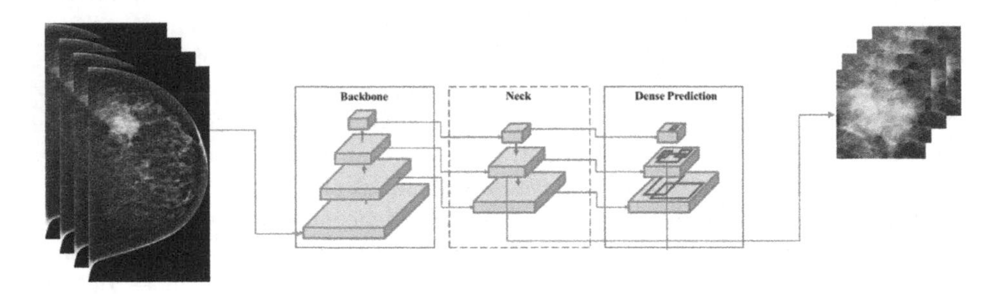

Fig. 1. Proposed method for breast mass detection

This model identifies the regions in the images containing potential breast masses. A typical YOLO model has three main components. The backbone, neck, and prediction head. Each of these components will be discussed later in detail. Once the areas of interest (potential breast masses) have been identified, these regions are cropped from the original images, preparing them for the next stage.

Following the object detection phase, the regions of interest, which are the cropped breast mass regions, are directed into the classification stage, as can be seen in Fig. 2. The classification stage aims to distinguish between benign and malignant breast masses. To do this, we used transfer learning, specifically through fine-tuning pre-trained models. We used a range of pre-trained models for the classification task, including DenseNet 121, Inception V3, VGG 16, AlexNet, ResNet 18, and ResNet 50.

All the experiments were carried out using two NVIDIA GeForce GTX Titan X GPUs with 12 GB and 16 GB VRAM.

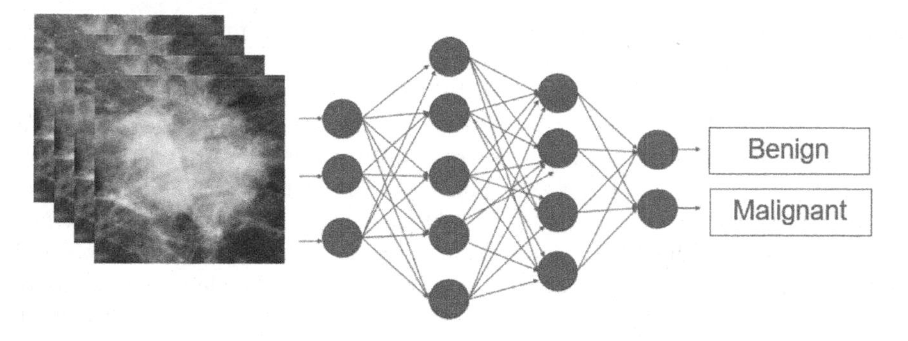

Fig. 2. Proposed method for breast mass classification followed by detection

2.3 YOLO

YOLO object detection algorithm is a widely used one-stage object detection algorithm. As opposed to two-stage detectors, YOLO performs object detection in one go. The input image is split into grids, and each cell in the grid is responsible for predicting objects within it. It simultaneously predicts the bounding boxes and class probabilities for these boxes [15]. Two-stage detectors, such as R-CNN, Fast R-CNN, and Faster R-CNN, approach object detection in two primary steps. The first stage involves generating a set of proposal regions within the image where the object might be located, and this is usually done using a region proposal network. Once these candidate regions are proposed, the second stage involves extracting features from them and classifying them to identify the object [16].

3 Results

In this section, we will discuss in detail the performance of the different YOLO models and versions that we have utilized to detect breast masses. We will deal with the specific performance of each of these models in our task and analyze the detection accuracy, classification precision, recall, and F1-score. The performance of these models will be compared with each other, providing us with a clear idea about which model performs the best in the context of breast mass detection and classification.

3.1 Breast Mass Detection

The results obtained from the various trained models of YOLOV5 for breast mass detection are summarized in the table below. The four performance metrics used in the analysis include Precision, Recall, mAP@50 at Intersection over Union (IoU) over 0.50), and TPR at FPPI of 0.85 (True Positive Rate at a False Positive Per Image rate of 0.85). In Table 1, YOLOV5m performed best in terms of precision, achieving a score of 0.694. Precision measures the proportion of correctly predicted positive observations to the total predicted positives. Regarding recall metrics, the DenseNet121 model trained on the RadImageNet dataset with the YOLOV5m architecture performed the best, achieving a recall of 0.713. Recall measures the proportion of correctly identified positive cases from all actual positive cases. Looking at mAP@50, the DenseNet121 model trained on the RadImageNet dataset with YOLOV5m again stands out, achieving a mAP@50 of 0.718. The TPR at an FPPI of 0.85 is highest for the DenseNet121 with a value of 0.97. In conclusion, based on the metrics, the DenseNet121 model trained on the RadImageNet dataset with YOLOV5m seems to outperform the other models in this task. This model appears to provide a good balance of precision and recall, leading to better overall performance in detecting and classifying breast masses.

Table 1. YOLOV5 all trained models result

	Precision	Recall	mAP@50	TPR at FPPI 0.85
YOLOV5s	0.619	0.537	0.556	0.93
YOLOV5m	**0.694**	0.609	0.600	0.88
YOLOV5l	0.616	0.544	0.533	0.90
YOLOV5m6	0.636	0.606	0.588	0.85
YOLOV6l6	–	0.601	0.625	–
YOLOV7-e6e	0.655	0.632	0.618	0.92
YOLOV8X	0.690	0.523	0.579	–
YOLOV8L	0.673	0.506	0.563	–
YOLOV5m6 Frozen Layer	0.641	0.529	0.534	0.93
YOLOV5l Transformer Head	0.558	0.540	0.510	0.89
YOLOV5m6 Transformer Head	0.652	0.551	0.568	0.87
RadImageNet ResNet50 YOLOV5m	0.605	0.606	0.557	0.88
RadImageNet InceptionV3 YOLOV5m	0.605	0.606	0.557	0.88
RadImageNet DenseNet121 YOLOV5m	0.678	**0.713**	**0.718**	**0.97**
RadImageNet DenseNet121 YOLOV7-e6e	0.677	0.585	0.600	0.91

Our experiments with the DenseNet121 model trained on the RadImageNet dataset and subsequently applied to breast mass detection have shown good results. A visual representation of these results is illustrated in Fig. 3.

The Transformer Prediction Head variant of YOLOv5 is another modified version of the YOLOV5 models' heads. In a nutshell, it integrates a transformer prediction head in place of the conventional convolutional layers in YOLOv5's prediction stage. This was inspired by the recent successes of transformers compared to CNN.

In our experiment, we examined the performance of the YOLOv5 model with a Transformer prediction head, coupled with RadImageNet DenseNet 121 weights, to evaluate its efficiency in detecting breast masses. We achieved a TPR of 0.89 at FPPI with a threshold of 0.85.

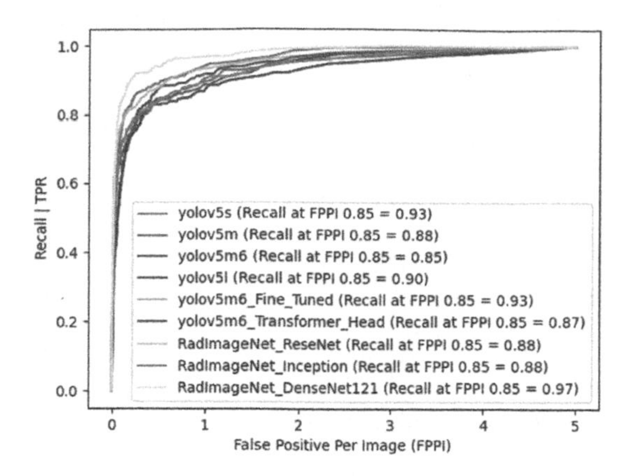

Fig. 3. FROC Curve for YOLOV5 all trial

Yolov5 transformer prediction head model result, when compared with another variant of the YOLOv5 model, which utilizes the DenseNet121 weights from RadImageNet, shows a notable performance difference, as can be seen in Fig. 4. The YOLOv5 model, when paired with DenseNet121 weights, outperformed the Transformer prediction head model variant by yielding the highest performance in our tests. It achieved a TPR of 0.97 at the same FPPI threshold of 0.85.

This performance gap highlights model architecture's significant impact on detection performance. It indicates that the YOLOv5 model combined with DenseNet121 weights performs more efficiently detecting breast masses, providing more accurate results than the variant that uses a transformer prediction head. It's important to note that while transformers have shown promising results, their performance in object detection tasks might vary depending on the dataset and problem context.

In our experiment with the YOLOv6 object detection model, we used the YOLOv6l6 model, which performs better due to its larger architecture. The original YOLOv6l6 had reached a mAP of 57.2 at an IoU of 0.5 on a COCO dataset. After fine-tuning the model and running it on for 100 epochs, we got an mAP of 0.625 at 0.5 IoU.

There are six models provided by YOLOv7, all of which were trained on the MS COCO dataset. Out of these, we choose the YOLOv7-E6E model. It's the largest as well as it is also more accurate compared to the rest of the models, which is significantly important when it comes to choosing a model for this study. It is important to highlight that the performance benefits of YOLOv7-E6E come at the expense of speed. As our work with the model demonstrated, it is relatively slower than the other options available.

We also trained the YOLOv7-E6E model with an initialized weight from RadImageNet - DenseNet121. This strategy was adopted to evaluate and compare the performance of the YOLOv7-E6E model against the RadIma-

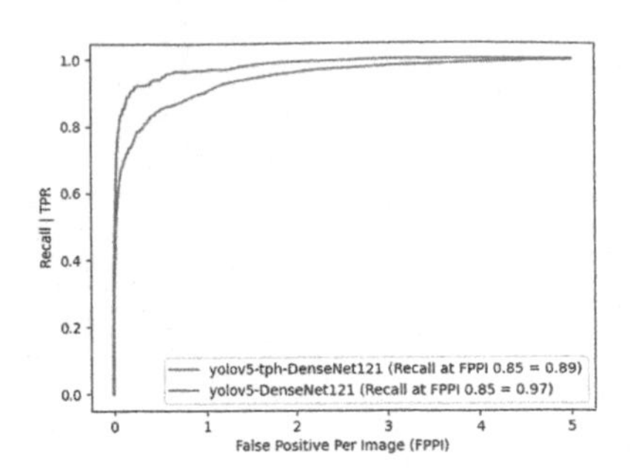

Fig. 4. Comparision between Yolov5 with Yolov5-tph

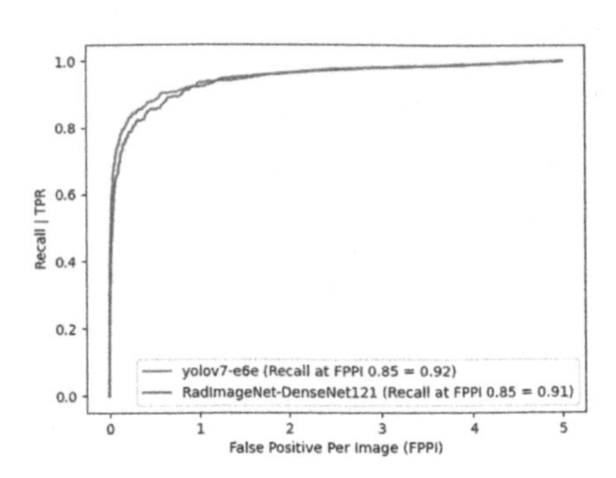

Fig. 5. FROC Curve for YOLOV7 trial

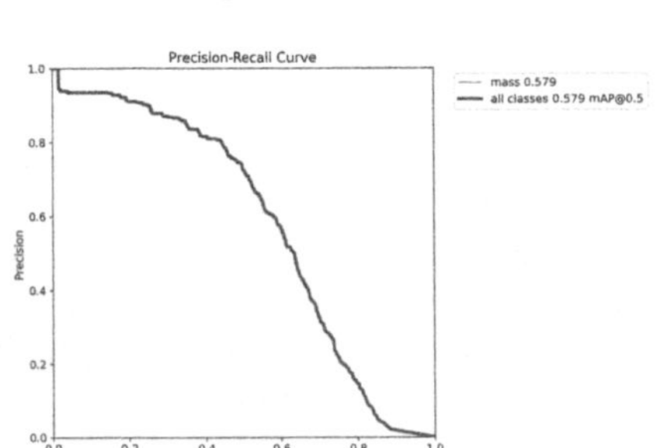

Fig. 6. Precision vs Recall curve for Yolov8x model

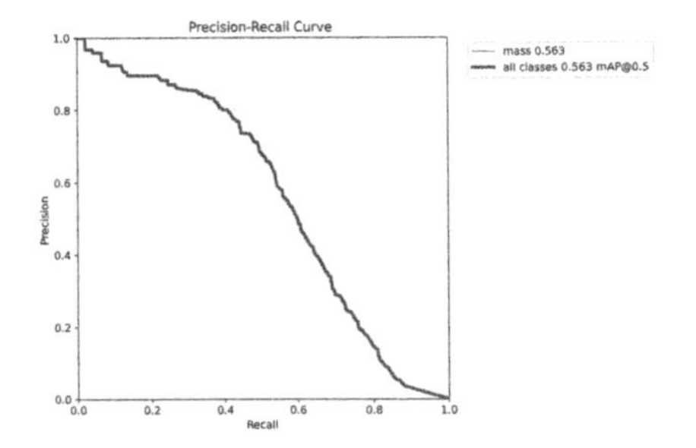

Fig. 7. Precision vs Recall curve for Yolov8l model

geNet dataset. The results show that the performances of the RadImageNet - DenseNet121 and the YOLOv7-E6E models were closely matched. However, the YOLOv7-E6E model exhibited slightly better performance.

After integrating the DenseNet121 weights trained on the RadImageNet dataset with the YOLOv7-E6E model, a slight decrease in the mAP value was observed during our experiments, as shown in Fig. 5.

For the yolov8 trial, we used a yolov8X and yolov8l models, yolov8x is an extra-large size, although it leads to a relatively slower computational speed, often providing higher accuracy. The YOLOv8X model's performance has been evaluated, and it has achieved a 0.53 mAP score at 0.5 IoU on the COCO validation dataset. When we applied this model to our dataset, the mAP results at 0.5 IoU reached 0.579. Figure 6 visually represents the Precision vs Recall curve for the YOLOv8X model on our dataset. Additionally, the yolov8l, a large-sized model, performs closely compared to the yolov8x model with an mAP of 0.563 at 0.5 IoU. The precision vs recall curve shows the results in Fig. 7.

Our experiments with the DenseNet121 model trained on the RadImageNet dataset and subsequently applied to breast mass detection have shown good results. A visual representation of these results is illustrated in Fig. 8. Additionally, we proceeded to feed the cropped breast mass ROI into a classifier for the consequent breast mass classification task.

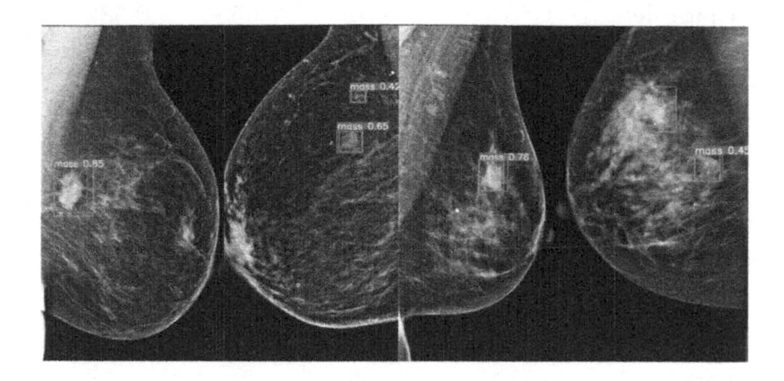

Fig. 8. Examples of detected breast masses with yolov5m with RadImageNet-DenseNet121 weights

3.2 Breast Mass Classification

Our approach used the YOLOV5m model, trained with RadImageNet DenseNet121 weights, to initially detect breast masses. This particular model was chosen due to its outstanding performance compared to other models. After the detection phase, the identified breast masses were extracted from the original mammogram images. This process involves cropping the image around the region identified as a mass by the YOLOV5m detector. These cropped sections, each containing a single mass, were prepared for the next process phase: classification.

The classification model has been trained to recognize the distinguishing features between benign and malignant breast masses, thereby accurately classifying new instances. By feeding the cropped images from the detection phase into this classifier, we generated a robust, two-step diagnostic tool that identifies and categorizes breast masses.

In our study, we faced a substantial class imbalance problem that could potentially impact the performance of our classification model. This issue is largely attributed to the nature of our dataset, which comprises a significant number of benign breast masses compared to malignant cases. Out of the 3363 detected breast masses, only 312 were benign, while the rest 3051 were malignant cases. Addressing this problem required implementing several strategies to ensure our model performed optimally despite the imbalance. One of the approaches we employed was applying class weights during the training phase.

By assigning higher weights to the minority class (benign cases in our dataset), we can increase their impact on the model's learning process, thereby reducing the bias towards the majority class. Another strategy we implemented was using a regularization method known as weight decay. The training phase used Stochastic Gradient Descent(SDG) optimizer, with a batch size of 8 for 300 epochs and a cross-entropy loss function.

Table 2. Transfer Learning for classification task

	Accuracy	F1-Score	Precision	Recall
DenseNet121	0.896	0.602	0.655	0.582
InceptionV3	0.893	0.634	0.661	0.617
AlexNet	0.901	0.556	0.647	0.544
VGG-16	0.924	0.731	0.793	0.695
ReseNet18	0.881	0.620	0.631	0.611
ResNet 50	0.888	0.606	0.634	0.591
EfficientNetB0	0.906	0.505	0.672	0.513

4 Discussion

In this study, we carried out a two-step approach for breast mass detection and classification. Primarily, we utilized a variety of YOLO-based architecture variants for our detection task, by effectively integrating pre-trained weights on the MS COCO dataset and leveraging the RadImageNet dataset weights across several CNN architectures, including InceptionV3, DenseNet121, and ResNet50. For the selection of the detection models, we evaluated an array of YOLO variants, including YOLOV5, YOLOV6, YOLOV7, and YOLOV8. Among the evaluated models, DenseNet121, initialized with RadImageNet weights and integrated with the YOLOV5m architecture, stood out in terms of performance. In this experiment, we got 0.97 TPR at 0.85 FPPI.

The best-performing detection model predicated the subsequent breast mass classification phase on the detected regions. We leveraged transfer learning architectures to train the classification task. This task posed a challenge due to a significant class imbalance between the benign and malignant classes. To resolve this issue, we implemented class weighting during the training process. This approach assigns different weights to the classes inversely proportional to their frequency.

Alongside class weighting, we also applied weight decay regularization, a common method to prevent overfitting in deep learning models. Despite these adjustments, the class imbalance issue wasn't fully resolved. Yet, we achieved a reasonable performance with the VGG16 model, which attained an accuracy of 0.924 a comparative result in other evaluation metrics.

5 Conclusion

In conclusion, our work showcases a two-stage approach for detecting and classifying breast masses, demonstrating the potential for integrating object detection models, such as YOLO and its variants for breast mass detection. Moreover, our work underlines the importance of using models trained specifically on medical images. In our study, the use of the RadImageNet model, which is specially designed and trained on radiological images, stands out as a particularly effective strategy. The positive impact of using such domain-specific models has significant implications for future medical imaging studies.

For our breast mass classification task, we used transfer learning. The models were fine-tuned for our dataset, allowing us to adapt these high-performing models into the specifics of our task. Although we faced a high-class imbalance in our breast mass detection dataset, where malignant cases significantly outnumbered the benign cases. We mitigate the problem using class weighting and regularization approaches.

Acknowledgements. 'The authors acknowledge the OPTIMAM project and Cancer Research Technology for providing the images used in this study, the staff at Royal Surrey NHS Foundation Trust who developed OMI-DB, and the charity Cancer Research UK which funded the OPTIMAM project. This work was supported by MUR (Italian Ministry for University and Research) funding to AB, CM, and MM through the DIEI Department of Excellence 2018–2022 (law 232/2016) and to FT through the DIEM Department of Excellence 2023–2027 (law 232/2016). Ruth Kehali Kassahun holds an EACEA Erasmus+ grant for the master in Medical Imaging and Applications.

References

1. Mattiuzzi, C., Lippi, G.: Current cancer epidemiology. J. Epidemiol. Global Health **9**, 217 (2019)
2. Siegel, R.L., Miller, K.D., Wagle, N.S., Jemal, A.: Cancer statistics. CA: Cancer J. Clinic. **73**, 17–48 (2023). https://doi.org/10.3322/caac.21763
3. Tabar, L., et al.: The incidence of fatal breast cancer measures the increased effectiveness of therapy in women participating in mammography screening. Cancer **125**, 515–523 (2018). https://doi.org/10.1002/cncr.31840
4. Evans, K.K., Haygood, T.M., Cooper, J., Culpan, A.M., Wolfe, J.M.: A half-second glimpse often lets radiologists identify breast cancer cases even when viewing the mammogram of the opposite breast. Proc. Natl. Acad. Sci. **113**, 10292–10297 (2016)
5. Sampat, M.P., Markey, M.K., Bovik, A.C., et al.: Computer-aided detection and diagnosis in mammography. Handbook Image Video Process. **2**, 1195–1217 (2005)
6. Lin, T.-Y., et al.: Microsoft COCO: common objects in context. In: Fleet, D., Pajdla, T., Schiele, B., Tuytelaars, T. (eds.) ECCV 2014. LNCS, vol. 8693, pp. 740–755. Springer, Cham (2014). https://doi.org/10.1007/978-3-319-10602-1_48
7. Mei, X., et al.: RadimageNet: an open radiologic deep learning research dataset for effective transfer learning. Radiol.: Artif. Intell. **4**, e210315 (2022)
8. Sechopoulos, I., Teuwen, J., Mann, R.: Artificial intelligence for breast cancer detection in mammography and digital breast tomosynthesis: state of the art, in: Seminars in Cancer Biology, Elsevier, pp. 214–225 (2021)
9. Akselrod-Ballin, A., et al.: Predicting breast cancer by applying deep learning to linked health records and mammograms. Radiology **292**, 331–342 (2019). https://doi.org/10.1148/radiol.2019182622
10. Yan, Y., Conze, P.H., Lamard, M., Quellec, G., Cochener, B., Coatrieux, G.: Towards improved breast mass detection using dual-view mammogram matching. Med. Image Anal. **71**, 102083 (2021). https://doi.org/10.1016/j.media.2021.102083

11. Agarwal, R., Dıaz, O., Yap, M.H., Llado, X., Martı, R.: Deep learning for mass detection in full field digital mammograms. Comput. Biol. Med. **121**, 103774 (2020). https://doi.org/10.1016/j.compbiomed.2020.103774
12. Betancourt Tarifa, A.S., Marrocco, C., Molinara, M., Tortorella, F., Bria, A.: Transformer-based mass detection in digital mammograms. J. Ambient. Intell. Humaniz. Comput. **14**, 2723–2737 (2023)
13. Ryspayeva, M., Molinara, M.: Breast mass detection and classification using transfer learning. Master's thesis. University of Cassino and Southern Lazio (2022)
14. Levy, D., Jain, A.: Breast mass classification from mammograms using deep convolutional neural networks. arXiv preprint arXiv:1612.00542 (2016)
15. Redmon, J., Divvala, S., Girshick, R., Farhadi, A.: You only look once: unified, real-time object detection. In: Proceedings of the IEEE Conference on Computer Vision and Pattern Recognition, pp. 779–788 (2016)
16. Du, L., Zhang, R., Wang, X.: Overview of two-stage object detection algorithms. J. Phys.: Conf. Ser., 012033. IOP Publishing (2020)
17. Cantone, M., Marrocco, C., Tortorella, F., Bria, A.: Convolutional networks and transformers for mammography classification: an experimental study. Sensors **23**, 1229 (2023). https://doi.org/10.3390/s23031229
18. Halling-Brown, M.D., et al.: Optimam mammography image database: a large-scale resource of mammography images and clinical data. Radiol.: Artif. Intell. **3**, e200103 (2020)

Prostate Cancer Detection: Performance of Radiomics Analysis in Multiparametric MRI

Muhammad Ali[1,2], Viviana Benfante[1,2](✉), Giuseppe Cutaia[3], Leonardo Salvaggio[2,3], Sara Rubino[3], Marzia Portoghese[3], Marcella Ferraro[3], Rosario Corso[4], Giovanni Piraino[2,5], Tommaso Ingrassia[5], Gabriele Tulone[6], Nicola Pavan[6], Domenico Di Raimondo[1], Antonino Tuttolomondo[1], Alchiede Simonato[6], and Giuseppe Salvaggio[3]

[1] Department of Health Promotion, Mother and Child Care, Internal Medicine and Medical Specialties, Molecular and Clinical Medicine, University of Palermo, 90127 Palermo, Italy
vbenfante@fondazionerimed.com
[2] Ri.MED Foundation, Via Bandiera 11, 90133 Palermo, Italy
[3] Department of Biomedicine, Neuroscience and Advanced Diagnostics, University of Palermo, 90100 Palermo, Italy
[4] Department of Mathematics and Computer Science, University of Palermo, ViaArchirafi 34, 90123 Palermo, Italy
[5] Department of Engineering, University of Palermo, 90100 Palermo, Italy
[6] Department of Surgical, Oncological and Stomatological Sciences, Urology Section, University of Palermo, 90100 Palermo, Italy

Abstract. The purpose of the study was to evaluate the performance of radiomics analysis of MR images for the detection of prostate cancer. The radiomics analysis was conducted using axial T2-weighted images from 49 prostate cancers. The study employs a sophisticated hybrid descriptive-inferential method for the meticulous selection and reduction of features, followed by discriminant analysis to construct a robust predictive model. Among 71 radiomics features, original_glrlm_ShortRunLowGrayLevelEmphasis demonstrated exemplary performance in differentiating between the whole prostate gland and prostate cancer. It had an AUROC of 68.46 (95% CI 0.544 – 0.824; p = 0.022), sensitivity of 76.25%, specificity of 73.15%, and accuracy of 71.02%. Radiomic analysis of T2 weighted MR images was demonstrated to have clinical application in prostate cancer detection, paving the way for improved diagnostic procedures and tailor-made treatment plans for prostate cancer patients.

Keywords: Magnetic Resonance Imaging · Prostate Cancer · Texture Analysis · Radiomics

1 Introduction

Prostate cancer (PCa) remains a significant health concern worldwide, with increasing incidence rates and a substantial impact on mortality. Early and accurate detection of prostate cancer is critical to improving patient outcomes and reducing the disease burden.

G. L. Foresti et al. (Eds.): ICIAP 2023 Workshops, LNCS 14366, pp. 83–92, 2024.
https://doi.org/10.1007/978-3-031-51026-7_8

In recent years, the advent of advanced imaging techniques has revolutionized oncology, offering new avenues for the detection and characterization of various cancers.

Multiparametric MRI (mpMRI) is an imaging modality that combines anatomical MR imaging with one or more functional MRI sequences and has become an important tool for detecting and characterizing PCa [1].

In the best hands, mpMRI of the prostate has a specificity approaching 90% and a negative predictive value of around 85% [2]. Prostate Imaging Reporting and Data System (PI-RADS) V2 was established to grade MRI findings from 1 to 5, correlating to the likelihood of clinically significant cancer [1].

With the growing speed of data generation and the increasing volume of imaging data per patient, development of computerised methods to process mpMRI data and extract useful information is highly demanded.

Among this methods, radiomics, an emerging field that extracts large amounts of features from radiographic medical images using data-characterization algorithms, has shown promise in enhancing diagnostic, prognostic, and predictive accuracy across various types of cancer, including PCa [3]. By converting medical images into mineable high-dimensional data [4], radiomics allows for the comprehensive quantification of tumor phenotypes, providing a more detailed picture of the disease.

This study investigates the potential of radiomics analysis of MR images as a non-invasive method for prostate cancer detection [3].

This study contributes to the growing body of literature on radiomics applications in oncology, with a specific focus on PCa. By leveraging radiomics, this study aims to uncover hidden patterns in MR images that could differentiate between healthy and cancerous prostate tissues [5]. The findings of this study could potentially lead to the development of a novel, non-invasive diagnostic tool for prostate cancer, improving patient care and outcomes.

2 Materials and Methods

This retrospective study was conducted at a single facility, the AOUP Paolo Giaccone, under the auspices of the University of Palermo. The study received approval from our institutional ethics committee, which did not require informed consent.

2.1 Population

Between January 2019 and June 2020, 503 patients visited our radiology department to undergo mpMRI. The examinations were performed for at least one of the following clinical reasons: onset of lower urinary tract symptoms (urinary flow obstruction, frequent urination, nocturia, straining, urgency, hematuria, and erectile dysfunction); suspicious EDAR and/or TRUS findings for prostate pathology; elevated PSA levels ($>$ 4 ng/mL); follow-up in patients undergoing radiotherapy; increase in serum PSA levels ($>$ 0.2 ng/ml) in patients with a history of PCa who had undergone radical prostatectomy.

Exclusion criteria for the study were: patients with incomplete MRI examination due to claustrophobic crises; subjects previously subjected to radical prostatectomy, TURP or radiotherapy; patients with negative mpMRI or with PI-RADS lesion < 3; lack of histopathological results obtained through fusion biopsy.

2.2 MRI Technique

The MRI examinations were performed using a 16-channel phase array coil (HD torso XL, Philips healthcare) on a 1.5 T magnet (Achieva, Philips Healthcare, Best, The Netherlands). The MRI examinations were performed by an experienced radiologist (>14 years), at least 6 weeks after prostate biopsy to allow the reabsorption of postoperative bleeding, thus avoiding possible diagnostic errors.

The MRI protocol for prostate imaging included axial, coronal and sagittal T2W sequences of the prostate and seminal vesicles. In addition, axial T1W was used to assess images with the presence of hyperintense residual bleeding after biopsy. Furthermore, diffusion weighted imaging (DWI) and dynamic contrast enhancement (DCE) sequences were incorporated. DWI sequences were acquired at b values of 0, 800, and 1400 s/mm^2 and post-processed to obtain the ADC maps. After unenhanced imaging, patients received a dose of 1 mmol/kg body weight of Gadoteric acid (Gd-DOTA, Dotarem, Guerbet) with a rate of 3 ml/sec, followed by infusion of 20–30 ml of saline solution at the same injection rate. Axial T1W 3-dimensional spoiled gradient-recalled echo volumetric interpolated images were obtained after contrast agent injection, to obtain a perfusion MRI of the prostate.

2.3 Manual Segmentation

Manual segmentation of the prostate and lesions is a critical step in radiomics analysis. This task was performed by a radiologist with 14 years of MRI experience. This ensured a high level of expertise and accuracy in the delineation of regions of interest.

For manual segmentation of the whole prostate glands and lesions, we took a similar approach to the one used by Cutaia et al. [3]. The process started with the importation of each MRI exam into an open-source DICOM viewer, specifically 3Dslicer[3] (https://slicer.org.org/). This software platform provides a wide range of tools for interactive medical image visualization, processing, and analysis. It supports a variety of imaging modalities, including MRI, and is widely used in the medical imaging community.

Axial T2W sequences were used for the aim of the study. Segmentation of prostate and lesion were performed manually (Fig. 1). This process involves the manual tracing of the whole prostate glands and of the lesions contours on each slice of the axial T2W sequence. Radiologist used a contouring tool, specifically a closed polygon, to accurately delineate the prostate and lesions boundaries. This tool allows precise control over the shape and size of the region of interest. This ensures that the segmentation accurately reflects the anatomical structure [6].

Following manual segmentation, the "compute volume" tool was used to calculate the volume of the entire prostate and lesions. This tool uses the segmented regions to compute the volume, providing a quantitative measure that can be used in radiomic analysis, successively. The volume of a lesion can provide valuable information about its size and growth rate.

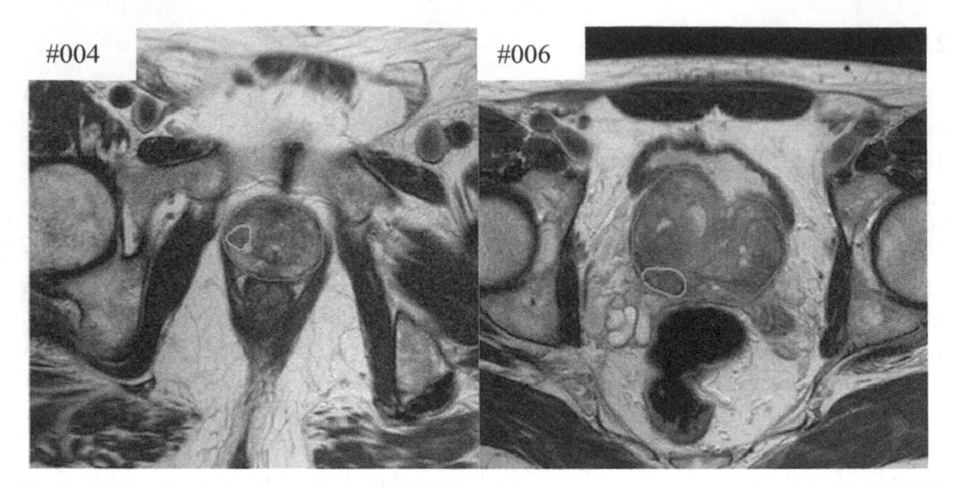

Fig. 1. Example of whole prostate gland (red) and prostate cancer segmentation (green) on T2W sequences of the prostate.

2.4 Radiomics Features Extraction

Radiomic parameters were extracted using Image Biomarker Standardization Initiative (IBSI) compliant analysis software, PyRadiomics. One of the key points in radiomic studies is to increase reproducibility of extracted textures. Through a broad panel of hard-coded engineered algorithms, reproducible radiomic features can be obtained. PyRadiomics [7] is implemented in Python, a popular open-source language for scientific computing, which also is installable on any system. PyRadiomics provides a flexible analysis platform with a back-end interface that allows automation in data processing, feature definition, and batch management.

This software extracts several feature classes, namely, shape descriptors, first-order statistics, and the following texture matrices: Gray Level Co-occurrence Matrix (GLCM), Gray Level Run Length Matrix (GLRLM), Gray Level Dependence Matrix (GLDM), Gray Level Size Zone Matrix (GLSZM), and Neighboring Gray Level Dependence Matrix (NGLDM).

Shape descriptors are independent of the image's gray-level intensity distribution. They describe geometric aspects such as volume, maximum diameter, surface, compactness, and sphericity. Specifically, the surface is calculated through a process that produces a mesh of triangles that completely cover the organ's surface (triangulation) to calculate the surface-volume ratio, while compactness and sphericity describe how the organ's shape (from a functional point of view) differs from that of a circle (2D) or a sphere (3D).

First-order statistical descriptors (or "histogram-based" features) describe the frequency distribution of voxels within an organ through the analysis of the histogram of gray-level intensity values. Among others, there are skewness, which indicates the asymmetry of the data distribution curve with respect to lower (negative) or higher (positive) values than the mean, kurtosis, which indicates the tail of the data distribution compared to a Gaussian distribution due to outlier values, entropy, and uniformity.

In addition, texture features are used to estimate the relative positions of voxels within the image. This provides information on the spatial organization of gray levels in the organ of interest. These features are classified according to the following texture classes from which they are obtained:

- GLCM quantifies the incidence of voxels with the same intensity at a predetermined distance along a fixed direction.
- GLRLM quantifies consecutive voxels with the same intensity in fixed directions.
- GLDM quantifies the number of voxel segments having the same intensity in a given direction.
- GLSZM quantifies the number of connected voxels with the same gray-level intensity.
- NGTDM quantifies the spatial interrelationships between three or more voxels.

Normalize was set to True and BinWidth to 0.27 for the extraction of radiomic features with pyradiomics, the others were kept by default.

2.5 Computational and Statistical Analyses

Computational interpretation of radiomic parameters was performed by a biomedical imaging senior researcher with eleven years of experience in image processing and data analysis. The performance outcome of the proposed approach was ascertained using a two-way analysis of variance (ANOVA) [5].

ANOVA is a statistical tool used to determine if there are statistically significant differences between the means of two or more independent groups. This is achieved by comparing the variability within these groups with the variability between the groups. In our case, it compares the means between the groups (the two different segmentations of the entire prostate and the lesion alone), determining if one of these means is statistically different from the others. It tests the null hypothesis: $H0 = \mu 1 = \mu 2 \cdots \mu k(1)$, where μ = group mean, k = number of groups. If the two-way ANOVA returns a statistically significant result (p-value less than 0.05), we accept the alternative hypothesis: there are at least two group means that are statistically significantly different from each other.

We considered the radiomic features with p-value < 0.05 in two-way ANOVA to find the radiomic features (out of 112 total radiomic features) different between the two groups (the entire prostate and the lesion alone). Once the different features between the two groups (the entire prostate and the lesion alone) were identified, a hybrid statistical method, independent operator, was adopted for the selection and reduction of different features between the two groups, while discriminant analysis was used to build the predictive model to discriminate the prostate from the tumor.

To identify the most discriminant features and reduce the redundancy of radiomic features with high correlation, an innovative sequential mixed descriptive-inferential approach was used [8]. Specifically, for each feature, the point-biserial correlation coefficient was calculated between features and the dichotomous variable assuming a value equal to 1 or 0 for the reference standard. Subsequently, the features were sorted in descending order based on the absolute value of the point-biserial correlation coefficient.

Then, a cycle began to add one column at a time performing a logistic regression analysis. Subsequently, the p-value of the current iteration was compared with the p-value of the previous iteration. The cycle stopped when the p-value of the current iteration did not decrease compared to the previous iteration. In this way, the most discriminant features were identified.

Discriminant analysis was used to create a predictive model for our dataset [9]. We trained the discriminant analysis with the most discriminant features to evaluate the differences between the prostate and the tumor. The training task had to be performed only once and, once completed, the discriminant analysis classified the newly encountered radiomic texture.

To generate the training input for the classifier, the expert radiologist's evaluation results were used as the gold standard. In our study, the k-fold (with $k = 5$) strategy was used to divide the data into training and validation sets. In this way, the studies were divided into k-folds. One of the sets was used as a validation set and the remaining sets were combined into the training set. This process was repeated k times using each set as a validation set and the remaining sets as a training set. To ensure disjoint validation sets, the leave-one-out approach was not adopted.

In this way, it is possible to obtain more robust results in the classification model [4, 10] implementation. ROC curves (ROC) with 95% confidence intervals (95% CI) and areas under the ROC curve (AUROC) were calculated to evaluate the diagnostic performance of the most discriminant selected parameters on the entire prostate and tumor images. Sensitivity, specificity, and accuracy were calculated.

The computational system used to provide the proposed computational statistical analysis was implemented in the Matlab R2019a simulation environment (MathWorks, Natick, MA, USA), running on an Intel Core i7 2.5 GHz computer, 16 GB iMac 1600 DDR3 MHz memory and OS X Sierra. This powerful computational environment facilitated the efficient processing and analysis of the intricate radiomic data.

3 Results

3.1 Population

Our final population consisted of 73 patients aged between 51 and 88 years (mean: 67 $\pm$ 7.6), accounting for a total of 80 lesions (7 patients had 2 lesions). In total, we had 5 PI-RADS 3 lesions (4 chronic inflammations and 1 ASAP), 45 PI-RADS 4 lesions [21 chronic inflammations, 1 PIN + ASAP, 11 GS 6, 3 GS 7 (3 + 4), 5 GS 7 (4 + 3), 3 GS 8 and 1 GS 9] and 30 PI-RADS 5 lesions [5 chronic inflammations, 1 acinar carcinoma, 7 GS 6, 4 GS 7 (3 + 4), 5 GS 7 (4 + 3), 5 GS 8, 3 GS 9 and 1 GS 10]. Non-tumor lesions (n°: 31) were excluded from the study, so that 49 PCa were used for radiomic analysis.

3.2 Performance of Radiomics

There were 71 radiomics features that were statistically different between the whole prostate gland and the tumor lesion. Among these, in the T2 sequences, the feature original_glrlm_ShortRunLowGrayLevelEmphasis ($p = 0.00072$) demonstrated good performance in differentiating between the whole prostate gland and the tumor lesion (Fig. 2), with an AUROC of 68.46 (95% CI 0.544 –0.824; $p = 0.022$), sensitivity of 67.25%, specificity of 73.15%, and accuracy of 71.02% (table 1).

Table 1. Selected radiomics features and their performance for the differentiation between the whole prostate and the tumor lesions

Feature	Sensitivity	Specificity	Accuracy	AUC (95% CI)	*p* value
Original glrlm_ShortRunLowGrayLevelEmphasis	67.25%	73.15%	71.02%	0.685 (0.545–0.825)	0.022

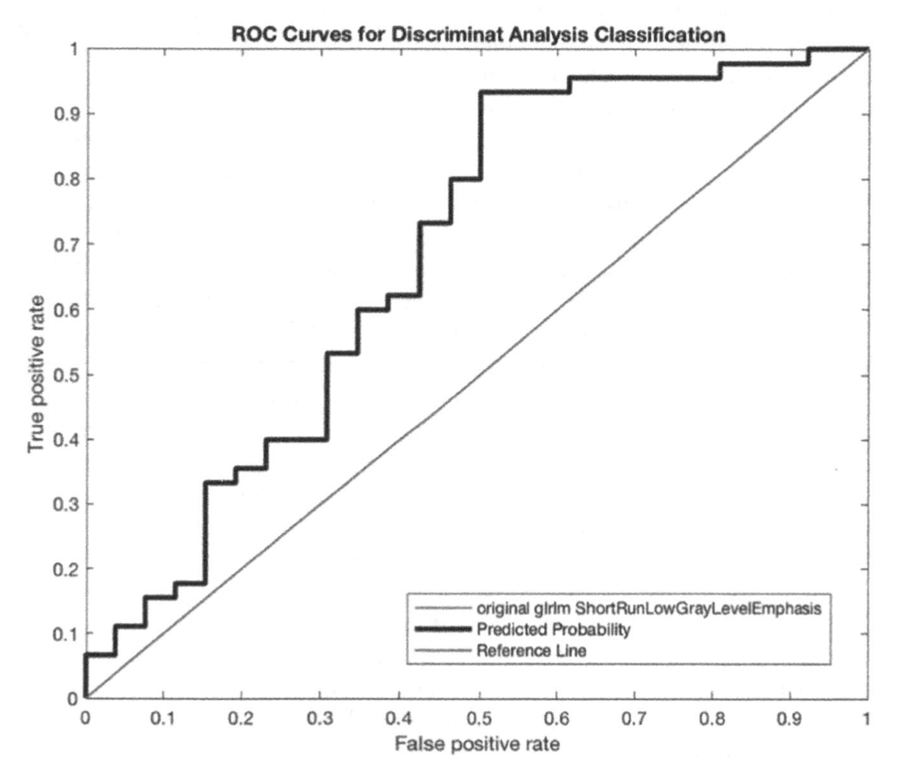

Fig. 2. Receiver operating characteristics curve of Original_glrlm_ShortRunLowGrayLevel Emphasis on T2W sequences of the prostate.

4 Discussion

Radiomics applied to prostate MRI has a broad range of applications, including tumor localization and detection, prognosis prediction, and post-treatment follow-up. Therefore, radiomics and artificial intelligence algorithms [10, 11] will help to limit discrepancies between different radiologists [12]. Specifically, radiomics applied to mpMRI can enhance the reliability of prostate carcinoma diagnosis by providing qualitative parameters (related to the diagnostic ability of the expert radiologist, capable of providing an accurate PI-RADS score) and quantitative measurements (such as tumor dimensions, prostate volume, and radiomic features, which are calculated directly by the computer, allowing for objective results independent of the radiologist's skill) [13–15].

In our study, we evaluated the performance of radiomics analysis applied to mpMRI in distinguishing between the whole prostate gland and tumor lesions for PCa detection. The results of our radiomic analysis demonstrated good performance, with a AUC of 0.685 (95% CI 0.544 - 0.824; $p = 0.022$), a sensitivity of 76.25%, a specificity of 73.15%, and an accuracy of 71.02%.

These results indicate that radiomics analysis of mpMRI images can be a promising method for distinguishing between the whole prostate gland and tumor lesion in the detection PCa. The significant AUC and high sensitivity suggest that this technique could be useful as a supportive tool for diagnosis and disease evaluation. The reasonable specificity and accuracy indicate that radiomic analysis may contribute to reducing false positives and improving the overall diagnostic accuracy.

These results are consistent with recent literature. Woznicki et al. [16] showed that features extracted from mpMRI images are reliable quantitative imaging biomarkers for PCa detection: their radiomics model achieved good predictive performance with a mean AUC of 0.783 for the differentiation of malignant versus benign prostate lesions. Moreover, they found that the radiomics model ensemble with PI-RADS, PSAD, and DRE achieved an excellent predictive performance with an AUC of 0.889. Bleker et al. [17] by extracting radiomics features from prostate mpMRI using auto-fixed VOIs have also achieved promising results on peripheral zone clinically significant PCa detection, with the highest AUC of 0.87. They used a combination of mpMRI features taken from T2-w, DWI, and DCE imaging and these factors could potentially account for the higher AUC value they achieved. Khalvati et al. [18] proposed a radiomics approach for automated localization and detection of PCa using clinical prostate mpMRI data from 30 patients. The authors demonstrated a sensitivity, specificity and accuracy of 0.82, 0.89, and 0.86, respectively in distinguishing between tumor and healthy tissue.

Our study has limitations. First, it is a retrospective study that only enrolls patients with a confirmed histological diagnosis, so there may be a selection bias. However, knowing the lesions we have to deal with is currently essential to test the potential of radiomic analyses and neural networks [19–21]. Secondly, our "gold standard" was the histological analysis obtained from fusion biopsy and not from prostatectomy, so there is an imperfect reference bias.

5 Conclusion

Our study demonstrated the clinical applicability of radiomic analysis in the detection of PCa using T2W MRI images of the prostate. The implementation of radiomic analysis holds the potential to enhance PCa detection and address the challenges associated with subjective interpretation of MRI images. By extracting quantitative features from the T2Wimages, radiomics analysis provides an objective and systematic approach to PCa diagnosis, potentially improving accuracy and reducing variability in interpretation. These findings suggest that radiomic analysis has a good performance in the field of PCa detection, offering a complementary and reliable method to overcome the limitations of a subjective MRI image interpretation.

References

1. Johnson, L.M., Turkbey, B., Figg, W.D., Choyke, P.L.: Multiparametric MRI in prostate cancer management. Nat. Rev. Clin. Oncol. **11**, 346–353 (2014). https://doi.org/10.1038/nrclinonc.2014.69
2. Thompson, J., Lawrentschuk, N., Frydenberg, M., Thompson, L., Stricker, P.: The role of magnetic resonance imaging in the diagnosis and management of prostate cancer. BJU Int. **112**, 6–20 (2013). https://doi.org/10.1111/bju.12381
3. Cutaia, G., et al.: Radiomics and prostate MRI: current role and future applications. J. Imaging **7**, 34 (2021). https://doi.org/10.3390/jimaging7020034
4. Stefano, A., et al.: Robustness of PET radiomics features: impact of co-registration with MRI. Appl. Sci. **11**, 10170 (2021). https://doi.org/10.3390/app112110170
5. Aerts, H.J.W.L., et al.: Decoding tumour phenotype by noninvasive imaging using a quantitative radiomics approach. Nat. Commun. **5**, 4006 (2014). https://doi.org/10.1038/ncomms5006
6. Cuocolo, R., et al.: Deep learning whole-gland and zonal prostate segmentation on a public MRI dataset. J. Magn. Reson. Imaging **54**, 452–459 (2021). https://doi.org/10.1002/jmri.27585
7. Gallotta, A., et al.: A novel algorithm for the prediction of prostate cancer in clinically suspected patients. Cancer Biomark. **13**, 227–234 (2013). https://doi.org/10.3233/CBM-130357
8. Barone, S., et al.: Hybrid descriptive-inferential method for key feature selection in prostate cancer radiomics. Appl. Stoch. Models Bus. Ind. **37**, 961–972 (2021). https://doi.org/10.1002/asmb.2642
9. Kumar, V., et al.: Radiomics: the process and the challenges. Magn. Reson. Imaging **30**, 1234–1248 (2012). https://doi.org/10.1016/j.mri.2012.06.010
10. Alongi, P., et al.: Choline PET/CT features to predict survival outcome in high-risk prostate cancer restaging: a preliminary machine-learning radiomics study. Q. J. Nuclear Med. Molecular Imaging 66 (2022). https://doi.org/10.23736/S1824-4785.20.03227-6
11. Cairone, L., et al.: Robustness of radiomics features to varying segmentation algorithms in magnetic resonance images. Presented at the (2022). https://doi.org/10.1007/978-3-031-13321-3_41
12. Vernuccio, F., Cannella, R., Comelli, A., Salvaggio, G., Lagalla, R., Midiri, M.: [Radiomics and artificial intelligence: new frontiers in medicine.]. Recenti Prog Med. **111**, 130–135 (2020). https://doi.org/10.1701/3315.32853
13. Giambelluca, D., et al.: PI-RADS 3 lesions: role of prostate MRI texture analysis in the identification of prostate cancer. Curr. Probl. Diagn. Radiol. **50**, 175–185 (2021). https://doi.org/10.1067/j.cpradiol.2019.10.009
14. Comelli, A., et al.: Deep learning-based methods for prostate segmentation in magnetic resonance imaging. Appl. Sci. **11**, 782 (2021). https://doi.org/10.3390/app11020782
15. Salvaggio, G., et al.: Deep learning network for segmentation of the prostate gland with median lobe enlargement in T2-weighted MR images: comparison with manual segmentation method. Curr. Probl. Diagn. Radiol. **51**, 328–333 (2022). https://doi.org/10.1067/j.cpradiol.2021.06.006
16. Woźnicki, P., et al.: Multiparametric MRI for prostate cancer characterization: combined use of radiomics model with PI-RADS and clinical parameters. Cancers (Basel). **12**, 1767 (2020). https://doi.org/10.3390/cancers12071767
17. Bleker, J., Kwee, T.C., Dierckx, R.A.J.O., de Jong, I.J., Huisman, H., Yakar, D.: Multiparametric MRI and auto-fixed volume of interest-based radiomics signature for clinically significant peripheral zone prostate cancer. Eur. Radiol. **30**, 1313–1324 (2020). https://doi.org/10.1007/s00330-019-06488-y

18. Khalvati, F., Zhang, J., Chung, A.G., Shafiee, M.J., Wong, A., Haider, M.A.: MPCaD: a multi-scale radiomics-driven framework for automated prostate cancer localization and detection. BMC Med. Imaging **18**, 16 (2018). https://doi.org/10.1186/s12880-018-0258-4
19. Lee, H., Hwang, S. Il, Lee, H.J., Byun, S.-S., Lee, S.E., Hong, S.K.: Diagnostic performance of diffusion-weighted imaging for prostate cancer: Peripheral zone versus transition zone. PLoS One. **13**, e0199636 (2018). https://doi.org/10.1371/journal.pone.0199636
20. Comelli, A., et al.: Tissue classification to support local active delineation of brain tumors. Presented (2020). https://doi.org/10.1007/978-3-030-39343-4_1
21. Agnello, L., Comelli, A., Ardizzone, E., Vitabile, S.: Unsupervised tissue classification of brain MR images for voxel-based morphometry analysis. Int. J. Imaging Syst. Technol. **26**, 136–150 (2016). https://doi.org/10.1002/ima.22168

Grading and Staging of Bladder Tumors Using Radiomics Analysis in Magnetic Resonance Imaging

Viviana Benfante[1,2], Giuseppe Salvaggio[3](✉), Muhammad Ali[1,2], Giuseppe Cutaia[3], Leonardo Salvaggio[2,3], Sergio Salerno[3], Gabriele Busè[3], Gabriele Tulone[4], Nicola Pavan[4], Domenico Di Raimondo[1], Antonino Tuttolomondo[1], Alchiede Simonato[4], and Albert Comelli[2]

[1] Department of Health Promotion, Mother and Child Care, Internal Medicine and Medical Specialties, Molecular and Clinical Medicine, University of Palermo, 90127 Palermo, Italy

[2] Ri.MED Foundation, Via Bandiera 11, 90133 Palermo, Italy

[3] Department of Biomedicine, Neuroscience and Advanced Diagnostics, University of Palermo, 90100 Palermo, Italy

p.salvaggio@libero.it

[4] Urology Section, Department of Surgical, Oncological and Stomatological Sciences, University of Palermo, 90100 Palermo, Italy

Abstract. Aim of this study was to evaluate the performance of MRI radiomics analysis in distinguishing low-grade (LG) versus high-grade (HG) bladder lesions and non-muscle-invasive bladder cancer (NMIBC) versus muscle-invasive bladder cancer (MIBC). We proposed a computational statistical analysis model that is standardized and reproducible, and identified predictive and prognostic models to facilitate the process of making medical decisions. Sixteen patients with bladder lesions and preoperative mpMRI were included for a total of 35 bladder lesions. Lesions were manually segmented from T2-weighted sequences. PyRadiomics software was used to extract radiomics features and a total of 120 radiomics features were obtained from each lesions. An operator-independent statistical system was adopted for the selection and reduction of the characteristics, while discriminant analysis was used for the construction of the predictive model. The performance in the discrimination between LG and HG lesions, with an AUROC of 0.84 (95% C.I. between 0.71 and 0.98), sensitivity of 65.6%, specificity of 81.5%, with $p < 0.001$. The performance in the discrimination between NMIBC and MIBC, with an AUROC of 0.7 (95% C.I. between 0.11 and 1), sensitivity of 100%, specificity of 86.7%, with p-value of 0.0031. Our results demonstrate the valuable contribution of radiomics analysis in improving the characterization and differentiation of bladder lesions, both in terms of differentiating LG from HG lesions and discriminating NMIBC from MIBC.

Keywords: Magnetic Resonance Imaging · Bladder Lesions · Texture Analysis · Radiomics · Machine Learning

G. L. Foresti et al. (Eds.): ICIAP 2023 Workshops, LNCS 14366, pp. 93–103, 2024.
https://doi.org/10.1007/978-3-031-51026-7_9

1 Introduction

Bladder cancer (BCa) is the ninth most common cancer in the world and one of the most expensive cancers to treat [1]. There are two types of bladder tumors according to their histology: low-grade (LG) and high-grade (HG) [1]. While most non-muscle invasive (NMIBC) neoplasms are relatively low-grade and indolent [2, 3], approximately one third of these neoplasms are HG, with a high risk of progression to muscle-invasive forms of neoplasms (MIBC) and the development of systemic disease and distant metastases in about 20–25% of patients [1–3]. BCa typically affects elderly patients, with a mean age at presentation of approximately 65 years [4]. The prevalence of this tumor is higher in men and it represents the fourth most common tumor in men, especially as historically most of these tumors are associated with industrial exposure to various chemical compounds [4–7]. The most common histological type (>90%) of BCa is urothelial cell carcinoma. The other types include squamous cell carcinoma, adenocarcinoma and small cell carcinoma [8]. Approximately 70% of diagnoses are associated with non-infiltrating and infiltrating urothelial neoplasms [9]. Therefore, in consideration of the clear difference in terms of both therapeutic management and prognosis, the clinical staging of bladder tumors distinguishes first of all, according to the TNM 2017 8th edition classification, between NMIBC (stages Ta, Tis and T1) and MIBC (stages T2 and above).

As far as grading is concerned, the reference classification is the one proposed by the WHO in 2004 and updated in 2016 [8] with the elimination of the intermediate grade G2, the subject of controversy, and therefore distinguishes between LG and HG neoplasm.

The first examination in the clinical suspicion of BCa is ultrasonography, which has high accuracy and specificity [10–12], followed by urethrocystoscopy, and cytological examination [13, 14].

In addition to imaging CT urography used for the detection and staging of urothelial carcinoma, MRI provides valuable insights into bladder and synchronous upper urinary tract disease, resulting in an alternative to CT urography in patients with contraindications to CT [14]. In terms of local T staging of BCa, MRI has a significant advantage over CT due to its high contrast resolution in soft tissue. In BCa staging, T2-weighted imaging (T2W) is the preferred method as it depicts the individual layers of the bladder wall in relation to the tumor.

Radiomics represents a new frontier in advanced image analysis [15], as it allows for the extraction of a large number of quantitative features from radiological images that are not easily measurable by humans [16]. In oncologic patients, radiomics has been widely applied, and the results have been promising for the differential diagnosis of focal lesions and correlation with histopathological characteristics of tumor aggressiveness, treatment response prediction after local or systemic treatments and prognosis prediction after complete surgical resection [4–7].

The aim of this study was to evaluate the performance of MRI radiomics analysis in distinguishing LG versus HG bladder lesions and MIBC versus NMIBC.

2 Materials and Methods

Our retrospective study was conducted at a single center. The study received approval from our institutional ethics committee, which did not require informed consent.

2.1 Population

A total of 30 consecutive patients with pathologically confirmed BCa who underwent multiparametric MRI examinations from March 2019 to May 2020 were enrolled in our study. Inclusion criteria were as follows: (1) postoperative pathologically confirmed BCa; and (2) 1.5 T MRI scans performed < 1 month before surgery.

The exclusion criteria were (1) a history of previous treatments, including chemotherapy, radiotherapy, TURBT, or BCG, before MRI examination (n = 11) and (2) incomplete pathological information (n = 3).

The acquired studies were evaluated by a single radiologist with 6 years of experience in MRI of the bladder. Patient-related variables (age, gender, etc.) and histopathologic findings were collected.

2.2 MRI Technique

MRI was performed using the same protocol for all patients using a 1.5T magnet (Achieva, Philips Healthcare, Best, The Netherlands) phased array coil (16 channel HD Torso XL). Table 1 provides a description of the sequences used.

Standard bladder MRI includes axial, coronal, and sagittal T2-weighted images, axial T1-weighted images, and diffusion weighted imaging (DWI) images with b values of 0, 800, and 1400 s/mm^2. Afterwards, the DWI images were processed to obtain the ADC maps. After acquiring images without contrast, patients were administered 1 mmol/kg body weight of contrast medium (Gd-DOTA, Dotarem®, Guerbet, USA) at a flow rate of 3 ml/s, followed by a infusion of 20–30 ml of saline solution at the same flow rate. Following the injection of the contrast medium, a volumetric 3D-T1-GRE sequence was acquired to obtain the perfusion study of the prostate (dynamic contrast-enhanced - DCE). DCE and ADC images were not analyzed in this study due to the potential variability of bladder lesion perfusion and the limited role of texture analysis applied to DCE images as compared to T2WI.

2.3 Qualitative Imaging Analysis

By combining and adapting several existing and new approaches, some of which have already been used in previous studies [17], we were able to overcome the limitations of a radiomics study, resulting in a new operational methodology designed to implement a radiomics strategy in patients undergoing magnetic resonance imaging with bladder lesions. Following the selection of the target lesions [18], a 3D segmentation of the lesions and the bladder wall adjacent to the lesions was performed by two radiologists in consensus using open-source software (Horos, LGPL license at Horosproject.org). Lesion segmentation was performed by manually tracing, on contiguous images, regions of interest (ROIs) along the lesion profile and the adjacent bladder wall (Fig. 1).

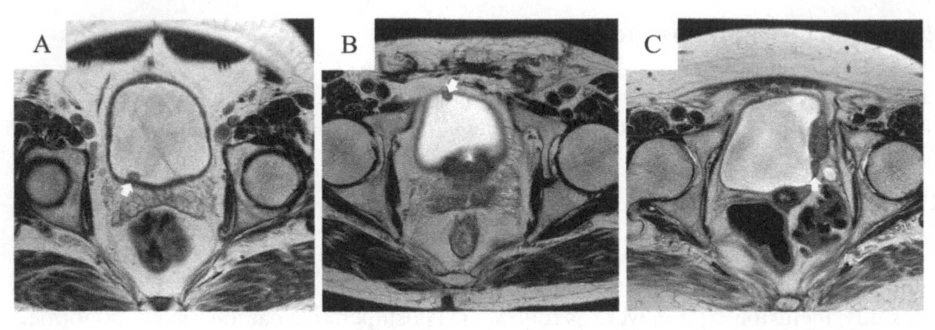

Fig. 1. A: LG bladder lesion in the right posterior wall (Red Arrow). B: HG bladder lesion in the right anterior wall (Red Arrow). C: MIBC in the left lateral wall (Red Arrow). (Color figure online)

2.4 Segmentation and Radiomics Features Extraction

For obtaining radiomics parameters, PyRadiomics (3.0, Computational Imaging & Bioinformatics lab - Harvard Medical School) was used. For each ROI, over 120 radiomics parameters are automatically obtained, including first-order texture parameters (which are related to gray level distribution within the ROI without taking into account the spatial relationships between voxels), parameters of second order (they consider the spatial relationships between voxels) and third order parameters (they evaluate the spatial relationships between three or more voxels).

From gray level histogram analysis, texture parameters were derived (e.g. mean, variance, skewness, kurtosis, and percentiles), and co-occurence matrix was calculated in five measurements (e.g. contrast, correlation, sums of squares, inverse difference moments, sums average, sum variances, sum entropies, difference variances, difference entropies). In addition, the "run-length matrix" was calculated in four directions (e.g., run length nonuniformity [RLN], grey-level nonuniformity normalized [GLNN], long run emphasis [LRE], short run emphasis [SRE]).

Radiomics parameters are described in detail in the PyRadiomics online manual [16].

2.5 Computational and Statistical Analyses

Computational analysis of radiomics features was conducted by a senior scientist in biomedical imaging (with 10 years of experience on image and data processing and analysis), which had access to both the extracted features and the reference standards.

The data obtained, collected as continuous variables, were expressed as mean and standard deviation (SD), and the categorical variables were expressed as numbers and percentages. Based on the histopathological findings, the analyzed lesions were grouped into "low grade" and "high grade" and "non-muscle invasive" and "muscle invasive".

Our method for MRI texture analysis included an operator-independent statistical system for feature reduction and selection in conjunction with a predictive method based on discriminant analysis, as performed in previous studies [16, 19].

The first step involved an approach that uses a mixed descriptive-inferential sequential approach to identify the most discriminating features and reduce redundant data. In particular, for each characteristic, the punctual biserial correlation index between continuous variables and dichotomous variables was calculated assuming a value of 1 or 0 (reference standard). The characteristics were then sorted in descending order of their absolute value of the point biserial correlation index. A cycle of adding one column at a time was then initiated by performing logistic regression analysis. Next, the p-value of the current iteration was compared with the p-value of the previous iteration. When the p-value of the current iteration has not decreased compared to the previous iteration, the cycle stops. In this way, the most discriminating characteristics for the differential diagnosis between "low grade" and "high grade" and between "non-muscle invasive" and "muscle-invasive" were identified.

In the second approach, we used discriminant analysis to develop a predictive model for our patient dataset [3]. The discriminant analysis was trained with the most discriminant features in order to predict the diagnosis between "low grade" and "high grade", as well as between "non-muscle invasive" and "muscle invasive". Once the training task had been completed, the discriminant analysis was able to classify the radiomics texture encountered earlier [20]. In order to generate training input for the classifier, histopathological examination results were used as a reference. Our study used 80% of the samples to train the classifier, while the remaining 20% tested the ability of the classifier to provide reliable classifications. To strengthen the reliability of the results and avoid common problems such as over-fitting, a 5-fold cross-validation was also performed [3]. As a point of clarification, the k-fold cross-validation technique was used in this study only to validate the two classification techniques (GLM and DA) [21, 22] and not to select the features. Receiver operating characteristics (ROC) with 95% confidence intervals (C.I.) and areas under the ROC curve (AUROC; 95% C.I.) were calculated to evaluate the diagnostic performance of selected most discriminating parameters on T2WI. The system used to provide the proposed computational statistical analysis was implemented in the Matlab R2019a simulation environment (MathWorks, Natick, MA, USA), running on a 2.5 GHz, 16 GB Intel Core i7-based iMac computer 1600 MHz DDR3 memory and OS X Sierra.

3 Results

3.1 Population

A total of 16 patients were eligible in the study [14 men, 2 women; mean age 72 years ± 8.5 (range 57 to 87 years)] and a total of 35 BCa lesions (mean diameter 13,9 × 12,4 mm; range 2–60 mm) were examined (1 patient had 2 lesions, one had 3 lesions, one had 4 lesions, one had 5 lesions, and one 8 lesions).

BCa lesions were divided into LG (12 lesions) and HG (23 lesions), and into MIBC (32 lesions) and NMIBC (3 lesions), according to pathological results.

3.2 Performance of Radiomics

There were two radiomics features extracted that were statistically significant in distinguishing LG from HG lesions, as shown in Table 2.

Table 1. Characteristics of the study population.

Characteristics	
Age (years), mean ± SD (range)	72 ± 8,5 (57–87)
Grading of the lesion, n (%)	35
Low Grade	12 (34.3%)
High Grade	23 (65.7%)
Muscular invasion	
NMIBC	32 (91.4%)
MIBC	3 (8.6%)
Diameter of lesion (mm), mean ± SD (range)	
All lesions	13.9 ± 12.4 (2–60)
Low Grade	8.2 ± 4.2 (4–16)
High Grade	15.7 ± 13.3 (2–60)
NMIBC	11.2 ± 7.9 (2–30)
MIBC	34.7 ± 22.1 (19–60)

Table 2. Discrimination of LG from HG lesions based on significant features.

Texture Features	p-value
SmallAreaEmphasis (SAE)	0.000064
SizeZoneNonUniformityNormalized (SZNN)	0.00031

The SAE is a measure of the distribution of small areas, with a higher value indicative of smaller areas and finer textures, while the SZNN indicates the degree to which volume sizes are variable throughout the image, with a lower value suggesting greater homogeneity between them. The combination of these features presents high diagnostic performance in the discrimination between LG and HG lesions, with an AUROC of 0.84, 95% C.I. between 0.71 and 0.98, sensitivity of 65.6%, specificity of 81.5%, with $p < 0.001$. The ROC curve related to the two features analyzed is shown in Fig. 2.a.

As shown in Table 3, two other radiomics features extracted were statistically significant in distinguishing MIBC from NMIBC.

Table 3. Significant features in the discrimination between NMIBCa and MIBCa lesions.

Texture Features	p-value
LargeAreaHighGrayLevelEmphasis (LAHGLE)	0.00063
LargeAreaEmphasis (LAE)	0.0018

LAHGLE measures the proportion of the joint distribution of larger areas with higher gray level values. LAE measures the distribution of large areas, with a higher value indicative of areas of higher dimensions and coarser plots.

The combination of these features presents high diagnostic performance in the discrimination between NMIBC and MIBC, with an AUROC of 0.7, 95% C.I. between 0.11 and 1, sensitivity of 100%, specificity of 86.7%, with p of 0.0031. Figure 2.b shows the ROC curve relating to the two features analysed.

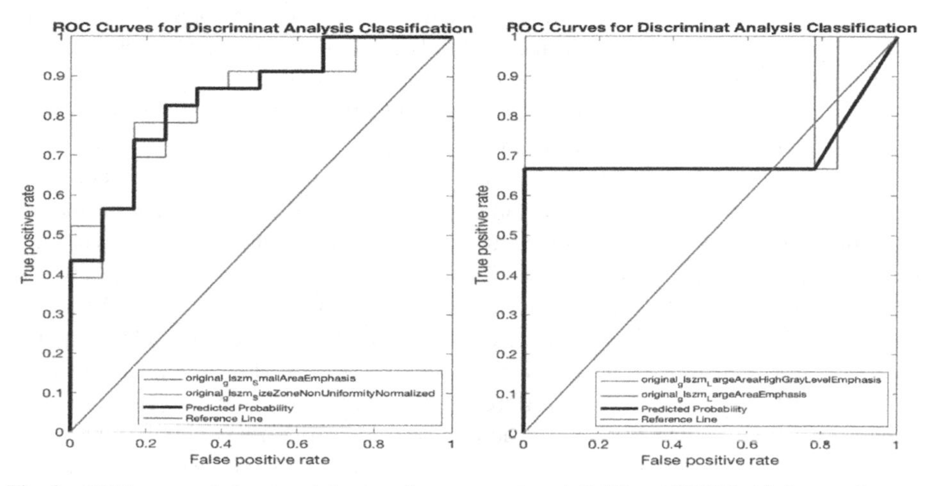

Fig. 2. ROC curve relating to: a) the two features analysed (SAE and SZNN); b) the two features analysed (LAHGLE and LAE).

4 Discussion

LG Versus HG Lesions

Our results showed that the combination of two specific radiomics features provided a significant discrimination value between low-grade and high-grade lesions. This suggests that these features, likely correlated to structural or morphological aspects of the lesions, can provide useful information to distinguish between bladder tumors of different malignancy grades.

In our study we observed that not all radiomics parameters have equal efficacy in discriminating between LG bladder lesions and HG bladder lesions: out of over 120 features extracted, only two showed a statistically significant correlation. We postulate that the texture heterogeneity of bladder tumor MR images reflects the underlying changes in tumor texture that occur histopathologically during carcinogenesis. It highlights the importance of carefully selecting and evaluating specific radiomics parameters to maximize the discriminatory power of the model. According to our results, texture parameters could be used in the grading of bladder lesions as a preliminary indication of their validity. The validation of radiomics parameters for multiparametric MRI of the bladder may have significant clinical implications for the management and diagnostic algorithm of these lesions in clinical practice.

The overall diagnostic performance of the model was evaluated using the AUC. In this case, an AUC of 0.84 indicates a good discriminative power of the model in distinguishing between LG and HG lesions. The 95% confidence interval ranging from 0.71 to 0.98 suggests some variability in the accuracy of the model, but its overall performance is still considerable.

The sensitivity of the model, representing its ability to correctly identify positive cases (high-grade lesions), was 65.6%. This means that the model recognized the majority of high-grade lesions present in the study. The specificity of 81.5% indicates the model's ability to correctly identify negative cases (low-grade lesions). These results suggest that the model is capable of discriminating between LG and HG lesions with a good balance of sensitivity and specificity.

Zhang et al. [11] employed Histogram and gray-level co-occurrence matrix (GLCM)-based radiomics to extract radiomics features from DWI and ADC images. They achieved an area under the receiver operating characteristic curve (AUC), accuracy, sensitivity, and specificity of 0.861, 82.9%, 78.4%, and 87.1%, respectively, in differentiating between LG and HG lesions. These results are similar to ours, even though we used T2WI. This suggests that the radiomics analysis using individual MRI sequences can yield comparable performance in distinguishing between LG and HG lesions. However, the integration of different MRI sequences can yield different results. For example, Wang et al. [12] evaluated the performance of an MRI-based radiomics strategy in distinguishing between LG and HG lesions. They initially assessed the features extracted from T2WI and obtained an AUC of 0.7933, with a 95% confidence interval (CI) ranging from 0.7471 to 0.8396. However, the performance significantly improved when they incorporated the features extracted from diffusion-weighted imaging (DWI) sequences and apparent diffusion coefficient (ADC) maps. The AUC values reached up to 0.9276, indicating a substantial enhancement in the discriminatory power of the radiomics analysis. This demonstrates the importance of integrating features from multiple MRI sequences in improving the performance of radiomics analysis.

NMIBC Versus MIBC Lesions

In the discrimination between NMIBC and MIBC, the radiomics analysis has also yielded promising results through the combination of the two extracted radiomics features. This combination has demonstrated good diagnostic performance in distinguishing between the two conditions. The area under the curve (AUC) is reported to be 0.7, indicating a moderate level of discriminatory power. The 95% confidence interval (CI) for the AUC ranges from 0.11 to 1, suggesting some variability in the performance estimate. The sensitivity of the model is 100%, implying that it correctly identifies all positive cases (MIBC), while the specificity is 86.7%, indicating its ability to accurately identify negative cases (NMIBC). These results are statistically significant with a p-value of 0.0031, indicating a strong likelihood of a true association between the features and the discrimination between NMIBC and MIBC.

Wang et al. [13] used radiomics features extracted from T2, DWI, and ADC sequences. They used a radiomics signature, namely Radscore, which was generated using 36 selected radiomics features. The Radscore demonstrated a favorable ability to predict the presence of MIBC in both the training cohort (AUC 0.880) and the validation cohort (AUC 0.813). Furthermore, the researchers integrated the Radscore with another independent predictor, the MRI-determined tumor stalk, into a nomogram. This integration resulted in improved discriminatory performance, with the AUC increasing to 0.924 in the training cohort and 0.877 in the validation cohort. These findings indicate that the combined use of the Radscore and MRI-determined tumor stalk enhances the predictive accuracy in distinguishing between non-muscle invasive bladder cancer (NMIBC) and muscle-invasive bladder cancer (MIBC). These results suggest that the integration of radiomics features extracted from different MRI sequences, along with other independent predictors, can significantly enhance the performance of radiomics analysis. This integration improves the ability to differentiate between NMIBC and MIBC. By incorporating information from various MRI sequences, the radiomics analysis becomes more comprehensive, capturing a broader range of tumor characteristics and improving the accuracy of classification. This finding highlights the importance of considering multiple imaging modalities and integrating diverse features to achieve more precise and reliable discrimination between NMIBC and MIBC.

Limits

Our study has several limitations that need to be reported. First, it was performed as a retrospective analysis, including only pathologically proven lesions, which could lead to selection bias. Despite this, this remains an essential criterion for validating the ability of radiomics to diagnose malignant lesions. Second, it is important to acknowledge that the small number of MIBC in our study, with only 3 out of 35 cases, represents a limitation in the statistical analysis conducted. This limited sample size reduces the generalizability and reliability of the results, despite the initial promising findings. The small sample of muscle invasive lesions may not accurately represent the true distribution and characteristics of these specific lesions in the population. Therefore, caution should be exercised in interpreting these results, and further studies with larger sample sizes are necessary to validate and confirm the findings observed in this study. Finally, the results of the texture parameters were specific to the available data based on our small population conducted in a single reference center, and therefore in the future it will be recommended to perform a confirmatory analysis with more data.

5 Conclusion

Despite the promising results obtained so far, it is important to emphasize that radiomics of bladder MRI for distinguishing between LG and HG bladder lesions, and MIBC and NMIBC is still in the developmental phase and requires further studies and validations. A standardized method for extracting radiomics features is crucial, as well as creating appropriate training and validation datasets to ensure result reproducibility. Additionally, integrating other clinical and molecular information may further enhance the accuracy of radiomics predictions.

References

1. Svatek, R.S., et al.: The economics of bladder cancer: costs and considerations of caring for this disease. Eur. Urol. **66**, 253–262 (2014). https://doi.org/10.1016/j.eururo.2014.01.006
2. Sylvester, R.J., et al.: Predicting recurrence and progression in individual patients with stage Ta T1 bladder cancer using EORTC risk tables: a combined analysis of 2596 patients from seven EORTC trials. Eur. Urol. **49**, 466–477 (2006). https://doi.org/10.1016/j.eururo.2005.12.031
3. Moch, H., Cubilla, A.L., Humphrey, P.A., Reuter, V.E., Ulbright, T.M.: The 2016 WHO classification of tumours of the urinary system and male genital organs—Part A: renal, penile, and testicular tumours. Eur. Urol. **70**, 93–105 (2016). https://doi.org/10.1016/j.eururo.2016.02.029
4. Cutaia, G., et al.: Radiomics and Prostate MRI: current role and future applications. J. Imaging **7**, 34 (2021). https://doi.org/10.3390/jimaging7020034
5. Cannella, R., Grutta, L.L., Midiri, M., Bartolotta, T.V.: New advances in radiomics of gastrointestinal stromal tumors. World J. Gastroenterol. **26**, 4729–4738 (2020). https://doi.org/10.3748/wjg.v26.i32.4729
6. Stefano, A., et al.: Performance of radiomics features in the quantification of idiopathic pulmonary fibrosis from HRCT. Diagnostics. **10**, 306 (2020). https://doi.org/10.3390/diagnostics10050306
7. Giambelluca, D., et al.: PI-RADS 3 Lesions: role of prostate MRI texture analysis in the identification of prostate cancer. Curr. Probl. Diagn. Radiol. **50**, 175–185 (2021). https://doi.org/10.1067/j.cpradiol.2019.10.009
8. Comelli, A., et al.: Active contour algorithm with discriminant analysis for delineating tumors in positron emission tomography. Artif. Intell. Med. **94**, 67–78 (2019). https://doi.org/10.1016/j.artmed.2019.01.002
9. Comelli, A., et al.: K-nearest neighbor driving active contours to delineate biological tumor volumes. Eng. Appl. Artif. Intell. **81**, 133–144 (2019). https://doi.org/10.1016/j.engappai.2019.02.005
10. Dibb, M.J., et al.: Ultrasonographic analysis of bladder tumors. Clin. Imaging **25**, 416–420 (2001). https://doi.org/10.1016/S0899-7071(01)00304-7
11. Zhang, X., et al.: Radiomics assessment of bladder cancer grade using texture features from diffusion-weighted imaging. J. Magn. Reson. Imaging **46**, 1281–1288 (2017). https://doi.org/10.1002/jmri.25669
12. Wang, H., et al.: Radiomics analysis of multiparametric MRI for the preoperative evaluation of pathological grade in bladder cancer tumors. Eur. Radiol. **29**, 6182–6190 (2019). https://doi.org/10.1007/s00330-019-06222-8
13. Wang, H., et al.: Elaboration of a multisequence MRI-based radiomics signature for the preoperative prediction of the muscle-invasive status of bladder cancer: a double-center study. Eur. Radiol. **30**, 4816–4827 (2020). https://doi.org/10.1007/s00330-020-06796-8
14. Mirmomen, S.M., Shinagare, A.B., Williams, K.E., Silverman, S.G., Malayeri, A.A.: Preoperative imaging for locoregional staging of bladder cancer. Abdom. Radiol. **44**, 3843–3857 (2019). https://doi.org/10.1007/s00261-019-02168-z
15. Benfante, V., et al.: A new preclinical decision support system based on PET radiomics: a preliminary study on the evaluation of an innovative 64Cu-labeled chelator in mouse models. J. Imaging. **8**, 92 (2022). https://doi.org/10.3390/jimaging8040092
16. Vernuccio, F., Cannella, R., Comelli, A., Salvaggio, G., Lagalla, R., Midiri, M.: Radiomics and artificial intelligence: new frontiers in medicine. Recenti Prog. Med. **111**, 130–135 (2020). https://doi.org/10.1701/3315.32853

17. Comelli, A., et al.: Radiomics: a new biomedical workflow to create a predictive model. Presented (2020). https://doi.org/10.1007/978-3-030-52791-4_22
18. Cairone, L., et al.: Robustness of radiomics features to varying segmentation algorithms in magnetic resonance images. Presented (2022). https://doi.org/10.1007/978-3-031-13321-3_41
19. Panebianco, V., et al.: An evaluation of morphological and functional multi-parametric MRI sequences in classifying non-muscle and muscle invasive bladder cancer. Eur. Radiol. **27**, 3759–3766 (2017). https://doi.org/10.1007/s00330-017-4758-3
20. Alongi, P., et al.: Artificial intelligence applications on restaging [18F]FDG PET/CT in metastatic colorectal cancer: a preliminary report of Morpho-functional radiomics classification for prediction of disease outcome. Appl. Sci. **12**, 2941 (2022). https://doi.org/10.3390/app12062941
21. Comelli, A., et al.: Tissue classification to support local active delineation of brain tumors. In: Zheng, Y., Williams, B.M., Chen, K. (eds.) MIUA 2019. CCIS, vol. 1065, pp. 3–14. Springer, Cham (2019). https://doi.org/10.1007/978-3-030-39343-4_1
22. Agnello, L., Comelli, A., Ardizzone, E., Vitabile, S.: Unsupervised tissue classification of brain MR images for voxel-based morphometry analysis. Int. J. Imaging Syst. Technol. **26**, 136–150 (2016). https://doi.org/10.1002/ima.22168

Combined Data Augmentation for HEp-2 Cells Image Classification

Gennaro Percannella, Umberto Petruzzello(✉), Francesco Tortorella, and Mario Vento

Department of Information and Electrical Engineering and Applied Mathematics, University of Salerno, Via Giovanni Paolo II 132,, Fisciano 84084 Salerno, Italy
upetruzzello@unisa.it

Abstract. The Antinuclear Antibody (ANA) test is a valuable diagnostic tool for autoimmune disorders that uses Indirect Immunofluorescence (IIF) microscopy with HEp-2 cells as the substrate to identify antibodies and their distinct staining patterns. Machine learning-based approaches have shown promise in automating this diagnosis process, with Data Augmentation (DA) techniques playing a crucial role in improving performance. Even though traditional DA methods have yielded positive results, generative techniques like Variational AutoEncoders (VAEs) have shown potential in exploring the input distribution and generating new images. To address the limitations of traditional DA and explore the potential of generative approaches, this paper focuses on applying Conditional Variational AutoEncoders (CVAEs) to HEp-2 cell image classification. A customized CVAE architecture is proposed, considering multiple labels during generation to enhance versatility. Extensive experiments were conducted with the largest publicly available dataset of HEp-2 cell images, the *I3A* dataset. The performance of traditional and generative data augmentation techniques were compared while investigating potential synergies between them. The findings highlight the benefits of combining these techniques, especially in scenarios with class imbalance. Thorough statistical analysis provides valuable insights from the experimental results.

Keywords: Data Augmentation · Variational Autoencoder · HEp-2 cells · Intensity Classification · Pattern Classification · Class Imbalance

1 Introduction

The Antinuclear Antibody (ANA) test is an essential diagnostic tool for individuals suspected to have an autoimmune disorder. The gold standard method for ANA screening utilizes Indirect Immunofluorescence (IIF) microscopy, with human epithelial type 2 (HEp-2) cells serving as the substrate. This technique allows for the identification of antibodies that bind to specific intracellular targets, resulting in the formation of distinct staining patterns. The classification of

G. L. Foresti et al. (Eds.): ICIAP 2023 Workshops, LNCS 14366, pp. 104–115, 2024.
https://doi.org/10.1007/978-3-031-51026-7_10

HEp-2 cell patterns and intensities plays a crucial role in the accurate diagnosis of patients [5].

A substrate comprised of HEp-2 cells and an ANA slide are used to incubate the patient's serum. The intensity level is then determined by measuring the intensity strength of the slide under a fluorescence microscope. Following that, segmentation and the identification of particular ANA patterns-that is, immunofluorescent staining patterns pertinent to diagnostic purposes-were proposed. The medical workflow is described in depth in [6].

In this research field, Data Augmentation (DA) has proven to be instrumental in improving performance. Common DA operations include rotations, mirroring, and flipping [16]. These operations generate transformed images while preserving the original image's semantic information. Several authors have successfully applied these techniques in their work [3,9,17]. While traditional data augmentation techniques are often effective and yield good results, it is worth noting that they are not the only options available. Figure 1 provides a taxonomy of DA methods, highlighting two main branches: basic image manipulations and Deep Learning approaches.

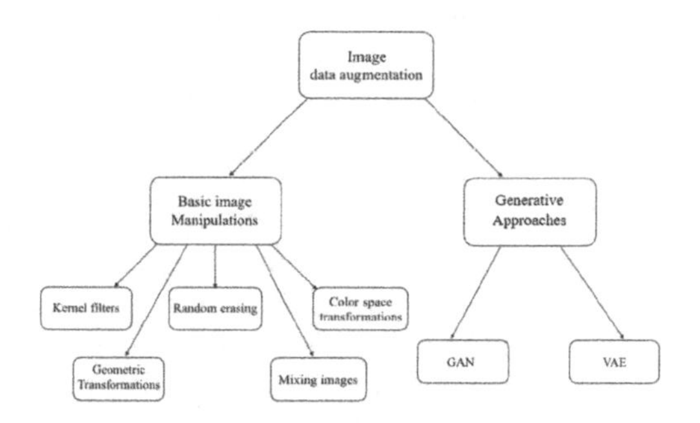

Fig. 1. A taxonomy of image data augmentations

Image manipulation approaches have demonstrated effectiveness in various domains. However, these approaches have certain limitations. Firstly, not all transformations can be applied universally as preserving sample semantics is crucial. For instance, in the case of HEp-2 cell images, only specific transformations like rotation and flipping are acceptable to avoid distorting sample information. Other transformations, such as variations in color and brightness, can compromise the ability to distinguish between positive and intermediate samples [16]. As a matter of fact, DA methods applied to HEp-2 cell images mostly rely on simple alterations to the images. For example, Li et al. [9] achieved notable improvements in classification performance by employing rotation and mirroring techniques while using the same Modified LeNet-5 network. Additionally, traditional approaches may not fully explore the distribution of inputs [14,18].

To address these limitations, generative approaches [19] can be considered. As shown in Fig. 1, they can be categorized into two main types: Generative

Adversarial Networks (GANs) and Variational AutoEncoders (VAEs). GANs aim to generate new images by training multiple networks with conflicting objectives, while VAEs learn a latent space to approximate the data distribution. Both approaches aim to better explore the input distribution and generate new images within it [2,4]. Generative approaches have been successful in generating new data in various medical imaging applications. For example, in the field of brain tumor imaging, combining different GAN models can better understand distributed features, as observed in the work of [13]. Pesteie et al. [15] used VAEs to synthesize new images and augment their datasets, incorporating sample labels into the architecture. Similarly, Biffi et al. [1] utilized generative architectures, specifically Conditional Variational AutoEncoders (CVAEs), for accurate segmentation of cardiac structures in MRI scans. On this basis, some generative data augmentation methods have already been proposed also to HEp-2 cell images. For example, in [11] GANs were used to generate synthetic HEp-2 cell images, but the attained results showed that traditional methods were more effective than GAN-based data augmentation. In another notable work, Li et al. [10] proposed a conditional GAN (cC-GAN) model for robust segmentation among different HEp-2 datasets. This model helped mitigate network overfitting and achieved better throughput for small-scale HEp-2 data. Majtner et al. [12] further improved the quality of GAN-generated synthetic samples by validating their quality with an independently fine-tuned neural network. However, all these GAN-based approaches did not achieve better results compared to traditional DA methods. In this paper, our objective is to investigate alternative generative methods for HEp-2 cell images, specifically focusing on Conditional Variational AutoEncoders (CVAEs). To the best of our knowledge, these methods have not been previously applied in the context of HEp-2 cell images. To address these research gaps, this paper aims to explore the synergies between traditional and CVAE-based DA techniques to enhance the classification performance of HEp-2 images. Specifically, the study aims to investigate the most effective augmentation techniques for classifying HEp-2 patterns and intensities, the importance of class balancing, and the potential benefits of combining different DA techniques.

The innovative contributions of this paper can be summarized as follows:

- This work introduces the application of CVAE for data generation in the HEp-2 literature. This approach offers a novel perspective on generative modeling for HEp-2 cell images.
- A customized CVAE architecture is implemented, specifically tailored for the HEp-2 cell image classification problem. Notably, this model considers multiple labels during the generation, enhancing its versatility and adaptability.
- The study includes thorough experimentation with statistical analysis, exploring the combined use of traditional and generative DA. This detailed analysis provides valuable insights into the efficacy of various DA strategies.

2 Materials and Method

In this section, we describe the dataset that was employed in our study. We also provide a description of the image manipulation techniques that will be utilized in the experiments. Furthermore, we describe the CVAE architecture that we propose for generative DA specifically tailored for HEp-2 cell images. Finally, we outline the experimental protocol that was adopted.

2.1 Dataset

For our experiments, we utilized the I3A dataset [5], which is the largest publicly available collection of HEp-2 cell images for research purposes. The dataset consists of two subsets: Task-1, which contains images of single-cell nuclei, and Task-2, which consists of whole specimens images. Our analysis focuses on the Task-1 subset, which comprises a total of 13,596 monochromatic images of single-cell nuclei extracted from 83 different specimen images. Each cell image in the Task-1 dataset is annotated by experts, providing bounding box annotations for each nucleus. For brevity, we will use the term "cell" to refer to the nucleus in the subsequent discussion.

The Task-1 dataset also includes annotations regarding the staining pattern, classified into six classes (Homogeneous, Speckled, Nucleolar, Centromere, Golgi, and Nuclear membrane - NuMem), as well as the label of the intensity level (Positive or Intermediate) and the segmentation mask.

Since the dataset provides several labels related to different issues associated with each image; in order to enable proper augmentation of the same, it is crucial that all related labels be generated with each augmented image.

2.2 Basic Image Manipulation

In terms of traditional data augmentation techniques, we primarily employ basic image manipulation operations such as rotation and flipping, as discussed in Sect. 1. These transformations are carefully chosen to increase the dataset size without introducing bias into the classification process. The selection of these transformations considers the characteristics of HEp-2 cell images, aiming to avoid image distortion while effectively augmenting the dataset. However, it is important to acknowledge the limitations of traditional data augmentation when applied to deep architectures. Although these techniques allows to increase the number of samples, they do not introduce new information or enhance the inherent variability of the dataset. In this paper, we have carefully chosen geometric transformations to apply to the original dataset samples. Our objective was to preserve the inherent cellular structures while avoiding any potential deformation. Our approach incorporates rotations of 90°, 180°, and 270°, along with horizontal and vertical flips, applied to the samples.

2.3 CVAE

To overcome the limitations of the basic image manipulation DA techniques with the aim of improving the performance of the classification models, we explore the use of *VAE* as generative approaches, specifically CVAEs.

As previously described, the analysis of HEp-2 images comprises several tasks, namely the staining pattern recognition according to six different classes, the scoring of the intensity according as intermediate or positive and the delineation of the boundary of single cells with respect to the surrounding cytoplasm. Consequently, the CVAE architecture has to incorporate multiple labels associated to each sample during both the learning of the CVAE itself and at inference time when the network is used to generate new samples to augment the dataset so that it can be used to address all the above tasks. To this aim, we enrich the traditional CVAE architecture in order to incorporate multiple labels for each sample (see Fig. 2). This design captures the relationships between input cell images and their associated labels, providing a versatile and adaptive generative approach.

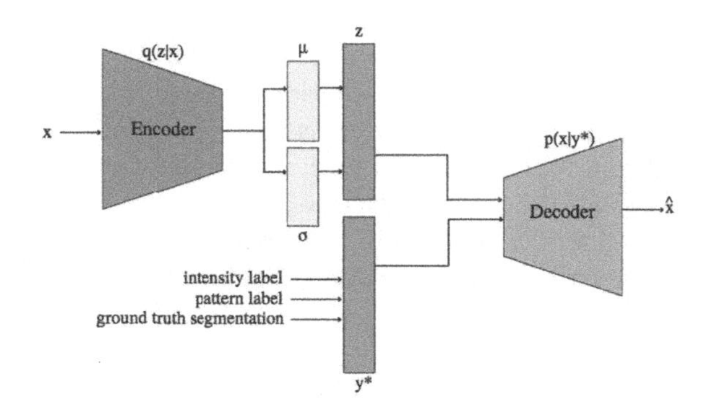

Fig. 2. CVAE Architecture Overview

The CVAE architecture consists of three main components: the Encoder module, the Linear layers, and the Decoder module. The Encoder module, depicted in Fig. 3(a), is composed of four Convolutional Layers (CL) with ReLU activation function. Each *CL* has a kernel size of 4×4 and a stride of 2. The module takes a cell image with dimensions $128 \times 128 \times 1$ as input and produces an encoded output with dimensions $6 \times 6 \times 256$. This output is then flattened into a vector with dimensions 9216. The purpose of the Encoder module is to map the input sample, denoted as x, to the corresponding distribution in the latent space. During the training phase, the encoder learns the distribution $q(z|x)$. This architecture enables the capture of relationships between input cell images and their associated labels. The resulting vector with dimensions 4096 serves as the input for two linear layers, as shown in Fig. 3(b). These layers process the input vector and generate two output vectors of size 512, representing the mean and variance of the distribution in the latent space associated with the sample x. Sampling of z from this distribution occurs once the mean and variance vectors are obtained. By employing linear layers, we transform the input vector into a latent space representation characterized by its mean and variance. This transformation enables the capture of statistical properties in the input data and facilitates the sampling process. Sampling from the latent space distribution is a crucial step in the generation process, as it allows the generation of new samples that align with the learned distribution.

The Decoder module, illustrated in Fig. 3(c), is comprised of three Transposed Convolutional Layers (*TCL*). Each *TCL* utilizes the ReLU activation function, except for the last one which uses the sigmoid activation function. The kernel size for the first three *TCLs* is set to 4×4, with a stride of 2. The last *TCL* has a kernel size of 6×6 and a stride of 2, ensuring that the output dimensions match the input image. The purpose of the Decoder module is to reconstruct the image based on the information encoded in the latent space representation, learning the distribution $p(x|y^*, z)$. Through their upscaling operations, the *TCs* progressively transform the latent space representation into a pixel-level representation, ultimately generating the reconstructed image as the decoding output. The activation functions within this module introduce non-linearity and enable the capturing of desired features and patterns in the reconstructed image.

In the training procedure, an error term is calculated between the input x and the output of the Decoder $\hat{x}$ using the Binary Cross Entropy loss function. This error term is then back-propagated through the network to update the model's parameters. By doing so, the Conditional Variational AutoEncoder (CVAE) can learn the relationships between the input samples, their associated labels, and their corresponding distributions in the latent space. Capturing these relationships enables the model to generate new samples by specifying their class labels, providing an effective generative approach for data augmentation in the HEp-2 cell image classification task.

2.4 Experimental Protocol

To analyze the results, we carefully selected widely used metrics in the existing literature [16]. Specifically, for both intensity classification and pattern classification, we adopted the Mean Class Accuracy (MCA) which is calculated by summing up the accuracies for each class and dividing it by the total number of classes. To ensure robust and reliable experimental analysis, we employed k-fold cross-validation. This approach allows for rigorous model evaluation and ensures the generalizability of findings across different partitions of the dataset. We divided the dataset into five folds while preserving the original class distribution in each fold (the distribution of samples per class for each fold is presented in Table 1).

For our CVAE training process, we employed specific hyperparameters. We used a learning rate of 0.0001 and a batch size of 32. The model was trained for a total of 500 epochs. To optimize the model's parameters during training, we utilized the Adam optimizer, a widely used optimization algorithm that adapts the learning rate based on the estimated first and second moments of the gradients.

Regarding the experiments, our first set of investigations focused on the importance of dataset balancing and the identification of optimal data augmentation techniques tailored to the specific tasks. The experiments in this group include:

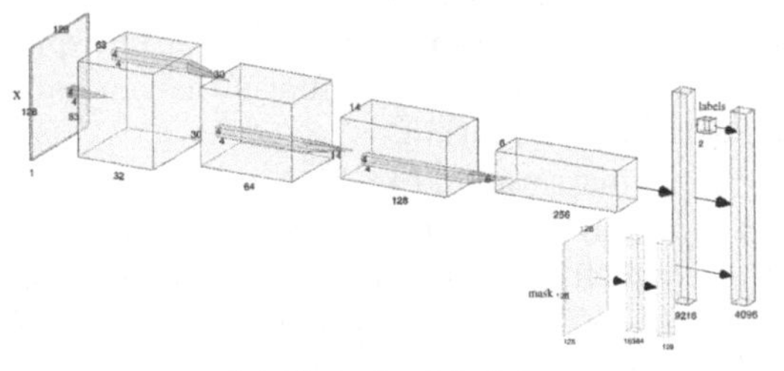

(a) Encoder Module

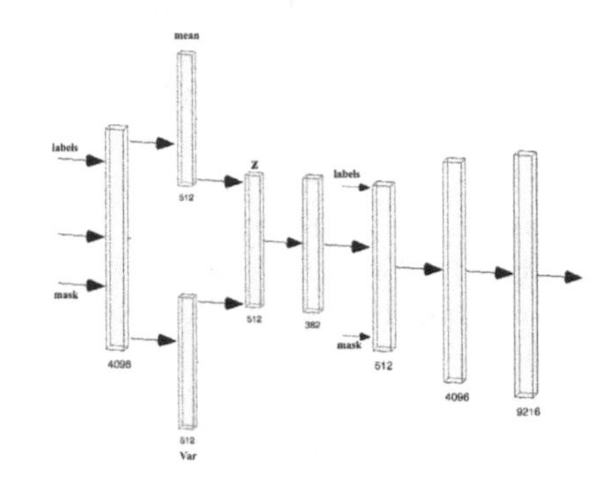

(b) Linear layers

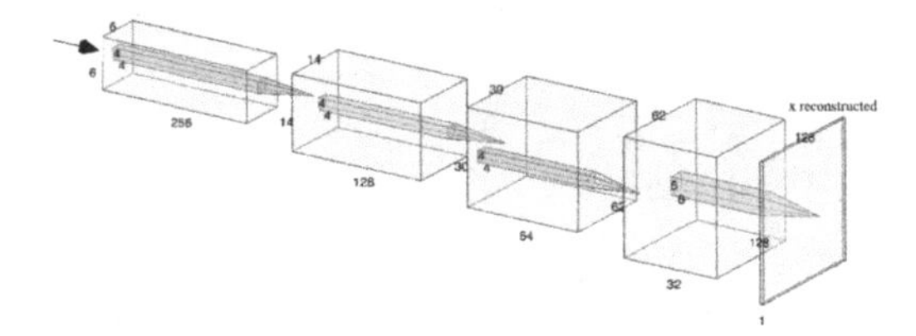

(c) Decoder module

Fig. 3. CVAE Detailed Architecture

- E1: Unbalanced original dataset I3A Cell Level.
- E2: Balanced dataset I3A Cell Level using undersampling technique.
- E3: Balanced dataset I3A Cell Level with traditional Data Augmentation.
- E4: Balanced dataset I3A Cell Level with samples generated by CVAE.
- E5: Balanced dataset I3A Cell Level with samples generated by CVAE and traditional DA techniques.

Except for E1 and E2, the sample number for each class in the defined experiments was adjusted to align with the count of the most representative class in the original dataset. This adjustment ensured fair and equitable representation for all classes during the analysis.

The subsequent set of experiments aimed to explore the impact of increasing the number of artificially generated samples using augmentation techniques. In

Table 1. 5-Fold cross-validation division of the I3A public dataset

Class	Fold 1	Fold 2	Fold 3	Fold 4	Fold 5
Homogeneous Positive	217	217	217	217	219
Homogeneous Intermediate	281	281	281	281	283
Speckled Positive	291	291	291	291	293
Speckled Intermediate	275	275	275	275	274
Nucleolar Positive	187	187	187	187	186
Nucleolar Intermediate	333	333	333	333	332
Centromere Positive	276	276	276	276	274
Centromere Intermediate	273	273	273	273	271
Golgi Positive	70	70	70	70	69
Golgi Intermediate	75	75	75	75	75
Nuclear Membrane Positive	188	188	188	188	191
Nuclear Membrane Intermediate	253	253	253	253	253
Total Samples	2719	2719	2719	2719	2720

this study, we increased the size of each fold by approximately five times while maintaining a balanced distribution of samples across different classes within each fold (12,000 samples per fold, 1,500 samples per class). The experiments in this group are:

- E6: Balanced dataset I3A Cell Level with traditional Data Augmentation and increased samples.
- E7: Balanced dataset I3A Cell Level with samples generated by CVAE and increased samples.
- E8: Balanced dataset I3A Cell Level with samples generated by CVAE and traditional DA techniques, with increased samples.

3 Results

To ensure a fair comparison among the considered data augmentation (DA) techniques, we utilized the ResNet50 deep network architecture for each of them. This architecture has demonstrated good performance on HEp-2 cell classification tasks in the literature [3,8]. In our 5-fold cross-validation procedure, the training set underwent the respective DA techniques, while the test set was used at the end of each training cycle to report the final performance. The performance results are summarized in Table 2. To assess the statistical significance of performance differences between each pair of experimental configurations in Table 2, we conducted a Wilcoxon test [20] within the 5-fold CV scheme. A significance level of 0.05 was considered for all tests, and the computed p-values were corrected using the Bonferroni-Holm method to account for multiple comparisons [7]. The results of the statistical analysis are presented in Table 3.

Based on the results shown in Tables 2 and 3, several remark can be made:

Dataset Balancing and Augmented: Experiment *E2* provides clear evidence of the ineffectiveness of undersampling techniques, as it showcases a significant performance drop of over 10% points compared to the original dataset. This decrease can be attributed to the decreased generalization capability resulting from the substantial reduction in the number of samples, which adversely affects

Table 2. Experimental results on *Intensity Classification*, *Pattern Classification*, and on *the combination of them*, compared among the experiments performed. The values are expressed as average ± standard deviation across the 5-fold CV.

Experiment	MCA	MCA Pattern	MCA Intensity
E1: Original dataset	71.60±5.50	78.60±3.71	85.87±4.56
E2: Balanced Dataset with undersampling	60.83±7.39	68.00±5.71	85.70±3.26
E3: Balanced Dataset with traditional DA	75.83±2.89	83.00±3.57	86.65±3.78
E4: Balanced Dataset with samples generated by CVAE	72.33±4.89	79.00±6.21	86.55±4.55
E5: Balanced Dataset with samples generated by CVAE and traditional DA	74.83±4.87	80.00±7.14	86.50±4.23
E6: Balanced Dataset traditional DA with more samples	78.00±2.96	83.35±2.70	87.60±3.03
E7: Balanced Dataset with CVAE with more samples	72.75±6.10	80.46±6.54	86.80±4.58
E8: Balanced Dataset the combination of traditional DA and CVAE with more samples	81.60±7.94	86.00±7.25	91.20±3.57

the model's ability to accurately learn and generalize patterns and intensity. On the other hand, the traditional augmentation technique, as seen in experiments *E3* and *E6*, shows a significant improvement in performance by balancing the classes and increasing the dataset size, achieving performance improvements of 4% and 6%, respectively. The effectiveness of this type of data augmentation approach is confirmed by the statistical analysis in Table 3, where traditional DA (*E2*) demonstrates both good performance and stability, exhibiting a statistically significant performance superiority over the original dataset.

Generative Approaches with CVAE: Experiments *E4* and *E7* demonstrate that generative approaches using CVAE achieve a slightly lower performance increase compared to traditional techniques with respect to the baseline configuration (*E1*), with a performance increase of only 1%. However, they also introduce an increase in instability compared to traditional DA. This suggests that generative approaches alone may not be as effective as traditional DA techniques.

Combined Approach: The combined approach of traditional and generative data augmentation, proposed in experiments *E5* and *E8*, yields better performance, particularly when the number of generated samples increases. It achieves a gain of 6.4% compared to the baseline. By increasing the number of generated samples and combining them with traditional techniques, advantageous components of both approaches are exploited. The table evidently shows that this approach greatly benefits the intensity classification. However, it would require a significant increase in the number of samples and their variability, like those found in the combined approach. The statistical significance analysis in Table 2 reveals that the configuration in experiment *E8* performs better than all its competitors.

To gain a deeper understanding of the outcomes achieved through different data augmentation techniques, we provide the performance per class of each experiment in Table 4. It is evident that the combined data augmentation technique (*E8*) leads to improvements on almost all classes. The performance

Table 3. Matrix of p-value among the experiments on the 12 classes.

	E1	E2	E3	E4	E5	E6	E7	E8
E1	–							
E2	0.776	–						
E3	0.115	**0.003**	–					
E4	0.153	**0.040**	0.548	–				
E5	0.138	**0.028**	0.294	0.186	–			
E6	**0.028**	**0.002**	0.093	0.052	0.105	–		
E7	0.316	**0.039**	0.584	0.381	0.527	0.673	–	
E8	**0.009**	**0.001**	**0.032**	**0.011**	**0.027**	**0.047**	**0.013**	–

obtained in (*E8*) is generally higher than the other experiments, with substantial differences observed in many classes. The only two classes that do not achieve the best performance are "Centromere Positive" and "Centromere Intermediate." However, even in these cases, the performance of the combined DA is still very close to the best experiment. The table clearly demonstrates how the combined approach leverages the advantages of both techniques. For example, the generative DA with CVAE excels in the "Centromere Positive" class, while the traditional DA achieves superior performance in the "Centromere Intermediate" class. The combined methodology effectively integrates the strengths of the two techniques and performs exceptionally well across all classes.

Table 4. Performances per class over the 12 combined classes

	E1	E2	E3	E4	E5	E6	E7	E8
H/P	$38.0_{\pm}29.9$	$38.0_{\pm}14.7$	$53.0_{\pm}27.1$	$46.0_{\pm}30.7$	$50.0_{\pm}29.0$	$58.0_{\pm}31.9$	$50.5_{\pm}29.5$	$\mathbf{67.5_{\pm}27.0}$
H/I	$60.0_{\pm}16.7$	$52.0_{\pm}11.7$	$60.0_{\pm}19.0$	$56.0_{\pm}10.2$	$66.0_{\pm}10.2$	$71.5_{\pm}18.7$	$66.0_{\pm}17.7$	$\mathbf{71.8_{\pm}25.2}$
S/P	$82.0_{\pm}17.2$	$66.0_{\pm}16.2$	$72.0_{\pm}21.4$	$80.0_{\pm}19.0$	$76.0_{\pm}18.5$	$75.9_{\pm}20.0$	$71.5_{\pm}20.7$	$\mathbf{81.9_{\pm}20.9}$
S/I	$66.0_{\pm}15.0$	$32.0_{\pm}11.7$	$68.0_{\pm}13.3$	$60.0_{\pm}20.0$	$70.0_{\pm}14.1$	$70.0_{\pm}11.1$	$61.6_{\pm}14.6$	$\mathbf{71.7_{\pm}14.3}$
N/P	$76.0_{\pm}4.9$	$58.0_{\pm}27.9$	$94.0_{\pm}12.0$	$90.0_{\pm}12.6$	$92.0_{\pm}7.5$	$84.7_{\pm}19.0$	$88.4_{\pm}12.0$	$\mathbf{95.6_{\pm}2.8}$
N/I	$70.0_{\pm}16.7$	$54.0_{\pm}18.5$	$72.0_{\pm}9.8$	$64.0_{\pm}13.6$	$64.0_{\pm}21.4$	$74.5_{\pm}10.8$	$70.5_{\pm}14.8$	$\mathbf{81.4_{\pm}19.9}$
C/P	$94.0_{\pm}8.0$	$76.0_{\pm}18.5$	$87.0_{\pm}7.5$	$\mathbf{91.0_{\pm}8.0}$	$91.0_{\pm}11.7$	$86.1_{\pm}12.4$	$90.5_{\pm}4.6$	$88.9_{\pm}16.8$
C/I	$52.0_{\pm}25.6$	$50.0_{\pm}21.9$	$64.0_{\pm}27.1$	$54.0_{\pm}28.0$	$64.0_{\pm}17.4$	$\mathbf{66.4_{\pm}28.4}$	$60.0_{\pm}26.8$	$66.2_{\pm}21.4$
G/P	$90.0_{\pm}8.9$	$96.0_{\pm}4.9$	$98.0_{\pm}2.0$	$98.0_{\pm}4.0$	$96.0_{\pm}4.9$	$98.4_{\pm}1.2$	$90.0_{\pm}8.6$	$\mathbf{99.1_{\pm}1.1}$
G/I	$76.0_{\pm}8.0$	$88.0_{\pm}14.7$	$90.0_{\pm}4.0$	$89.0_{\pm}7.5$	$90.0_{\pm}0.0$	$89.1_{\pm}2.7$	$85.3_{\pm}4.9$	$\mathbf{91.2_{\pm}8.5}$
NM/P	$82.0_{\pm}11.7$	$76.0_{\pm}19.6$	$82.0_{\pm}5.3$	$74.0_{\pm}4.0$	$79.0_{\pm}15.5$	$87.7_{\pm}4.3$	$73.2_{\pm}16.0$	$\mathbf{92.2_{\pm}4.7}$
NM/I	$68.0_{\pm}27.1$	$44.0_{\pm}25.8$	$70.0_{\pm}16.9$	$66.0_{\pm}26.5$	$60.0_{\pm}12.6$	$73.8_{\pm}16.6$	$65.6_{\pm}24.1$	$\mathbf{78.0_{\pm}13.3}$

Examining the images generated in Figs. 4 and 5, it can be observed that the images produced using CVAE closely resemble the original ones but exhibit slight variations. While maintaining a fundamental similarity to the original images, these generated images have been subtly adjusted to enhance the distribution of samples within the input range. The CVAE-generated images demonstrate the model's ability to generate samples that align with the original dataset while introducing beneficial variations for improved performance and exploration of the input distribution. This advantageous characteristic is evident in the achieved performance, as described and elaborated above.

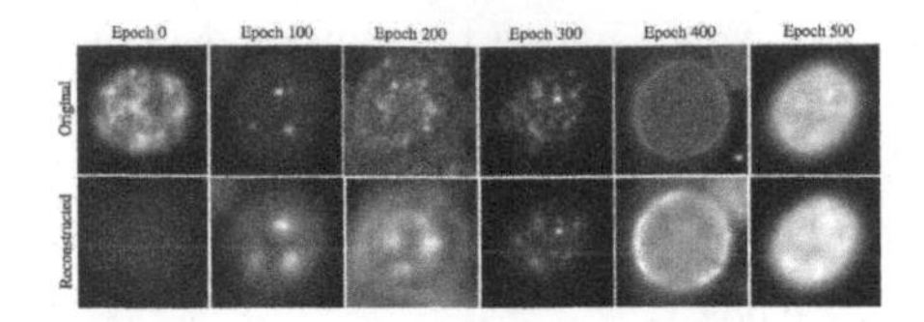

Fig. 4. Comparison among Original and Reconstructed samples

Fig. 5. Examples of samples generated with CVAE with their labels

4 Conclusions

This study focused on analyzing and evaluating HEp-2 image augmentation techniques to enhance classification performance. A learning-based approach utilizing Conditional Variational Auto-Encoders (CVAEs) was proposed to generate new samples tailored to specific classification tasks. The performance of classical and generative data augmentation techniques, including CVAE, was compared, and the potential of combining these techniques was explored.

The experiments were conducted on the *I3A* dataset, which is the largest and most widely used dataset of HEp-2 cell images. The results revealed several key findings. Firstly, incorporating samples generated by CVAEs into the training set resulted in improved performance compared to using the original training set. However, combining generated and modified images with basic image transformations showed interesting potential. The combination of traditional and generative data augmentation techniques led to a statistically significant performance increase. This can be attributed to the generative techniques' ability to explore the sample distribution by introducing variability, while traditional data augmentation increased the sample size while maintaining the reference distribution, enhancing stability. Additionally, the experiments consistently demonstrated the effectiveness of traditional data augmentation techniques for HEp-2 image classification. The promising results obtained through the combination of CVAE-generated samples and basic manipulation indicate a compelling alternative that merits further investigation. The potential benefits of combining these techniques can also be leveraged in other research fields.

In conclusion, the combined use of traditional and generative data augmentation techniques, especially for less represented classes, opens up possibilities for developing more robust classification systems, particularly in scenarios with class imbalance. This work contributes to advancing the understanding and application of DA techniques in the context of HEp-2 image classification.

References

1. Biffi, C., et al.: 3D High-resolution cardiac segmentation reconstruction from 2D views using conditional variational autoencoders. In: 2019 IEEE 16th International Symposium on Biomedical Imaging, pp. 1643–1646 (2019)
2. Chou, J.: Generated loss and augmented training of MNIST VAE. ArXiv (2019)
3. Cascio, D., Taormina, V., Raso, G.: Deep convolutional neural network for HEp-2 fluorescence intensity classification. Appl. Sci. **9**(3), 408 (2019)
4. Goodfellow, I.J., et al.: Generative adversarial networks (2014)
5. Hobson, P., et al.: Competition on cells classification by fluorescent image analysis. Proceedings of 20th IEEE International Conference on Image Processing (2013)
6. Hobson, P., et al.: Computer aided diagnosis for anti-nuclear antibodies HEp-2 images: progress and challenges. Pattern Recogn. Lett. **82**, 3–11 (2016)
7. Holm, S.: A simple sequentially rejective multiple test procedure. Scand. J. Statist. **6**(2), 65–70 (1979)
8. Lei, H., et al.: Cross-modal transfer learning for HEp-2 cell classification based on deep residual network. In: IEEE International Symposium on Multimedia (2017)
9. Li, H., Zhang, J., Zheng, W.: Deep CNNs for HEp-2 cells classification: A cross-specimen analysis. ArXiv (2016)
10. Li, Y., Shen, L.: cC-GAN: a robust transfer-learning framework for HEp-2 specimen image segmentation. IEEE Access **6**, 14048–14058 (2018)
11. Majtner, T., et al.: On the Effectiveness of Generative Adversarial Networks as HEp-2 Image Augmentation Tool. Image Analysis (2019)
12. Majtner, T.: HEp-2 cell image recognition with transferable cross-dataset synthetic samples. In: Computer Analysis of Images and Patterns, pp. 215–225 (2021)
13. Mukherkjee, D., et al.: Brain tumor image generation using an aggregation of GAN models with style transfer. Sci. Rep. **2**, 9141 (2022)
14. Mumuni, A., Mumuni, F.: Data augmentation: a comprehensive survey of modern approaches. Array **16**, 100258 (2022)
15. Pesteie, M., Abolmaesumi, P., Rohling, R.N.: Adaptive Augmentation of Medical Data using Independently Conditional Variational Auto-Encoders. IEEE Trans. Med, Imaging (2019)
16. Rahman, S., Wang, L., Sun, C., Zhou, L.: Deep learning based HEp-2 image classification: a comprehensive review. Med. Image Anal. **65**, 101764 (2020)
17. Rodrigues, L.F., Naldi, M.C., Mari, J.F.: HEp-2 cell image classification based on convolutional neural networks. Workshop of Computer Vision (2017)
18. Shorten, Connor, Khoshgoftaar, Taghi M..: A survey on image data augmentation for deep learning. J. Big Data **6**(1), 1–48 (2019). https://doi.org/10.1186/s40537-019-0197-0
19. Sohn, K., Lee, H., Yan, X.: Learning structured output representation using deep conditional generative models. In: Cortes, C., Lawrence, N., Lee, D., Sugiyama, M., Garnett, R. (eds.) Advances in Neural Information Processing Systems. vol. 28. Curran Associates, Inc. (2015)
20. Wilcoxon, F.: Individual comparisons by ranking methods. Biometrics Bull. **1**(6), 196–202 (80–83)

Multi-modal Medical Imaging Processing (M3IP)

Harnessing Multi-modality and Expert Knowledge for Adverse Events Prediction in Clinical Notes

Marco Postiglione[1](✉), Giovanni Esposito[2], Raffaele Izzo[2], Valerio La Gatta[1], Vincenzo Moscato[1], and Raffaele Piccolo[2]

[1] Department of Electrical Engineering and Information Technology, University of Naples Federico II, Naples, Italy
marco.postiglione@unina.it

[2] Department of Advanced Biomedical Sciences, University of Naples Federico II, Naples, Italy
{giovanni.esposito,raffaele.izzo,raffaele.piccolo}@unina.it

Abstract. Recent advancements in machine learning and deep learning techniques have revolutionized the field of adverse event prediction, which plays a vital role in healthcare by enabling early identification and intervention for high-risk patients. Traditionally, researchers have relied on structured data, including demographic information, vital signs, laboratory results, and medication records. However, the widespread adoption of electronic health records (EHRs) has introduced a substantial amount of unstructured information in the form of clinical notes, which have been largely underutilized. Natural Language Processing (NLP) techniques have emerged as a powerful tool for extracting valuable insights from these clinical notes and incorporating them into machine learning frameworks. Additionally, multimodal machine learning, which integrates structured and unstructured data, has gained considerable attention to enhance the accuracy of adverse event prediction. This research focuses on the application of multimodal machine learning for predicting adverse events such as atrial fibrillation, heart failure, and ischemic myocardial infarction. The study aims to compare the performance of a Machine Learning specialist without domain knowledge would obtain with an approach guided by physicians, that includes an information retrieval step using unstructured clinical notes. The analysis is carried out using a dataset provided by the Hospital of Naples Federico II. The results not only shed light on the importance of leveraging different aspects of a patient's medical history and extracting information from unstructured notes but also highlight the added value of domain expertise.

Keywords: Adverse events prediction · Multi-modal Machine Learning · EHRs · Natural Language Processing

G. L. Foresti et al. (Eds.): ICIAP 2023 Workshops, LNCS 14366, pp. 119–130, 2024.
https://doi.org/10.1007/978-3-031-51026-7_11

1 Introduction

Adverse event prediction plays a crucial role in healthcare as it enables early identification and intervention for patients at risk. In recent years, the healthcare field has witnessed significant advancements in the application of machine learning and deep learning techniques. For example, Tomašev et al. [11] propose a workflow for early prediction of patient outcomes, focusing their attention on acute kidney injuries, mortality, length of stay and 30-day hospital readmission; similarily, Sheu et al. [10] use AI models to predict the differential response to antidepressant classes.

Traditionally, researchers have primarily relied on structured data, such as demographic information, vital signs, laboratory results, and medication records, to predict adverse events. For example, Mortazavi et al. [8] leverage information such as demographics (e.g. age, gender, insurance), patient history (i.e. problems list and admission diagnosis codes), visit information (e.g. primary principal procedure, admission time), medical information (e.g. medications, laboratory results, patient vitals) and Rothman Index scores to predict adverse events, such as respiratory failure or infection, in patients undergoing major cardiovascular procedures.

However, with the proliferation of electronic health records (EHRs), there is a vast amount of unstructured information available in the form of clinical notes that remains largely untapped. Clinical notes, containing free-text descriptions of patients' medical history, symptoms, treatments, and physician's observations, provide valuable insights that are not captured in structured data. Natural Language Processing (NLP) techniques have been employed to extract relevant information from unstructured text and incorporate it into machine learning frameworks [2,5,12]. NLP methods, such as named entity recognition, entity linking, and sentiment analysis, have been applied to identify key concepts, clinical events, and temporal relationships, enhancing the predictive capabilities of machine learning models. BEHRT [7] is a transformer-based architecture able to process EHR data to predict future disorders; similarily, G-BERT [9] augments the standard transformer architecture with graph data structures for medication recommendation.

As a consequence of the ever-increasing performance of NLP systems, the utilization of multimodal machine learning techniques, that aim to leverage the complementary nature of structured and unstructured data to improve predictive performance and enable more accurate adverse event prediction, has gained substantial attention. Several prior studies have explored the use of multimodal machine learning for adverse event prediction in healthcare. For example, Hernandez et al. [4] use ECG, HRV, arterial blood pressure (ABP) waveform from an arterial line, pulse plethysmography (PPG) waveform from a pulse oximeter, and EHR data to predict hemodynamic decompensation. Similarly, Krix et al. [6] integrate various data sources on protein functions, gene expression, chemical compound structures and more, into the prediction of drug adverse events.

In this work, we focus on the use of multimodal machine learning for predicting the occurrence of adverse events, such as atrial fibrillation, heart failure, and IMA (ischemic myocardial infarction). Given the available dataset, provided by the Hospital of Naples Federico II, we will compare (1) the performance that a Machine Learning specialist, without any domain knowledge, would reach, with (2) the performance obtained with a pre-processing guided by physicians and (3) with an information retrieval step from the unstructured text of clinical notes. Not only do results show the importance of combining different aspects of a patient's medical history and the information retrieved from unstructured notes, but they also provide insights about the importance of the additional domain expertise. It is worth to note that we used Italian clinical notes, where the availability of Natural Language Processing resources, such as publicly available datasets and pre-trained models, is limited, but the help of domain experts has allowed us to surpass the low-resource language obstacle.

The remainder of this work is structured as follows: in Sect. 2 we formalize the adverse events prediction task, providing details about the events of interest; in Sect. 3 we provided details about the available dataset, the features extracted and the methodology applied to extract them from the structured and the unstructured parts of EHRs; in Sect. 4 we summarize the training methodology and provide information about the models used in our study; results are shown and discussed in Sect. 5; finally, we conclude our work and provide insights for future studies in Sect. 6.

2 Adverse Events Prediction: Task Formulation

In this study, our objective is to utilize the comprehensive patient history and information acquired during their initial visit as feature inputs for our machine learning model. The dataset encompasses a diverse collection of both structured and unstructured data, which will be subjected to appropriate processing techniques to derive patient-specific features. Let $\mathbf{X}$ represent the feature matrix, where each row $\mathbf{x}_i$ denotes the extracted features from the patient's initial visit. The primary focus of our investigation revolves around addressing various machine learning tasks pertaining to specific adverse events, such as heart attacks, acute myocardial infarctions, and aortic diseases. Specifically, we endeavor to model the conditional probability $P(event|\mathbf{x}_i)$, which characterizes the likelihood of future occurrences of these events in a given patient, based on the recorded information from their initial visit.

3 Data and Information Extraction

The dataset utilized for this work has been provided by the Department of Cardiology at the Federico II University Hospital in Naples. It encompasses a comprehensive collection of information pertaining to patients who have undergone visits at the cardiology department. Each medical event is recorded by a physician, rendering it a valuable source of data for our analysis. The dataset comprises several variables, including patients' demographic data (such as age, gender, and geographical origin), clinical information (such as physical examination results, presented symptoms, diagnoses, prescriptions, and treatments), as well as follow-up information regarding patients over time. The overall size of the dataset is significant, encompassing 54462 patients. The utilization of this dataset enables us to deepen our understanding of cardiac conditions and develop machine learning models capable of predicting diagnoses, clinical outcomes, and therapeutic recommendations. The collaboration with the Department of Cardiology at the Federico II University Hospital has facilitated access to a high-quality clinical data source, thereby contributing to the success and validity of our study.

In the remainder of this section, we first describe the features of interest that are used to characterize a patient, and then detail their retrieval process using both the unstructured and structured information stored in the available dataset.

3.1 Features of Interest

In this section, we provide a detailed description of the features characterizing a patient's medical history, as defined with the support of domain experts. Each feature is binary, indicating the presence or absence of a specific event or condition in the patient's medical record. They capture various aspects of a patient's medical history, providing essential information for predicting future adverse events and enabling more informed clinical decision-making. A comprehensive overview is provided in Table 1.

3.2 Features Extraction from Structured Data

By working closely with the medical experts, a systematic approach was followed to identify and extract relevant information from the structured fields. This collaborative effort ensured that the features were accurately represented and aligned with the medical domain knowledge.

In many instances, binary features were derived from columns containing categorical values. Therefore, mapping criteria were established in collaboration with the medical professionals. Additionally, certain feature values were obtained from measurements and laboratory tests for the retrieval of specific features. Some examples are reported as follows:

- A fasting plasma glucose level greater than 126 mg/dL is indicative of the presence of *diabetes*
- A Body Mass Index (BMI) value greater than 30 is commonly used as an indicator of obesity.
- The CKD-EPI (Chronic Kidney Disease Epidemiology Collaboration) equation is used to estimate glomerular filtrate rate (GFR) from serum creatinine and other readily available clinical parameters. GFR is then leveraged to retrieve the stage of CKD.

Table 1. Summary of the features of interest considered in this work.

Feature	Description
Diabetes	The patient has a diagnosis of diabetes.
Insulin-treated	The patient requires insulin therapy for diabetes management.
Oral therapy	The patient is undergoing oral therapy for diabetes management.
Dyslipidemia	This feature indicates the presence of abnormal lipid levels in the patient's medical record.
Statins	The patient is receiving statin medication for dyslipidemia management.
Fibrates	The patient is receiving fibrate medication for dyslipidemia management.
Hypercholesterolemia	This feature indicates the presence of hypercholesterolemia in the patient's medical record.
Hypertriglyceridemia	This feature indicates the presence of hypertriglyceridemia in the patient's medical record.
Mixed	This feature indicates the presence of mixed dyslipidemia, characterized by abnormalities in both cholesterol and triglyceride levels.
Arterial Hypertension	The patient has a diagnosis of high blood pressure.
Current Smoker	The patient is currently a smoker.
Ex-Smoker	The patient is an ex-smoker.
Obesity	The patient is classified as obese based on their body mass index (BMI).
CKD	This feature indicates the presence of chronic kidney disease (CKD) in the patient's medical record.
CKD Stage 1, 2, 3A, 3B, 4, 5	These features indicate the specific stage of chronic kidney disease experienced by the patient (ranging from Stage 1 to Stage 5)
CAD	The patient has a diagnosis of coronary artery disease (CAD).
IMA	The patient has experienced a previous episode of acute myocardial infarction (IMA).
PCI	The patient has undergone a previous percutaneous coronary intervention (PCI).
CABG	The patient has undergone previous coronary artery bypass grafting (CABG) surgery.
CAD Familial History	There is a family history of coronary artery disease.
Cerebral	There is a family history of cerebral vascular disease.
Cardiac	There is a family history of cardiac-related coronary artery disease.
Atrial Fibrillation	The patient has a diagnosis of atrial fibrillation.
Stroke	The patient has a history of stroke.
Heart Failure	The patient has a diagnosis of heart failure.
Aortic Disease	The patient has a diagnosis of aortic disease.
Ascending Aortic Dilation	This feature indicates the presence of ascending aortic dilation in the patient's medical record.
Ascending Aortic Aneurysm	This feature indicates the presence of an ascending aortic aneurysm in the patient's medical record.
Abdominal Aortic Aneurysm	This feature indicates the presence of an abdominal aortic aneurysm in the patient's medical record.
COPD	The patient has a diagnosis of chronic obstructive pulmonary disease (COPD).
Peripheral Revascularization	The patient has undergone peripheral revascularization surgery.
Peripheral Artery Disease	The patient has a diagnosis of peripheral artery disease.
SVA_SVM	This feature indicates the presence of supraventricular arrhythmias (SVA_SVM) in the patient's medical record.
ICD	The patient has an implantable cardioverter-defibrillator (ICD).
PM	The patient has a pacemaker (PM).
CRT-D	The patient has received cardiac resynchronization therapy with a defibrillator (CRT-D).

3.3 Features Extraction from Unstructured Data

In the extraction process of unstructured data, guided by medical experts, we have defined sets of aliases for each target feature. When a match is found between the aliases and the clinical records, the corresponding feature is associated with the patient. However, an additional step of *assertion classification* is required to verify the actual presence of the concept. This is necessary because the clinical records may contain statements such as "the patient *denies* familiarity with CAD." In this case, we should not consider the "familiarity with CAD" as a feature to be associated with the patient. To address this, under the guidance of domain experts, we proceeded with the annotation of a dataset comprising approximately 8000 randomly selected instances from available clinical records. Subsequently, a transformer-based model was trained to classify the assertion levels into three categories: present, absent, and familiarity.

The annotation process involved labeling the instances to reflect the accurate assertion status of the targeted features. This dataset served as training data for the transformer-based model, enabling it to learn and classify assertions accurately. The model's architecture, based on a pre-trained transformer model[1], was selected due to their effectiveness in capturing contextual information and understanding the relationships between words and phrases. By training the model on this annotated dataset, we aimed to develop a reliable system capable of correctly classifying the assertion levels associated with each feature in the clinical records.

3.4 Multi-modality: Early and Late Fusion

We apply and compare two different fusion methods, i.e. early and late fusion. Specifically, *early fusion* is obtained by combining features extracted from structured and unstructured data with an OR operation (i.e. if the concept appears at least in one of the two feature sets, we consider it as present). Conversely, *late fusion* is obtained by combining predictions of models trained on the two sets of features, separately. For a given prediction task, each model returns a score denoting the probability of an adverse event happening in the future. If P_s and P_u are the score returned by the models trained on structured and unstructured features, respectively, then the final score P is obtained by late fusion of named models:

$$P = P_s \cdot w_s + P_u \cdot w_u, \tag{1}$$

where late fusion weights w_i are obtained by random-search and cross-validation, and their sum is equal to 1, i.e. $w_s + w_u = 1$.

4 Training

In this section, we provide a detailed description of how our classification settings as well as the training procedure.

4.1 Datasets and Metrics

We address multiple binary classification tasks, where each task corresponds to a dataset with specific disease listed in Sect. 3. All tasks involve analyzing the same set of patients. However, the definition of positive and negative classes varies depending on the disease. Specifically, patients experiencing the adverse event are considered as the positive class, while the remaining patients are classified as the negative class.

To characterize each patient, we extract features from their EHRs. We consider four distinct feature groups for each patient:

[1] https://huggingface.co/dbmdz/bert-base-italian-xxl-cased.

- *Without domain knowledge* features: These are raw tabular features extracted directly from the patient's EHR.
- *Unstructured* features: These are domain-enhanced features extracted from unstructured text within the EHR.
- *Structured* features: These are domain-enhanced features extracted from structured data in the EHR.
- *Multi-modal (early)* features: These are domain-enhanced features that combine both "text" and "no text" features.

It is worth to note that domain-enhanced features undergo processing based on medical guidelines. For instance, diabetes features involve numeric parameters (e.g., insulin, glycemia), which are further processed using medical thresholds (e.g., presence of diabetes if a patient takes insulin or if glycemia exceeds 126).

We aim to evaluate the contribution of features from each group to the overall performance of the classification tasks. Additionally, we investigate whether combining this multimodal data leads to improved predictive power.

Regarding the classification nature of the tasks, we employ standard classification metrics, including accuracy, precision, recall, and F1-score. While accuracy is widely recognized for evaluation, we emphasize the importance of utilizing other metrics in our specific use case due to factors such as class imbalance and importance.

4.2 Classification Suite

In our evaluation, we compared various machine learning models:

- *Random Forest* is an ensemble learning method that combines multiple decision trees to make predictions. It creates a collection of decision trees and aggregates their results to obtain the final prediction.
- *Logistic Regression* is a statistical model used for binary classification. It estimates the probability of an instance belonging to a certain class using a logistic function.
- *Support Vector Machine (SVM)* is a powerful classification algorithm that separates data points by finding an optimal hyperplane that maximizes the margin between classes.
- *Gaussian Naive Bayes* is a probabilistic classifier based on Bayes' theorem.
- *Multi-Layer Perceptron (MLP)* is a type of artificial neural network with multiple layers of interconnected nodes (neurons). It can learn complex nonlinear relationships between features and the target variable.

For each model and task, we conducted 10-fold stratified cross-validation with default hyper-parameters and calculated the 95% confidence interval for the observed performance. Since hyper-parameter optimization was not performed, we divided the datasets into training (80%) and testing (20%) sets, without utilizing a validation set.

4.3 Imbalance Learning

All the datasets analyzed in this study exhibit class imbalance, where the positive class occurs much more frequently than the negative class. This imbalance arises due to the fact that there are significantly fewer patients who have experienced the adverse event compared to those who are healthy in relation to that particular disease.

To address this issue, we employed various augmentation strategies:

- *Random Oversampling* is a technique used to address class imbalance by randomly duplicating instances from the minority class until a balanced distribution is achieved. This technique aims to increase the representation of the minority class in the training set.
- *Random Undersampling* involves randomly removing instances from the majority class to achieve a balanced distribution.
- *SMOTE* [1] generates synthetic instances for the minority class by interpolating between existing instances. In particular, the algorithm selects a minority class instance and creates synthetic instances along the line segments connecting it to its nearest neighbors.
- *Adasyn* [3] is an extension of SMOTE that focuses on generating synthetic instances for the minority class based on their difficulty of classification. It assigns higher weights to those instances that are more challenging to classify correctly.

Specifically, we augmented the training set to create a balanced distribution and then evaluated the models on the original unbalanced testing set. This approach allowed us to train the models under balanced settings while assessing their performance on real-world imbalanced data.

5 Results

Table 2 shows the optimal configurations and corresponding performance metrics achieved for each disease, sorted by F1-score. Notably, the optimal configurations vary significantly across different diseases. Interestingly, the SVM model and the random undersampling strategy does never achieve the best results on average.

When considering the classification metrics, it becomes evident that accuracy alone is not a reliable performance indicator. For instance, while the accuracy for CAD exceeds 90%, the precision, recall, and F1-score values are considerably lower at 58.4%, 68.2%, and 57%, respectively. This behavior can be attributed to class imbalance and indicates a bias in the model towards predicting patients as healthy with respect to this disease. Moreover, it is worth noting that the performance for hypertension and CKD stands out significantly compared to other diseases. This outcome is likely influenced by the higher number of positive instances (patients diagnosed with these diseases) available in the dataset.

Table 2. Best training configuration and performance per disease. The *models* are as follows: "RF" Random Forest, "LR" Logistic Regression, "MLP" Multi-Later Perceptron, "GNB" Gaussian Naive-Bayes. The *imbalance* learning techniques are as follows: "ROS" Random OverSampling, "OSS" One Side Selection. The features are as follows: "Unstructured" are features extracted from textual anamnesis, "Structured" are tabular features extracted from patient's medical history, "Multi-modal (early)" results from the combination of "Unstructured" and "Structured" features, "Multi-modal (late)" results from late fusion of structured- and unstructured-based models.

Disease	Model	Imb.	Features	Accuracy	Precision	Recall	F1
Hypertension	RF	ROS	Multi-modal (early)	0.798	0.754	0.763	0.714
CKD	LR	SMOTE	Structured	0.707	0.723	0.714	0.678
Dyslipidemia	MLP	OSS	Multi-modal (early)	0.889	0.665	0.570	0.585
CAD	LR	ADASYN	Unstructured	0.901	0.584	0.682	0.570
Diabetes	GNB	ROS	Multi-modal (late)	0.689	0.583	0.597	0.576
Heart failure	LR	SMOTE	Unstructured	0.817	0.586	0.603	0.562
Atrial Fibrillation	RF	SMOTE	Multi-modal (early)	0.776	0.562	0.654	0.538
Stroke	GNB	OSS	Structured	0.901	0.534	0.561	0.532
IMA	MLP	OSS	Multi-modal (early)	0.956	0.563	0.521	0.527
Aortic Disease	RF	SMOTE	Multi-modal (late)	0.671	0.538	0.542	0.521

Shifting our focus on the comparison of the types of features we can use for training, Fig. 1 illustrates the F1-score performance, along with their corresponding 95% confidence intervals, that can be obtained with different configurations. For example, the use of *unstructured* features yields the best performance for CAD, but significantly underperforms for diabetes compared to other feature groups. Conversely, *structured* features exhibit the best performance for diabetes. Interestingly, we observe that the combination of *unstructured* and *structured* features, referred to as *multi-modal (early)* features, does not consistently improve classification accuracy except for a few cases. This result stems from variations in doctors' practices, where they may record relevant information either in the unstructured textual part or the structured tabular part of the patient's EHR. Conversely, *late fusion* consistently leads to high performance. Furthermore, we observe that incorporating domain medical knowledge consistently leads to better performance, on average, compared to directly utilizing the extracted features. This improvement can be attributed to the training process and the inherent understanding of the medical domain incorporated into the domain-enhanced features.

A focus on late fusion is presented in Fig. 2. Here, we show the best weights for structured and unstructured features, obtained during the fusion process. Results highlight that different tasks need different weights, due to the different ways physicians apply to store information related to different diseases.

We now present a comparison of classification models and imbalance learning techniques. For the sake of conciseness, we present the results for only five diseases. Notably, there is no single configuration that consistently achieves the

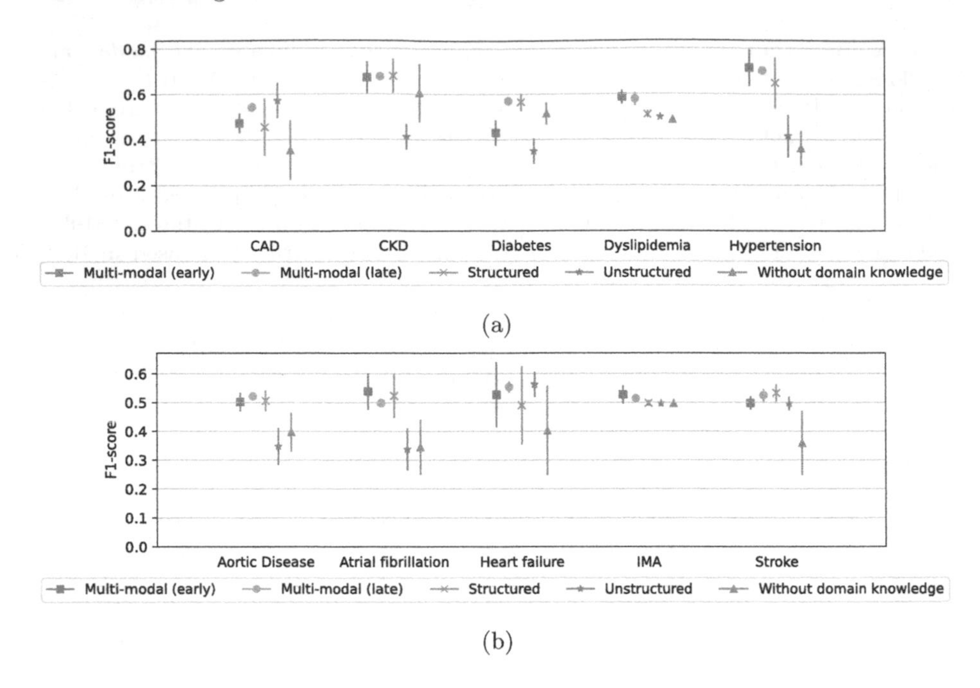

Fig. 1. Performance (and their 95% confidence interval) by varying the type of features used for training

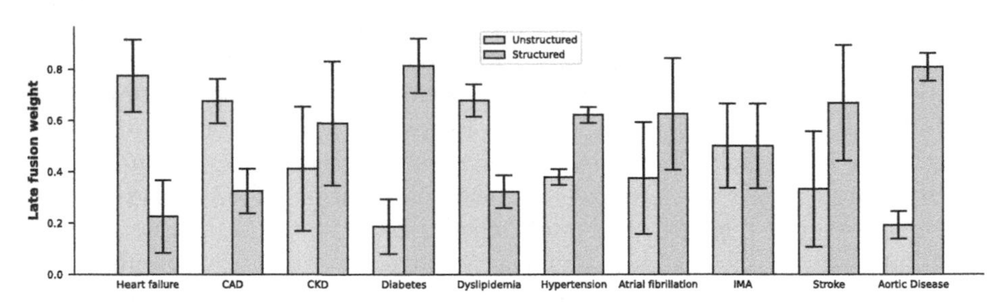

Fig. 2. Weights assigned during the late-fusion process, averaged over cross-validation splits.

best performance across all diseases. Figure 3 shows the effects the classification model in term of F1-performance. Interestingly, we observe that confidence intervals for each disease are completely overlapped, thus highlighting there is no statistical difference across different models. Finally, Fig. 4 shows the effects of the imbalance learning techniques. We observe no statistical difference in most of the cases except for Stroke where undersampling techinques (random undersampling and one-sided selection) generally yields better performance.

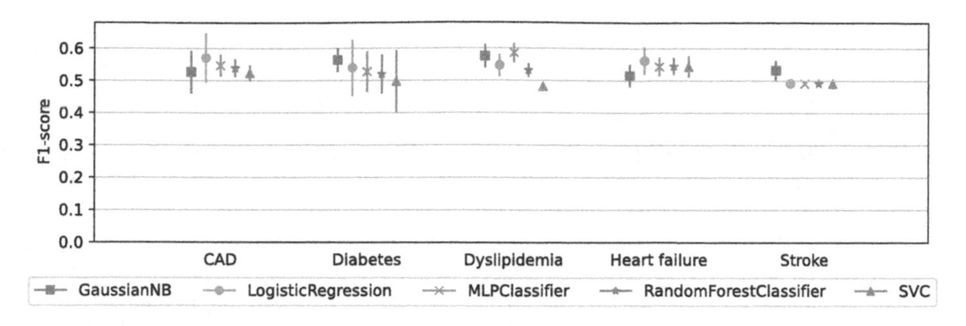

Fig. 3. Performance (and their 95% confidence interval) by varying the model used for training

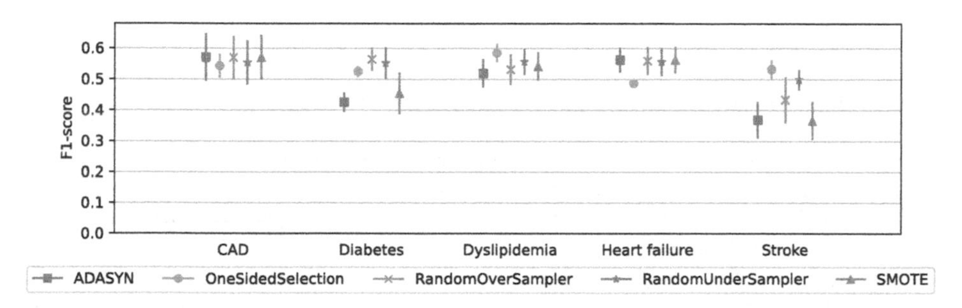

Fig. 4. Performance (and their 95% confidence interval) by varying the technique used for imbalanced learning

6 Conclusion and Future Work

Our study demonstrates the efficacy of multimodal machine learning techniques in the field of adverse event prediction in healthcare. Not only do we have evaluated the contribution of structured data and unstructured clinical notes, but we have also quantified the value of domain expertise in the methodological workflow. Further research can be carried out to refine feature selection methods, ensure data quality and integrity and enhance interpretability of the models. By addressing these challenges, we can improve the early identification and intervention for patients at risk of adverse events, thereby significantly improving healthcare outcomes.

References

1. Chawla, N.V., Bowyer, K.W., Hall, L.O., Kegelmeyer, W.P.: SMOTE: synthetic minority over-sampling technique. J. Artif. Int. Res. **16**(1), 321–357 (2002)
2. French, E., McInnes, B.T.: An overview of biomedical entity linking throughout the years. J. Biomed. Inf. **137**, 104252 (2023). https://doi.org/10.1016/j.jbi.2022.104252

3. He, H., Bai, Y., Garcia, E.A., Li, S.: ADASYN: adaptive synthetic sampling approach for imbalanced learning. In: 2008 IEEE International Joint Conference on Neural Networks (IEEE World Congress on Computational Intelligence), pp. 1322–1328 (2008). https://doi.org/10.1109/IJCNN.2008.4633969
4. Hernandez, L., et al.: Multimodal tensor-based method for integrative and continuous patient monitoring during postoperative cardiac care. Artif. Intell. Med. **113**, 102032 (2021). https://doi.org/10.1016/j.artmed.2021.102032
5. Huang, M., Lai, P., Lin, P., You, Y., Tsai, R.T., Hsu, W.: Biomedical named entity recognition and linking datasets: survey and our recent development. Briefings Bioinform. **21**(6), 2219–2238 (2020). https://doi.org/10.1093/bib/bbaa054
6. Krix, S., et al.: MultiGML: Multimodal graph machine learning for prediction of adverse drug events. bioRxiv (2022)
7. Li, Y., et al.: BEHRT: transformer for electronic health records. CoRR abs/1907.09538 (2019)
8. Mortazavi, B., et al.: Prediction of adverse events in patients undergoing major cardiovascular procedures. IEEE J. Biomed. Health Informatics **21**(6), 1719–1729 (2017). https://doi.org/10.1109/JBHI.2017.2675340
9. Shang, J., Ma, T., Xiao, C., Sun, J.: Pre-training of graph augmented transformers for medication recommendation. In: Kraus, S. (ed.) Proceedings of the Twenty-Eighth International Joint Conference on Artificial Intelligence, IJCAI 2019, Macao, China, 10–16 August 2019, pp. 5953–5959. ijcai.org (2019). https://doi.org/10.24963/ijcai.2019/825
10. han Sheu, Y., Magdamo, C.G., Miller, M., Das, S., Blacker, D., Smoller, J.W.: AI-assisted prediction of differential response to antidepressant classes using electronic health records. NPJ Digital Med. **6**, 73 (2023)
11. Tomavsev, N., et al.: Use of deep learning to develop continuous-risk models for adverse event prediction from electronic health records. Nat. Protoc. **16**, 2765–2787 (2021)
12. Yadav, V., Bethard, S.: A survey on recent advances in named entity recognition from deep learning models. In: Bender, E.M., Derczynski, L., Isabelle, P. (eds.) Proceedings of the 27th International Conference on Computational Linguistics, COLING 2018, Santa Fe, New Mexico, USA, 20–26 August 2018, pp. 2145–2158. Association for Computational Linguistics (2018)

A Multimodal Deep Learning Based Approach for Alzheimer's Disease Diagnosis

Adriano De Simone(✉) and Carlo Sansone

Department of Electrical Engineering and Information Technologies, University of Naples Federico II, Via Claudio 21, 80125 Naples, Italy
adriano.desimone2@unina.it

Abstract. Alzheimer's Disease is among the most common causes of death worldwide, and it is expected to have a greater impact in the years to come. Currently, there are no effective means to halt its progression, but researchers are actively exploring prevention, diagnosis, prognosis, and treatment options to find better solutions in each domain. Notably, extensive studies have shown that early detection plays a crucial role in developing more accurate prognoses and appropriate treatments. Presently, the primary diagnostic tests employed in this regard are images derived from Positron Emission Tomography (PET) and Magnetic Resonance Imaging (MRI). PETs are mainly used for obtaining functional information from the produced image, while MRIs reveal structural impairment. As artificial intelligence (AI) is increasingly used to support diagnostics for the development of ever more performing classifiers, in this paper we intend to show a study related to a Multimodal Deep Learning (MDL) approach that could guarantee better classification performance, integrating MRI structural information with PET functional information. The classifiers we are going to introduce is based on 3D Deep Convolutional Neural Networks (CNN). Here we will focus on Early Fusion (EF) and Late Fusion (LF) approaches on unbalanced and incomplete datasets exported from the Californian ADNI project.

Keywords: MDL · Early Fusion · Late Fusion · Alzheimer's Disease

1 Introduction

According to the World Health Organization (WHO), Alzheimer's Disease (AD) is one of the most frequent causes of death in the world and its impact is going to scale up even more in the next future [10]. More specifically, this neuro-pathology is actually recognized as the first typology of dementia diagnosed in the world and its appearance is going to grow even more in the next 25 years, passing from the actual number of 74 millions diagnoses to 131 millions in 2050 [10].

Alzheimer's can be described as a chronic neurodegenerative and progressive pathology that provides progressive decay of cognitive faculties, manifested in loss of memory, orientation, feeling perceptions, and speaking capabilities [1,8].

G. L. Foresti et al. (Eds.): ICIAP 2023 Workshops, LNCS 14366, pp. 131–139, 2024.
https://doi.org/10.1007/978-3-031-51026-7_12

Even though there are no ways of stopping this morbus, researchers are trying to deeply explore prevention, diagnosis, prognosis, and treatments to reach better solutions in every domain. In particular, significant studies have demonstrated that an early diagnosis can better help in developing prognosis and appreciable treatments. Even if there are different diagnostic systems to confirm the presence of AD, over the years, radiomics have acquired a more relevant role in medical science. By radiomics, we mean the analysis of medical images aimed at being obtained in a quicker way through appropriate mathematical methods and the use of computers; quantitative information that cannot always be detected through simple visual observation by the operator [9]. Radiomics is going to facilitate the domain expert's work thanks to the spread of Artificial Intelligence (AI) techniques. Indeed, researchers have been exploring AI-based approaches for image analysis, reaching interesting results, especially with the use of Deep Learning (DL) techniques based on Convolutional Neural Networks (CNN).

In AD the most used diagnostic tests could be recognized in Positron Emission Tomography (PET) and Magnetic Resonance Image (MRI) [3,4]. In particular, PETs are used to distinguish and monitor the growing impairments of functional areas in the brain while MRIs are mainly used for looking at the growth of structural atrophy provoked by Alzheimer's evolution [3]. AI approaches based on Multimodal Deep Learning allow the use of both these two information sources to empower the classification capability of the network [5,11,14]. Multimodal stays just to indicate that more than one single modality (or source) has been used. In literature, the most used Multimodal Deep Learning (MDL) [11] techniques are Early Fusion (EF), Intermediate Fusion (IF), and Late Fusion (LF), which differ for the timing in which modalities fusion is realized. EF provides to merge information coming from different input modalities from the early stages of the model. This is a robust solution for having a unique input for training the neural network and it could be advantageous when input modalities are closely related. IF implies that the fusion of different modalities occurs at an intermediate stage of the model architecture. This approach is more convenient for structuring a more plastic and adaptable model. Finally, LF requires independent input modalities to be processed separately. Thus, LF just provides a fusion at the decision level and it could be advantageous when input modalities provide complementary information that can be better exploited by specialized models or when processing each modality requires specific approaches.

The most widely used approach seen in the literature is the IF. Indeed, appreciable results have been reached by Heinerics et al. [6] in a classification based on the early stages of the morbus by Y. Asim et al. [2] in approaches based on segmentation. Moreover, an interesting work has been developed by D. Zhang et al. [15] about the analysis of the transition between different stages of dementia and AD.

However, the main limitation of IF is the need for a paired dataset, in which for each patient all the image modalities are present. As a consequence, in this paper, we aim to introduce a comparative study on the EF and LF approaches evaluated on an uncompleted and unbalanced dataset, analyzing the contribution given by the multimodal approach in comparison with unimodal ones.

All the approaches are developed on images coming from the Alzheimer's Disease Neuroimaging Initiative (ADNI) [7] public database project from the University of Southern California. ADNI collects AD images, thanks to the contribution of laboratories and experts from all over the world, and among all the different types of images provided, the study we propose in this paper is focused on MRI with magnetic fields at 1 T (MRI T1-w) and Fludeoxyglucose PET (PET-FDG), the most commonly used diagnostic tools actually used for diagnosis and evaluation of dementia [12,13].

The rest of the paper is organized as follows: Sect. 2 introduces the proposed materials and methods; Sect. 3 describes the experimental setup; Sect. 4 shows the obtained results; finally Sect. 5 provides some conclusions.

2 Materials and Methods

2.1 The Population

All processed images come from the Alzheimer's Disease Neuroimaging Initiative (ADNI) [7] of the University of Southern California (USC). The dataset is public and every laboratory or authorized researcher can contribute to its growth. This dataset collects different image typologies classified by the stadium of the morbus recognizable in the acquisition. Our attention has been focused on acquisition classified as "AD" (Alzheimer's Disease), "CN" (Cognitive Normal) and "MCI" (Mild Cognitive Impairments). AD and MCI instances have been put together for evaluating the classification capability on the binary task considering sick and healthy patients. From the whole dataset, we acquired 2542 MRI T1-w and 750 FDG PET. Table 1 reports the distribution of MRI e PET volumes for each class, highlighting the unbalanced nature of the dataset, while Table 2 shows the same information in terms of patients. The medium age of patients could be considered around 75 years old and, specifically, man ages move in the range [55–97] while women move in the range [51–96].

Table 1. MRI and PET volumes distribution for each class

Class	Women		Men		Total	
	MRI	PET	MRI	PET	MRI	PET
AD	141	18	158	13	299	31
CN	507	136	466	100	973	236
MCI	553	279	717	204	1270	483
Total	1201	433	1341	317	2542	750

2.2 Data Preprocessing

All data have been preprocessed to standardize the network's input. Both image modalities have been submitted to the same processes. Firstly, a padding operation has been performed for centering the brain in the 3D volume. All images have

Table 2. MRI and PET patients distribution for each class

Class	Women		Men		Total	
	MRI	PET	MRI	PET	MRI	PET
AD	82	12	94	15	176	27
CN	176	57	152	89	328	146
MCI	175	98	226	132	401	230
Total	433	167	472	236	905	403

then been submitted to a dimensioning fixed to the standard of $128 \times 128 \times 128$. Moreover, intensity normalization has been provided to scale image intensity in a range [0,1]. Finally, a skull stripping session has been provided to extract just the informative content, concentrated in the center of the image.

2.3 The Neural Network

In this paper, we propose a 3D CNN based on two main cores: the first one is structured with convolutional layers that provide the features extraction while the second core is composed by fully connected layers that provide the final decision (Fig. 1). To be more specific, we're going to indicate as Convolutional Core (CC) the first core composed of many Reduction blocks. Every single reduction block is structured with a 3D convolutional layer, followed by a batch normalization, followed by a ReLU function. The output of this Core corresponds to the input for the second one. The second core provides two main operations: a global average pooling followed by a fully connected layer. Its output is the predicted classification label.

2.4 The Proposed Multimodal Approach

With Multimodal Learning we refer to the process of learning and understanding information from multiple modalities or sources of input simultaneously [11]. In the context of artificial intelligence and machine learning, modalities can include different types of data, such as text, images, speech, video, and sensor data. The goal of multimodal learning is to leverage the complementary nature of different modalities to enhance the understanding and representation of information. By combining multiple modalities, the model can capture a richer and more comprehensive understanding of the data, as each modality provides unique and complementary information [11].

When Multimodal Learning recurs to the use of Deep Neural Networks, we talk about Multimodal Deep Learning (MDL) [11] approaches and, as already mentioned, MDL techniques for data fusion could be classified and differentiated by the timing in which fusion is concretely realized. As mentioned in the previous section we're now going to concentrate on EF and LF approaches.

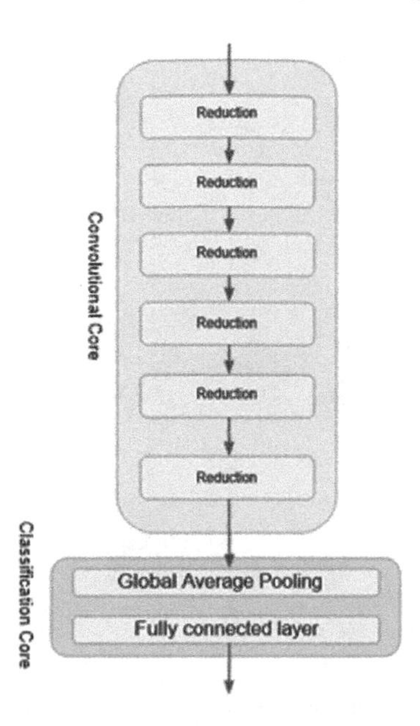

Fig. 1. Architecture Model of the proposed 3D Network

Early Fusion. Early Fusion, or Fusion at the Feature Level, provides to combine or concatenate information coming from different input modalities from the early stages of the model [11]. In this way, a unique multi-channel volume representation is assembled and can be used as input for training the neural network. Even though this kind of integration in a high-level representation could not be easy to recreate, the EF approach could be advantageous when input modalities are closely related and information from each modality is necessary from the beginning to achieve good predictive performance. Figure 2 shows the implemented EF approach.

Late Fusion. Late Fusion, also known as Decision Level Fusion, is the preferred approach when the input modalities are independent from each other, allowing them to be processed separately [11]. In this approach, separate models are trained for each input modality, and their predictions are then combined or fused to obtain a final decision, as reported in Fig. 3. LF can be beneficial when the input modalities offer distinct and complementary information that can be effectively utilized by specialized models. It is also advantageous when each modality necessitates specific techniques or approaches for processing. In our case decision in the LF approach will be operated with a weighted majority voting (WMV) where a weight is assigned in relation to the probability of the output.

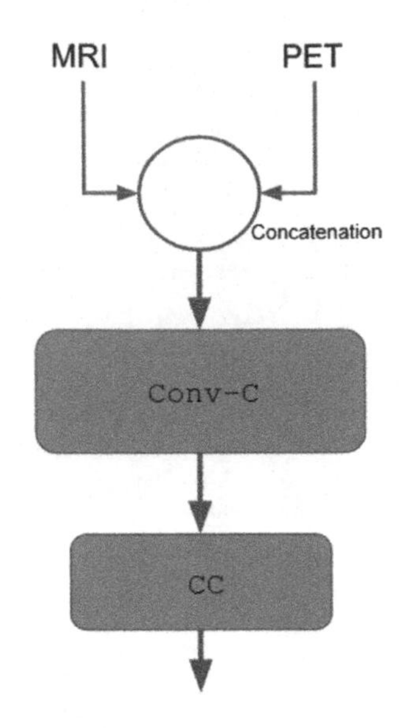

Fig. 2. 3D Network proposed in the EF approach

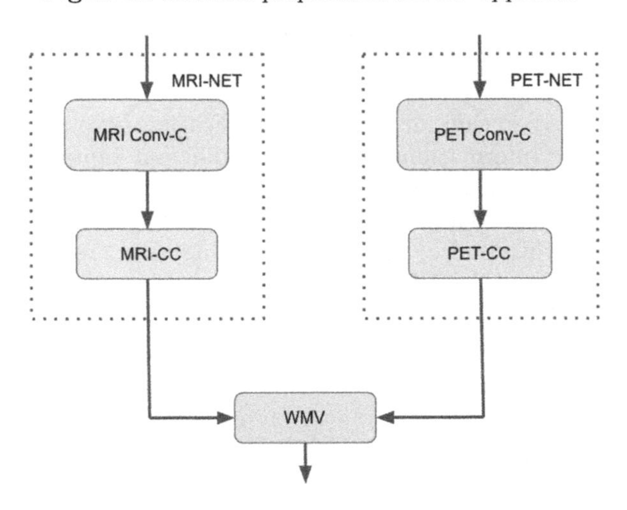

Fig. 3. 3D Network proposed in the LF approach

3 Experimental Set-Up

In this paper we focus on the dementia analysis in AD considering two different tasks:

- Binary classification: that considers healthy and sick patients. We can indicate a sick patient if it is labeled as AD or MCI, while we just consider a healthy patient labeled as CN. So the task could be summarized as NO-CN (AD + MCI) vs CN.

- Three classes classification: that considers CN, MCI and AD as three different classes.

We evaluated both Unimodal (U) and multimodal experiments with Early Fusion (EF) and Late Fusion (LF). In particular, in the U approach, we train the 3D CNN considering MRI and PET volumes separately, while in the LF we combine the prediction obtained with the unimodal method using a WMV strategy.

Data have been finally split in 70% for the training set, 15% for the validation set, and 15% for the test set. It is worth noting that we implement a patient-based hold-out method to avoid inserting in different sets MRI or PET belonging to the same subject.

Training has reached better performances around 100 Epochs with batch dimension fixed at 32; while 10^{-4} has been the best learning rate value. The Loss function used is the Cross-Entropy Loss.

Performance is evaluated in terms of Accuracy (ACC), Precision, Recall, and the Area Under the Curve (AUC). Due to the incomplete and unbalanced nature of the dataset involved in our study, we assess the results considering different sets of data. More in detail, while the groups denoted with MRI and PET contain the images acquired with a specific modality, the PAIRED set includes the MRI-PET pairs belonging to the same subject obtained within a time interval of one year. Moreover, the COMPLETE set is the result of the combination of the MRI, PET, and PAIRED groups. In this case, the network trained in the U approach is used to handle the incomplete acquisitions.

4 Results

Tables 3 and 4 summarizes the results of the implemented experiments, detailing the considered set of data, the used approach in columns *Set* and *App.* and the performance in terms of ACC, Precision, Recall, and AUC.

Table 3 reports the results of the experiments implemented for the binary classification task. As shown in the table, the Unimodal Approach (U) has already given a good result for PET processing, but comparing its percentage of accuracy with the U-PET one, based on the single subset of PET in the PAIRED set, we can notice how the multimodal approach has given a good contribution improving the results. Specifically, we have an improvement of around 2 precentual points, passing from 94,83% to 96,51%. This characteristic could be even more visible looking at the MRI accuracy that shows a stronger improvement, reaching 81,39%. On the same test set, EF and LF approaches have demonstrated an appreciable result reaching the levels of 97,04% and 92,44%, respectively. The behavior is more appreciable in the EF approach rather than in LF, since the fusion managed only at the decisional level involves a greater impact of the MRIs, which have shown poor performance results in the unimodal approach. Contrary to this, the EF approach provides for an initial fusion, and therefore, the entire training is more performant.

Table 4 reports the results of the experiments implemented for the three classes classification task. As shown in the table, even in this case we can notice

Table 3. Performance of the experiments considering the binary classification task

Set	App.	ACC	Precision		Recall		AUC
			CN	NO-CN	CN	NO-CN	
MRI	U	67.05%	52.90%	76.16%	64.06%	66.81%	74.05%
PET	U	94.83%	88.09%	98.65%	97.37%	93.59%	95.44%
PAIRED	U-MRI	81.39%	83.34%	81.32%	13.89%	**99.26%**	78.08%
	U-PET	96.51%	85.71%	**100.00%**	**100.00%**	95.59%	**99.84%**
	EF	**97.04%**	89.74%	99.23%	97.22%	96.99%	98.66%
	LF	92.44%	**90.95%**	95.59%	85.22%	97.74%	98.98%
COMPLETE	EF	73.60%	60.20%	84.50%	72.69%	74.82%	80.53%
	LF	73.06%	60.40%	82.40%	71.69%	73.82%	80.93%

Table 4. Performance of the experiments considering the three classes classification task

Set	App.	ACC	Precision			Recall			AUC
			CN	MCI	AD	CN	MCI	AD	
MRI	U	48.05%	52.63%	59.04%	18.86%	40.81%	57.51%	32.00%	65.26%
PET	U	91.52%	84.00%	98.61%	50.00%	92.10%	93.42%	50.00%	95.60%
PAIRED	U-MRI	80.25%	82.85%	85.87%	57.38%	78.91%	81.86%	77.78%	84.73%
	U-PET	91.73%	83.34%	98.62%	71.43%	92.10%	93.42%	71.43%	95.78%
	EF	**95.93%**	83.72%	**100.00%**	0.00%	**100.00%**	**96.99%**	0.00%	**98.69%**
	LF	90.25%	**85.88%**	92.33%	**76.51%**	89.20%	87.74%	**86.92%**	91.25%
COMPLETE	EF	82.90%	82.96%	89.45%	59.92%	81.62%	85.13%	75.51%	88.01%
	LF	83.10%	83.42%	89.45%	57.58%	81.62%	85.13%	77.55%	88.18%

an improvement given by the multimodal approach in comparison to unimodal values. Indeed, we can see a little improvement in the performance of PETs against a much greater improvement margin on MRIs. Evaluating the multimodal performance on EF approach we can notice appreciable values of general accuracy but in LF results can be pejorative for PETs. Moreover, we can notice that AD is very badly recognized. All these phenomena are due to the imbalance of the dataset in the subdivision of instances and patients by class. However, EF approach can still be considered the best one in comparison to the LF on the PAIRED set and we can recognize the appreciable contribution given by the multimodal approach.

5 Conclusion

In this paper, we aimed to investigate the role of MDL for dementia analysis in AD. In particular, we focused on the publicly available ADNI [7] dataset comparing the performance of the unimodal and multimodal approaches in early and late fusion settings. Indeed, the high unbalanced and incomplete nature of the dataset involved in our experiments made the implementation of fusion at an intermediate level very hard. The obtained results showed the promising performance of MDL highlighting the need to further explore the integration of heterogeneous and complementary diagnostic tools in medical imaging for diagnostic purposes.

However, future works will include the analysis of the intermediate fusion method considering different network architectures and efficient training strategies.

References

1. Malattie neurologiche: un confrontro tra l'Alzheimer e demenze vascolari. https://www.humanitas.it/news/malattie-neurologiche-un-confronto-tra-alzheimer-e-demenze-vascolari/
2. Asim, Y., Raza, B., Malik, A.K., Rathore, S., Hussain, L., Iftikhar, M.A.: A multimodal, multi-atlas-based approach for Alzheimer detection via machine learning. Int. J. Imaging Syst. Technol. **28**(2), 113–123 (2018)
3. Frisoni, G.B., Fox, N.C., Jack, C.R., Jr., Scheltens, P., Thompson, P.M.: The clinical use of structural MRI in Alzheimer disease. Nat. Rev. Neurol. **6**(2), 67–77 (2010)
4. Gambassi, G., et al.: Impiego delle tecniche di imaging nelle demenze. In: Impiego delle tecniche di imaging nelle demenze, pp. 1–63. Istituto Superiore di Sanità (2010)
5. Hermessi, H., Mourali, O., Zagrouba, E.: Multimodal medical image fusion review: theoretical background and recent advances. Sig. Process. **183**, 108036 (2021)
6. Hinrichs, C., Singh, V., Xu, G., Johnson, S.C., Initiative, A.D.N., et al.: Predictive markers for ad in a multi-modality framework: an analysis of mci progression in the ADNI population. Neuroimage **55**(2), 574–589 (2011)
7. Jack Jr, C.R., et al.: The Alzheimer's disease neuroimaging initiative (ADNI): MRI methods. J. Magn. Resonan. Imaging Off. J. Int. Soc. Magn. Resonan. Med. **27**(4), 685–691 (2008)
8. KNOW, W.D.W.: What is Alzheimer's disease? (1986)
9. Lambin, P., et al.: Radiomics: extracting more information from medical images using advanced feature analysis. Eur. J. Cancer **48**(4), 441–446 (2012)
10. Prince, P.M., Wimo, P.A., Guerchet, D.M., Ali, S.G.C., Wu, D.Y.T., Prina, D.M.: L'impatto globale della demenza. un'analisi di prevalenza, incidenza, costi e dati di tendenza. World Alzheimer Report 2015 (2015)
11. Ramachandram, D., Taylor, G.W.: Deep multimodal learning: a survey on recent advances and trends. IEEE Sig. Process. Mag. **34**(6), 96–108 (2017). https://doi.org/10.1109/MSP.2017.2738401
12. Scheltens, P., Fox, N., Barkhof, F., De Carli, C.: Structural magnetic resonance imaging in the practical assessment of dementia: beyond exclusion. Lancet Neurol. **1**(1), 13–21 (2002)
13. Vernooij, M., et al.: Dementia imaging in clinical practice: a European-wide survey of 193 centres and conclusions by the ESNR working group. Neuroradiology **61**, 633–642 (2019)
14. Xu, Y.: Deep learning in multimodal medical image analysis. In: Wang, H., Siuly, S., Zhou, R., Martin-Sanchez, F., Zhang, Y., Huang, Z. (eds.) HIS 2019. LNCS, vol. 11837, pp. 193–200. Springer, Cham (2019). https://doi.org/10.1007/978-3-030-32962-4_18
15. Zhang, D., Wang, Y., Zhou, L., Yuan, H., Shen, D., Initiative, A.D.N., et al.: Multimodal classification of Alzheimer's disease and mild cognitive impairment. Neuroimage **55**(3), 856–867 (2011)

A Systematic Review of Multimodal Deep Learning Approaches for COVID-19 Diagnosis

Salvatore Capuozzo(✉) and Carlo Sansone

Department of Electrical Engineering and Information Technologies, University of Naples Federico II, Via Claudio 21, 80125 Naples, Italy
salvatore.capuozzo@unina.it

Abstract. During and after the years of the COVID-19 pandemic, researchers and domain experts put all their effort into the discovery of accurate and reliable techniques for the detection and diagnosis of this disease in potentially sick patients. In the meanwhile, Deep Learning (DL) techniques are continuously improving and expanding, becoming more and more efficient and compatible in several fields of study and with different kinds of data. This huge but heterogeneous set of data cannot be fully exploited if DL models are not designed to be compatible with different sources of data at the same time, therefore multimodal approaches were designed and adopted, resulting in better prediction results than the classic approaches. Given these premises, several multimodal solutions for COVID-19 diagnosis were built in these years, but it may result hard to have a complete overview of the current state-of-the-art. For this reason, this paper wants to be a useful review of multimodal approaches and related adopted datasets, and therefore a starting point to quickly check what to improve to bring more accurate solutions.

Keywords: Biomedical Imaging · Convolutional Neural Network · COVID-19 · Deep Learning · Multimodal

1 Introduction

In December 2019, in the Chinese city of Wuhan an unknown virus started to spread. This virus led to the beginning of the COVID-19 pandemic, the fifth deadliest one in human history. In less than four years, this disease brought to about 7 million confirmed deaths and more than 750 million cases [30].

Once exposed to this virus, people can have one of the following symptoms: fever, cough, shortness of breath, fatigue, body aches, headache, loss of taste and smell, sore throat, vomiting. As these are the possible symptoms, it is unrecognizable from a flu. The only differences are regarding the severity of the symptoms and the times to show them and of contagiousness. This information cannot be adopted for diagnosis since it can be obtained after reaching an advanced stage

G. L. Foresti et al. (Eds.): ICIAP 2023 Workshops, LNCS 14366, pp. 140–151, 2024.
https://doi.org/10.1007/978-3-031-51026-7_13

of the disease, when having a diagnosis may be too late. That is the meaning why it is crucial to find other ways to diagnose this virus.

This unexpected and catastrophic event forced researchers from all around the world to find out any kind of solution which can slow down, prevent or block the consequences of this disease, even after the administration of vaccines. During these years, lots of datasets from different clinical entities have been generated, containing any kind of data, from Computed Tomography (CT) scans to sounds, from X-ray images to self-reported symptoms, from ultrasounds to genetic information. Given this huge amount of information, domain experts can analyze it to understand common and recurrent trends in sick patients data features and, therefore, identify potential biomarkers for COVID-19 detection, so that they can give quicker and more accurate diagnoses to new patients.

It is crucial to remark that, in the recent years, the fast evolution of Machine Learning (ML) and Deep Learning (DL) techniques enabled the design and the development of classification and detection solutions in any field of application, which are extremely useful when the object to identify is still quite unknown to the scientific community. The promising results obtained with these techniques are related to the available amount of data, despite not having a complete knowledge of the chosen research topic, since these techniques are data-driven and not model-driven, therefore features are extracted from data distributions.

Since the pandemic required to have quick responses despite the little amount of knowledge regarding it, DL, in particular with Convolutional Neural Networks (CNNs), was the perfect approach for the diagnosis task. Several solutions adopting CNNs for COVID-19 diagnosis already exist, as well as literature reviews collecting them for a faster overview of the current state-of-the-art [1,8,11].

However, despite having a huge amount of data, this collection is composed of different kinds of data (X-rays, CT scans, cough sound profiles, clinical notes), and so far most of researchers focused on adopting only one kind of data, limiting the power of a data-driven approach like DL. For this reason, in the recent years a new DL approach came out and revolutionized the way to adopt data: Multimodal Deep Learning (MDL) [50]. Thanks to this approach, neural network models can capture complex features from the fusion of information coming from different data sources and, at the same time, take advantage of a bigger amount of data. According to when information from modalities is merged, we can have three types of data fusion: Early Fusion (EF), Intermediate Fusion (IF) and Late Fusion (LF). Figure 1 will detail the three architecture schemas.

EF is when modalities are merged before putting them as input in the neural network, IF is when, inside the neural network itself, there are more subnets which somehow merge before the output, LF is when there are more neural networks running in parallel and their results are evaluated by a voting classifier.

Currently, there are only literature reviews on COVID-19 data but without prediction models [2,25,53], but there is a lack of reviews on MDL solutions. Therefore, the purpose of this review is to collect all the most relevant articles and studies regarding MDL approaches on COVID-19 data for diagnosis task.

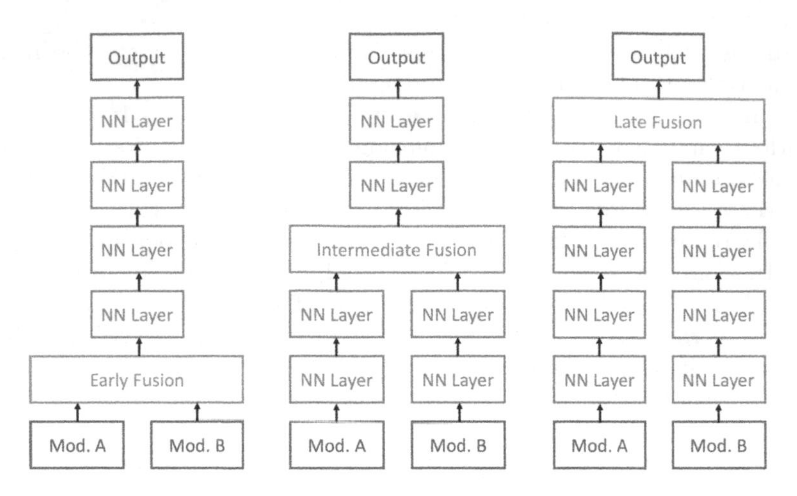

Fig. 1. Architecture schemas for EF, IF and LF

The paper is organised as follows: Sect. 2 briefly analyzes the existing literature reviews; Sect. 3 provides all the available data sources for the chosen task, the related articles adopting these sources, the search strategy adopted to find them and an overview of the proposed solutions; Sect. 4 reports and discusses the results obtained from the collected works, making a comparison among them; finally, Sect. 5 provides some conclusions from the works analysis.

2 Existing Literature Reviews

In order to have a quick summary of available solutions at the moment, literature reviews are crucial. Therefore, even for COVID-19 diagnosis solutions, there are some reviews which collect, analyze and compare them. The approach to identify literature reviews about the AI approaches for COVID-19 diagnosis, regardless the multimodal approach, is the same adopted for identifying studies related to the MDL approach for COVID-19 diagnosis, detailed in the next section.

In the literature, a total of five literature reviews can be found [1,8,11,29, 45]. As it is noticeable in these reviews, apart from the first one [29], all the other literature reviews have not the specific focus on the multimodal approach. Indeed, even though in the recent years the importance of multimodal data has been remarked in any medicine field [6], this kind of approach is not widely adopted yet, primarily for dataset limited dimensions.

Focusing on the only existing work making a literature review on multimodal models for COVID-19 tasks, this one collects several solutions, mathematical models, probabilistic models, machine learning models and so on, but evaluation info of the collected solutions are missing, as well as a reference and a focus on data adopted by these solutions, making the existing work an incomplete review.

3 Materials and Methods

Before listing the most relevant MDL studies, it is useful to understand the available multimodal data and the search strategy adopted for studies collection. In this way, this review will be compliant to the guidelines of the Preferred Reporting Items for Systematic reviews and Meta-Analyses (PRISMA) statement [38], which enable to make a useful systematic review with the purpose of helping researchers to bring new solutions which outperform the existing ones.

3.1 Data Sources

COVID-19 available data, both from public and private sources, can be retrieved both with Google and Google Scholar results with combinations of keywords like "COVID-19 multimodal dataset", "Coronavirus biomedical images", "COVID-19 CT and X-rays", and "Coronavirus clinical data", and by searching sources reported in the related articles listed in the next section.

These datasets can have one or more of different kinds of modalities, starting from images, where we can have CT scans, chest X-ray images and lung ultrasounds, moving to audio, from cough profiles to speech and breathing samples, and ending to tabular data like clinical notes and reported symptoms. Most of the public datasets contains only one modality, while there are only few ones which contain at least two modalities, and, as reported later, multimodal studies adopted all of the latter ones or a combination of the first ones.

In Table 1 there is a list of the available public datasets, all of them adopted by the studies regarding multimodal imaging for COVID-19. These datasets

Table 1. Available public datasets regarding COVID-19.

Dataset Name	Modalities	Data Count
AIforCOVID Imaging Archive [49]	Clinical Data, Chest X-Ray	820
Coswara Project [46]	Cough Profile, Speech, Breathing, Self-Reported Info	2746
UCSD-AI4H/COVID-CT [55]	CT Scans	2849
RunwenHu/COVID-19 [43]	CT Scans	802
SARS-CoV-2 CT-scan Dataset [48]	CT Scans	2481
Curated Covid CT [28]	CT Scans	14486
COVID-19 Radiography Database [14]	Chest X-Ray	3886
Covid19 Pneumonia Normal Chest XrayPA Dataset [5]	Chest X-Ray	6939
COUGHVID Dataset [36]	Cough Profile	8868
COVID-19 Image Data Collection [13]	Chest X-Ray	2905
POCOVID-Net Dataset [9]	Lung Ultrasounds	545
An Aggregated Dataset of Clinical Outcomes for COVID-19 Patients [7]	Clinical Data	
NIH Chest X-Ray [17]	Chest X-Ray	11680
Virufy COVID-19 Open Cough Dataset [23]	Cough Profile	
CoronaHack-Chest X-Ray-Dataset [39]	Chest X-Ray	5910
N-CLAHE Lung Medical Images [21]	Chest X-ray, CT scans, Lung Ultrasounds	2940
COVIDGR Dataset [51]	Chest X-Ray	852

have modalities which can have a different type of format (images, audio, tabular data), but only with images format it is possible to have large (>10000) datasets.

3.2 Search Strategy and Related Articles

Once found out the available datasets, the next step was to identify the existing works related to the COVID-19 diagnosis using a multimodal approach.

Several literature reviews related to COVID-19 diagnosis using a unimodal approach are already available, and since the purpose of this review is to identify only the more rare multimodal approaches, the classic ones will not be listed. As for the datasets research task, the approach adopted for studies discovery is a keyword search strategy through Google Scholar, where sentences with combinations of keywords like "multimodal", "COVID-19", "diagnosis", "DL", "multi-imaging", "model", "AI", "coronavirus" and "detector" have been searched.

All the results have been filtered by inherence to both the biomedical purpose and the multimodality and reported in Table 2, where the order of the papers is correlated to the appearance in the results (first ones are the top ones by searching the previous sentences through Google Scholar). This table reports the study names, which types of modalities have been adopted, and consequently the datasets adopted, and the data fusion approach. In particular, the last one defines if the modalities are merged in the preprocessing step (EF), in the neural network architecture (IF) or in the end as an ensemble/voting system of the results (LF). These fusion subgroups are already defined in the research community [10]. In this case, the task performed is omitted since the task (detection and diagnosis) is fixed. The reason of this choice is due to the lack of other cases (prognosis, feature explainability, etc.) during the studies research.

4 Results and Discussion

Given all the articles reported in this literature review, it is possible to compare them by their fusion strategy, network architecture and evaluation metrics. In Table 3, studies are reported again along with input types, network architectures and results on test sets coming from their respective studies, with the available evaluation metrics.

There can be made several observations from the studies collected. Firstly, it is noticeable that most solutions adopt multiple kinds of images for multimodality, but there are even cases where images are combined with tabular data.

Another curious aspect is that half of the studies adopted the IF and the other half adopted the LF, therefore there is no study involving the EF approach. The reason could be that, given the fact that most studies adopted multiple unimodal datasets in order to retrieve a multimodal one, these ones are hard to match during a preprocessing step, since they come from different sources.

By observing the submission year of these works, it is possible to understand that, even though datasets were available since 2020, most of them have been

Table 2. Studies on multimodal DL approaches for COVID-19 diagnosis.

Study	Modalities	Datasets	Fusion
M Hammad et al., 2022 [19]	Tabular, Images	Private	LF
SA Almutairi et al., 2023 [4]	Tabular, Audio, Images	[14], Private	IF
V Guarrasi et al., 2022 [18]	Tabular, Images	[49]	IF
K Nguyen-Trong et al., 2023 [34]	Tabular, Audio	[46]	IF
GA Fahmy et al., 2022 [16]	Images	[43,55], [28,48], [14], Private	LF
S Almuayqil et al., 2023 [3]	Images	[5,48] Private	LF
S Tang et al., 2022 [52]	Tabular, Audio	[36,46]	IF
MDA Rahman et al., 2021 [41]	Tabular, Audio, Images	Several	LF
SE Mukhi et al., 2023 [32]	Images	[14]	IF
A Qayyum et al., 2022 [40]	Images	[9,13]	IF
M Effati et al., 2021 [15]	Tabular	[7]	LF
T Padmapriya et al., 2022 [37]	Images	Not reported	IF
V Mayya et al., 2021 [31]	Tabular, Images	Not reported	IF
Y Li et al., 2022 [26]	Tabular, Images	Private	IF
MJ Horry et al., 2020 [21]	Images	[13,17], [9,55], [21]	LF
MY Sikkandar et al., 2023 [47]	Images	[13]	LF
VP Jayachitra et al., 2021 [22]	Audio, Images	[23,46], [13,48]	LF
U Sait et al., 2021 [44]	Audio, Images	[14], Private	LF
N Hilmizen et al., 2020 [20]	Images	[39,48]	IF
N Nasir et al., 2023 [33]	Tabular, Images	[13]	LF
S Kumar et al., 2022 [24]	Audio, Images	[51], Private	LF
S R Chetupalli et al., 2021 [12]	Tabular, Audio	[46]	LF
T Xia et al., 2021 [54]	Audio	Private	LF
H Rashid et al., 2022 [42]	Tabular, Audio	[36]	IF

published starting from 2022, even though other articles, using the same datasets but with the classical approach, have been published earlier (2020 and 2021).

The most common modality type adopted in analyzed works is the image (in 18 works out of 24), followed by tabular data (12/24) and audio (10/24). Regarding works using images, almost any of them (16/18) use CNNs, with the exception of [18,26], using respectively autoencoders and attention modules. The others adopt either fine tuned CNNs, in particular 8 works [16,21,22,24,31,32,47] or ad hoc CNNs, in particular 10 works [4,19,20,22,33,37,40,41,44]. Two of the works use both. Among the works adopting tabular data, the most common architecture is the MLP, in particular in 7 works [4,18,34,41,42,52], but there are also works adopting Long Short-Term Memories (LSTMs) [18,19] and decision trees [12,15,22]. Most of solutions adopting audio profiles have MLPs/DNNs for Mel-Frequency Cepstral Coefficients (MFCCs) features extraction, in particular 4 works [24,34,42,52], 2 of which use even CNNs for spectrogram analysis [24,52].

Every work adopts its own set of metrics for evaluation. In particular, the adopted metrics are accuracy, precision, recall, sensitivity and specificity. The accuracy is the ratio between true predictions (positive and negative) and all the predictions. Precision and recall are a couple of metrics typically adopted

Table 3. Architectures and results obtained with the reported MDL approaches.

Study	Modalities	Network Architecture	Evaluation on Test Set
M Hammad et al., 2022 [19]	Tabular, Images	Ad hoc CNN or ConvLSTM	3 classes CNN: Acc: 98.2% - Spec: 99.2% - Sens: 97.2% 3 classes ConvLSTM: Acc: 96.8% - Spec: 99.6% - Sens: 94.0%
SA Almutairi et al., 2023 [4]	Tabular, Audio, Images	Ad hoc DNN	Audio layer: Acc: 96.55% - Spec: 99.72% - Sens: 93.39% Images layer: Acc: 96.67% - Spec: 99.70% - Sens: 93.63%
V Guarrasi et al., 2022 [18]	Tabular, Images	Autoencoder, Convolutional AE, MLP	Tabular set: Acc: 75.78% - Spec: 74.74% - Sens: 76.63% Images set: Acc: 74.14% - Spec: 73.96% - Sens: 74.57%
K Nguyen-Trong et al., 2023 [34]	Tabular, Audio	CNN BiLSTM, MFCC features extraction, EfficientNet, MLP	Acc: 96.88% - Prec: 96.25% - Rec: 97.47%
GA Fahmy et al., 2022 [16]	Images	Ad hoc CNN, Fine tuned CNN	CT scan: Acc: 98.5% X-ray: Acc: 98.6%
S Almuayqil et al., 2023 [3]	Images	Fine tuned CNN	MobileNetV2: Acc: 99.19% - Spec: 99.18% - Sens: 99.18%
S Tang et al., 2022 [52]	Tabular, Audio	MFCC features extraction, MLP, Transformer	Acc: 81.32%
MDA Rahman et al., 2021 [41]	Tabular, Audio, Images	Ad hoc CNN, MLP	
SE Mukhi et al., 2023 [32]	Images	Fine tuned CNN	VGG-19 CT scan: Acc: 93.00% VGG-19 X-ray: Acc: 94.17%
A Qayyum et al., 2022 [40]	Images	Ad hoc CNN	X-ray: Prec: 85% - Rec: 89% Ultrasounds: Prec: 93% - Rec: 94%
M Effati et al., 2021 [15]	Tabular	Decision Tree	
T Padmapriya et al., 2022 [37]	Images	Ad hoc CNN	Acc: 87.5% - Spec: 94.0% - Sens: 72.0%
V Mayya et al., 2021 [31]	Tabular, Images	Word2Vec, Fine tuned CNN	Acc: 99%
Y Li et al., 2022 [26]	Tabular, Images	Reciprocal Attention Module, Ad hoc DNN	Acc: 92.75% - Prec: 82.92% - Rec: 80.95%
MJ Horry et al., 2020 [21]	Images	Fine tuned CNN	VGG-19 X-ray: F1-score: 87% VGG-19 CT scan: F1-score: 78% VGG-19 Ultrasounds: F1-score: 99%
MY Sikkandar et al., 2023 [47]	Images	Fine tuned CNN + MLP	Acc: 99.21% - Spec: 99.21% - Sens: 99.21%
VP Jayachitra et al., 2021 [22]	Audio, Images	Random Forest, Fine tuned CNN, Ad hoc CNN	Cough: Acc: 97.12% - Prec: 96.96% - Rec: 94.11% Speech: Acc: 96.08% - Prec: 100.00% - Rec: 82.87% Breathing: Acc: 97.45% - Prec: 100.00% - Rec: 93.53% X-ray: Acc: 98.40% - Prec: 100.00% - Rec: 89.00% CT scan: Acc: 99.19% - Prec: 100.00% - Rec: 98.32%
U Sait et al., 2021 [44]	Audio, Images	Ad hoc CNN, MLP	Acc: 91.1% - Prec: 91.4% - Rec: 91.1%
N Hilmizen et al., 2020 [20]	Images	Ad hoc CNN	Acc: 99.87% - Spec: 100.00% - Sens: 99.74%
N Nasir et al., 2023 [33]	Tabular, Images	Ad hoc CNN	Acc: 96.30% - Spec: 88.89% - Sens: 100.00%
S Kumar et al., 2022 [24]	Audio, Images	Fine tuned CNN, MFCC features extraction	Acc: 98.90% - Prec: 96.70% - Rec: 98.90%
S R Chetupalli et al., 2021 [12]	Tabular, Audio	SVM Classifier, Decision Tree	Spec: 95.00% - Sens: 69.00%
T Xia et al., 2021 [54]	Audio	SVM Classifier, Fine tuned CNN	OpenSMILE+SVM: Spec: 67.00% - Sens: 54.00% Pre-trained VGGish: Spec: 63.00% - Sens: 50.00% Fine-tuned VGGish: Spec: 69.00% - Sens: 65.00%
H Rashid et al., 2022 [42]	Tabular, Audio	Ad hoc DNN, MFCC features extraction	Prec: 86.30% - Spec: 85.80% - Sens: 85.60%

together which, respectively, are the ratio between true positive predictions and total positive predictions and true positive predictions and total positive labels. Sensitivity and specificity are another couple of metrics adopted together, where the first is the same of the recall, while the second is the ratio between true negative predictions and total negative labels.

As noticeable, most of the reported results have accuracies higher than the 90%, even though some of the studies, in particular the ones with less evaluation metrics reported, are not totally reliable, since they are biased by the distributions of the adopted datasets, and there are no information which guarantees the model robustness. Some studies rely on multimodality over different kinds of multimedia, and they show better results since solutions which adopt different kinds of media at the same time (images, sounds, notes), can bring more useful information for the inference. Moreover, the customization of architectures in order to be compatible with multiple data at the same time (IF) has been proven to be a more efficient solution for medical diagnosis [27,35].

5 Conclusions

By reading all the articles it can be noticed that, even though several models have been trained with the same dataset, none of them is an evolution of already existing models, since probably no author contacted others in order not to start from the ground up. A good reason would be that, actually, no systematic literature reviews on MDL for COVID-19 diagnosis were made before.

Unfortunately, almost none of the authors have made any Risk of Bias assessment, however, for solutions which adopt more than one data source per modality (for instance the one proposed by Fahmy et al. [16]), there should be less likelihood of incurring bias risks.

Considering all the information and results reported in this review, it is possible to say that there are already several methods involving the multimodal approach, most of which show extremely good results and therefore could be adopted for accurate COVID-19 diagnoses. However, there are some limitations which constraint the adoption of these models. All the models, even if they adopt the innovative approach of multimodality, do not use state-of-the-art or cutting-edge architectures (for instance transformers, attention models or LLMs), therefore they may be outdated quickly in the next months from this review drafting. Then, Datasets are not collected in a unique collection, which may help researchers to quickly generate new solutions. Finally, few studies use a designed to be multimodal dataset, indeed in most cases multimodality is brought by the adoption of multiple sources, which typically extracts data in different ways, and this may lead to the detection of misleading features during the training phase.

The studies reported are the one available publicly, therefore this review wants to give the opportunity to both improve the performances of solutions with similar architectures and test new approaches which reflect the current state-of-the-art adopted for similar fields.

In this way, new and different solutions can be developed in order to fight the spread of this virus, even though the apex of its mortality is already passed.

Indeed, a multimodal solution can be an extremely powerful tool of doctors which will help them to understand as soon as possible if a patient has been infected, thanks to clinical images, symptoms, reports, and even sounds.

Acknowledgements. This work is part of the POR FESR CAMPANIA 2014-2020 Synergy for COVID project (CUP H69I22000710002).

References

1. Adamidi, E.S., Mitsis, K., Nikita, K.S.: Artificial intelligence in clinical care amidst COVID-19 pandemic: a systematic review. Comput. Struct. Biotechnol. J. **19**, 2833–2850 (2021)
2. Agricola, E., et al.: Heart and lung multimodality imaging in COVID-19. JACC Cardiovasc. Imaging **13**(8), 1792–1808 (2020)
3. Almuayqil, S., Abd El-Ghany, S., Shehab, A.: Multimodality imaging of COVID-19 using fine-tuned deep learning models. Diagnostics **13**(7), 1268 (2023)
4. Almutairi, S.A.: A multimodal AI-based non-invasive COVID-19 grading framework powered by deep learning, manta ray, and fuzzy inference system from multimedia vital signs. Heliyon **9**(6) (2023)
5. Asraf, A.: Covid19-pneumonia-normal-chest-xray-pa-dataset (kaggle) (2020)
6. Behrad, F., Saniee Abadeh, M.: An overview of deep learning methods for multimodal medical data mining. Expert Syst. Appl. **200**, 117006 (2022). https://doi.org/10.1016/j.eswa.2022.117006, https://www.sciencedirect.com/science/article/pii/S0957417422004249
7. Bertsimas, D., et al.: An aggregated dataset of clinical outcomes for COVID-19 patients (2020). https://www.covidanalytics.io/datasetdocumentation
8. Born, J., et al.: On the role of artificial intelligence in medical imaging of COVID-19. Patterns **2**(6) (2021)
9. Born, J., et al.: Pocovid-net: automatic detection of COVID-19 from a new lung ultrasound imaging dataset (pocus). arXiv preprint arXiv:2004.12084 (2020)
10. Boulahia, S.Y., Amamra, A., Madi, M.R., Daikh, S.: Early, intermediate and late fusion strategies for robust deep learning-based multimodal action recognition. Mach. Vis. Appl. **32**(6), 121 (2021)
11. Cenggoro, T.W., Pardamean, B., et al.: A systematic literature review of machine learning application in COVID-19 medical image classification. Procedia Comput. Sci. **216**, 749–756 (2023)
12. Chetupalli, S.R., et al.: Multi-modal point-of-care diagnostics for COVID-19 based on acoustics and symptoms. IEEE J. Transl. Eng. Health Med. **11**, 199–210 (2023)
13. Cohen, J.P., Morrison, P., Dao, L.: COVID-19 image data collection. arXiv preprint arXiv:2003.11597 (2020)
14. COVID, K.: Radiography database. Radiol. Soc. North Am. (2019). https://www.kaggle.com/tawsifurrahman/covid19-radiography-database. Accessed 1 Oct 2021
15. Effati, M., Sun, Y.C., Naguib, H.E., Nejat, G.: Multimodal detection of COVID-19 symptoms using deep learning & probability-based weighting of modes. In: 2021 17th International Conference on Wireless and Mobile Computing, Networking and Communications (WiMob), pp. 151–156. IEEE (2021)
16. Fahmy, G.A., Abd-Elrahman, E., Zorkany, M.: COVID-19 detection using multimodal and multi-model ensemble based deep learning technique. In: 2022 39th National Radio Science Conference (NRSC), vol. 1, pp. 241–253. IEEE (2022)

17. Filice, R.W., et al.: Crowdsourcing pneumothorax annotations using machine learning annotations on the NIH chest x-ray dataset. J. Digit. Imaging **33**, 490–496 (2020)
18. Guarrasi, V., et al.: Multimodal explainability via latent shift applied to COVID-19 stratification. arXiv preprint arXiv:2212.14084 (2022)
19. Hammad, M., et al.: Efficient multimodal deep-learning-based COVID-19 diagnostic system for noisy and corrupted images. J. King Saud Univ.-Sci. **34**(3), 101898 (2022)
20. Hilmizen, N., Bustamam, A., Sarwinda, D.: The multimodal deep learning for diagnosing COVID-19 pneumonia from chest CT-scan and x-ray images. In: 2020 3rd International Seminar on Research of Information Technology and Intelligent Systems (ISRITI), pp. 26–31. IEEE (2020)
21. Horry, M.J., et al.: COVID-19 detection through transfer learning using multimodal imaging data. IEEE Access **8**, 149808–149824 (2020)
22. Jayachitra, V., Nivetha, S., Nivetha, R., Harini, R.: A cognitive IoT-based framework for effective diagnosis of COVID-19 using multimodal data. Biomed. Sig. Process. Control **70**, 102960 (2021)
23. Khanzada, A., Wilson, T.: Virufy COVID-19 open cough dataset, github (2020) (2021)
24. Kumar, S., et al.: A novel multimodal fusion framework for early diagnosis and accurate classification of COVID-19 patients using x-ray images and speech signal processing techniques. Comput. Methods Programs Biomed. **226**, 107109 (2022)
25. Larici, A.R., et al.: Multimodality imaging of COVID-19 pneumonia: from diagnosis to follow-up. A comprehensive review. Eur. J. Radiol. **131**, 109217 (2020)
26. Li, Y., et al.: Automated multi-view multi-modal assessment of COVID-19 patients using reciprocal attention and biomedical transform. Front. Publ. Health **10**, 886958 (2022)
27. Lu, Y., Niu, K., Peng, X., Zeng, J., Pei, S.: Multi-modal intermediate fusion model for diagnosis prediction. In: 2022 the 6th International Conference on Innovation in Artificial Intelligence (ICIAI). ICIAI 2022, pp. 38–43. Association for Computing Machinery, New York, NY, USA (2022). https://doi.org/10.1145/3529466.3529496
28. Maftouni, M., Law, A.C.C., Shen, B., Grado, Z.J.K., Zhou, Y., Yazdi, N.A.: A robust ensemble-deep learning model for COVID-19 diagnosis based on an integrated CT scan images database. In: IIE Annual Conference. Proceedings, pp. 632–637. Institute of Industrial and Systems Engineers (IISE) (2021)
29. Mahalle, P.N., Sable, N.P., Mahalle, N.P., Shinde, G.R.: Data analytics: COVID-19 prediction using multimodal data. In: Intelligent Systems and Methods to Combat Covid-19, pp. 1–10 (2020)
30. Mathieu, E., et al.: Coronavirus pandemic (COVID-19). Our World in Data (2020). https://ourworldindata.org/coronavirus
31. Mayya, V., Karthik, K., Sowmya, K.S., Karadka, K., Jeganathan, J.: COVIDdx: AI-based clinical decision support system for learning COVID-19 disease representations from multimodal patient data. In: HEALTHINF, pp. 659–666 (2021)
32. Mukhi, S.E., Varshini, R.T., Sherley, S.E.F.: Diagnosis of COVID-19 from multimodal imaging data using optimized deep learning techniques. SN Comput. Sci. **4**(3), 212 (2023)
33. Nasir, N., et al.: Multi-modal image classification of COVID-19 cases using computed tomography and x-rays scans. Intell. Syst. Appl. **17**, 200160 (2023)
34. Nguyen-Trong, K., Nguyen-Hoang, K.: Multi-modal approach for COVID-19 detection using coughs and self-reported symptoms. J. Intell. Fuzzy Syst. (Preprint), 1–13 (2023)

35. Niu, K., Zhang, K., Peng, X., Pan, Y., Xiao, N.: Deep multi-modal intermediate fusion of clinical record and time series data in mortality prediction. Front. Mol. Biosci. **10**, 1136071 (2023)
36. Orlandic, L., Teijeiro, T., Atienza, D.: The coughvid crowdsourcing dataset, a corpus for the study of large-scale cough analysis algorithms. Sci. Data **8**(1), 156 (2021)
37. Padmapriya, T., Kalaiselvi, T., Priyadharshini, V.: Multimodal COVID network: multimodal bespoke convolutional neural network architectures for COVID-19 detection from chest x-ray's and computerized tomography scans. Int. J. Imaging Syst. Technol. **32**(3), 704–716 (2022)
38. Page, M.J., et al.: The Prisma 2020 statement: an updated guideline for reporting systematic reviews. Int. J. Surg. **88**, 105906 (2021)
39. Praveen: Coronahack -chest x-ray-dataset (kaggle) (2020)
40. Qayyum, A., Ahmad, K., Ahsan, M.A., Al-Fuqaha, A., Qadir, J.: Collaborative federated learning for healthcare: multi-modal COVID-19 diagnosis at the edge. IEEE Open J. Comput. Soc. **3**, 172–184 (2022)
41. Rahman, M.A., Hossain, M.S., Alrajeh, N.A., Gupta, B.: A multimodal, multimedia point-of-care deep learning framework for COVID-19 diagnosis. ACM Trans. Multimidia Comput. Commun. Appl. **17**(1s), 1–24 (2021)
42. Rashid, H.A., Sajadi, M.M., Mohsenin, T.: Coughnet-v2: a scalable multimodal DNN framework for point-of-care edge devices to detect symptomatic covid-19 cough. In: 2022 IEEE Healthcare Innovations and Point of Care Technologies (HI-POCT), pp. 37–40. IEEE (2022)
43. Hu, R., Ruan, G., Xiang, S., Huang, M., Liang, Q., Li, J.: Automated diagnosis of COVID-19 using deep learning and data augmentation on chest CT. medRxiv (2020)
44. Sait, U., et al.: A deep-learning based multimodal system for COVID-19 diagnosis using breathing sounds and chest x-ray images. Appl. Soft Comput. **109**, 107522 (2021)
45. Sekaran, K., Gnanasambandan, R., Thirunavukarasu, R., Iyyadurai, R., Karthick, G., Doss, C.G.P.: A systematic review of artificial intelligence-based COVID-19 modeling on multimodal genetic information. Progr. Biophys. Mol. Biol. (2023)
46. Sharma, N., et al.: Coswara - a database of breathing, cough, and voice sounds for COVID-19 diagnosis. arXiv preprint arXiv:2005.10548 (2020)
47. Sikkandar, M.Y., et al.: Leveraging multimodal ensemble fusion-based deep learning for COVID-19 on chest radiographs. Comput. Syst. Sci. Eng. **47**(1) (2023)
48. Soares, E., Angelov, P., Biaso, S., Froes, M.H., Abe, D.K.: Sars-cov-2 CT-scan dataset: a large dataset of real patients CT scans for sars-cov-2 identification. medRxiv, pp. 2020–04 (2020)
49. Soda, P., et al.: AiforCOVID: predicting the clinical outcomes in patients with COVID-19 applying AI to chest-x-rays. An Italian multicentre study. Med. Image Anal. **74**, 102216 (2021)
50. Stahlschmidt, S.R., Ulfenborg, B., Synnergren, J.: Multimodal deep learning for biomedical data fusion: a review. Brief. Bioinform. **23**(2), bbab569 (2022)
51. Tabik, S., et al.: CovidGR dataset and COVID-SDNET methodology for predicting COVID-19 based on chest x-ray images. IEEE J. Biomed. Health Inform. **24**(12), 3595–3605 (2020). https://doi.org/10.1109/JBHI.2020.3037127
52. Tang, S., Hu, X., Atlas, L., Khanzada, A., Pilanci, M.: Hierarchical multi-modal transformer for automatic detection of COVID-19. In: Proceedings of the 2022 5th International Conference on Signal Processing and Machine Learning, pp. 197–202 (2022)

53. Varadarajan, V., Shabani, M., Ambale Venkatesh, B., Lima, J.A.: Role of imaging in diagnosis and management of COVID-19: a multiorgan multimodality imaging review. Front. Med. **8**, 765975 (2021)
54. Xia, T., et al.: COVID-19 sounds: a large-scale audio dataset for digital respiratory screening. In: Thirty-Fifth Conference on Neural Information Processing Systems Datasets and Benchmarks Track (round 2) (2021)
55. Zhao, J., Zhang, Y., He, X., Xie, P.: COVID-CT-dataset: a CT scan dataset about COVID-19. arXiv preprint arXiv:2003.13865 490(10.48550) (2020)

A Multi-dimensional Joint ICA Model with Gaussian Copula

Oktay Agcaoglu[1](✉), Rogers F. Silva[1], Deniz Alacam[1,2], and Vince Calhoun[1]

[1] Translational Research in Neuroimaging and Data Science, Atlanta, USA
oagcaoglu@gsu.edu
[2] Uludag University, Bursa, Turkey

Abstract. Different imaging modalities can provide complementary information and fusing those can leverage their unique views into the brain. Independent component analysis (ICA) and its multimodal version, joint ICA (jICA), have been useful for brain imaging data mining. Conventionally, jICA assumes a common mixing matrix and independent latent joint components with independent and identical marginals. Thus, jICA maximizes a (melded) 1D distribution for each joint component, by either maximum likelihood or the infomax principle. In this study, we propose a joint ICA method that relaxes these assumptions by allowing samples from same voxels (in this case, fMRI and sMRI) to originate from a non-factorial bivariate distribution. We then maximize the likelihood of this joint 2D distribution. The full 2D bivariate distribution is defined by two marginal distributions linked with a copula. Several ICA-based studies on neuroimaging data have successfully modeled independent sources with a logistic distribution, providing robust and replicable results across modalities. This is because neuroimaging data often consists of rapid fluctuations around a baseline, resulting in super-Gaussian distributions. For consistency with prior literature, we choose the logistic distribution to model the marginals, combined with a Gaussian copula to model linkage via simple correlation. However, it should be noted that the proposed algorithm can easily adapt to different types of copulas and alternative marginal distributions. We demonstrated the performance of the proposed method on a simulated dataset and applied the proposed method to analyze structural and functional magnetic resonance imaging dataset from the Alzheimer's disease neuroimaging initiative (ADNI) dataset.

Keywords: Data fusion · copula · joint ICA · multi-modal · Alzheimer's

1 Introduction

Independent component analysis (ICA) is a robust data analysis technique for brain imaging data with a wide range of successful applications in different modalities, such as functional magnetic resonance imaging (fMRI), structural MRI (sMRI), magnetoencephalography (MEG) and electroencephalograms (EEG) [1–8]. ICA assumes that the observed data is a mixture of the sources and aims to blindly estimate these sources

G. L. Foresti et al. (Eds.): ICIAP 2023 Workshops, LNCS 14366, pp. 152–163, 2024.
https://doi.org/10.1007/978-3-031-51026-7_14

by assuming they are statistically independent from one another and have a known distribution. In other words, the joint distribution of sources factorizes as the products of their marginals. For this study, our focus is specifically on the case of ICA with linear mixing. Joint ICA (jICA) extends ICA to data fusion scenarios by assuming that multiple imaging modalities share the same mixing process. Its objective is to estimate the common unmixing matrix, utilizing the complementary information from these modalities to promote discovery of links from the joint datasets.

There are several successful applications of jICA [9–11] where meaningful results and clinical biomarkers were identified. However, jICA relies on some strong assumptions, namely the same linear mixing in different modalities and identical independent source distributions at different modalities (Fig. 1). Different algorithms have been proposed and utilized to modify these assumptions, such as parallel ICA [12], canonical correlation analysis (CCA) [13], independent vector analysis (IVA) [14], Multiview ICA [15] and Bayesian linked tensor ICA (LICA) [16]. Except for LICA, they all assume different linear mixing per modality. Apart from IVA and CCA, these algorithms do not utilize topological information between different modalities. In the case of CCA, the sources are only uncorrelated and not statistically independent. Also, neither CCA nor IVA allow different marginal distributions per modality in linked sources.

In this study, we propose a solution that leverages topological information by modeling the joint distribution of linked sources explicitly, while allowing for arbitrary marginal distributions per modality. Specifically, we align the data from the same voxel location across modalities, such as in fMRI and sMRI data, to model topographical similarities while also estimating a different set of scaling parameters for each modality individually. We achieve this by utilizing a copula to capture topographical linkage and an arbitrary (marginal) distribution for each modality (here, logistic distributions with different scale parameters for consistency with prior literature). We also leverage the constraint that modalities share the same linear mixing to reduce the parameter search space and emphasize patterns with highly similar expression levels across modalities.

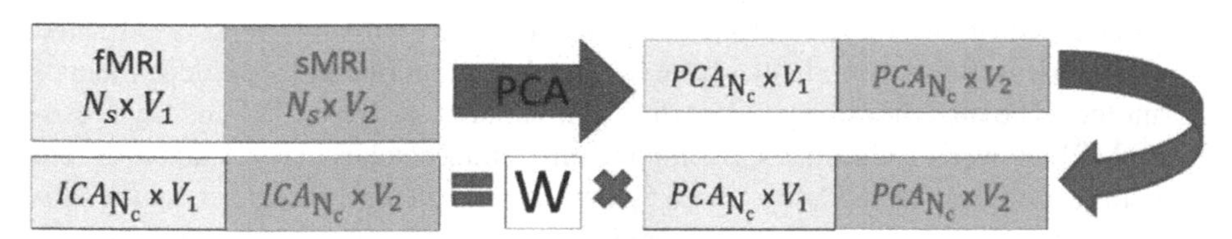

Fig. 1. The block diagram of the conventional jICA (1D-jICA). Here N_S is the number of subjects, V_1 and V_2 are the number of voxels in each modality, which do not need to be equal. N_C is the number of components and **W** is the common unmixing matrix.

2 Dataset

We tested the performance of our proposed methods and compared with the conventional jICA method by analyzing fMRI and sMRI data from the ADNI dataset [17]. The fMRI data went into a unimodal NeuroMark ICA pipeline [18] and subject-specific spatial maps for 53 intrinsic cognitive networks (ICNs) were estimated [19]. We combined these subject-specific maps by taking the maximum intensity value in each voxel, reducing them to a single map per subject. These combined ICN maps are the first modality we used in jICA. The second modality consisted of sMRI gray matter (GM) probability density maps segmented from subject-specific T1 scans. GM images were down sampled to match fMRI spatial resolution.

The datasets were normalized to have the same total sum of squares and, additionally, the sMRI dataset had the sign of every other voxel flipped [11]. All datasets were demeaned per subject, and jointly reduced to 53 principal components. A total of 1943 scans from 872 participants were included (some subjects have multiple scans). Among them, 170 participants have Alzheimer's disease (ad), 756 participants are cognitively normal (cn), 394 participants have early mild cognitive impairment (emci), 296 of them were diagnosed with mild cognitive impairment (mci), 206 participants have late mild cognitive impairment and 118 of the participants have significant memory concern (smc).

Apart from this, we tested the algorithm on a special case, where the left hemisphere gray matter volumes and right hemisphere gray matter volumes are treated as different modalities. In this scenario, the modalities are expected to be highly dependent, and because of homotopic regions were linked with a gaussian copula, the estimated dependency can be interpreted as a laterality metric of the components.

3 Methods

3.1 Conventional Joint ICA

Joint ICA (linear) assumes that there is a common unmixing matrix **W** that can separate the concatenated data $\mathbf{X} = [\mathbf{X}1\ \mathbf{X}2]$ into its independent components $\mathbf{Y} = \mathbf{WX} = [\mathbf{Y}1\ \mathbf{Y}2]$. Existing unimodal ICA algorithms can be used for jICA provided that the data is concatenated and normalized properly [11]. The modalities are normalized separately, concatenated, and then their principal component analysis (PCA) is jointly calculated and retained with the desired model order. Then, a common unmixing matrix is estimated the sources at the different modalities have identical distributions and are independent of one another. The modalities may reflect different features, such as fMRI and high resolution sMRI, or even FNC features. During training, mini-batches may contain different numbers of samples per modality, and originate from different voxels across modalities. The training can be based on maximum likelihood estimation or infomax principle, which are same algorithms [20] (Please note that Infomax algorithm with sigmoid nonlinear transformation function also implies logistic marginal distribution.) We refer to this method as 1D-jICA. The block diagram of 1D-jICA is presented in Fig. 1.

The probability of observing x for a given **W** can be written as:

$$p_x(x) = p_y(y) * |\det W| = \prod_k^{N_c} p_y\left(y_{k_{fMRI}}\right) * p_y\left(y_{k_{sMRI}}\right) * |\det W| \quad (1)$$

3.2 Joint ICA with Different Variances

As mentioned earlier, one issue with conventional jICA is that it implicitly expects the sources to have the same variance across different modalities. This potentially leads to a suboptimal solution because a single unmixing matrix **W** may not be able to accommodate two sets of scales, and it can hinder convergence when more than 2 modalities are analyzed. To overcome these issues, we adapted the model so that they can have different scaling. The block diagram of the modified algorithm is shown in Fig. 2, where **W** is the shared unmixing matrix and **S** is a learnable diagonal scaling matrix and "0"s are N_c by N_c square matrices filled with 0s. We refer to this method as 1Dscale-jICA. The block diagram of the 1Dscale-jICA is presented in Fig. 2.

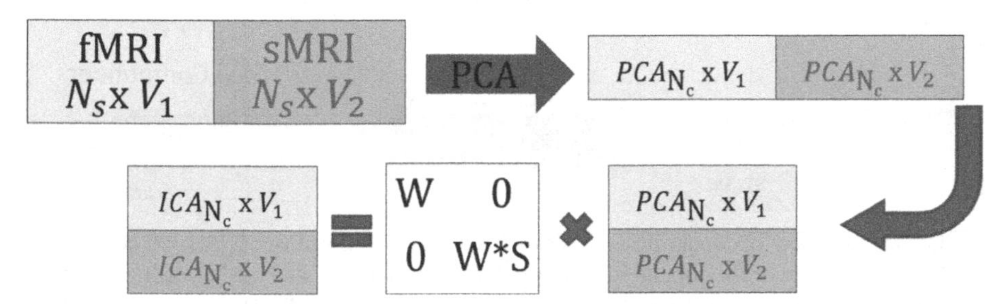

Fig. 2. The block diagram of the proposed jICA (1Dscale-jICA), which learns different scaling for each modality. The matrix **S** is diagonal and enables the second source pairs to have different variance from its multimodal counterpart. **S** is estimated during training. V_1 and V_2 do not need to be equal. N_c is the number of components and **W** is the common unmixing matrix.

The probability of observing x for a given **W** and **S** can be written as:

$$p_x(x) = \prod_{k}^{N_c} p_y\left(y_{k_{fMRI}}\right) * p_y\left(y_{k_{sMRI}}\right) * |\det W|^2 * |\det S| \tag{2}$$

3.3 Proposed Copula Joint ICA

The proposed copula joint ICA algorithm finds a shared unmixing matrix **W** that maximizes the likelihood of the data in (1), assuming a 2D joint distribution which implements a gaussian copula link. For each joint independent component, we use a 2D gaussian copula to link the 1D logistic distribution (marginal) of each modality [1]. A gaussian copula is defined in terms of univariate Uniform variables that are then processed through the bivariate Normal inverse cumulative distribution function. Based on Sklar's theorem [21], any multivariate distribution can be expressed as the product of their marginals and a copula. While we assume the unmixing matrixes are same for both modalities (like 1D-jICA), we also add a scaling layer to enable different variances across modalities (like 1Dscale-jICA). Importantly, during training, batches contained data from corresponding voxels from both modalities. Both the 1D logistic and 1D normal distribution are assumed to be zero mean and unit variance. The copula correlation parameter ρ and the scaling parameter σ (S) are tunable per joint component and estimated during training. A block diagram of the proposed algorithm is shown in Fig. 3, we call this algorithm as 2Dcopula-jICA.

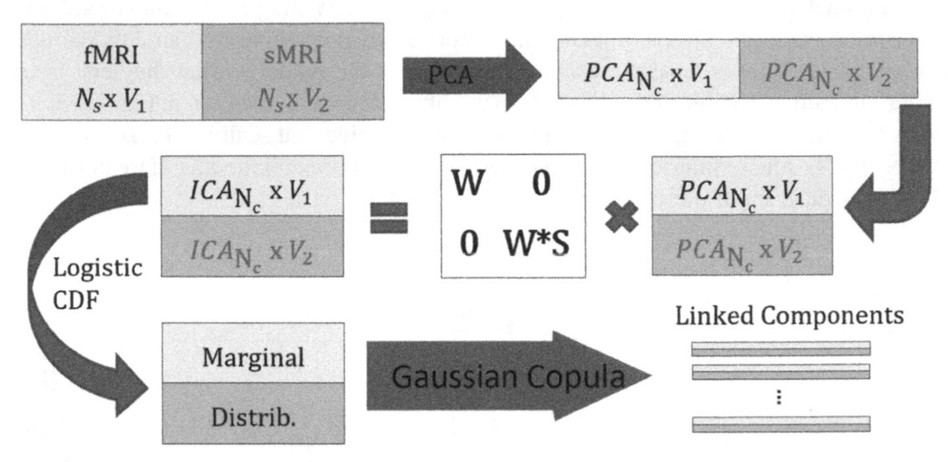

Fig. 3. The proposed 2Dcopula-jICA, where the sources are linked by a gaussian copula. The correlation parameter of the copula is estimated during training, as well as the scaling parameter **S**. V_1 and V_2 must be equal. N_c is the number of components and **W** is the common unmixing matrix.

The probability of observing x for a given **W** and **S** can be written by combining Eq. 2 and Sklar's theorem [21] as:

$$p_x(x) = \prod_k^{N_c} p_y\left(y_{k_{fMRI}}\right) * p_y\left(y_{k_{sMRI}}\right) * |\det W|^2 * |\det S|$$

$$= \prod_k^{N_c} p_{copula\left(\sigma\left(y_{k_{fMRI}}\right),\sigma\left(y_{k_{sMRI}}\right)\right)} * p_{logist\left(y_{k_{fMRI}}\right)} * p_{logist\left(y_{k_{sMRI}}\right)} * |\det W|^2 * |\det S| \quad (3)$$

where the Gaussian Copula function will be:

$$p_{copula}\left(\sigma\left(y_{k_{fMRI}}\right), \sigma\left(y_{k_{sMRI}}\right)\right) =$$

$$= \frac{1}{\sqrt{\det R}} \cdot \exp\left(-\frac{1}{2}\begin{pmatrix}\phi^{-1}\left(u_{k_{fMRI}}\right) \\ \phi^{-1}\left(u_{k_{sMRI}}\right)\end{pmatrix}^T\right) * \left(R^{-1} - I\right) * \begin{pmatrix}\phi^{-1}\left(u_{k_{fMRI}}\right) \\ \phi^{-1}\left(u_{k_{sMRI}}\right)\end{pmatrix}$$

where, $R = \begin{pmatrix} 1 & \sigma_k \\ \sigma_k & 1 \end{pmatrix}$, and

$$u_{k_{fMRI}} = CDF_{logistic}\left(y_{k_{fMRI}}\right), u_{k_{fMRI}} = CDF_{logistic}\left(y_{k_{sMRI}}\right) \quad (4)$$

4 Implementation

We implemented these three algorithms using python and pytorch. All three algorithm were initialized randomly with the same initial unmixing matrix and data from corresponding voxels were included in the training batches, and the optimization were conducted using Adam optimizer in the sense of minimizing negative log-likelihood. After each training epoch, total loss is calculated for the whole sample and parameters that minimized the total loss were used in the results. For the 1Dscale and 2Dcopula, scaling parameters were initialized as 1 and dependence parameters for each component, ρ, are randomly initialized between 0 and 0.9.

4.1 Simulation

We simulated 2D imaging dataset using the simulation toolbox [22], to test and validate the proposed 2Dcopula algorithm. We generated five independent components that have logistic distribution as first modality. We generated the second modality by sampling other pairs of logistic distribution that are correlated with first modality. For simplicity, we set all the components to have same correlation between first and second modality and set to 0.95. The simulation components for first and second modality are displayed in Fig. 4.

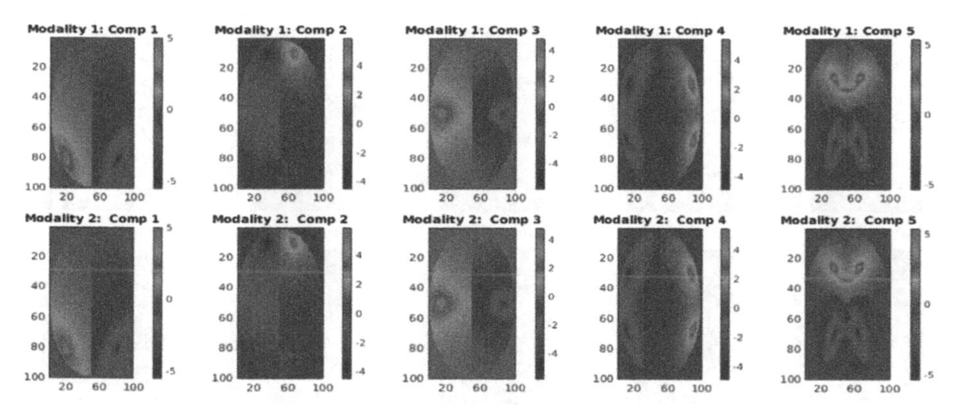

Fig. 4. Simulated ground truth components for modality one and modality 2, each components have 0.95 correlation among different modalities.

Later, we randomly generated a 5 by 5 mixing matrix, that was used as the mixing matrix of the first modality. In order to make the simulation more realistic, we added random gaussian noise with signal to noise ratio equal to 23 db.

After generating the mixing samples, we applied the PCA and then applied ICA with a 1D and a 2D Gaussian copula model. Results are presented in Fig. 5 and 6. The Gaussian copula performed better at recovering the original components, specifically first component couldn't be separated at the infomax, however 2D copula was able to extract it using its leveraging the correlation between modalities.

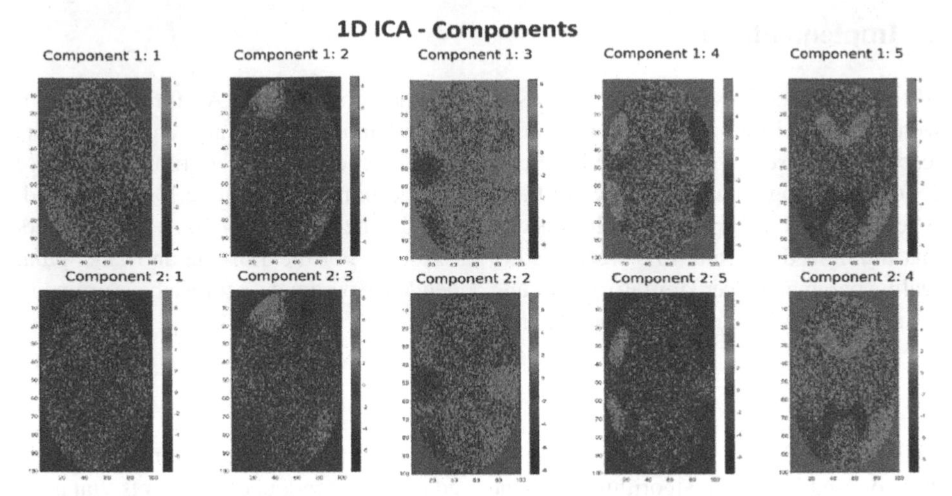

Fig. 5. Simulation results using 1D jICA algorithm. 1D jICA failed to reconstruct some components under high noise. To facilitate easy comparison, we sorted the components to best match the ground truth components. Please note that due to the sign ambiguity of ICA, the components do not have matching positive and negative directions.

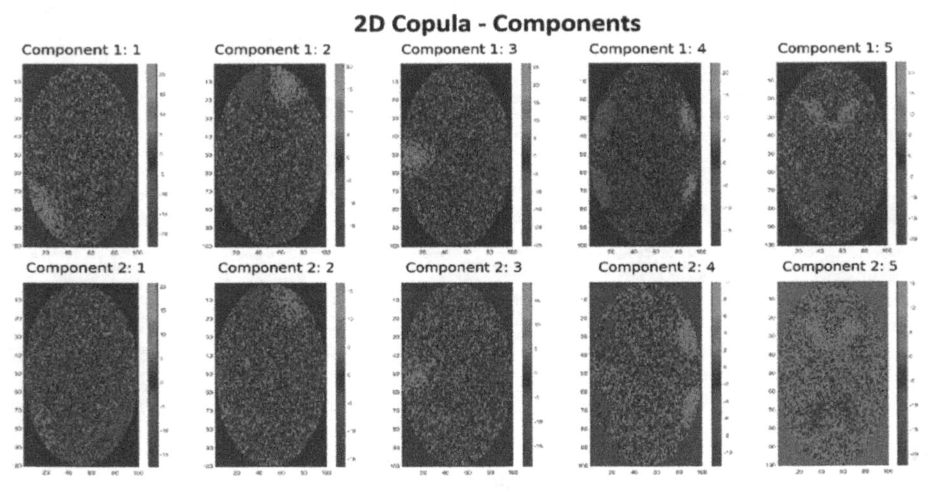

Fig. 6. Simulation results using the proposed 2D copula jICA algorithm showed that our proposed algorithm successfully managed to separate and extract components, performing better compared to the 1D jICA in this case. To facilitate easy comparison, we sorted the components to best match the ground truth components. Please note that due to the sign ambiguity of ICA, the components do not have matching positive and negative directions.

5 Results

Our proposed 2Dcopula jICA models find covarying sources in both modalities while accounting for the possible correlation between sources of different modalities at corresponding voxels. After training, to facilitate better comparison, the estimated unmixing matrixes normalized to have unit power in each row and sorted according to Pearson correlation between different algorithm results to get best matching component. Note that the sorting first done between 1D and 2Dcopula, and then between 1Dscale and 1D, using 1D as reference (Fig. 7).

For the sMRI and ICNs case, the estimated ρ values were low, ranging between 0.0112 and −0.0177. For the left and right hemisphere case, the estimated ρ values are ranging from 0.85 to −0.99, and with minimum absolute value of 0.13 (Fig. 8). This indicates that the correlation between sMRI and ICNs at the corresponding voxels are low, and their dependence is statistically low.

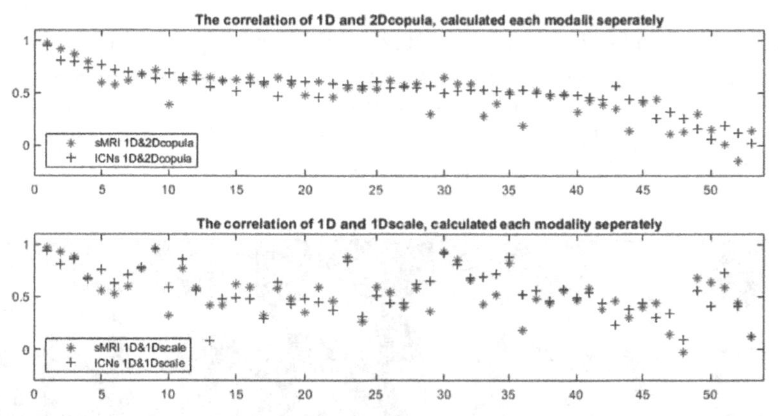

Fig. 7. Pearson correlation values between estimated components sorted from high to low. At the top figure, between 1D and 2Dcopula, at the bottom figure between 1D and 1D scale.

Some of the estimated components are displayed in Fig. 9, for ICNs and sMRI. For better comparison, the differences are also shown. For this component, the three algorithm gives very different brain regions, 1D scale for sMRI and 2D copula for ICNs have very small region. sMRI of 2D copula includes region from cuneus and superior frontal gyrus. We presented average loading parameters for ad and cn clinical groups in Fig. 10. Separately, we also checked for groups differences between ad and cn with a two-sample t-test on loading parameters and corrected for multiple comparison with false discovery rate. Those who showed FDR significant group differences are marked in Fig. 10. 1D scale and 2D copula have more components with significant group difference, comparing to 1D. The component presented in Fig. 9 shows significant group differences only in 2D copula case. 1D components do not show any significant differences after components #18; note that components are ordered from high correlated to low correlated. This can be interpreted as forcing different modalities to same distribution may cause some biomarker not detected.

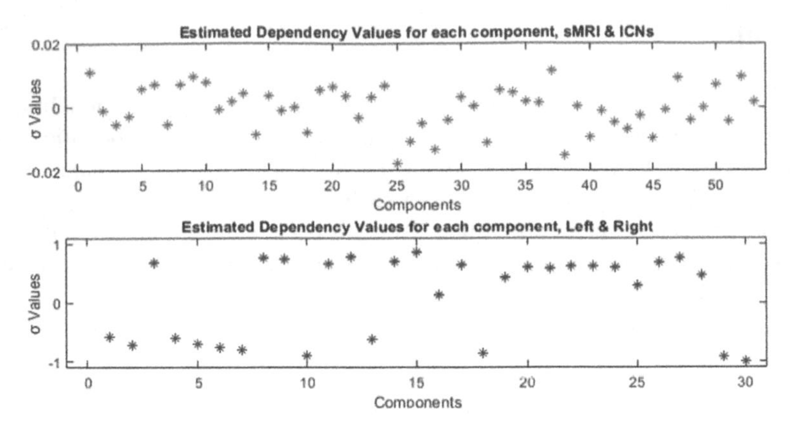

Fig. 8. Estimated dependencies for each component, ρ values, between modalities, these ρ values are low for sMRI&ICNs case, and as high as 0.99 for Left & Right case.

Estimated left and right components 26,28 and 29 are displayed in Fig. 11; left and right hemisphere components are fused after jICA run for display purposes. Estimated sigma values are also shown in the Fig. 8. While 1D and 1Dscale components look mostly similar, 2D copula components show differences. The sigma value is correlated with how bilateral the 2Dcopula components looks.

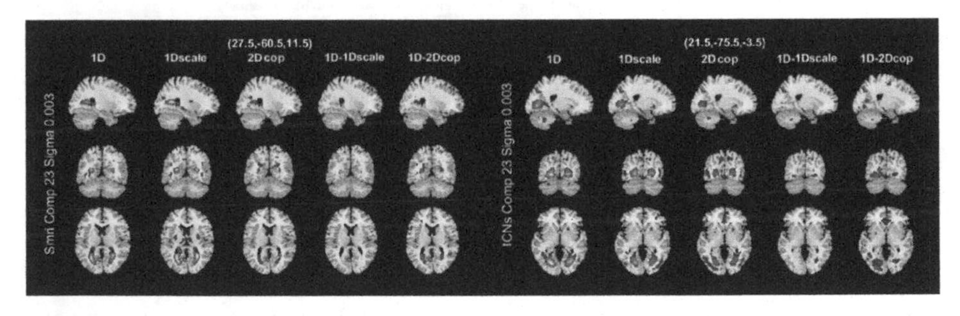

Fig. 9. ICNs and sMRI components, for 3 algorithms and their difference are displayed.

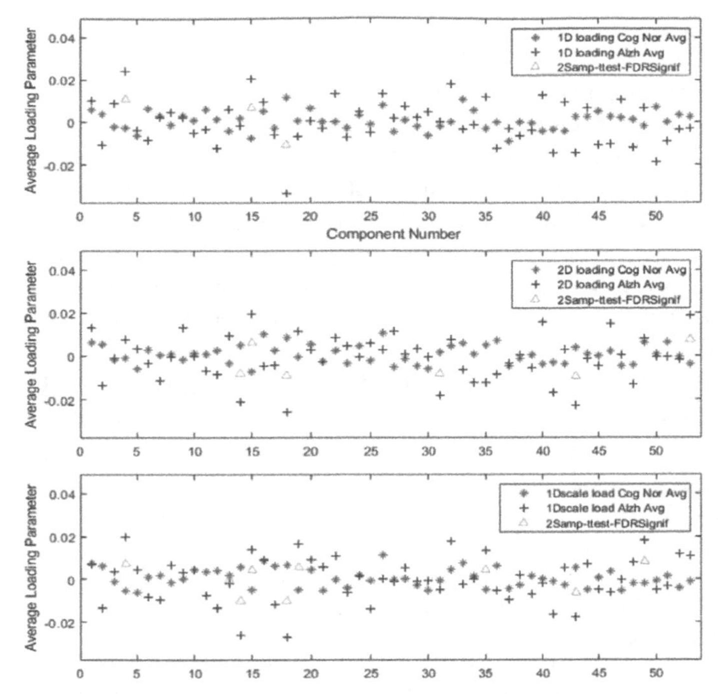

Fig. 10. Loading parameters averaged over cognitively normal and Alzheimer's patients are presented. Separately, two-sample t-tests were conducted to compare clinical groups and those that showed FDR significant difference was marked. 1Dscale and 2Dcopula have more components with significant differences. 1D does not show any significant differences after component #20, which has a correlation with 2Dcopula 0.6 (Fig. 7). Therefore, we do not observe significant group difference among the components showing less correlation with 2Dcopula.

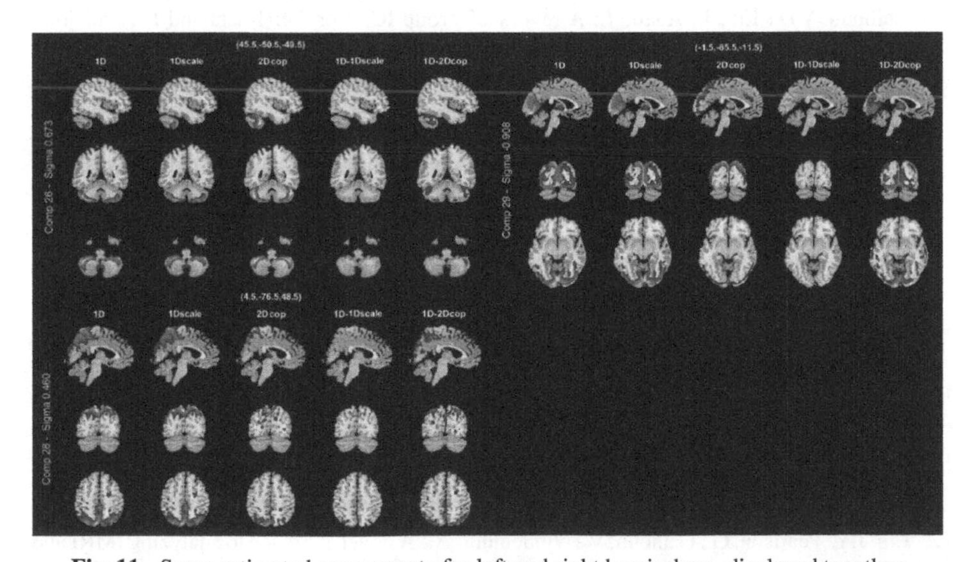

Fig. 11. Some estimated components for left and right hemisphere, displayed together.

6 Conclusion

In this study, we proposed two joint ICA algorithms that provide softer coupling option by altering the independence assumption between different modalities and estimating the dependence between corresponding components among different modalities, as well as enabling modalities to have different distributions and different variance. We presented results for sMRI & ICNs modalities and generated meaningful components. We found more components with significant group differences between Alzheimer's and cognitively normal participants using the proposed 1D scale and 2D copula algorithms. We also presented results for a special case where brain hemispheres are considered as different modalities and illustrated that 2Dcopula jICA can recover meaningful components for this highly dependent structure. In future work, this can be adapted to investigate hemispheric lateralization by determining the statistical dependencies among homotopic regions.

References

1. Wu, L., Eichele, T., Calhoun, V.D.: Reactivity of hemodynamic responses and functional connectivity to different states of alpha synchrony: a concurrent EEG-fMRI study. Neuroimage **52**(4), 1252–1260 (2010)
2. Allen, E.A., Damaraju, E., Eichele, T., Wu, L., Calhoun, V.D.: EEG signatures of dynamic functional network connectivity states. Brain Topogr. **31**(1), 101–116 (2018)
3. Cetin, M.S., et al.: Multimodal classification of schizophrenia patients with MEG and fMRI data using static and dynamic connectivity measures. Front. Neurosci. Methods **10**(466), 466 (2016). (in English)
4. Agcaoglu, O., Silva, R.F., Calhoun, V.: Multimodal fusion of brain imaging data with joint non-linear independent component analysis. In: 2022 IEEE 14th Image, Video, and Multidimensional Signal Processing Workshop (IVMSP), pp. 1–5 (2022)
5. Calhoun, V.D., Liu, J., Adali, T.: A review of group ICA for fMRI data and ICA for joint inference of imaging, genetic, and ERP data. Neuroimage **45**(1 Suppl.), S163–S172 (2009)
6. Agcaoglu, O., Miller, R., Mayer, A.R., Hugdahl, K., Calhoun, V.D.: Increased spatial granularity of left brain activation and unique age/gender signatures: a 4D frequency domain approach to cerebral lateralization at rest. Brain Imaging Behav. **10**(4), 1004–1014 (2016)
7. Calhoun, V.D., Adali, T.: Multisubject independent component analysis of fMRI: a decade of intrinsic networks, default mode, and neurodiagnostic discovery. IEEE Rev. Biomed. Eng. **5**, 60–73 (2012)
8. Rashid, B., et al.: A framework for linking resting-state chronnectome/genome features in schizophrenia: a pilot study. Neuroimage **184**, 843–854 (2019)
9. Moosmann, M., Eichele, T., Nordby, H., Hugdahl, K., Calhoun, V.D.: Joint independent component analysis for simultaneous EEG-fMRI: principle and simulation. Int. J. Psychophysiol. **67**(3), 212–221 (2008)
10. Eichele, T., et al.: Unmixing concurrent EEG-fMRI with parallel independent component analysis. Int. J. Psychophysiol. **67**(3), 222–234 (2008)
11. Calhoun, V.D., Adali, T., Giuliani, N.R., Pekar, J.J., Kiehl, K.A., Pearlson, G.D.: Method for multimodal analysis of independent source differences in schizophrenia: combining gray matter structural and auditory oddball functional data. Hum. Brain Mapp. **27**(1), 47–62 (2006)
12. Liu, J.Y., Pearlson, G., Calhoun, V., Windemuth, A.: A novel approach to analyzing fMRI and SNP data via parallel independent component analysis. In: Proceedings of Spie, vol. 6511 (2007). (in English)

13. Correa, N., Li, K.O., Adali, T., Calhoun, V.D.: Examining associations between fMRI and EEG data using canonical correlation analysis. I S Biomed. Imaging, 1251+ (2008). (in English)
14. Kim, T., Lee, I., Lee, T.W.: Independent vector analysis: definition and algorithms. In: Conf. Rec. Asilomar C, 1393+ (2006). (in English)
15. Richard, H., Gresele, L., Hyvärinen, A., Thirion, B., Gramfort, A., Ablin, P.: Modeling shared responses in neuroimaging studies through multiview ICA. arXiv:2006.06635. https://doi.org/10.48550/arXiv.2006.06635. https://ui.adsabs.harvard.edu/abs/2020arXiv200606635R. Accessed 01 June 2020
16. Groves, A.R., Beckmann, C.F., Smith, S.M., Woolrich, M.W.: Linked independent component analysis for multimodal data fusion. Neuroimage **54**(3), 2198–2217 (2011). (in English)
17. Jack, C.R., Jr., et al.: The Alzheimer's disease neuroimaging initiative (ADNI): MRI methods. J. Magn. Reson. Imaging **27**(4), 685–691 (2008)
18. Du, Y., et al.: NeuroMark: an automated and adaptive ICA based pipeline to identify reproducible fMRI markers of brain disorders. Neuroimage Clin. **28**, 102375 (2020)
19. Hassanzadeh, R., Abrol, A., Calhoun, V.: Classification of Schizophrenia and Alzheimer's disease using resting-state functional network connectivity. In: 2022 IEEE-EMBS International Conference on Biomedical and Health Informatics (BHI), pp. 01–04 (2022)
20. Hyvarinen, A., Oja, E.: Independent component analysis: algorithms and applications. Neural Netw. **13**(4–5), 411–30 (2000)
21. Sklar, M.J.: Fonctions de repartition a n dimensions et leurs marges (1959)
22. Simulation Toolbox. https://github.com/trendscenter/

Federated Learning in Medical Imaging and Vision (FEDMED)

Federated Learning for Data and Model Heterogeneity in Medical Imaging

Hussain Ahmad Madni[1(✉)], Rao Muhammad Umer[2], and Gian Luca Foresti[1]

[1] Department of Mathematics, Computer Science and Physics (DMIF), University of Udine, 33100 Udine, Italy
madni.hussainahmad@spes.uniud.it

[2] Institute of AI for Health, Helmholtz Zentrum München - German Research Center for Environmental Health, 85764 Neuherberg, Germany

Abstract. Federated Learning (FL) is an evolving machine learning method in which multiple clients participate in collaborative learning without sharing their data with each other and the central server. In real-world applications such as hospitals and industries, FL counters the challenges of data heterogeneity and model heterogeneity as an inevitable part of the collaborative training. More specifically, different organizations, such as hospitals, have their own private data and customized models for local training. To the best of our knowledge, the existing methods do not effectively address both problems of model heterogeneity and data heterogeneity in FL. In this paper, we exploit the data and model heterogeneity simultaneously, and propose a method, MDH-FL (Exploiting Model and Data Heterogeneity in FL) to solve such problems to enhance the efficiency of the global model in FL. We use knowledge distillation and a symmetric loss to minimize the heterogeneity and its impact on the model performance. Knowledge distillation is used to solve the problem of model heterogeneity, and symmetric loss tackles with the data and label heterogeneity. We evaluate our method on the medical datasets to conform the real-world scenario of hospitals, and compare with the existing methods. The experimental results demonstrate the superiority of the proposed approach over the other existing methods.

Keywords: Federated Learning · Medical Imaging · Heterogeneous Data · Heterogeneous Model

1 Introduction

Federated Learning (FL), initially introduced by [20], has become a popular machine learning technique because of distributed model training without sharing the private data of participating hosts. In FL, participants (i.e., clients) including organizations and devices generally have heterogeneous data and heterogeneous models that are customized according to the tasks and local data. In real-world applications, data from multiple sources are heterogeneous and contain non-independent and identically and distributed data (non-IID). Moreover,

G. L. Foresti et al. (Eds.): ICIAP 2023 Workshops, LNCS 14366, pp. 167–178, 2024.
https://doi.org/10.1007/978-3-031-51026-7_15

data from multiple source may produce diverse labels and classes that is more challenging for the convergence of FL model. Traditional training methods based on centralized data cannot be used in practical applications due to privacy concerns and data silos at multiple locations [10]. FL has the ability to train a global model by allowing multiple participants to train collaboratively with their decentralized private data. In this way, private data of an individual participant are never shared with the central server and other participant in FL environment. Most common FL algorithms are FedProx [14] and FedAvg [20] that aggregate the model parameters obtained from the participating clients. Most of the existing methods [13,28] using these algorithms consider the homogeneous data and same architecture of the local model used by all participants.

In practical applications, each participant has its own data and might need to design its own customized model [8,24] due to specific and personalized requirements [14]. Such heterogeneity in data and model is natural in healthcare organizations that design custom models for specific tasks as illustrated in Fig. 1. In such environment, hospitals are hesitant to reveal their data and model architecture due to privacy concerns and business matters. Thus, numerous methods have been proposed to perform FL with such heterogeneous data [7,30] and clients [12,15,16]. FedMD [12] is a method that implemented knowledge distillation based on class scoring calculated by local models of clients trained on public dataset. FedDF [16] is another method that performs ensemble distillation by leveraging the unlabeled data for every model architecture. Such existing methods are dependent on shared models and mutual consensus. However, a mutual consensus is another challenge in which each client is unable to set its learning direction to adjust the deviations among all participants. Moreover, designing additional models enhance the processing overhead, and eventually affect the performance. Thus, FL containing heterogeneous data and models without depending on global sharing and consensus is critical and challenging.

The methods discussed above are mostly dependent on the assumption that every participant has homogeneous, independent and identically distributed (IID) data that is not possible in real-world scenarios. More specifically, in collaborative learning each participant has its own data and requires a customized model for the specific nature of data and task. As FL has many participants with heterogeneous models and data, each model suffers data diversity effecting the overall performance of the global model. Existing methods such as [4,27] have been presented that designed robust loss function to minimize the negative impact of heterogeneous labels in data. The current existing methods either tackle the data heterogeneity or model heterogeneity. In FL, it is required that a model should be robust and learn sufficiently from the data during local update.

To tackle with the heterogeneous participants and data containing diverse labels are the prominent challenges in FL. Model heterogeneity in FL causes the diverse noise patterns and decision boundaries. Moreover, data heterogeneity based on non-IID and label diversity creates difficulty in convergence of the global model during global learning phase in FL. It is required for each client to concentrate on the contribution of other participants and align the learning

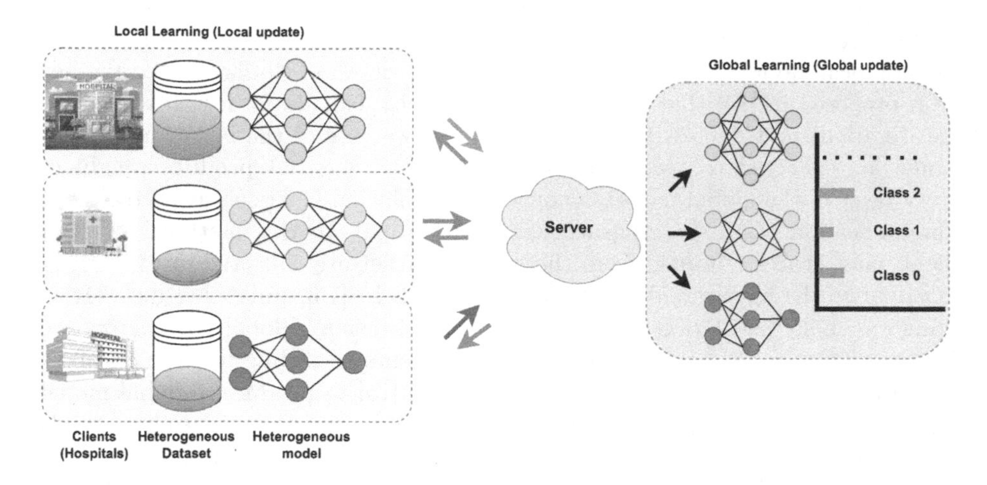

Fig. 1. Participating hospitals (i.e., clients) contain heterogeneous local models trained on heterogeneous data and diverse labels. Each client has its own data and custom model as per requirements and tasks.

to produce a robust global model. In this paper, we propose a solution for the heterogeneous data and model in FL. 1) For the model heterogeneity, a model distribution (i.e., logits output) is aligned by learning the knowledge distribution and feedback from other clients using public data. In this way, each participant learns with its own strategy without depending on the public model. 2) To tackle with the data heterogeneity having diverse labels, an additional symmetric loss function as proposed in [29], is used to minimize the diversity impact on model learning.
Our main contribution are as follows.

- We explore the real-world scenario of data and model heterogeneity in hospitals implementing decentralized collaborative model training.
- We use knowledge distillation for the alignment of model output (i.e., logits) to solve the problem of model heterogeneity and to produce an efficient global model in FL.
- We utilize an additional symmetric loss function to optimize the model learning based-on heterogeneous data containing diverse labels.
- We evaluate the proposed method on hematological cytomorphology clinical datasets with heterogeneous model and data scenarios, and experimental results show the supremacy of the proposed method over the existing FL methods.

2 Related Work

2.1 Federated Learning

Federated Learning (FL), firstly proposed by [20] is a machine learning method in which multiple clients train a global model without sharing their private local data to preserve the privacy. Initially, FedAvg was used to aggregate the parameters of local models trained on local data [20]. A method similar to FedAvg has been proposed in [14] that can customize the local calculations with respect to the iterations and devices used in FL. In [28], weights of the layers in a client model are collected to accomplish one-layer matching that produce weight of every layer in the global model. Knowledge distillation has been utilized for the communication of FL heterogeneous models in [12]. In this method, for each client, class scores obtained from the public dataset are collected on the server to calculate the aggregated value to be updated. In [16], unlabeled data leveraging ensemble distillation is used for the model fusion. Global parameters are dynamically assigned as a subset to the local clients according to their capabilities in [2]. An algorithm has been introduced in [15] to produce a global model from the learning of local representations. We summarize that existing methods assume that all clients have homogeneous data without consideration of any type of heterogeneity. No research have been conducted for the mitigation of diverse impact of data and model heterogeneity simultaneously during the collaborative learning in FL.

2.2 Model and Data Heterogeneity

Numerous methods have been presented to tackle with data heterogeneity, but not much research have been conducted for the model heterogeneity and label diversity in the scenario of FL. Some existing methods use loss functions for the optimization such as [4,27]. A convex classification calibration loss has been proposed by [27] that is robust for incorrect classes and labels. Some loss functions are evaluated by [4] that prove the robustness of MAE to perturbed classes in deep learning. Estimation of the probability for every class flipped to some other class has been utilized in existing methods [22,25,32]. In [32], corrupted data is transformed into Dirichlet distribution space and label regression technique is used to infer the correct classes, and finally data modeling and classifier are trained together. Some existing methods extract clean samples, re-weight each instance, or apply some transformation on the heterogeneous data for model training [5,9,31].

A method JoCoR has been proposed in [31] that uses Co-Regularization for the joint loss estimation. In this method, the samples with minimum loss are selected to update the model parameters. MentorNet is another method proposed by [9] that comes up with a technique used to weight a sample such as used in StudentNet and MentorNet. Co-teaching method has been proposed in [5] that selects data for cross training of the two deep networks simultaneously. To avoid the model overfitting specific samples, robust regularization is used in [1,21,33]. A method Mixup has been proposed in [33] to regularize the deep network by training the convex pairs of instance and their corresponding labels. Regularization is used by [1] to minimize the impact of corrupted data while not affecting the training of actual samples. A regularization method has been introduced in [21] that depends on the virtual adversarial loss and adversarial direction that do not require any label information. Most of the existing methods that solve the problem of data heterogeneity and corrupted data are based on centralized data and a single model. However, server is not able to access the local data of a client directly in FL environment. Moreover, heterogeneous clients have diverse patterns and decision boundaries.

3 Federated Learning with Heterogeneous Data and Models

In FL with heterogeneous participants P and a server, we consider C as the number of all clients where $|C| = P$. Thus, the pth participant $c_p \in C$ has its local data $d_p = \{(x_i^p, y_i^p)\}_{i=1}^{N_p}$ where $|x|^p = N_p$. Moreover, $y_i^p \in \{0,1\}^{N_p}$ is a one hot vector containing ground truth labels. Furthermore, a local model Θ_p owned by a client c_p has different architecture and $f(x^p, \Theta_p)$ represents the logits output produced by the network $f(.)$ using input x^p calculated with Θ_p. The server has a public dataset $d_0 = \{x_i^0\}_{i=1}^{N_0}$ that may belongs to the client data for different classification tasks. In FL, overall process is divided into local training and collaborative learning in which local training is performed by E_l rounds and collaborative learning is performed by E_c rounds. Our purpose is to perform FL with heterogeneous (i.e., non-IID) data containing diverse labels and heterogeneous clients, so a client has its heterogeneous data $\tilde{d} = \{(x_i^p, \tilde{y}_i^p)\}_{i=1}^{N_p}$ in which $\tilde{y}_i^p$ denotes the heterogeneous annotations. Each client has different noise patterns and decision boundaries due to model heterogeneity that can be expressed as $f(x, \Theta_{p_1}) \neq f(x, \Theta_{p_2})$. Thus, each client c_p must also consider the heterogeneity of other clients $c_{p_0} \neq p$, other than heterogeneity in its own dataset. The overall objective is to find an optimal solution for model parameters $\Theta_p = argmin\ \mathcal{L}(f(x^p, \Theta_p), y^p)$. The architecture of the proposed method is shown in Fig. 2. Each client is trained on its private dataset and subsequently on public dataset to use the knowledge distillation and alignment of knowledge distribution as given in Eq. (3). Moreover, each local client is updated and optimized using symmetric loss given in Eq. (8).

3.1 Model Heterogeneity

The knowledge distribution represented as $D_p^{t_c} = f(d_0, \Theta_p^{t_c})$ is produced for the client c_p. To estimate the variance in knowledge distribution, Kullback-Leibler ($\mathcal{KL}$) divergence is used by each client as proposed in [3]. KL divergence represents the deviation between two probability distributions. If there are two clients c_{p_1} and c_{p_2} having knowledge distributions $D_{p_1}^{t_c} = f(d_0, \Theta_{p1}^{t_c})$ and $D_{p_2}^{tc} = f(d_0, \Theta_{p_2}^{t_c})$ respectively, then the difference between their knowledge distributions can be formulated as:

$$\mathcal{KL}(D_{p_1}^{e_c} || D_{p_2}^{e_c}) = \sum D_{p_1}^{e_c} \log(\frac{D_{p_1}^{e_c}}{D_{p_2}^{e_c}}) \tag{1}$$

If the difference between two knowledge distributions $D_{p_1}^{t_c}$ and $D_{p_2}^{t_c}$ is higher, there is more opportunity for the clients c_{p_1} and c_{p_2} to learn from each other, and

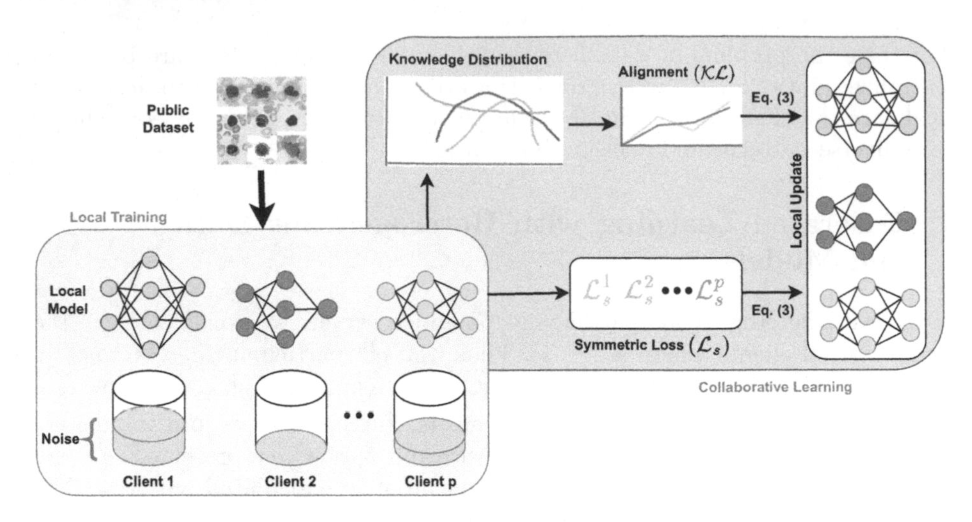

Fig. 2. Proposed approach containing local training and global learning in FL. Local models are updated with Kullback-Leibler loss based on knowledge distribution (Eq. 3), and Symmetric loss (Eq. 8). In local training phase, private models are individually trained on private datasets, and in global or collaborative learning, local clients are updated through loss functions (i.e., $\mathcal{KL}$ and $\mathcal{L}_f$).

vice versa. If the $\mathcal{KL}$ difference is minimized due to the probability distributions $D_{p_1}^{t_c}$ and $D_{p_2}^{t_c}$, it is assumed a technique that allows a client c_{p_1} for learning from the client c_{p_2}. Thus, knowledge distribution difference for a client c_p can be expressed as:

$$\mathcal{L}_{pl}^{p,e_c} = \sum_{p_0=1, p_0 \neq p}^{P} \mathcal{KL}(D_{p_0}^{e_c} || D_p^{e_c}) \tag{2}$$

where p_0 is a participant other than c_p. Moreover, knowledge distribution difference is calculated for a client c_p, so other participants can access the knowledge of c_p without leakage of model architecture and data privacy. All participants are prompted for the collaborative learning due to significant difference in their knowledge distributions. Thus, each participant aligns its knowledge distribution by learning from other participants. This process can be mathematically formulated as follows.

$$\Theta_p^{e_c} \leftarrow \Theta_p^{e_c-1} - \alpha \nabla_{\Theta} \left(\frac{1}{P-1} \cdot \mathcal{L}_{pl}^{p,e_c-1} \right) \tag{3}$$

where α represents the learning rate.

3.2 Data and Labels Heterogeneity

We utilize the Symmetric Cross Entropy proposed in [29] to minimize the effect of local noise in model learning. Cross Etnropy (CE) is a very common loss function used in most of the classification tasks. CE is deformation of $\mathcal{KL}$ divergence, so

$\mathcal{KL}$ can be formulated in term of CE. For example, if p and g are predicted and label class distribution respectively, the $\mathcal{KL}$ divergence can be formulated as:

$$\mathcal{KL}(g||p) = \underbrace{\sum g(x)\log(g(x))}_{\text{entropy of g}} - \underbrace{\sum g(x)log(p(x))}_{\text{cross entropy}} \tag{4}$$

The Eq. (4) contains entropy of g and cross entropy terms. Thus, CE loss for the input x is represented as:

$$\mathcal{L}_c = -\sum_{i=1}^{N} g(x_i)\log(p(x_i)) \tag{5}$$

Cross Entropy loss ($\mathcal{L}_c$) has limitations due to label noise. It does not make overall classes to learn enough from all categories due to various simplicity levels in the classes. To converge the model for such difficult classes, extra communication rounds are required for additional learning. In such scenario, there is a possibility of overfitting to the heterogeneous labels that eventually reduces the overall efficiency of the model.

Generally, a model has limited ability for some categories to classify correctly. Moreover, a model prediction is reliable up to some extent due to label noise. Thus, if g is not a real class distribution, reliability of prediction p as a true class distribution is limited. To solve this problem, a Reverse Cross Entropy (RCE) loss function proposed in [29], on the basis of p is exploited to align the predicted class distribution by the model. RCE loss for the input x is formulated as:

$$\mathcal{L}_{rc} = -\sum_{i=1}^{N} p(x_i)\log(g(x_i)) \tag{6}$$

It is feasible to learn the difficult classes for the model if both $\mathcal{L}_c$ and $\mathcal{L}_{rc}$ are combined, and overfitting can be avoided. This combined loss is named as Symmetric loss [29] that can be expressed as:

$$\mathcal{L}_s = \lambda\mathcal{L}_c + \mathcal{L}_{rc} \tag{7}$$

where λ is used to control the overfitting to noise. Thus, $\mathcal{L}_c$ fits the model on each class and $\mathcal{L}_{rc}$ tackles with the label noise.

A client aligns its local knowledge with the knowledge of other participants using a local learning process. A local model updated with its local data to prevent the local knowledge forgetting. In the process of local training, label noise redirects the model to wrong learning that causes convergence failure for the model. To solve this problem, symmetric loss ($\mathcal{L}_s$) is used to compute the loss between given label and the predicted pseudo-label by the model. Local update for a model can be expressed as:

$$\Theta_p^{el} \leftarrow \Theta_p^{e_l-1} - \alpha\nabla_\Theta\mathcal{L}_s^{p,e_l-1}(f(x^p, \Theta_p^{e_l-1}), \tilde{y}^p) \tag{8}$$

where $e_p \in E_l$ denotes the e_p-th epoch in model training. A client leverages $\mathcal{L}_s$ to update its model that strengthens the local knowledge, and avoids the overfitting to label noise. Thus, model learning is promoted with the $\mathcal{L}_s$ loss.

Table 1. Federated Learning hyperparameters.

Hyperparameter	Value(s)
E_c (global epochs for collaborative learning)	40
E_l (local epochs for local training)	$\frac{N_p}{N_0}$
Optimizer	Adam [11]
α (Initial learning rate for optimizer)	0.001
b (batch size)	16
λ	0.1
μ (labels diversity rate)	$\{0.1, 0.2, 0.3\}$
flip %age in data $\tilde{d}$	20
γ (data heterogeneity rate)	0.5

4 Experimental Results

4.1 Datasets and Models

In the experiments, two hematological cytomorphology clinical datasets, INT_20 dataset [26] and Matek_19 dataset [19] are used for the single-cell classification in Leukemia (i.e., cancer detection). INT_20 dataset [26] is used as a public dataset on the server, and Matek_19 dataset [19] is distributed to the clients as their local private datasets. INT_20 dataset has 26379 samples of 13 classes containing 288×288 colored blood images. Matek_19 dataset contains a total of 14681 samples of 13 classes having blood images with resolution of 400×400. In each experiment, four clients are set up for the collaborative learning and Matek_19 dataset is equally divided to these clients using Dirichlet distribution (i.e., Dir (γ)) to make non-IID dataset [17]. The size of public data on the server and private data on each client is $N_0 = 26379$ and $N_p = 3670$ respectively.

For the homogeneous clients, ResNet-12 [6] is used for the training of all clients, and for the heterogeneous scenario, ShuffleNet [34], ResNet10 [6], Mobilenetv2 [23], and ResNet12 [6] are assigned to the clients for local training on the private datasets.

To produce labels diversity in data, a matrix $\mathcal{M}$ for the label transition is used represented as $M_{ij} = flip(\tilde{y} = j|y = i)$ that shows that label y is moved to a heterogeneous class j from a class i. As the real-word scenario, a client c_p selects N_p examples randomly from the private data (Matek_19), so each client has different noise proportion in its local data. Pair flip [5] and Symmetric flip [27] are the two common categories of matrix $\mathcal{M}$. In Pair flip, a label of original class is swapped with a same wrong category, and in Symmetric flip, a class label is swapped with a wrong class label having same probability. Other implementation configuration is given in Table 1.

In Table 1, E_c is used as epoch for global or collaborative learning, and E_l is used as local epoch in local training. Adam [11] is used as optimizer with the learning rate α. In each experiment, $\lambda = 0.1$ is fixed to control the overfitting to

Table 2. Results produced by FL training with heterogeneous models trained on homogeneous data containing homogeneous labels.

Method	Θ_1	Θ_2	Θ_3	Average
SL-FedL [17]	74.34	77.45	79.23	77.01
FedDF [16]	75.54	79.45	77.96	77.65
Swarm-FHE [18]	76.58	72.98	73.89	74.48
FedMD [12]	77.83	78.11	78.05	78.00
Ours	**82.27**	**79.24**	**83.57**	**81.69**

label diversity. Different diversity rate μ is used to check the performance of the model with varying data and label heterogeneity. Moreover, label flip percentage for the heterogeneous data $\tilde{d}$ is fixed as 20 in all the experiments, where $\gamma = 0.5$ is the data heterogeneity rate.

4.2 Comparison with State-of-the-Art Methods

We perform experiments to evaluate and compare the proposed method with existing methods on the basis of accuracy. Table 2 shows the results of different methods using non-heterogeneous training models with $\mu = 0$ (i.e., no label diversity) in local datasets. Performance of each individual client is given in terms of accuracy (%age), and in the last column average accuracy is given for each method. It is evident that the proposed method performs better when using non-heterogeneous models and homogeneous data without label diversity for model training.

Table 3 shows the comparison of the proposed method with similar existing methods. We use different labels-diversity techniques for the datasets used with heterogeneous models for the training. Performance of each method is decreased with the increasing labels-diversity rate. Moreover, there is a remarkable difference among all methods when the type of labels diversity is changed. This is because heterogeneous data or labels lead to wrong learning and communication of participating clients. Moreover, heterogeneous models produce different noise patterns that eventually decrease the model performance. Results from the Table 2 and Table 3 are computed when using heterogeneous models. However, these results are computed from the experiments without using additional loss functions.

We compare the proposed method with similar baseline methods, SL-FedL [17], FedDF [16], Swarm-FHE [18], and FedMD [12]. In the experiments, different diversity rate (i.e., $\mu = \{0.1, 0.2, 0.3\}$) and types are used for the fair comparison. Symmetric loss is used to optimize the model against data and label diversity. Moreover, knowledge distribution (i.e., $\mathcal{KL}$ loss) is implemented to align the output of all heterogeneous participants. Table 4 shows the comparison of proposed method with the existing methods, and demonstrates the outperforming of the proposed method compared to the existing similar methods.

Table 3. FL training results computed on heterogeneous models and heterogeneous data for different methods.

		Symmetric flip				Pair flip			
Noise Rate (μ)	Method	Θ_1	Θ_2	Θ_3	Average	Θ_1	Θ_2	Θ_3	Average
0.1	SL-FedL [17]	69.48	71.89	74.59	71.99	70.04	71.98	75.55	72.52
	FedDF [16]	71.77	74.45	72.47	72.90	73.02	74.58	73.24	73.61
	Swarm-FHE [18]	71.15	67.08	67.54	68.59	71.89	67.66	71.93	70.49
	FedMD [12]	72.88	75.76	73.37	74.00	73.17	75.91	74.42	74.50
	Ours	**79.38**	**76.95**	**79.36**	**78.56**	**80.04**	**77.22**	**79.78**	**79.01**
0.2	SL-FedL [17]	66.53	68.23	71.15	68.64	66.27	68.76	71.88	68.97
	FedDF [16]	68.65	70.14	68.64	69.14	69.94	70.09	69.60	69.88
	Swarm-FHE [18]	65.35	62.89	62.45	63.56	65.82	63.56	68.26	65.88
	FedMD [12]	67.22	70.32	69.10	68.88	69.18	70.65	71.84	70.56
	Ours	**74.06**	**71.77**	**73.94**	**73.26**	**78.27**	**75.44**	**76.67**	**76.79**
0.3	SL-FedL [17]	62.14	65.06	66.85	64.68	62.16	64.78	66.76	64.57
	FedDF [16]	62.87	66.23	62.97	64.02	66.09	67.35	67.58	67.01
	Swarm-FHE [18]	59.44	55.91	54.34	56.56	59.96	58.75	61.16	59.96
	FedMD [12]	61.87	63.78	64.93	63.53	67.45	66.33	67.48	67.09
	Ours	**66.54**	**65.60**	**66.20**	**66.11**	**73.25**	**68.12**	**69.12**	**70.16**

Table 4. Training results computed with different methods. Heterogeneous models and data are used for each experiment. Two losses (i.e., $\mathcal{L}_s$ and $\mathcal{KL}$) are used to minimize the heterogeneity impact and to improve the overall performance of the global model.

Method	Symmetric flip			Pair flip		
	$\mu = 0.1$	$\mu = 0.2$	$\mu = 0.3$	$\mu = 0.1$	$\mu = 0.2$	$\mu = 0.3$
SL-FedL [17]	76.21	72.16	67.02	78.24	73.87	68.44
FedDF [16]	78.53	74.47	68.77	78.91	74.22	69.65
Swarm-FHE [18]	72.44	66.88	59.94	73.68	67.20	60.72
FedMD [12]	79.78	74.18	68.11	80.85	76.15	73.26
Ours	**83.69**	**79.82**	**72.93**	**84.06**	**80.10**	**73.94**

5 Conclusion

In this paper, a real-world problem of model and data heterogeneity in medical imaging has been explored. To solve the problem of heterogeneous data and labels diversity, an additional symmetric loss has been used to optimize the model trained on local and private data. To tackle with the heterogeneous participants in FL, Kullback-Leibler has been exploited to align the different noise patterns produced by the heterogeneous participants. Moreover, each participating client uses the knowledge distribution of other participants to improve the performance of global FL model. Experimental results conclude that the proposed method outperforms the existing similar methods.

Acknowledgements. This work was supported by the Departmental Strategic Plan (PSD) of the University of Udine Interdepartmental Project on Artificial Intelligence (2020–2025).

References

1. Arpit, D., et al.: A closer look at memorization in deep networks. In: International Conference on Machine Learning, pp. 233–242. PMLR (2017)
2. Diao, E., Ding, J., Tarokh, V.: Heterofl: computation and communication efficient federated learning for heterogeneous clients. arXiv preprint arXiv:2010.01264 (2020)
3. Fang, X., Ye, M.: Robust federated learning with noisy and heterogeneous clients. In: Proceedings of the IEEE/CVF Conference on Computer Vision and Pattern Recognition, pp. 10072–10081 (2022)
4. Ghosh, A., Kumar, H., Sastry, P.S.: Robust loss functions under label noise for deep neural networks. In: Proceedings of the AAAI Conference on Artificial Intelligence, vol. 31 (2017)
5. Han, B., et al.: Co-teaching: robust training of deep neural networks with extremely noisy labels. In: Advances in Neural Information Processing Systems 31 (2018)
6. He, K., Zhang, X., Ren, S., Sun, J.: Deep residual learning for image recognition. In: Proceedings of the IEEE Conference on Computer Vision and Pattern Recognition, pp. 770–778 (2016)
7. Huang, Y., Gupta, S., Song, Z., Li, K., Arora, S.: Evaluating gradient inversion attacks and defenses in federated learning. Adv. Neural. Inf. Process. Syst. **34**, 7232–7241 (2021)
8. Jeong, W., Yoon, J., Yang, E., Hwang, S.J.: Federated semi-supervised learning with inter-client consistency & disjoint learning. arXiv preprint arXiv:2006.12097 (2020)
9. Jiang, L., Zhou, Z., Leung, T., Li, L.J., Fei-Fei, L.: Mentornet: learning data-driven curriculum for very deep neural networks on corrupted labels. In: International Conference on Machine Learning, pp. 2304–2313. PMLR (2018)
10. Kairouz, P., et al.: Advances and open problems in federated learning. Found. Trends Mach. Learn. **14**(1–2), 1–210 (2021)
11. Kingma, D.P., Ba, J.: Adam: a method for stochastic optimization. arXiv preprint arXiv:1412.6980 (2014)
12. Li, D., Wang, J.: Fedmd: heterogenous federated learning via model distillation. arXiv preprint arXiv:1910.03581 (2019)
13. Li, Q., He, B., Song, D.: Model-contrastive federated learning. In: Proceedings of the IEEE/CVF Conference on Computer Vision and Pattern Recognition, pp. 10713–10722 (2021)
14. Li, T., Sahu, A.K., Zaheer, M., Sanjabi, M., Talwalkar, A., Smith, V.: Federated optimization in heterogeneous networks. Proceedings of Machine learning and systems **2**, 429–450 (2020)
15. Liang, P.P., et al.: Think locally, act globally: federated learning with local and global representations. arXiv preprint arXiv:2001.01523 (2020)
16. Lin, T., Kong, L., Stich, S.U., Jaggi, M.: Ensemble distillation for robust model fusion in federated learning. Adv. Neural. Inf. Process. Syst. **33**, 2351–2363 (2020)
17. Madni, H.A., Umer, R.M., Foresti, G.L.: Blockchain-based swarm learning for the mitigation of gradient leakage in federated learning. IEEE Access **11**, 16549–16556 (2023)

18. Madni, H.A., Umer, R.M., Foresti, G.L.: Swarm-fhe: fully homomorphic encryption based swarm learning for malicious clients. Int. J. Neural Syst. (2023)
19. Matek, C., Schwarz, S., Spiekermann, K., Marr, C.: Human-level recognition of blast cells in acute myeloid leukaemia with convolutional neural networks. Nature Mach. Intell. **1**(11), 538–544 (2019)
20. McMahan, B., Moore, E., Ramage, D., Hampson, S., y Arcas, B.A.: Communication-efficient learning of deep networks from decentralized data. In: Artificial Intelligence and Statistics, pp. 1273–1282. PMLR (2017)
21. Miyato, T., Maeda, S.i., Koyama, M., Ishii, S.: Virtual adversarial training: a regularization method for supervised and semi-supervised learning. IEEE Trans. Pattern Anal. Mach. Intell. **41**(8), 1979–1993 (2018)
22. Patrini, G., Rozza, A., Krishna Menon, A., Nock, R., Qu, L.: Making deep neural networks robust to label noise: a loss correction approach. In: Proceedings of the IEEE Conference on Computer Vision and Pattern Recognition, pp. 1944–1952 (2017)
23. Sandler, M., Howard, A., Zhu, M., Zhmoginov, A., Chen, L.C.: Mobilenetv 2: inverted residuals and linear bottlenecks. In: Proceedings of the IEEE Conference on Computer Vision and Pattern Recognition, pp. 4510–4520 (2018)
24. Shen, T., et al.: Federated mutual learning. arXiv preprint arXiv:2006.16765 (2020)
25. Sukhbaatar, S., Bruna, J., Paluri, M., Bourdev, L., Fergus, R.: Training convolutional networks with noisy labels. arXiv preprint arXiv:1406.2080 (2014)
26. Umer, R.M., Gruber, A., Boushehri, S.S., Metak, C., Marr, C.: Imbalanced domain generalization for robust single cell classification in hematological cytomorphology. In: ICLR 2023 Workshop on Domain Generalization (2023)
27. Van Rooyen, B., Menon, A., Williamson, R.C.: Learning with symmetric label noise: The importance of being unhinged. In: Advances in neural information processing systems 28 (2015)
28. Wang, H., Yurochkin, M., Sun, Y., Papailiopoulos, D., Khazaeni, Y.: Federated learning with matched averaging. arXiv preprint arXiv:2002.06440 (2020)
29. Wang, Y., Ma, X., Chen, Z., Luo, Y., Yi, J., Bailey, J.: Symmetric cross entropy for robust learning with noisy labels. In: Proceedings of the IEEE/CVF International Conference on Computer Vision, pp. 322–330 (2019)
30. Warnat-Herresthal, S., et al.: Swarm learning for decentralized and confidential clinical machine learning. Nature **594**(7862), 265–270 (2021)
31. Wei, H., Feng, L., Chen, X., An, B.: Combating noisy labels by agreement: a joint training method with co-regularization. In: Proceedings of the IEEE/CVF Conference on Computer Vision and Pattern Recognition, pp. 13726–13735 (2020)
32. Yao, J., Wu, H., Zhang, Y., Tsang, I.W., Sun, J.: Safeguarded dynamic label regression for noisy supervision. In: Proceedings of the AAAI Conference on Artificial Intelligence, vol. 33, pp. 9103–9110 (2019)
33. Zhang, H., Cisse, M., Dauphin, Y.N., Lopez-Paz, D.: mixup: beyond empirical risk minimization. arXiv preprint arXiv:1710.09412 (2017)
34. Zhang, X., Zhou, X., Lin, M., Sun, J.: Shufflenet: an extremely efficient convolutional neural network for mobile devices. In: Proceedings of the IEEE Conference on Computer Vision and Pattern Recognition, pp. 6848–6856 (2018)

Experience Sharing and Human-in-the-Loop Optimization for Federated Robot Navigation Recommendation

Morteza Moradi(✉), Mohammad Moradi, and Dario Calogero Guastella

Department of Electrical, Electronic and Computer Engineering, University of Catania, Catania, Italy
morteza.moradi@phd.unict.it, dario.guastella@unict.it

Abstract. Mobile robot navigation in unknown, dynamic, hostile and/or crowded environments is a challenging task, especially when it comes to multi-robot systems. Looking at the issue from the lens of human-robot interaction (HRI) and artificial general intelligence, the keys are enabling (mobile) robots to deal with such a problem as human beings do by learning from (themselves or) counterparts' experience and providing them with freedom of choice. Besides traditional solutions in the field, learning from historical knowledge, in the form of experience sharing and experience replay gained momentum within the recent years. Extending these notions, the idea of taking advantages of the robot's collected information and perception of the environment (and the obstacles within); previous (case-specific) decisions and their consequences as well as complementary information provided by human-in-the-loop optimization, such as human-generated suggestions and advice, to provide the robotic agent with context-aware navigation recommendations is introduced in this paper. More specifically, the conceptual architecture of a robot navigation recommender system (RoboRecSys) is proposed to provide the agent with several options (based on different criteria) for making more efficient decisions in finding the most appropriate path towards its goal. Moreover, in order to preserve the privacy of both agents' data and environmental perception information and their decisions (and feedback) based on the received recommendations, the federated learning approach is employed.

Keywords: Mobile Robot Navigation · Experience Sharing · Robot Navigation Recommendation · Human-in-the-Loop · Human-Robot Interaction

1 Introduction

Autonomous mobile robot navigation in outdoor, dynamic, partially unknown and potentially crowded environments is considered a challenging and computa-

M. Moradi and M. Moradi—Equal contribution.

G. L. Foresti et al. (Eds.): ICIAP 2023 Workshops, LNCS 14366, pp. 179–188, 2024.
https://doi.org/10.1007/978-3-031-51026-7_16

tionally expensive task. Therefore, to handle such an important and intelligent-intensive problem, it is of high importance to develop more comprehensive solutions rather than only relying on the robot's sensors, equipment and current knowledge. Looking at robots as intelligent agents which try to mimic humans' behavior and decision-making approaches to cope with problems; relying on nature-inspired strategies to tackle intelligence-intensive problems can be considered as the key. More specifically, in addition to benefiting from collective machine intelligence [16], adopting the idea of taking advantages of previous experiences and best practices for solving similar problems in the future, like what humans do in their life is the silver bullet to cope with a wide variety of challenges in the robot control and decision-making. While collective machine intelligence refers to the concept that collaboration of several machines, i.e., intelligent agents, will yield in more efficient problem solving, reusing previous experiences of facing with decision-making problems in a particular context, e.g., robot navigation, to facilitate coping with partially similar situations, has been formulated as experience sharing [22] and experience replay [21]. Although robots can leverage the experience and successful efforts of themselves in the previous episodes and other agents' records in the same environment or other similar configurations, the multivariable nature of the task (and the context) may not be addressed completely. In other words, regarding several high-level/case-specific criteria and constraints or unexpected requirements, the priorities may be changed over time without any known pattern. Therefore, and to be well prepared to handle such uncertainties while preserving the performance, the robot should be provided with multiple options for every single step in the process to select among them the most appropriate one according to the situation. From another point of view, despite recent advancements in artificial intelligence, such as ChatGPT and Bard, looking for machines to present (be prepared with) human-level intelligence (maybe in the form of human-imitative AI [30]) in the near future is not a realistic expectation [15]. To fill this gap, bringing humans into the loop has been experienced as a working approach to provide robots with instructions and feedback or teaching them how to perform tasks properly. In this regard, and with the aim of introducing an adaptive, human-in-the-loop solution for robot navigation, in this work, the concept of a navigation recommendation system, RoboRecSys, is proposed. In addition to generate different context-aware suggestions for improving the mobile robot navigation in a hostile environment based on various criteria and considerations, the recommender system is, in fact, a part of the model predictive control component of the robot. Another important aspect of this solution is preserving the privacy of the involved agents, the information collected by them and their decisions and feedback. Depending on different application scenarios, such information may be of confidential or sensitive nature that sharing can put them in danger of disclosure and misuse. To deal with this indispensable concern in the distributed context of multi-robot collaboration [23,35] introduced in this work, the privacy-preserved collaborative learning, i.e., federated learning [2], approach is considered as a working solution.

In this paper, the underlying idea, implications and conceptual architecture of the RoboRecSys is delineated.

2 Learning from Experience

Although robot navigation in unknown environments is essentially based on exploration and making decisions when facing unexpected obstacles or untraversable paths; incorporating previous similar knowledge and experiences can improve the process in some extent. As a matter of fact, however, neither all obstacles in different environments are similar in shape nor all locations could be traversed and treated equally; those share some features and elements that can be used for the sake of generalization and pattern matching. Following this interesting motivation, several concepts in the context of multi-agent (reinforcement) learning, including experience sharing [6] and experience replay [11], have been proposed and employed.

The underlying idea for these notions, regardless of implementation and deployment details, was a simple one: enabling robots (artificial agents) to learn from best practices of other agents which obtained in the same environment or adaptable experience from their operations in different situations. Also, taking advantages of the recorded decisions of the same agent for its future challenges or sharing such information in a collaborative manner with counterparts are other forms of realization of this idea. This relatively new and bio-inspired strategy that is regarded a breakthrough in the field of robotics and multi-agent systems facilitates the process of continual learning and efficient real-time decision making in a wide variety of applications. More specifically, when it comes to robot navigation (in unknown environments), the power of sharing experience and relying on the previous operation records will be revealed more than other scenarios. This is due to the complexity of the navigation task in such environments that faces with many uncertainties. Therefore, taking benefits of such an approach can provide several highly precious opportunities to improve the environmental awareness, enhancing accuracy of obstacle detection, path planning, etc. and totally leveling up the performance of the agent in performing the designated task. Based upon the case, the agent can incorporate the knowledge of how to recognize and avoid moving/stationary objects of different kinds according to their geometrical shape, position and location or how to select the most appropriate path towards reaching the target location while maximizing the reward based on historical decisions and records of (itself or) other agents' operation. Nonetheless, with respect to different unpredictable situations that the agent may face with, the process of synthesizing a decision based on the previous experiences or integrating various information from different sources to generate a specific pattern or rule are computationally expensive and can affect the agent's performance and primary tasks the agent must perform during the operation. In this regard, in spite of its benefits, not all of robotic agents could leverage such a mechanism and to make it generally available, there is need to migrate from a stand-alone solution to the cloud-based one.

3 Recommendation as the Silver Bullet

As previously mentioned, the overhead of additional computational tasks related to inferring from shared experiences and integrating them into the robot's current workflow may put the agent into trouble both from energy efficiency and computing capability perspectives. To cope with such challenges and also make it possible for the agent to take the maximum benefits of the experience shared by other agents, providing the agent with some concrete recommendations seems to be a working strategy. In other words, instead of feeding the robots with raw experience records and expect it to extract the pattern and turning that into the control command(s), it is more feasible to provide the robot with the final decision(s) and required actions to fulfill that. Recommender systems as a useful tool for providing user-specific suggestions to help users make better decisions are around for years [37]. The general concept of recommender systems is to match user's preferences, previous choices, relationships and similarity to other users and (based on the context) his/her individual-specific characteristics with the most appropriate items, which may be a book to buy, a travel to go and so on. Besides common applications of recommender systems in e-commerce, education, healthcare, etc., with the development of location-based services, navigation recommendation systems, such as [29], has gained momentum. Inspired by such systems, we introduce the RoboRecSys as the first ever robot navigation recommender system. The RoboRecSys as a hybrid, deep learning-based recommendation engine works by integrating various information, including satellite information (e.g., GPS), shared experience of other agents, the robot's perception of the environment, its operation and performance logs as well as its live status. Then, taken into account the robot's goal, it generates several case-specific navigation recommendations with the aim of helping it reach its goal safely. To further facilitate the usage of these recommendations, they are generated in a machine-readable format, namely in the form of low-level control signals to feed the model predictive controller. The RoboRecSys, also, incrementally improves its knowledge base by considering its suggestions' feedback on the robots' performance.

An important consideration for the realization of the RoboRecSys is dealing with privacy concerns. Since in different application scenarios robots may collect sensitive information about, e.g., secure places, industrial processes and so on, sharing such information with other agents and making them available on the cloud can lead to confidentiality breach and misuse. Moreover, the robots' decisions that made based on the generated recommendations and their feedback can also disclose confidential information or unintentionally facilitate adversarial intentions. Regarding the distributed nature of the RoboRecSys, employing federated learning approach can be regarded as a state-of-the-art efficient solution to preserve the privacy of the agents, their data and decision-related information.

However, using federated learning is a good fit for the RoboRecSys through facilitating the continual learning process and robotic agents' communication with the cloud server; limited resources and computational power of the mobile robots may cause some essential problems, especially when it comes to performing resource-intensive tasks during long missions. To cope with such a situation and make using the federated learning possible and viable, the edge-based network architecture [34] is used. Such architecture, by its very nature, provides the opportunity of faster and more efficient information processing for the robotic agents. Doing so, to train their own local models, the agents' need to on-device computing power is compensated, and on other side, the risky storage of all of the information in the centralized cloud is no longer required. Therefore, this decentralization workflow, in addition to preserve the privacy of the agents' data, generates the navigation recommendations in a more efficient way.

To provide a clear representation of the RoboRecSys, its high-level conceptual architecture is illustrated in Fig. 1.

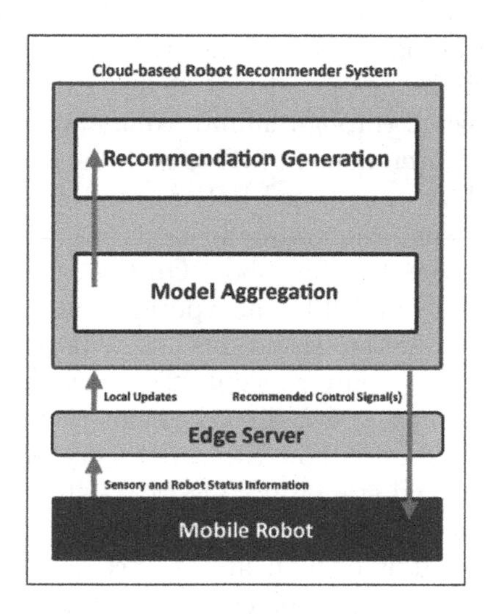

Fig. 1. The conceptual architecture of the RoboRecSys.

One of the most challenging issues with recommender systems is the lack of adequate information for the first-time users (cases), or simply cold-start problem. This is the case for the robot navigation scenarios in both directions when a robot should operate in a (new) unknown environment, where there is no information on the environment, including its traversability level and obstacles within. Also, the robot's performance in an unknown situation with unknown obstacles introduces some degree of uncertainties. To fill this gap, leveraging sim-to-real transfer approaches [17] can be regarded as a working solution. In fact, following this strategy helps to have some initial estimation on what robot may do and what it may face with. The recommendation generation process can also be extended to obstacle detection and other visual tasks. For example, type of an obstacle and how it should be dealt with may be recommended to the robot based on similarity (shape) of the perceived/sensed object to the known and recognized objects by other robots in the (similar) environment.

4 Human-in-the-Loop Optimization

With the aim of filling the gap between machine and human intelligence, or, in other words, to compensate the lack of human-level intelligence in machine-driven decisions, incorporating humans into the loop is known as a cost-effective strategy. Such an inclusion that sometimes is referred to as hybrid intelligence [9] has gained popularity in a wide variety of intelligence- and cognitive-intensive applications, ranging from semantic segmentation [26] to autonomous agents navigation [24] and control [7]. More specifically, leveraging human participation, e.g., in the form of crowdsourcing, in addition to preparing rich data for training artificial agents, can play a more active role by teaching the robot through demonstration [12], providing feedback on robot's actions and decisions [20], tuning reinforcement learning parameters [13] and so on. In the case of RoboRecSys, taking advantages of human participation is considered as a major pillar. In order to acquire human-generated navigation advice, their suggestions and preferences are collected through human computation game and simulation studies, such as what has been done in MIT's moral machine [1] and DeepTraffic [13] experiments as well as game-based HRI research [5]. Being integrated with contextual information, such human-made decisions will be used to generate navigation recommendations for the robot. On the other side, humans' intelligent and evaluative judgments will be incorporated into the process of assessing robots' selected choices in different situations [36]. This approach is even one step further than the shared autonomy concept [28], since instead of providing the robot with (raw) human inputs, it indirectly helps the agent with human-level judgement. In fact, to determine how the robot's decision (choice) is effective, in addition to traditional and standard statistical criteria, human feedback and scores are considered as complementary measures to assess how much the decisions are human-friendly. It is of high importance, specifically, when it comes to socially-aware robot navigation problem in crowded environments. To further improve the future recommendations, both types of feedback will be collected and analyzed to generate more accurate and human-friendly suggestions for the similar situations. The interaction among different components of the proposed system is represented in the workflow diagram in Fig. 2.

5 Security-Related Considerations

When it comes to real-world scenarios, the integrity and performance of the RoboRecSys may be threatened from two major sources, rather than common technical and unintentional problems. The first type of threats can be conducted by adversarial human participants. Generally, taking advantage of humans' intelligence through crowdsourcing is prone to adversarial attacks by malicious workers [4,14]. For example, some participants (possibly in an organized attack) may provide wrong (inaccurate) feedback on the robot's actions and decisions with the aim of influencing the recommendation process. To cope with human-initiated challenges, several consideration and regulations for worker selection [10], malicious worker detection [27,32] and quality control [8,18] should be taken into

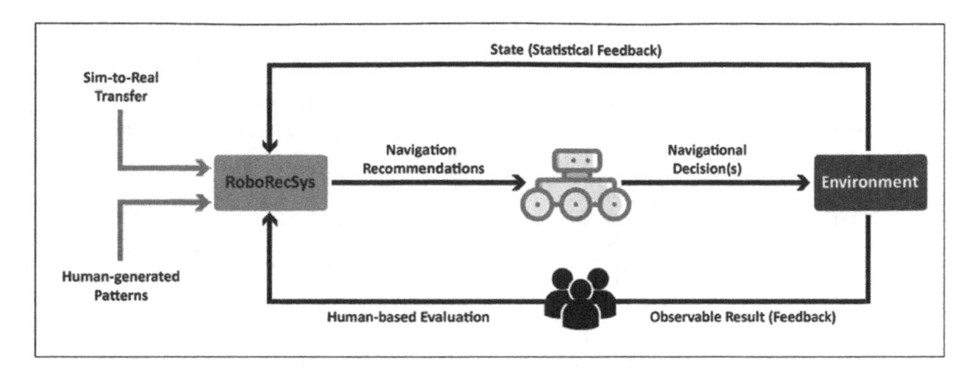

Fig. 2. The workflow of the RoboRecSys and the involved components.

account. On the other side, and as the second type of threats, robots may be targeted by cyber-attacks [33] for wide range of adversarial intentions, including impacting the agent's level of trust. Such type of threats makes robotic agents to provide misleading and unreliable information to consequently affect the recommendation generation process. Considering the trust of federated learning agents [19] and designing robust cooperative algorithms against attacks [31] and even in presence of malicious robots [3] are some of the state-of-the-art solutions in this regard. Moreover, the security of the recommender system [25] and communication infrastructure against must be preserved to mitigate other types of possible attacks.

6 Conclusion

In this work-in-progress, RoboRecSys, the first robot navigation recommender system is introduced. The underlying idea of the RoboRecSys is to provide the robot with several concrete options for dealing with navigation challenges, including obstacle recognition and finding safe paths, in an unknown environment. Such suggestions are generated based on the agent's previous decisions and experience shared by other agents as well as satellite information. Moreover, to incorporate human-level intelligence into the process of generating and evaluating recommendations and decisions, human participants are involved by providing feedback on the robot's choices and their consequences. From another point of view, in order to preserve the privacy of the agents' data and decisions they will make based on the received recommendation, the federated learning approach is employed. In this regard, and to realize this strategy, the edge-cloud architecture is used. Doing so, besides compensating the resource limitation of the robotic agents for performing computing tasks, the privacy preserving recommendation generation will be achieved.

References

1. Awad, E., Dsouza, S., Kim, R., Schulz, J., Henrich, J., Shariff, A., Bonnefon, J.F., Rahwan, I.: The moral machine experiment. Nature **563**(7729), 59–64 (2018)
2. Bonawitz, K., et al.: Towards federated learning at scale: System design. Proceedings of Machine Learning and Systems **1**, 374–388 (2019)
3. Cavorsi, M., Akgün, O.E., Yemini, M., Goldsmith, A.J., Gil, S.: Exploiting trust for resilient hypothesis testing with malicious robots. In: 2023 IEEE International Conference on Robotics and Automation (ICRA), pp. 7663–7669. IEEE (2023)
4. Checco, A., Bates, J., Demartini, G.: Adversarial attacks on crowdsourcing quality control. J. Artif. Intell. Res. **67**, 375–408 (2020)
5. Chernova, S., Orkin, J., Breazeal, C.: Crowdsourcing HRI through online multiplayer games. In: Dialog with Robots, Papers from the 2010 AAAI Fall Symposium, Arlington, Virginia, USA, November 11–13, 2010. AAAI Technical Report, vol. FS-10-05. AAAI (2010). https://www.aaai.org/ocs/index.php/FSS/FSS10/paper/view/2212
6. Christianos, F., Schäfer, L., Albrecht, S.: Shared experience actor-critic for multiagent reinforcement learning. Adv. Neural. Inf. Process. Syst. **33**, 10707–10717 (2020)
7. Dani, A.P., Salehi, I., Rotithor, G., Trombetta, D., Ravichandar, H.: Human-in-the-loop robot control for human-robot collaboration: human intention estimation and safe trajectory tracking control for collaborative tasks. IEEE Control Syst. Mag. **40**(6), 29–56 (2020)
8. Daniel, F., Kucherbaev, P., Cappiello, C., Benatallah, B., Allahbakhsh, M.: Quality control in crowdsourcing: a survey of quality attributes, assessment techniques, and assurance actions. ACM Comput. Surv. (CSUR) **51**(1), 1–40 (2018)
9. Dellermann, D., Ebel, P., Söllner, M., Leimeister, J.M.: Hybrid intelligence. Bus. Inf. Syst. Eng. **61**, 637–643 (2019)
10. Feng, W., Yan, Z., Yang, L.T., Zheng, Q.: Anonymous authentication on trust in blockchain-based mobile crowdsourcing. IEEE Internet Things J. **9**(16), 14185–14202 (2020)
11. Foerster, J., et al.: Stabilising experience replay for deep multi-agent reinforcement learning. In: International Conference on Machine Learning, pp. 1146–1155. PMLR (2017)
12. Forbes, M., Chung, M., Cakmak, M., Rao, R.: Robot programming by demonstration with crowdsourced action fixes. In: Proceedings of the AAAI Conference on Human Computation and Crowdsourcing, vol. 2, pp. 67–76 (2014)
13. Fridman, L., Terwilliger, J., Jenik, B.: Deeptraffic: Crowdsourced hyperparameter tuning of deep reinforcement learning systems for multi-agent dense traffic navigation. arXiv preprint arXiv:1801.02805 (2018)
14. Gadiraju, U., Kawase, R., Dietze, S., Demartini, G.: Understanding malicious behavior in crowdsourcing platforms: The case of online surveys. In: Proceedings of the 33rd Annual ACM Conference on Human Factors in Computing Systems, pp. 1631–1640 (2015)
15. Grace, K., Salvatier, J., Dafoe, A., Zhang, B., Evans, O.: Viewpoint: When will AI exceed human performance? evidence from AI experts. J. Artif. Intell. Res. **62**, 729–754 (2018). https://doi.org/10.1613/jair.1.11222. https://doi.org/10.1613/jair.1.11222
16. Halmes, M.: Measurements of collective machine intelligence. CoRR abs/1306.6649 (2013). https://arxiv.org/abs/1306.6649

17. Hu, H., Zhang, K., Tan, A.H., Ruan, M., Agia, C., Nejat, G.: A sim-to-real pipeline for deep reinforcement learning for autonomous robot navigation in cluttered rough terrain. IEEE Robot. Autom. Lett. **6**(4), 6569–6576 (2021)
18. Hu, Q., Wang, S., Ma, P., Cheng, X., Lv, W., Bie, R.: Quality control in crowdsourcing using sequential zero-determinant strategies. IEEE Trans. Knowl. Data Eng. **32**(5), 998–1009 (2019)
19. Imteaj, A., Amini, M.H.: Fedar: activity and resource-aware federated learning model for distributed mobile robots. In: 2020 19th IEEE International Conference on Machine Learning and Applications (ICMLA), pp. 1153–1160. IEEE (2020)
20. Jain, A., Das, D., Gupta, J.K., Saxena, A.: Planit: a crowdsourcing approach for learning to plan paths from large scale preference feedback. In: 2015 IEEE International Conference on Robotics and Automation (ICRA), pp. 877–884. IEEE (2015)
21. Jiang, L., Huang, H., Ding, Z.: Path planning for intelligent robots based on deep q-learning with experience replay and heuristic knowledge. IEEE/CAA J. Automatica Sinica **7**(4), 1179–1189 (2019)
22. Levine, S., Shah, D.: Learning robotic navigation from experience: principles, methods and recent results. Philos. Trans. R. Soc. B **378**(1869), 20210447 (2023)
23. Li, L., Bayuelo, A., Bobadilla, L., Alam, T., Shell, D.A.: Coordinated multi-robot planning while preserving individual privacy. In: 2019 International Conference on Robotics and Automation (ICRA), pp. 2188–2194 (2019). https://doi.org/10.1109/ICRA.2019.8794460
24. Liu, C., et al.: Human-machine cooperation research for navigation of maritime autonomous surface ships: a review and consideration. Ocean Eng. **246**, 110555 (2022)
25. O'Mahony, M.P., Hurley, N.J., Silvestre, G.C.: Recommender systems: attack types and strategies. In: AAAI, pp. 334–339 (2005)
26. Qiao, N., Sun, Y., Liu, C., Xia, L., Luo, J., Zhang, K., Kuo, C.: Human-in-the-loop video semantic segmentation auto-annotation. In: IEEE/CVF Winter Conference on Applications of Computer Vision, WACV 2023, Waikoloa, HI, USA, January 2–7, 2023, pp. 5870–5880. IEEE (2023). https://doi.org/10.1109/WACV56688.2023.00583
27. Qiu, C., Squicciarini, A.C., Carminati, B., Caverlee, J., Khare, D.R.: Crowdselect: increasing accuracy of crowdsourcing tasks through behavior prediction and user selection. In: Proceedings of the 25th ACM International on Conference on Information and Knowledge Management, pp. 539–548 (2016)
28. Reddy, S., Dragan, A.D., Levine, S.: Shared autonomy via deep reinforcement learning. In: Kress-Gazit, H., Srinivasa, S.S., Howard, T., Atanasov, N. (eds.) Robotics: Science and Systems XIV, Carnegie Mellon University, Pittsburgh, Pennsylvania, USA, June 26–30, 2018 (2018). https://doi.org/10.15607/RSS.2018.XIV.005. https://www.roboticsproceedings.org/rss14/p05.html
29. Shilov, N.: Recommender system for navigation safety: Requirements and methodology. TransNav: Int. J. Marine Navigation Saf. Sea Transp. **14**(2) (2020)
30. Silver, D., et al.: Mastering the game of go without human knowledge. Nature **550**(7676), 354–359 (2017)
31. Tasooji, T.K., Marquez, H.J.: A secure decentralized event-triggered cooperative localization in multi-robot systems under cyber attack. IEEE Access **10**, 128101–128121 (2022)
32. Wang, G., Wang, T., Zheng, H., Zhao, B.Y.: Man vs. machine: practical adversarial detection of malicious crowdsourcing workers. In: 23rd USENIX Security Symposium (USENIX Security 14), pp. 239–254 (2014)

33. Yaacoub, J.P.A., Noura, H.N., Salman, O., Chehab, A.: Robotics cyber security: Vulnerabilities, attacks, countermeasures, and recommendations. Int. J. Inf. Secur., 1–44 (2022)
34. Ye, Y., Li, S., Liu, F., Tang, Y., Hu, W.: Edgefed: optimized federated learning based on edge computing. IEEE Access **8**, 209191–209198 (2020). https://doi.org/10.1109/ACCESS.2020.3038287
35. Zhang, K., Li, Z., Wang, Y., Louati, A., Chen, J.: Privacy-preserving dynamic average consensus via state decomposition: case study on multi-robot formation control. Automatica **139**, 110182 (2022). https://doi.org/10.1016/j.automatica.2022.110182
36. Zhang, R., Torabi, F., Guan, L., Ballard, D.H., Stone, P.: Leveraging human guidance for deep reinforcement learning tasks. In: Kraus, S. (ed.) Proceedings of the Twenty-Eighth International Joint Conference on Artificial Intelligence, IJCAI 2019, Macao, China, August 10–16, 2019. pp. 6339–6346. ijcai.org (2019). https://doi.org/10.24963/ijcai.2019/884. https://doi.org/10.24963/ijcai.2019/884
37. Zhang, S., Yao, L., Sun, A., Tay, Y.: Deep learning based recommender system: a survey and new perspectives. ACM Comput. Surv. **52**(1), 5:1–5:38 (2019). https://doi.org/10.1145/3285029. https://doi.org/10.1145/3285029

FeDETR: A Federated Approach for Stenosis Detection in Coronary Angiography

Raffaele Mineo[1,2(✉)], Amelia Sorrenti[1,2], and Federica Proietto Salanitri[2]

[1] Department of Engineering, University Campus Bio-Medico of Rome, Rome, Italy
{amelia.sorrenti,raffaele.mineo}@unicampus.it
[2] PeRCeiVe Lab, University of Catania, Catania, Italy
federicaproietto.salanitri@phd.unict.it

Abstract. Assessing the severity of stenoses in coronary angiography is critical to the patient's health, as coronary stenosis is an underlying factor in heart failure. Current practice for grading coronary lesions, i.e. *fractional flow reserve* (FFR) or *instantaneous wave-free ratio* (iFR), suffers from several drawbacks, including time, cost and invasiveness, alongside potential interobserver variability. In this context, some deep learning methods have emerged to assist cardiologists in automating the estimation of FFR/iFR values. Despite the effectiveness of these methods, their reliance on large datasets is challenging due to the distributed nature of sensitive medical data. Federated learning addresses this challenge by aggregating knowledge from multiple nodes to improve model generalization, while preserving data privacy. We propose the first federated detection transformer approach, *FeDETR*, to assess stenosis severity in angiography videos based on FFR/iFR values estimation. In our approach, each node trains a detection transformer (DETR) on its local dataset, with the central server federating the backbone part of the network. The proposed method is trained and evaluated on a dataset collected from five hospitals, consisting of 1001 angiographic examinations, and its performance is compared with state-of-the-art federated learning methods.

Keywords: federated learning · coronary angiography · medical imaging analysis · stenosis detection

1 Introduction

One of the main causes of heart failure is *coronary stenosis*, which occurs when the blood vessels narrow and prevent the normal pumping of blood [13,27]. Depending on the severity of the stenosis, the first step in the evaluation of a coronary angiography is to decide whether medical or surgical treatment is required.

Raffaele Mineo and Amelia Sorrenti: Equal contribution.

G. L. Foresti et al. (Eds.): ICIAP 2023 Workshops, LNCS 14366, pp. 189–200, 2024.
https://doi.org/10.1007/978-3-031-51026-7_17

Invasive assessment of coronary physiology, using either *fractional flow reserve* (FFR) or *instantaneous wave-free ratio* (iFR), is an established guideline method for grading coronary lesions [13,27]. However, these methods have certain limitations, such as time consumption, high costs, and potential complications due to their invasive nature. In addition, inter-observer variability in clinical decision making may occur depending on the expertise of the examiner in locating major stenosis [33].

Despite these drawbacks, quantification of stenoses severity using FFR and iFR has gained popularity in guiding revascularization strategies for multivessel disease. Studies suggest that FFR values below 0.80 and iFR values below 0.89 are indicative of hemodynamically-significant stenosis [6,8,27,35], while patients with values above these thresholds may not benefit significantly from revascularization compared with optimal treatment alone. Therefore, automating the estimation of FFR or iFR values would assist clinicians in making accurate decisions and reducing the number of treatments patients need to undergo.

Over the last decade, the field of medical image analysis has seen the emergence of several deep learning methods to assist cardiologists in a variety of tasks (e.g., cardiovascular imaging analysis and risk assessment [18,26], and automated/semi-automated quantification of artery stenosis in coronary angiography [38,41,42]). In the context of stenoses quantification, the most promising approaches [41,42] take advantage of multiple angiography views in conjunction with the concept of *key frame*, that is the highest quality video frame characterized by complete penetration of the contrast agent. Although these techniques have been proven to be effective, they require datasets consisting of a large number of patient examinations, besides the need for physicians to manually identify key frames.

However, due to the sensitive nature of medical data, collecting and sharing large amounts of patient data distributed across multiple healthcare institutions and research centers is challenging. Federated Learning [14,16,24] (FL) supports the medical domain by offering approaches that leverage the potential of distributed healthcare data and ensure that this data remains local. FL aggregates knowledge from multiple nodes, even when each individual dataset is small. This collective learning approach improves the generalization and performance of the model by taking advantage of the diversity of data from different sources.

To overcome the above-mentioned limitations, we propose an approach named **FeDETR** for the assessment of stenosis severity from angiography videos by taking advantage of the federated setting. In our training procedure, each node in the centralized federation trains a detection transformer (DETR) [4] on its own small heterogeneous dataset, while the central server federates the backbone network. The proposed method detects hemodynamically significant stenosis in key frames using both direct and indirect estimation of FFR or iFR values. To train and evaluate the performance of our approach, the dataset used has been collected in five prominent hospitals and includes a total of 1001 angiographic examinations. We assess the validity of the proposed approach by comparing it to state-of-the-art federated methods (i.e., FedAvg [24] and FedBN [16]).

2 Related Work

Federated Learning [24] aims to develop models that can be used to exploit the collective knowledge of multiple nodes, while preserving data privacy and security. In a classic FL setting, multiple client nodes fine-tune a base model received from a central server using local data and send back local model updates. The central server then aggregates these updates into a global model, which is iteratively sent back to the nodes until convergence is reached. This aggregation can be performed using several methods.

FedAvg [24] averages the model updates received from all participating devices to create the updated global model. FedProx [14] extends such an approach by introducing a proximal term into the optimization objective, maintaining similarity between the global model and local models during aggregation. Federated Median [29], as an alternative to FedAvg [24], performs the aggregation by taking the median value of the model updates received from participating nodes, reducing the impact of outliers or noisy updates. In FedBN [16], the central node aggregates model parameters while keeping batch normalization layers private.

In the context of FL, most of the methods used to perform object detection involve the use of YOLOv3 [30] and Faster R-CNN [31], as proposed in [12,22]. FedVision [20] is a federated proprietary algorithm based on YOLOv3 [30] that enables end-to-end joint training of object detection models with locally stored datasets from multiple clients. Similarly, Yu and Liu [40] applied the FedAvg [24] algorithm to the Single Shot MultiBox Detector (SSD) [19], enhancing it with the Abnormal Weights Clip in order to reduce the influence of non-IID data.

Similarly, existing classical methods are used and adapted for federated image segmentation tasks. Tedeschini et al. [34] and Yi et al. [39] proposed two different modified versions of the U-NET [32] architecture to perform brain tumor segmentation. Li et al. [15] compared several weight sharing strategies to mitigate the effects of data imbalance across different hospitals and to reduce the potential risk of training data being reverse-engineered from the model parameters. FedDis [3] is a federated method for segmentation that disentangles the parameter space into shape and appearance, and shares only the shape parameter across clients, mitigating domain shifts among individual clients.

Since the medical field is characterized by distributed and heterogeneous data, it can be used as test bench for federated learning methods [7,9,15]. FL provides a privacy-preserving approach that enables healthcare institutions to collaborate and build robust and generalized models by training on diverse data from different patient populations.

Over the past few years, several deep learning methods have been employed to detect and quantify the severity of stenosis from imaging data. These methods can be based on *2D* or *3D approaches*. The former analyzes individual frames previously extracted from an angiography video and then performs the final prediction. The latter, which has been less researched, extracts the spatio-temporal features directly from the whole video [38,41,42].

The vast majority of the 2D approaches perform stenosis classification by grading into two or three severity levels or by thresholding FFR/iFR values in order to detect hemodynamically significant stenosis. These methods typically rely on CNN architectures [1,25] or a combination of convolutional and recurrent networks [5,23,28] to automatically identify *key frames* and then employ a stenosis classification module for final classification.

Alternatively, some approaches locate stenosis by focusing only on blood vessels in the key frame, i.e., by highlighting them in a pre-processing segmentation step [2,37] or by analyzing their shape and visual appearance [36,43,44]. Zhao et al. [44] perform automatic vessel segmentation, followed by keypoint extraction and classification to identify segments with a high likelihood of stenosis.

In addition, some 2D methods apply interpretability techniques on frame-based stenosis classification models to generate activation maps that help guide the stenosis detection process [5,25].

Differently from the aforementioned methods, our approach combines detection transformers (DETR) [4] with federated learning to assess stenosis severity from angiographic key frames.

3 Method

An overview of the proposed approach is shown in Fig. 1. In this scenario, a federation consists of a set of n peer nodes managed by a central server. Each node owns a private dataset consisting of *key frames* extracted from coronary angiography videos. This dataset is used to train a detection transformer (DETR) [4], as shown in Fig. 2.

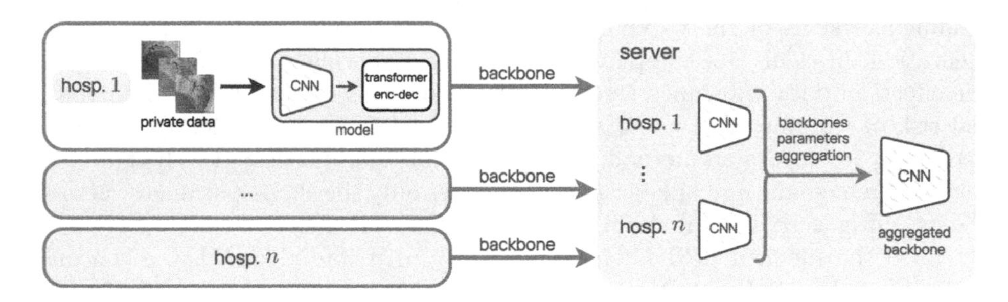

Fig. 1. Each node uses its own dataset, consisting of key frames extracted from coronary angiographies videos, to train a detection transformer (DETR). At the beginning of each round, the central server sends the aggregated backbone to each node, which uses it to extract features from the key frames. After each round, each node sends its locally trained backbone back to the central server, which aggregates them all.

During each round of centralized training, each node receives the backbone of the model from the server. The backbone is used to extract features from the images in the private dataset. After being flattened by the model and enhanced with position encoding, these features are fed into the encoder-decoder transformer. Each of the N output embeddings of the transformer decoder is then passed to a shared feed-forward network, which independently decodes it into bounding boxes and class labels, resulting in N final predictions.

At the end of each round, which involves multiple training iterations, the locally trained backbone from each device is sent back to the central server. The server performs the aggregation of the received backbones, and the entire process is repeated for the next round of training. This iterative process ensures collaboration between nodes while preserving data privacy and security.

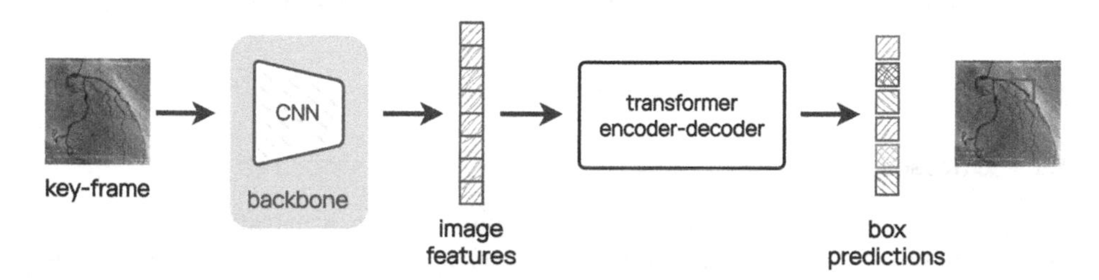

Fig. 2. Given a key frame, the CNN backbone learns a 2D representation, which is then flattened by the model. Before passing these learned features to the transformer encoder, the model also extends them with positional encoding. The transformer decoder outputs some learned positional embeddings and feeds them into a feed-forward network that predicts classes and their corresponding bounding boxes.

3.1 Problem Formulation

The proposed method operates in a federated setting, where K nodes jointly train for R rounds. At the end of each round, which consists of T local epochs, a communication with a central server is performed.

Each node i trains a local model M_i on its own private dataset $\mathcal{D}_i = \{(\mathbf{x}_1, y_1, \mathbf{b}_1), (\mathbf{x}_2, y_2, \mathbf{b}_2), \ldots, (\mathbf{x}_n, y_n, \mathbf{b}_n)\}$, where $\mathbf{x}_j \in \mathcal{X}$ represents a sample image within the dataset, $y_j \in \mathcal{Y}$ represents the target object class (low-severity or mild/high-severity stenosis), and $\mathbf{b}_j \in \mathbb{R}^4$ represents bound box coordinates. Each local model $M_i = \langle g, \boldsymbol{\theta}_i \rangle$ is parameterized by $\boldsymbol{\theta}_i$, with $g : \mathcal{X} \to \mathcal{Y}$. Specifically, in the proposed approach, the K nodes train a local DETR model [4] whose architecture is illustrated in Fig. 2. Since the model consists of two parts, the model parameters $\boldsymbol{\theta}_i$ can be written as the union of the backbone and the encoder-decoder transformer parameters, $\boldsymbol{\phi}_i$ and $\boldsymbol{\omega}_i$ respectively.

Hence, at training time, each node i in the federation, locally optimizes a finite sum objective in the form:

$$\min_{\boldsymbol{\theta}_i} f(\boldsymbol{\theta}_i) \quad \text{with} \quad f(\boldsymbol{\theta}_i) \equiv f(\boldsymbol{\phi}_i, \boldsymbol{\omega}_i), \tag{1}$$

defining $f(\boldsymbol{\phi}_i, \boldsymbol{\omega}_i)$ as follows:

$$f(\phi_i, \omega_i) = \frac{1}{n} \sum_{j=1}^{n} f_j(\phi_i, \omega_i), \tag{2}$$

where $f_j(\phi_i, \omega_i) = \ell(g(\mathbf{x}_j), y_j, \mathbf{b}_j; \phi_i, \omega_i)$ denotes an object detection loss for $(\mathbf{x}_j, y_j)$ using the model parameters $\boldsymbol{\theta}_i$.

In contrast to current federated approaches, which are mostly based on the averaging of all model parameters, our method only federates the backbone part of the model. Indeed, at the end of each training round, the nodes in the federation apply the FedAvg algorithm [24] on the subset of parameters $\{\phi_1, \ldots, \phi_K\}$, thus receiving an updated version of the backbone and resuming local training.

4 Experimental Evaluation

4.1 Dataset

The proposed method was trained and evaluated using a private dataset comprising of 1001 coronary angiographies from patients with chronic coronary syndrome (CCS) and acute coronary syndrome (ACS). The dataset was collected in five large hospitals between 2020 and 2022, and included 737 male and 264 female patients with an average age of 66.8 ± 9.26 years.

The coronary angiographies and the physiological measurements were carried out in accordance with the standard clinical practice. Then, two experienced cardiologists annotated the key frames in the angiography data. Each angiography was evaluated using invasive physiological assessment with iFR, FFR, or both. Specifically, FFR values were available for 613 patients (61.2%), iFR values for 667 patients (66.6%), and a subset of 279 patients (27.8%) had both FFR and iFR data. Cardiologists identified the major stenosis in each exam and labeled it as hemodynamically significant if the FFR was less than 0.80 [8,27,35] or if the iFR was less than 0.89 [6,27]. Consequently, the 40.5% of patients has been labeled as positive, resulting in a significant class imbalance.

4.2 Training Procedure

Each node trains a DETR model [4], consisting of a ResNet-50 [11] backbone and an encoder-decoder transformer. The model was pre-trained on the COCO 2017 dataset [17]. The DETR models were trained by minimizing the Hungarian loss [4] using the AdamW [21] optimizer with initial learning rates set to 10^{-4} and 10^{-5} for the transformer and the backbone, respectively, and weight decay set to 10^{-4}. The learning rate is then updated using a scheduler with the multiplicative factor γ set to 0.1. Gradient clipping was also applied with a maximum gradient norm of 0.1.

Both the encoder and the decoder consist of six layers characterized by a standard architecture. Their self- and cross-attention modules have eight heads, each receiving input embeddings of size 256. In addition, the feed-forward network (FFN) of these modules has a dimensionality of 2048. An additive dropout of 0.1 is applied after each multi-head attention and FFN and before performing the layer normalization. The output of the encoder is supplemented with a sinusoidal positional encoding before being passed to the decoder. The number of object queries of the transformer decoder is 20.

Data augmentation is performed by consecutively and randomly applying horizontal and vertical flips, and 90-degree rotations. Scale augmentation is then applied to the images, resizing them so that the shortest side is 512 pixels and the longest side is 1333 pixels or less. Furthermore, zero padding is also added to the images, making them rectangular in order to shift the detection area. Since bounding boxes are usually positioned in the upper left corner or centrally, as demonstrated by the heatmap in Fig. 3b, the additional padding should prevent the model from learning a location bias.

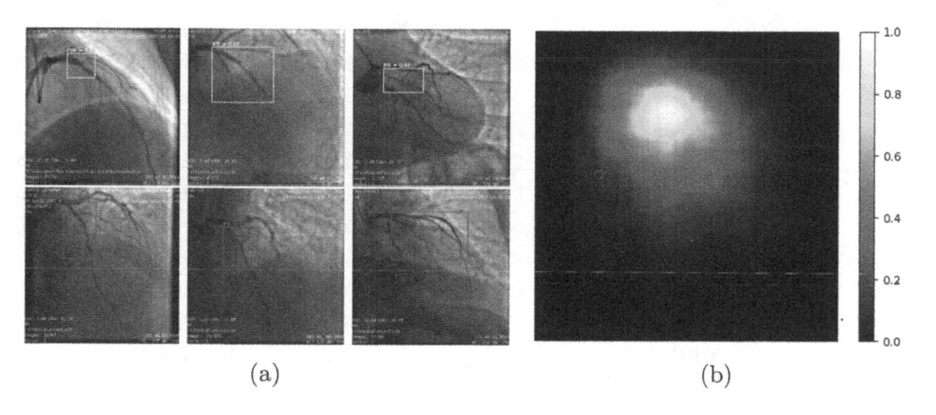

Fig. 3. Figure 3a shows representative examples of *low severity stenoses* (illustrated in the top row) and *mild or high severity stenoses* (depicted in the bottom row). Figure 3b presents an averaged heatmap, which is derived from the collective bounding box positions across all samples.

4.3 Results

In order to evaluate the performance of our method, FeDETR, we compare its performance to baseline approaches (FedAvg [24] and FedBN [16]) on standard object detection metrics, namely, accuracy (ACC), precision (PPV), recall (TPR) and intersection over union (IoU). Results are presented in Tab. 1.

The federated training approach consistently achieves significant accuracy gains over baseline DETR models that utilize a combined dataset from all nodes (*joint*), and even more compared to local training alone (*standalone*). This outcome underscores that training on smaller individual node datasets outperforms a unified dataset in a single node. Furthermore, when comparing federated training on individual node datasets to training on a single node's dataset, there is a notable 12-point increase in accuracy. This suggests that federated training on

Table 1. Performance comparison between the proposed approach, FedAvg and FedBN, for different combinations of number of rounds and epochs, on the target metrics. Standard deviations refer to inter-node variability.

Rounds	Epochs	Methods	PPV		TPR		IoU		ACC
			low	*high*	*low*	*high*	*low*	*high*	
—	200	*joint training*	69.8	61.9	80.2	48.1	52.2	50.8	67.3
—	200	*standalone*	64.8	58.1	68.7	43.2	43.9 ± 2.3	42.7 ± 3.8	58.5 ± 3.2
25	5	FedAvg [24]	55.4	29.5	78.0	25.9	53.1 ± 4.1	42.0 ± 11.1	72.2 ± 4.4
		FedBN [16]	57.5	27.9	73.3	34.1	51.8 ± 5.2	54.3 ± 1.4	71.8 ± 4.1
		Ours	57.9	33.5	73.4	33.9	51.2 ± 1.8	51.3 ± 6.8	70.9 ± 3.5
	10	FedAvg [24]	55.2	28.1	77.3	25.9	51.0± 2.6	40.8 ± 11.2	71.8 ± 4.6
		FedBN [16]	59.1	47.1	80.5	25.3	54.5 ± 3.2	52.1 ± 3.8	69.0 ± 1.9
		Ours	62.0	47.9	77.8	31.2	53.3 ± 4.0	37.3 ± 9.4	70.2 ± 3.0
	20	FedAvg [24]	57.8	31.7	77.3	32.9	51.4 ± 2.9	44.6 ± 8.5	74.3 ± 4.5
		FedBN [16]	58.3	29.3	74.1	34.4	55.7 ± 3.9	52.1 ± 3.6	72.4 ± 4.0
		Ours	59.8	38.7	68.1	41.1	51.1 ± 3.4	40.2 ± 5.2	71.0 ± 3.4
50	5	FedAvg [24]	58.3	28.9	74.0	35.3	53.2 ± 4.3	54.9 ± 2.9	73.1 ± 4.3
		FedBN [16]	63.0	73.3	85.2	32.8	52.8 ± 3.4	54.0 ± 1.3	71.1 ± 2.3
		Ours	64.4	66.3	78.9	39.8	46.2 ± 4.2	44.2 ± 5.2	69.4 ± 1.6
	10	FedAvg [24]	62.0	56.1	83.4	28.3	55.1 ± 4.3	47.5 ± 10.2	69.2 ± 2.0
		FedBN [16]	63.5	72.2	94.9	30.4	55.5 ± 3.7	54.3 ± 4.4	69.5 ± 3.2
		Ours	66.9	66.8	81.1	46.5	51.2 ± 2.9	46.9 ± 6.9	73.6 ± 3.8
	20	FedAvg [24]	59.3	49.8	74.1	37.3	56.6 ± 3.1	52.8 ± 2.1	73.9 ± 4.1
		FedBN [16]	61.2	47.6	85.4	27.8	58.5 ± 5.4	48.3 ± 4.8	66.0 ± 1.5
		Ours	67.6	49.8	69.5	47.9	50.3 ± 3.5	50.0 ± 2.9	70.4 ± 3.7

node-specific data surpasses training on a single consolidated dataset, possibly due to large data distribution shifts between nodes.

The comparison between the three federated approaches, i.e. FeDETR, FedAvg [24] and FedBN [16], shows that all of them achieve similar performance in both classification accuracy and object detection metrics. In particular, FeDETR shows generally improved performance at a lower number of rounds, highlighting better convergence properties. Also, while FeDETR seems to achieve better scores in precision rather than recall, we can observe a general trend of improvement on mild/high–severity stenosis, which can be a critical factor for diagnostic purposes. On the other hand, FeDETR shows worse performance on bounding box prediction, based on the IoU metric.

In particular, FeDETR employs a model that shares only 15% of the total model's parameters, specifically the backbone, omitting the transformer component. This strategic parameter sharing maintains privacy concerns (since weight sharing can be subject to malicious attacks aimed at reconstructing model inputs [10, 45]) while still delivering competitive performance.

These results collectively underscore the effectiveness of federated training in enhancing accuracy and performance across various scenarios, showcasing its potential for collaborative and privacy-preserving machine learning.

5 Conclusion

In this paper, we propose a novel federated learning approach for the assessment of coronary stenosis. Leveraging the Detection Transformer (DETR) model [4], the proposed method includes *fractional flow reserve* (FFR) and *instantaneous wave-free ratio* (iFR) values to accurately evaluate stenosis severity.

In our approach, each node in the federation trains a DETR model to detect high severity stenoses in key frames extracted from angiography videos. At the end of each round, the central server receives the backbones of the models from the nodes and aggregates them.

By taking advantage of federated learning, we address the challenges of privacy concerns and data sharing in the medical domain, ensuring the security and integrity of patient data. Moreover, we provide cardiologists with a reliable tool for accurate assessment of coronary stenosis, contributing to improved patient care and informed medical decision-making. Our evaluation goes beyond traditional benchmarks by using real data from a dataset collected from five hospitals: we demonstrate that the proposed model is able to achieve competitive performance to state-of-the-art approaches, and even yields better results in the identification of mild/high–severity stenoses, thus showing the robustness and practical applicability of our approach in real clinical settings.

To the best of our knowledge, this work stands as a pioneering endeavor in employing a federated detection transformer, holding applicability within computer vision for detection tasks that hinge upon datasets comprised of non-i.i.d. sub-datasets. In the future, we aim at improving our method by automating the procedure for the selection of the key frames from angiography videos. Furthermore, we envisage investigating broader applications, such as using our approach for medical examinations that use contrast agents to improve the quality of imaging.

Acknowledgments. Raffaele Mineo and Amelia Sorrenti are PhD students enrolled in the National PhD in Artificial Intelligence, cycle XXXVII and XXXVIII respectively, course on Health and life sciences, organized by University Campus Bio-Medico of Rome.

References

1. Antczak, K., Liberadzki, Ł.: Stenosis detection with deep convolutional neural networks. In: MATEC Web of Conferences, vol. 210, p. 04001. EDP Sciences (2018)
2. Au, B., et al.: Automated characterization of stenosis in invasive coronary angiography images with convolutional neural networks. arXiv preprint arXiv:1807.10597 (2018)

3. Bercea, C.I., Wiestler, B., Rueckert, D., Albarqouni, S.: Feddis: Disentangled federated learning for unsupervised brain pathology segmentation. arXiv preprint arXiv:2103.03705 (2021)
4. Carion, N., Massa, F., Synnaeve, G., Usunier, N., Kirillov, A., Zagoruyko, S.: End-to-end object detection with transformers. In: European Conference on Computer Vision, pp. 213–229. Springer (2020)
5. Cong, C., Kato, Y., Vasconcellos, H.D., Lima, J., Venkatesh, B.: Automated stenosis detection and classification in x-ray angiography using deep neural network. In: 2019 IEEE BIBM (2019)
6. Davies, J.E., et al.: Use of the instantaneous wave-free ratio or fractional flow reserve in pci. N. Engl. J. Med. **376**(19), 1824–1834 (2017)
7. Dayan, I., et al.: Federated learning for predicting clinical outcomes in patients with covid-19. Nat. Med. **27**(10), 1735–1743 (2021)
8. De Bruyne, B., et al.: Fractional flow reserve-guided pci versus medical therapy in stable coronary disease. N. Engl. J. Med. **367**(11), 991–1001 (2012)
9. Feki, I., Ammar, S., Kessentini, Y., Muhammad, K.: Federated learning for covid-19 screening from chest x-ray images. Appl. Soft Comput. **106**, 107330 (2021)
10. Geiping, J., et al.: Inverting gradients-how easy is it to break privacy in federated learning? NeurIPS (2020)
11. He, K., Zhang, X., Ren, S., Sun, J.: Deep residual learning for image recognition. In: Proceedings of the IEEE Conference on Computer Vision and Pattern Recognition, pp. 770–778 (2016)
12. Jallepalli, D., Ravikumar, N.C., Badarinath, P.V., Uchil, S., Suresh, M.A.: Federated learning for object detection in autonomous vehicles. In: 2021 IEEE Seventh International Conference on Big Data Computing Service and Applications (BigDataService), pp. 107–114. IEEE (2021)
13. Knuuti, J., Revenco, V.: 2019 esc guidelines for the diagnosis and management of chronic coronary syndromes. Eur. Heart J. **41**(5), 407–477 (2020)
14. Li, T., Sahu, A.K., Zaheer, M., Sanjabi, M., Talwalkar, A., Smith, V.: Federated optimization in heterogeneous networks. Proc. Mach. Learn. Syst. **2**, 429–450 (2020)
15. Li, W., et al.: Privacy-preserving federated brain tumour segmentation. In: Suk, H.-I., Liu, M., Yan, P., Lian, C. (eds.) MLMI 2019. LNCS, vol. 11861, pp. 133–141. Springer, Cham (2019). https://doi.org/10.1007/978-3-030-32692-0_16
16. Li, X., Jiang, M., Zhang, X., Kamp, M., Dou, Q.: Fedbn: federated learning on non-iid features via local batch normalization. arXiv preprint arXiv:2102.07623 (2021)
17. Lin, T.-Y., Maire, M., Belongie, S., Hays, J., Perona, P., Ramanan, D., Dollár, P., Zitnick, C.L.: Microsoft COCO: common objects in context. In: Fleet, D., Pajdla, T., Schiele, B., Tuytelaars, T. (eds.) ECCV 2014. LNCS, vol. 8693, pp. 740–755. Springer, Cham (2014). https://doi.org/10.1007/978-3-319-10602-1_48
18. Litjens, G., et al.: State-of-the-art deep learning in cardiovascular image analysis. JACC: Cardiovascular Imaging **12**(8 Part 1), 1549–1565 (2019)
19. Liu, W., Anguelov, D., Erhan, D., Szegedy, C., Reed, S., Fu, C.-Y., Berg, A.C.: SSD: single shot MultiBox detector. In: Leibe, B., Matas, J., Sebe, N., Welling, M. (eds.) ECCV 2016. LNCS, vol. 9905, pp. 21–37. Springer, Cham (2016). https://doi.org/10.1007/978-3-319-46448-0_2
20. Liu, Y., et al.: Fedvision: an online visual object detection platform powered by federated learning. In: Proceedings of the AAAI Conference on Artificial Intelligence, vol. 34, pp. 13172–13179 (2020)

21. Loshchilov, I., Hutter, F.: Decoupled weight decay regularization. arXiv preprint arXiv:1711.05101 (2017)
22. Luo, J., Wu, X., Luo, Y., Huang, A., Huang, Y., Liu, Y., Yang, Q.: Real-world image datasets for federated learning. arXiv preprint arXiv:1910.11089 (2019)
23. Ma, H., Ambrosini, P., van Walsum, T.: Fast prospective detection of contrast inflow in X-ray angiograms with convolutional neural network and recurrent neural network. In: Descoteaux, M., Maier-Hein, L., Franz, A., Jannin, P., Collins, D.L., Duchesne, S. (eds.) MICCAI 2017. LNCS, vol. 10435, pp. 453–461. Springer, Cham (2017). https://doi.org/10.1007/978-3-319-66179-7_52
24. McMahan, B., Moore, E., Ramage, D., Hampson, S., Arcas, B.A.: Communication-efficient learning of deep networks from decentralized data. In: Artificial Intelligence and Statistics, pp. 1273–1282. PMLR (2017)
25. Moon, J.H., Cha, W.C., Chung, M.J., Lee, K.S., Cho, B.H., Choi, J.H., et al.: Automatic stenosis recognition from coronary angiography using convolutional neural networks. Comput. Methods Programs Biomed. **198**, 105819 (2021)
26. Motwani, M., et al.: Machine learning for prediction of all-cause mortality in patients with suspected coronary artery disease: a 5-year multicentre prospective registry analysis. Eur. Heart J. **38**(7), 500–507 (2017)
27. Neumann, F.J., et al.: 2018 ESC/EACTS Guidelines on myocardial revascularization. Europ. Heart J. **40**(2), 87–165 (2018). https://doi.org/10.1093/eurheartj/ehy394
28. Ovalle-Magallanes, E., Avina-Cervantes, J.G., Cruz-Aceves, I., Ruiz-Pinales, J.: Hybrid classical-quantum convolutional neural network for stenosis detection in x-ray coronary angiography. Expert Syst. Appl. **189**, 116112 (2022)
29. Pillutla, K., Kakade, S.M., Harchaoui, Z.: Robust aggregation for federated learning. IEEE Trans. Signal Process. **70**, 1142–1154 (2022)
30. Redmon, J., Farhadi, A.: Yolov3: An incremental improvement. arXiv preprint arXiv:1804.02767 (2018)
31. Ren, S., He, K., Girshick, R., Sun, J.: Faster r-cnn: towards real-time object detection with region proposal networks. Advances in neural information processing systems 28 (2015)
32. Ronneberger, O., Fischer, P., Brox, T.: U-Net: convolutional networks for biomedical image segmentation. In: Navab, N., Hornegger, J., Wells, W.M., Frangi, A.F. (eds.) MICCAI 2015. LNCS, vol. 9351, pp. 234–241. Springer, Cham (2015). https://doi.org/10.1007/978-3-319-24574-4_28
33. Singh, K., Jacobsen, B., Solberg, S., Bønaa, K., Kumar, S., Bajic, R., Arnesen, E.: Intra-and interobserver variability in the measurements of abdominal aortic and common iliac artery diameter with computed tomography. the tromsø study. Europ. J. Vascular Endovascular Surg. **25**(5), 399–407 (2003)
34. Tedeschini, B.C., et al.: Decentralized federated learning for healthcare networks: a case study on tumor segmentation. IEEE Access **10**, 8693–8708 (2022)
35. Tonino, P.A., et al.: Fractional flow reserve versus angiography for guiding percutaneous coronary intervention. New England J. Med. **360**(3), 213–224 (2009)
36. Wan, T., Feng, H., Tong, C., Li, D., Qin, Z.: Automated identification and grading of coronary artery stenoses with x-ray angiography. Comput. Methods Programs Biomed. **167**, 13–22 (2018)
37. Wu, W., Zhang, J., Xie, H., Zhao, Y., Zhang, S., Gu, L.: Automatic detection of coronary artery stenosis by convolutional neural network with temporal constraint. Comput. Biol. Med. **118**, 103657 (2020)
38. Xue, W., Brahm, G., Pandey, S., Leung, S., Li, S.: Full left ventricle quantification via deep multitask relationships learning. Med. Image Anal. **43**, 54–65 (2018)

39. Yi, L., Zhang, J., Zhang, R., Shi, J., Wang, G., Liu, X.: SU-net: an efficient encoder-decoder model of federated learning for brain tumor segmentation. In: Farkaš, I., Masulli, P., Wermter, S. (eds.) ICANN 2020. LNCS, vol. 12396, pp. 761–773. Springer, Cham (2020). https://doi.org/10.1007/978-3-030-61609-0_60
40. Yu, P., Liu, Y.: Federated object detection: optimizing object detection model with federated learning. In: Proceedings of the 3rd International Conference on Vision, Image and Signal Processing, pp. 1–6 (2019)
41. Zhang, D., et al.: Direct quantification of coronary artery stenosis through hierarchical attentive multi-view learning. IEEE Trans. Med. Imaging **39**(12), 4322–4334 (2020)
42. Zhang, D., Yang, G., Zhao, S., Zhang, Y., Zhang, H., Li, S.: Direct quantification for coronary artery stenosis using multiview learning. In: Shen, D., Liu, T., Peters, T.M., Staib, L.H., Essert, C., Zhou, S., Yap, P.-T., Khan, A. (eds.) MICCAI 2019. LNCS, vol. 11765, pp. 449–457. Springer, Cham (2019). https://doi.org/10.1007/978-3-030-32245-8_50
43. Zhao, C., et al.: A new approach to extracting coronary arteries and detecting stenosis in invasive coronary angiograms. arXiv preprint arXiv:2101.09848 (2021)
44. Zhao, C., et al.: Automatic extraction and stenosis evaluation of coronary arteries in invasive coronary angiograms. Comput. Biol. Med. **136**, 104667 (2021)
45. Zhu, L., et al.: Deep leakage from gradients. NeurIPS (2019)

FeDZIO: Decentralized Federated Knowledge Distillation on Edge Devices

Luca Palazzo(✉), Matteo Pennisi, Giovanni Bellitto, and Isaak Kavasidis

PeRCeiVe Lab, University of Catania, Catania, Italy
{luca.palazzo,matteo.pennisi,giovanni.bellitto}@phd.unict.it, isaak.kavasidis@unict.it

Abstract. In recent years, the proliferation of edge devices and distributed sensors has fueled the need for training sophisticated deep learning models directly on resource-constrained nodes, in order to guarantee data locality and prevent the transmission of private information to centralized training infrastructures. However, executing large-scale models on edge devices poses significant challenges due to limited computational power, memory constraints and energy consumption limitation. Federated Learning (FL) has emerged as a promising approach to partially address these issues, enabling decentralized model training across multiple devices without the need to exchange local data. At the same time, Knowledge Distillation (KD) has demonstrated its efficacy in compressing complex models by transferring knowledge from a larger teacher model to a smaller student model.

This paper presents a novel framework combining Federated Learning with Knowledge Distillation, specifically tailored for accelerating training on edge devices. The proposed approach leverages the collaborative learning capabilities of federated learning to perform knowledge distillation in a privacy-preserving and efficient manner. Instead of relying on a central server for aggregation, edge devices with localized data collaboratively exchange knowledge with each other, enabling transmission of minimal quantities of data without compromising data privacy and model performance. The distributed nature of this approach allows edge devices to leverage collective intelligence while avoiding the need for sharing raw data across the network.

We conduct extensive experiments on diverse edge device scenarios using state-of-the-art deep learning architectures. The results demonstrate that our approach achieves substantial model compression while maintaining competitive performance compared to traditional knowledge distillation methods. Additionally, the federated nature of our approach ensures scalability and robustness, even in dynamic edge device environments.

Keywords: Federated learning · Knowledge distillation · Edge machine learning

G. L. Foresti et al. (Eds.): ICIAP 2023 Workshops, LNCS 14366, pp. 201–210, 2024.
https://doi.org/10.1007/978-3-031-51026-7_18

1 Introduction

In recent years, the world has witnessed an unprecedented surge in data generation, fueled by the proliferation of smart devices with low computation capabilities, such as Internet of Things (IoT) technologies, and the increasing adoption of digital services. This massive influx of data presents a unique opportunity to accelerate the development of Artificial Intelligence (AI) models and algorithms, promising groundbreaking advancements in various domains, ranging from healthcare and finance to transportation and natural language processing. However, with this exponential growth in data comes a considerable challenge regarding how to process and utilize it efficiently, securely, and in a privacy-preserving manner.

Traditional machine learning paradigms heavily rely on centralized data repositories, where all the training data is aggregated in a single location for model training. However, this approach raises significant concerns related to data privacy, security and sovereignty. To address these challenges and tap into the full potential of distributed data, centralized [6] and decentralized [7] federated learning (FL) has emerged as a promising paradigm that allows AI models to be trained collaboratively while preserving data privacy and control. Decentralized federated learning leverages the principles of federated learning and decentralization to distribute the model training process across multiple devices without the need to centralize the raw data. In this approach, individual devices participate in the model training process by collaboratively learning from their local data and sharing only model updates or gradients amongst themselves. As such, there is no need for a central aggregation server as in classical centralized federation learning.

However, in such scenario, it is usually implied that nodes participating in the federation have significant local computational power, allowing them to both train a local model and process incoming updates from other devices. This hypothesis may not be valid when federated training is carried at *at the edge*, i.e., directly on sensor nodes or devices with limited resources. In this context, Knowledge Distillation (KD) [2] has emerged as a technique used in machine learning to transfer the knowledge from one, more complex model (the "teacher") to another, less powerful model (the "student") in order to improve the performance of the student model.

This paper presents *FeDZIO, Federated Distillation with Zonal Interaction Optimizer*, a method that combines decentralized federated learning and knowledge distillation techniques to effectively train machine learning models at the edge, taking into account multiple *zones* characterized by different computational power and ensuring data privacy guarantees. To this aim, we train local models on *strong* and *weak* nodes (the latter representing edge devices) and perform federated knowledge distillation by sharing, in a decentralized way, class logits from strong nodes to weak nodes, thus enforcing the transmission of class-discriminative information while avoiding actual data sharing. Such integration leads to a robust and scalable learning ecosystem that can be deployed in scenarios where large datasets must be kept separated and private, but less capable nodes must be able to exploit top-level information from such datasets. We extensively evaluate the proposed approach, by comparing it to local (i.e., non-federated) training, varying different aspects of the proposed approach, as well as the proportion of each type of node. Our results show that our approach effectively allows for model compression on edge devices, while improving performance with respect to local training and highlighting the advantages of knowledge distillation even from a small number of nodes.

2 Related Work

The adoption of federated learning on edge devices has been the subject of extensive research due to its potential to revolutionize the field of distributed machine learning while addressing the challenges posed by centralized data processing [8]. A seminal work [5] laid the groundwork for decentralized federated learning, introducing a novel approach that allows edge devices to collaboratively learn from their local data and contribute to model updates without sharing raw data. The authors proposed a secure aggregation mechanism that ensures privacy and integrity during the model aggregation process. Their results demonstrated the feasibility of training deep learning models on resource-constrained edge devices, opening up new avenues for privacy-preserving AI [4], under the term *edge intelligence.*

In [1] the authors present an architecture that efficiently coordinates model aggregation and communication between edge devices and a central server. Their work emphasized the importance of addressing communication overheads and latency concerns for successful federated learning at the edge but however, highlighting the potential of edge intelligence in enabling federated learning on resource-constrained edge devices. However, in many scenarios the presence of an aggregation server is not guaranteed.

In such scenarios, completely decentralized federated learning methods, such as [11] offer the advantage of peer-to-peer model updates eventually converging to an optimum global model. However, in such cases, transmitting model updates is still a cumbersome process for devices with limited capabilities and energy availability.

Model partitioning is also used for distributing the learning process to edge devices without the need of a central aggregation server. In particular, each device is responsible for a small portion of the of the total network (e.g., a convolutional or fully-connected layer) [10,12] and the training load is divided uniformly among these devices. However, in such works there is the need for very large transmission overheads in order to achieve the training of a full model.

Knowledge distillation is the ability of extracting characteristic information from data using strong and complex neural networks and use such information in order to train weaker neural networks possibly minimizing overheads. In this context, the words *strong* and *weak* refer to the architecture of the network (although other possibilities may be taken into account, e.g., different amounts of available data). Works such as [9] use dropout in order to iteratively exclude neurons from the training process until memory constraints are met. Others, such as [3], use periodic updates starting from the edge devices to transmit data to a central server for aggregation, but a central server is still required.

In this work, we present an approach for completely decentralized federated learning that can accommodate heterogeneous architectures of networks. Through knowledge distillation, data privacy is guaranteed by sending only aggregated top-level features (i.e., logits) to weak nodes, ensuring optimal performance and efficiency for the whole federation.

3 Method

FeDZIO is based on the assumption that top-level features from stronger nodes (which act as teachers, in the knowledge distillation framework) can be exploited to enhance the training performance of weaker networks (which act as students), while guaranteeing data privacy (Fig. 1).

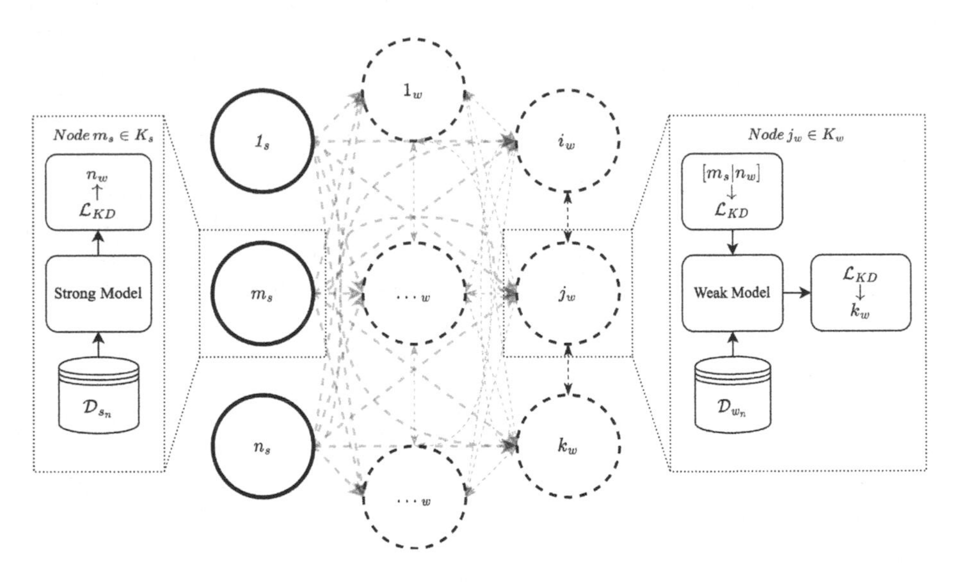

Fig. 1. Overview of the proposed federated approach. Strong nodes in the federation share average class logits with weak nodes, which train on a local dataset optimizing a classification and a knowledge distillation loss.

In this scenario, the main distinction we consider between strong and weak nodes is in the architecture of the local model. We assume that strong nodes are equipped with a model $f_{\boldsymbol{\theta}}$ while weak nodes employ a model $g_{\boldsymbol{\psi}}$, where model capacities differ in the amount of trainable parameters, i.e., $|\boldsymbol{\theta}| > |\boldsymbol{\psi}|$. Of course, each strong node will train its own set of parameters: for a family of K_s strong nodes, we will take into account independent parameters vectors $\boldsymbol{\theta}_1, \boldsymbol{\theta}_2, \ldots, \boldsymbol{\theta}_{K_s}$; similarly for K_w weak nodes.

We take into consideration a classification problem on the same subset of tasks for all nodes; for this reason, both strong and weak nodes include an output classification layer with the same number of neurons, equal to the number of classes C. Each node i is provided with a local dataset $\mathcal{D}_i = \{(\mathbf{x}_1, y_1), (\mathbf{x}_2, y_2), \ldots, (\mathbf{x}_N, y_N)\}$. Labels y are identically and independently distributed (*i.i.d.*: all nodes include samples for all classes, and the label distribution is the same on all nodes).

At each round of the proposed federated knowledge distillation approach, a node updates its parameters based on the following training objectives.

Local Training. Each node i updates its parameters ($\boldsymbol{\theta}_i$ or $\boldsymbol{\psi}_i$, depending on the type of the node) by optimizing a cross-entropy loss $\mathcal{L}_{\text{CE}}$ on its own dataset:

$$\mathcal{L}_{CE}(\mathbf{p}, y) = -\sum_{i=1}^{C} p_y \log p_y, \tag{1}$$

where $\mathbf{p} = [\, p_1, \ldots, p_C]$ = softmax$(m_i(\mathbf{x}))$ is the softmax distribution associated to class logits for input $\mathbf{x}$ with class y, and m_i is $f_{\boldsymbol{\theta}_i}$ or $g_{\boldsymbol{\psi}_i}$, for strong or weak nodes, respectively.

Federated Distillation. At the beginning of each training round, each *weak* node i receives a vector $\mathbf{k}_j$ from a random node j in the federation, which can be either a weak or a strong node. Vector $\mathbf{k}_j$ represents the average class logits computed by node j on $\mathcal{D}_j$ at the end of the previous round:

$$\mathbf{k}_j = \mathbb{E}_{(\mathbf{x},y)\sim\mathcal{D}_j}\Big[m_j(\mathbf{x})\Big] \tag{2}$$

Then, during the training round, node i introduces a second loss term, performing knowledge distillation on the received logits:

$$\mathcal{L}_{\text{KD}}(\mathbf{x}, \mathbf{k}) - \lambda \left\| g_{\psi}(\mathbf{x}) - \mathbf{k} \right\|_2^2, \tag{3}$$

with $\|\cdot\|$ being the L_2 norm and λ a weighing factor. In this scenario, we do not perform distillation on strong nodes, based on the assumption that weak nodes may provide less informative contributions due to their inferior model capacity.

Of course, the federated distillation stage is only carried out starting from the second training round of the protocol. Also, since strong nodes are in principle fewer than weak nodes, the same $\mathbf{k}_j$ vector from strong node j will be shared by multiple weak nodes; care is taken to ensure a uniform distribution of distilled vectors, as well as to guarantee that each weak node has the chance to perform knowledge distillation from as many strong nodes as possible, during the entire training procedure.

By executing multiple communication rounds, the decentralized federated learning process for each student device aims to converge to a model that captures the collective knowledge of all participating devices and performs uniformly across all nodes.

The overall procedure is described in pseudo-code in Algorithm 1.

Algorithm 1: Pseudo-code of the proposed federated learning approach.

```
Input: K: number of nodes, K_s: number of strong nodes, K_w: number of weak
       nodes
Parameters: θ_1, ..., θ_{K_s}, ψ_1, ..., ψ_{K_w}: model parameters
Data: D_1, D_2, ..., D_K: local datasets
Distilled vectors: k_1, k_2, ..., k_{K_s}: distilled vectors on strong nodes
Options: R: number of rounds, E: epochs per round
for r = 1...R                                         // For each round
do
    for k = 1...K_s                                   // For each strong node
    do
        // Local training
        for e = 1...E                                 // For each epoch
        do
            θ_k ← optimize(L_CE, θ_k, D_k)            // Optimize cross-entropy loss
        end
        // Distill knowledge
        k_k ← average class logits
    end
    for k = 1...K_w                                   // For each weak node
    do
        // Local training + knowledge distillation
        Receive a random k
        for e = 1...E                                 // For each epoch
        do
            θ_k ← optimize(L_CE + λL_KD, θ_k, D_k, k) // Optimize joint loss
        end
    end
end
```

4 Performance Evaluation

We conduct experiments that mirror real-world scenarios in which edge nodes with limited resources play a role in training neural networks. In the context of our research, weak nodes are characterized by their low computational capabilities and smaller models. Strong nodes represent more powerful entities that are equipped with higher computational resources and larger models. A variable percentage of strong nodes is incorporated into the federated network, with the remaining nodes in the network being weak nodes. The variable composition of strong and weak nodes allows us to explore the effects of different levels of computational resources and data availability on the overall performance of the federated network during the neural network training process.

4.1 Dataset

In our performance evaluation, we utilize the CIFAR10 dataset, a widely-used benchmark for machine learning tasks, comprising 60,000 color images of 32×32

pixels, divided into 10 classes, with 6,000 images per class, representing real-world objects. Each class features a diverse set of images, providing a realistic challenge for federated learning scenarios.

The CIFAR10 dataset is divided equally among all participating nodes, regardless of whether they were categorized as strong or weak. Each node received an equal share of data, with an identical number of images from each of the 10 classes, ensuring a balanced class distribution for each node. This setup allowed us to investigate the performance of federated learning with equal data distribution across nodes, without introducing any bias due to data skewness or imbalance.

4.2 Training Procedure

Our experiments consist of two distinct training scenarios to simulate the behavior of strong and weak nodes in a federated learning environment. For the strong nodes, we employ the ResNet-18 architecture, which is characterized by its relatively deep layers and high computational complexity, making it suitable for nodes with higher computational resources. Instead, for the weak nodes, we opt for the MobileNetV3 model, a lightweight neural network architecture optimized for mobile and edge devices with limited computational power.

The training procedure for both strong and weak nodes follows a federated learning approach. Each node is responsible for training a local model on its share of the CIFAR-10 dataset; additionally, each weak node carries out knowledge distillation on averaged class logits received by strong nodes.

We utilize the stochastic gradient descent (SGD) optimizer for training both ResNet-18 and MobileNetV3 models. A batch size of 64 is employed for each local update on the participating nodes. In our default settings, the federated learning process consists of 20 rounds, with each round comprising the following steps: local training, knowledge distillation and logit distribution from strong to weak nodes. During each federated learning round, every node trains its local model for ten epochs. This setup allows us to assess the performance of the federated learning process under different node capabilities and neural network architectures while ensuring consistency in the training procedure.

4.3 Experimental Results

We conduct experiments to assess the performance of the proposed FeDZIO algorithm against local training (i.e., when no federation learning is applied and each node's model in trained on its local dataset only) on CIFAR10. The default experimental setup considers the presence of 5 strong nodes and 5 weak nodes. Results are reported in Figs. 2 and 3 in terms of average test accuracy over the set of nodes in the federation, when using not-pretrained and pretrained models, respectively (when enabled, pretraining is carried out on ImageNet).

Comparing FeDZIO with local training, our algorithm exhibits superior performance in terms of average accuracy across all nodes. The advantage of FeDZIO becomes particularly evident in the not-pretrained case. This can be attributed

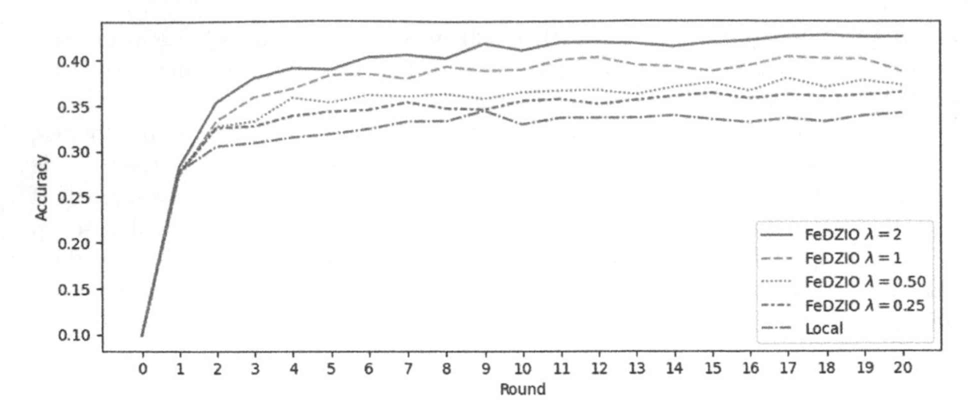

Fig. 2. Average accuracy on a 10-node federation (five strong nodes and five weak nodes), using not-pretrained models, for different λ values.

to the fact that the knowledge distillation process in FeDZIO facilitates weak nodes in learning meaningful representations from the logits shared by strong nodes, thereby compensating for their lack of pre-training. As the λ hyperparameter is increased, further enhancements in performance were observed, which is consistent with the expected behavior since a higher λ amplifies the contribution of distillation loss and reinforces knowledge transfer from strong to weak nodes.

We then assess the impact of the number of strong nodes on accuracy for both the pretrained and not-pretrained cases, by varying the federation structuring keep the overall number of nodes (ten) while changing the number of strong nodes from 0 to 8. Figure 4 reports the corresponding results. We notice that adding just one strong node to the federation substantially boosts performance. However, additional increases in the number of strong nodes brings only

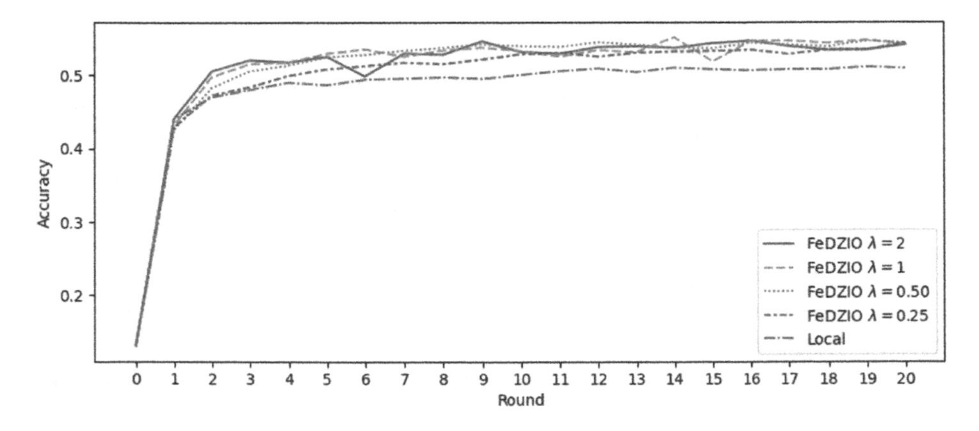

Fig. 3. Average accuracy on a 10-node federation (five strong nodes and five weak nodes), using pretrained models, for different λ values.

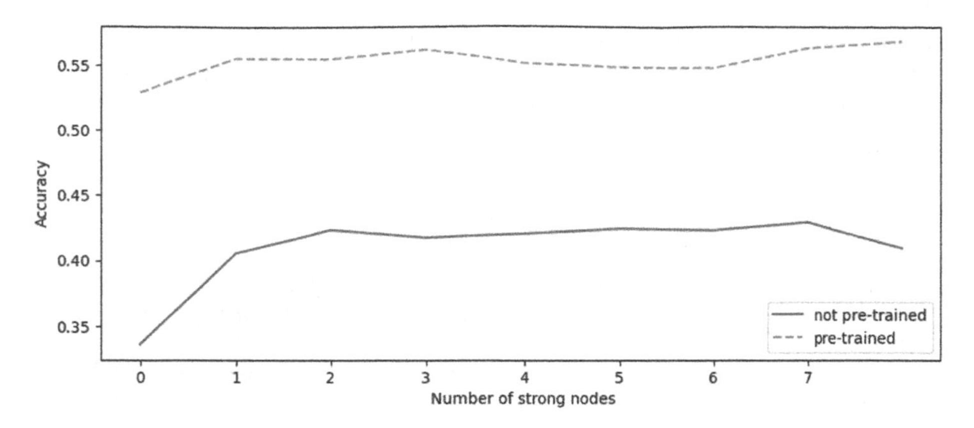

Fig. 4. Average accuracy on a 10-node federation for different numbers of strong nodes. Results are reported for both pretrained and not-pretrained models, after 20 training rounds, with $\lambda = 2$.

marginal improvements in performance. This can be attributed to the diminishing returns of increasing the number of strong nodes: the first strong node significantly augments the collective computational capacity of the federation, while subsequent nodes add less incremental value due to the redundancy of information.

5 Conclusions

We introduced FeDZIO, a novel federated learning approach that leverages knowledge distillation to enhance the performance of nodes in a decentralized network, specifically optimizing for heterogeneous nodes with varying computational resources. By categorizing nodes as strong or weak, FeDZIO manages to effectively transfer knowledge among nodes, resulting in higher performance than local training, especially for not-pretrained models. The superior performance of FeDZIO showcases the potential of our approach in effectively harnessing the computational capabilities of nodes in a decentralized network.

While the approach has demonstrated promising results, there are some limitations that should be noted. Firstly, the choice of the hyperparameter λ is critical to regulate the strength of knowledge distillation. An inappropriate value may hinder the optimal transfer of knowledge from strong to weak nodes. Additionally, while FeDZIO performs well with a single strong node, there is diminishing returns with the addition of further strong nodes. This suggests that an ideal balance between strong and weak nodes needs to be struck for optimal performance.

Future directions for this work include the exploration of adaptive methods to select the best value for λ, depending on the specific distribution and characteristics of the data. Furthermore, the potential exists for a more fine-grained categorization of nodes beyond just strong and weak. Lastly, it would be interesting to assess the generalizability of the FeDZIO algorithm to different datasets and to extend the approach to more complex tasks, such as image segmentation or object detection.

References

1. Deng, S., Zhao, H., Fang, W., Yin, J., Dustdar, S., Zomaya, A.Y.: Edge intelligence: the confluence of edge computing and artificial intelligence. IEEE Internet Things J. **7**(8), 7457–7469 (2020)
2. Gou, J., Yu, B., Maybank, S.J., Tao, D.: Knowledge distillation: a survey. Int. J. Comput. Vision **129**, 1789–1819 (2021)
3. He, C., Annavaram, M., Avestimehr, S.: Group knowledge transfer: federated learning of large CNNs at the edge. Adv. Neural. Inf. Process. Syst. **33**, 14068–14080 (2020)
4. Kaissis, G.A., Makowski, M.R., Rückert, D., Braren, R.F.: Secure, privacy-preserving and federated machine learning in medical imaging. Nat. Mach. Intell. **2**(6), 305–311 (2020)
5. Lalitha, A., Shekhar, S., Javidi, T., Koushanfar, F.: Fully decentralized federated learning. In: Third Workshop on Bayesian Deep Learning (NeurIPS), vol. 2 (2018)
6. Li, T., Sahu, A.K., Talwalkar, A., Smith, V.: Federated learning: challenges, methods, and future directions. IEEE Signal Process. Mag. **37**(3), 50–60 (2020)
7. Li, Y., Chen, C., Liu, N., Huang, H., Zheng, Z., Yan, Q.: A blockchain-based decentralized federated learning framework with committee consensus. IEEE Network **35**(1), 234–241 (2020)
8. Lim, W.Y.B., et al.: Federated learning in mobile edge networks: a comprehensive survey. IEEE Commun. Surv. Tutorials **22**(3), 2031–2063 (2020)
9. Mishra, R., Gupta, H.P.: Designing and training of lightweight neural networks on edge devices using early halting in knowledge distillation. IEEE Trans. Mob. Comput. (2023)
10. Pappas, C., Chatzopoulos, D., Lalis, S., Vavalis, M.: IPLS: a framework for decentralized federated learning. In: 2021 IFIP Networking Conference (IFIP Networking), pp. 1–6. IEEE (2021)
11. Roy, A.G., Siddiqui, S., Pölsterl, S., Navab, N., Wachinger, C.: BrainTorrent: a peer-to-peer environment for decentralized federated learning. arXiv preprint arXiv:1905.06731 (2019)
12. Yu, R., Li, P.: Toward resource-efficient federated learning in mobile edge computing. IEEE Network **35**(1), 148–155 (2021)

A Federated Learning Framework for Stenosis Detection

Mariachiara Di Cosmo[1(✉)], Giovanna Migliorelli[2], Matteo Francioni[3], Andi Muçaj[3], Alessandro Maolo[3], Alessandro Aprile[4], Emanuele Frontoni[5], Maria Chiara Fiorentino[1], and Sara Moccia[6]

[1] Department of Information Engineering, Universitá Politecnica delle Marche, Ancona, Italy
m.dicosmo@pm.univpm.it
[2] Department of Law, Universitá degli Studi di Macerata, Macerata, Italy
[3] U.O.C. Cardiology and Hemodynamics - Department of Cardiovascular Sciences, Azienda Ospedaliero Universitaria delle Marche, Ancona, Italy
[4] U.O.C. Cardiology-UTIC-Hemodynamics, Ospedale del Mare, ASL NA 1, Naples, Italy
[5] Department of Political Sciences, Communication and International Relations, University of Macerata, Macerata, Italy
[6] The BioRobotics Institute and Department of Excellence in Robotics and AI, Scuola Superiore Sant'Anna, Pisa, Italy

Abstract. This study explores the use of Federated Learning (FL) for stenosis detection in coronary angiography images (CA). Two heterogeneous datasets from two institutions were considered: Dataset 1 includes 1219 images from 200 patients, which we acquired at the Ospedale Riuniti of Ancona (Italy); Dataset 2 includes 7492 sequential images from 90 patients from a previous study available in the literature. Stenosis detection was performed by using a Faster R-CNN model. In our FL framework, only the weights of the model backbone were shared among the two client institutions, using Federated Averaging (FedAvg) for weight aggregation. We assessed the performance of stenosis detection using Precision ($Prec$), Recall (Rec), and F1 score ($F1$). Our results showed that the FL framework does not substantially affects clients 2 performance, which already achieved good performance with local training; for client 1, instead, FL framework increases the performance with respect to local model of +3.76%, +17.21% and +10.80%, respectively, reaching $Prec = 73.56$, $Rec = 67.01$ and $F1 = 70.13$. With such results, we showed that FL may enable multicentric studies relevant to automatic stenosis detection in CA by addressing data heterogeneity from various institutions, while preserving patient privacy.

Keywords: Federated Learning · Coronary Angiography · Stenosis detection · Computer Assisted Diagnosis

1 Introduction

Coronary artery disease (CAD) provokes stenoses, coronary segments with narrowed lumen, causing blood flow reduction and eventually leading to ischemia

G. L. Foresti et al. (Eds.): ICIAP 2023 Workshops, LNCS 14366, pp. 211–222, 2024.
https://doi.org/10.1007/978-3-031-51026-7_19

and heart attacks. CAD, which is a leading cause of mortality worldwide, is currently assessed through coronary angiography (CA), an imaging technique that uses X-rays and contrast dye to visualize coronary arteries and assess blood flow dynamics [17]. Precise and timely stenosis detection is crucial for effective CAD diagnosis and treatment. CA interpretation relies on clinician's expertise and requires tackling challenges such as complex vessel anatomy, stenosis variability (in shape, pattern and severity), presence of movement and shadowing artifacts, as well as varying imaging equipment and contrast agent levels [5].

Computer-aided decision support systems for stenosis detection from CA have the potential to improve CAD assessment and reduce clinician variability. In recent years, deep learning (DL) has shown great potential in automating stenosis detection from CA images. Many approaches propose a multi-stage framework, in which stenoses are identified following vessel enhancement [11] or segmentation [24]. In [24], a U-Net++ model with a feature pyramid network is proposed to automatically segment coronary arteries and from the artery centerline, diameters are calculated and stenotic levels are measured. In [11], after key frame detection using vessel extraction, a classification model identifies the stenosis presence in the key frame and through Class Activation Mapping (CAM) stenosis is qualitatively localized. The study considers right CA only.

To avoid error accumulation though image pre-processing steps, several studies develop DL methods to localize stenoses directly from CA images without relying on vessel analysis [1,2,14]. In [1], similarly to [11], CAM is employed to localize stenosis on top of an image-level stenosis classification performed by an Inception-v3. Several object detection models are trained in [2] and tested over one-vessel stenotic CA series to explore trade off between accuracy and performance efficiency. Inspired by quantum computing, [14] incorporates a quantum network in a ResNet network to detect stenosis by performing binary classification on fixed-size patches obtained from CA images. Some approaches [4,15,20,23] consider sequence of images to take advantage of temporal information intrinsic in CA. The work in [23] proposes a hierarchical attentive multi-view learning model to capture the pixel correlation from CA sequences for quantifying stenosis. In [20], a method is proposed that detects candidate stenoses using a deconvolutional single-shot detector (DSSD) on frames of a CA sequence selected by a U-Net model. Then, the seq-fps module takes advantage of the temporal information within the X-ray sequence to suppress false positives and generate the final stenosis detection results. The work in [15] includes sequence feature fusion and sequence consistency alignment into stenosis detection network to capture the spatio-temporal information using 166 CA sequence. To gather spatio-temporal features, a transformer-based module is used in [4] and to learn long-range context a feature aggregation network is developed. Most of these studies use proprietary datasets which have their own acquisition and annotation processes, or open-source datasets which are made available for research community (for stenosis detection only [2] shares dataset). This limits generalizability as DL model results relies on the dataset-specific characteristics such as imaging equipment, protocols, and patient variability.

When multicentric studies are considered [1,15,23], datasets are treated separately, simply testing their model on each dataset overlooking the potential influence of one dataset over the other, or as a single dataset, without considering issues related to sharing medical imaging data across multiple centers. Data privacy concerns, regulatory compliance, technical compatibility, and varying imaging protocols are addressed among the main challenges to face for DL-based decision support tool development for medical image analysis [3,13]. Federated learning (FL) [18] provides an opportunity for data-private multi-institutional collaborations, leveraging model learning sharing among datasets interactions while preserving their privacy. Each client's raw data are stored locally and remain under control of and private to that institution, while only model updates leave the client enabling the aggregation of learned patterns into a single global model. In addition, different FL strategies can be adopted to increase data privacy further, like partial model sharing [22], to proactively avoid data leakage during communication between the server and multiple clients. FL has already shown to be successful in medical image analysis [9,21]: it has been used to alleviate small sample size and lack of annotation problems, facilitate domain adaptation and mitigate domain shift. Despite its potential, no applications in stenoses detection have been proposed so far.

The present work explores FL for stenosis detection from two datasets with the aim of addressing challenges posed by data heterogeneity, variable acquisition protocols and inter-patient variability. Our contributions include:

- FL framework to address stenosis detection from CA;
- Privacy-preserving FL by partial aggregation of the detection model to capture and share intrinsic CA features;
- Collection of a new dataset of CA, composed by 1219 images from 200 patients and acquired in the clinical practice from three clinicians, who performed image annotation.

2 Material and Methods

Our FL framework relies on a learning paradigm between two hospitals (i.e., the local clients of the federation) and one central server.

To perform stenosis detection a Faster R-CNN model [16] is deployed. Faster R-CNN integrates a Convolutional Neural Network (CNN) as backbone for feature extraction, a Region Proposal Network (RPN) for proposal generation, and subsequent layers for accurate object classification and localization within proposed Region of Interest (ROI). In the present study, the model developed is a Faster R-CNN [16] with ResNet50 Feature Pyramid Network (FPN) as backbone network [7]. This version of Faster R-CNN benefits from the hierarchical features extracted by the ResNet50, the multi-scale feature representation provided by FPN, an extension of the RPN with two convolutional layers on top of the ROI heads for classification and regression of the stenotic regions. The Faster R-CNN, pre-trained on Coco dataset [8], is trained to minimize a multi-task loss as a weighted combination of cross-entropy loss for the identification of

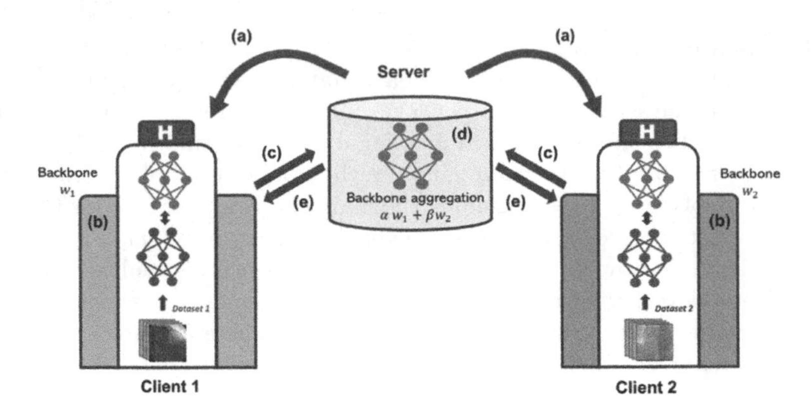

Fig. 1. Representation of proposed Federated Learning (FL) framework: (a) the server initially sends the Faster R-CNN model pre-trained on Coco dataset to both clients; (b) for each round, the model is trained locally on the training set of the specific client; (c) updated weights of the model backbone are sent back to the server; (d) the server receives and aggregates the updated weights from both clients; (e) the server sends the result of aggregation back to the two clients. Communication happens between server and clients aggregating at each round the backbone weights (w_1 for client 1 and w_2 for client 2) with Federate Averaging (FedAvg) defined as $\alpha w_1 + \beta w_2$, where α and β represent two parameters that determine the clients contribution to the aggregation process.

stenosis presence and regression loss for stenosis localization. Since the backbone is responsible for feature extraction and representation, we consider that sharing its parameters in the FL framework could allow the clients to collectively learn generalized and meaningful features of the CA images.

As shown in Fig. 1, the server initiates the federated process sending the Faster R-CNN model to both the clients, including the pre-trained backbone weights. Then, for each round of the federated computation, each client trains the model received on its local private data and sends back to the server only the updated backbone weights. The server receives the weight updates from each client and aggregates them to create a global model. The aggregation technique adopted relies on Federated Averaging (FedAvg) [10], which computes a weighted average of the backbone parameters collected by the server. Since only the backbone weights are shared, the server combines them without interacting with the rest of the Faster R-CNN model.

FedAvg aggregation strategy is defined as $\alpha w_1 + \beta w_2$, where α and β represent two parameters that determine the clients' contribution to the aggregation process and w_1 and w_2 are the backbone weights of client 1 and client 2, respectively.

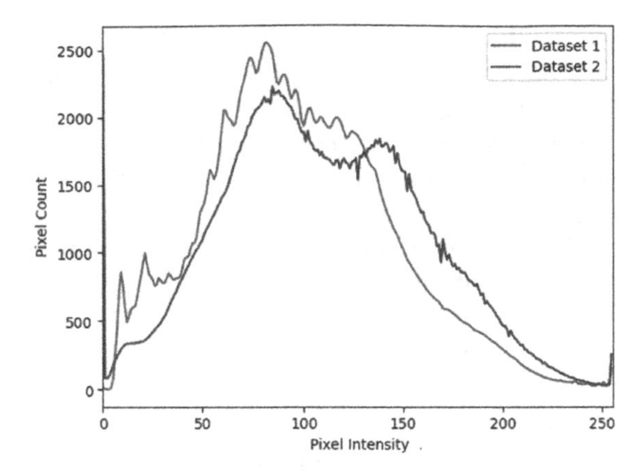

Fig. 2. Domain shift between the two coronary angiography (CA) datasets in terms of pixel value distribution.

2.1 Datasets

The datasets used in this study exhibit inherent dissimilarities in the acquisition protocols and annotation approaches. This results in a visible data heterogeneity: as depicted in Fig. 2, the intensity distributions at the two institutions underline variation in imaging characteristics posing domain shifts challenges, thus leading to reduced model generalization, increased bias and ineffective transfer learning. Domain shift arises especially from image characteristics, such as imaging equipment, image quality, frame selection process, and patient demographics.

Both datasets are made of gray-scale images with 512×512 pixels and all images acquired from the same patient are carefully considered as part of the same set (training or testing).

Dataset 1. The dataset consists of 1219 CA images, which we acquired at the Ospedali Riuniti of Ancona (Italy). The images are provided by 200 patients, who signed informed consent, underwent CA procedure and presented one or more stenotic regions along the coronary arteries (up to 4 stenoses per image). From each patient exam, a few relevant frames are selected by the clinicians of this study taking into account the presence of high contrast dye, various viewpoints, and potentially the diastolic phase. The data acquisition is conducted in compliance with the Helsinki Declaration and under the supervision of three expert clinicians, which evaluate the presence of the stenotic regions and provide the annotations. For model training and testing, 1106 (90%) images from 175 patients compose the training set and 112 (10%) images from 25 patients the test set.

Dataset 2. The dataset is provided by Danilov et al. [2] and contains CA series comprising a total of 7492 images. The images are obtained from 100 patients, who underwent CA at the Research Institute for Complex Problems of Cardiovascular Diseases (Kemerovo, Russia). All patients had confirmed one-vessel CAD. From each patient exam, images containing contrast passage through a stenotic vessel are extracted in sequences, discarding non-informative images. The manual annotation of the presence or absence of stenotic lesions for each image was performed by a single operator, as described in [2]. For model training and testing, 6660 (90%) images from 80 patients are used as training set, and 832 (10%) images from 10 patients as test set. The split was given by [2].

2.2 Experimental Protocol

In Faster R-CNN training and for all experiments, we used Adam optimizer with a constant learning rate set to 0.0001 and batch size equal to 16. Pixel values in the input images were normalized in the range [0,1], and the images were randomly augmented via horizontal flip. To handle the different sizes of Dataset 1 and Dataset 2 and ensure comparable number of training steps, the FL framework performed 20 rounds. In each round, the training process of the Faster R-CNN was performed locally for 20 epochs for Dataset 1 and 4 epochs for Dataset 2. For the first round, a warm-up strategy was implemented to promote weight stabilization and the local training was performed more extensively for 40 and 16 epochs for Dataset 1 and Dataset 2, respectively. FedAvg aggregation is performed by assigning weights proportionally to the dataset sizes, defining $\alpha = 1$ and $\beta = 6$. Even though our clients strongly differ in terms of number of images, we consider that giving more importance to the larger dataset could be beneficial for the smaller one.

We compared the performance of proposed FL framework with that obtained by training the Faster R-CNN model locally. The local models were trained for 200 epochs for client 1 and 50 epochs for client 2. We further performed the following ablation study:

FL1: *Faster R-CNN weight aggregation*
To probe the effectiveness of sharing only the weights of the backbone, focusing on extracting and sharing relevant features, we performed also the FL of the whole Faster R-CNN model to evaluate the impact on stenosis detection performance.

FL2: *Faster R-CNN backbone weight aggregation with equal client contribution*
Considering the strong difference in terms of size between the two datasets, we explored also the effect of applying weighting techniques to the aggregation process to ensure a fair and balanced representation of both clients ($\alpha = 1$ and $\beta = 1$). In this way, we examine whether the discrepancy in terms of size between the two datasets may lead to any performance penalty in detriment of smaller datasets.

FL3: *Faster R-CNN backbone weight aggregation with the exclusion of Batch Normalization layers parameters*
Based on the study of [6], which demonstrated that keeping local Batch Normalization parameters not synchronized with the global model reduces feature

Table 1. Mean values of Precision (*Prec*), Recall (*Rec*) and F1 score (*F*1): from top to bottom single clients' local model, **FL1**, **FL2**, **FL3** and proposed FL framework performances are reported.

	Dataset 1				Dataset 2			
	Prec	*Rec*	*F*1	*Round*	*Prec*	*Rec*	*F*1	Round
Local	70.89	57.14	63.28	–	**93.50**	82.83	87.84	–
FL1	76.25	62.24	68.54	2	92.61	82.71	87.38	2
FL2	**77.33**	59.18	67.05	3	92.54	83.43	87.75	1
FL3	69.32	62.24	65.59	1	92.91	83.43	**87.92**	5
Proposed	73.56	**67.01**	**70.13**	2	91.62	**84.03**	87.66	1

shifts in non-Independent and Identically Distributed (IID) data as in our case (see Fig. 2), we evaluated if the exclusion of the statistical non-trainable parameters of the Batch Normalization layers of the backbone could mitigate the discrepancy between the clients.

To assess the performance of stenosis detection, we computed Precision (*Prec*), Recall (*Rec*) and F1 score (*F*1) over each client test set. We considered a prediction as True Positive (*TP*) if it achieved an Intersection over Union (*IoU*) value with respect to the ground truth annotation greater than or equal to 0.5. Conversely, a predicted bounding box with a *IoU* value less than 0.5 was considered a False Positive (*FP*). When a stenosis did not have any corresponding prediction, it is regarded as a False Negative (*FN*). To ensure the lowest number of missing predictions, for each FL framework training and for each model training best configuration was selected in terms of average *Rec* over the client' test set.

The overall implementation was performed in Python 3.8.10 with PyTorch v.2.0.0, Torchvision v.0.15.1 and Flower v.2.0.0 libraries. Our model training and all experiments conducted were performed via 8 GPU bank where one or more GPUs were assigned to each client and one GPU was assigned to the server.

3 Results and Discussion

Prec, *Rec* and *F*1 values achieved from proposed FL framework are reported in Table 1 in comparison with performance obtained from single client's local training and from **FL1**, **FL2** and **FL3**. Table 1 shows that client 1 performance improved significantly within the FL framework: the *Prec*, *Rec* and *F*1 values increased considerably compared to local training of +3.76%, +17.21% and +10.80% respectively, demonstrating the positive effect of the interaction with client 2. The sharing of the backbone weights only is noteworthy for client 1: a higher *Rec* value compared to **FL1** (+7.66%) is achieved, suggesting a reduction of *FN* predictions. Client 2 had a greater contribution in the backbone aggregation process and this introduced additional insights and increased the

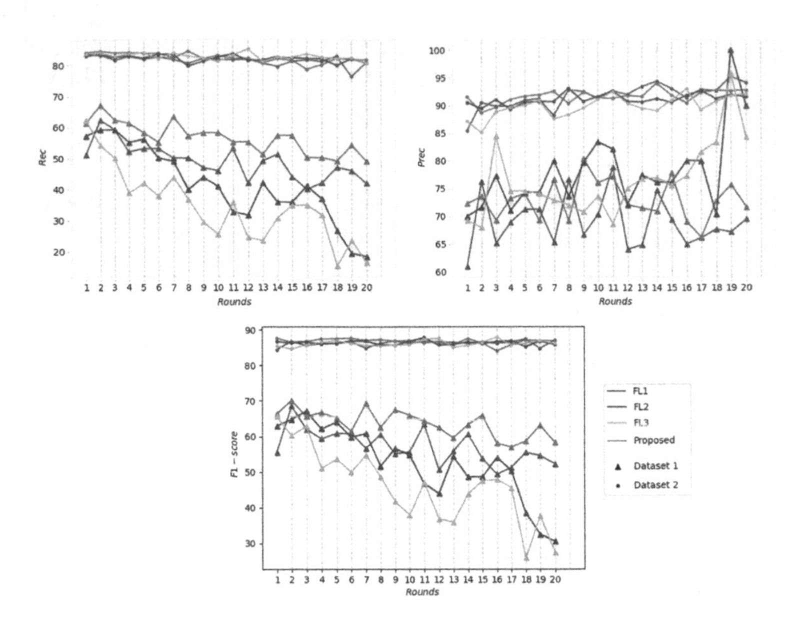

Fig. 3. Mean values of Precision ($Prec$), Recall (Rec) and F1 score ($F1$) at each round of **FL1**, **FL2**, **FL3** and proposed FL framework.

extraction of intrinsic features which boosted the ability of client 1 to detect stenotic regions. In Table 1, it is also evident that client 2 performance remained relatively stable and unaffected within the FL process: it consistently achieved comparable results across **FL1**, **FL2**, **FL3** and proposed FL framework. The $Prec$ value was highest when the model was trained locally, indicating that the introduction of a FL framework may not significantly improve performance for client 2. For what concerns **FL3**, differently from [6], client 1 did not benefit from the exclusion from the aggregation process of the Batch Normalization layers parameters, while client 2 exhibited good performance under this setting. With **FL3** experiment, we focused on investigating the specific impact of Batch Normalization exclusion as addressed in [6]; however, we are aware that in recent research [19] alternatives to Batch Normalization have been explored to mitigate challenges associated with non-IID data distribution in FL scenarios. In future work, we plan to expand our study exploring also alternative normalization techniques.

Figure 3 shows the trends of $Prec$, Rec and $F1$ obtained from the proposed FL framework and from **FL1**, **FL2** and **FL3** considering different number of rounds. The proposed FL framework confirmed to enhance client 1 performance. On the other hand, client 2 demonstrated remarkable results throughout the different rounds with minimal variation, suggesting that FL framework did not significantly impact over its performance. By increasing the number of rounds, the proposed FL framework and also **FL1**, **FL2** and **FL3** exhibited sings of overfitting. It is evident from the declining performance of client 1, especially in terms

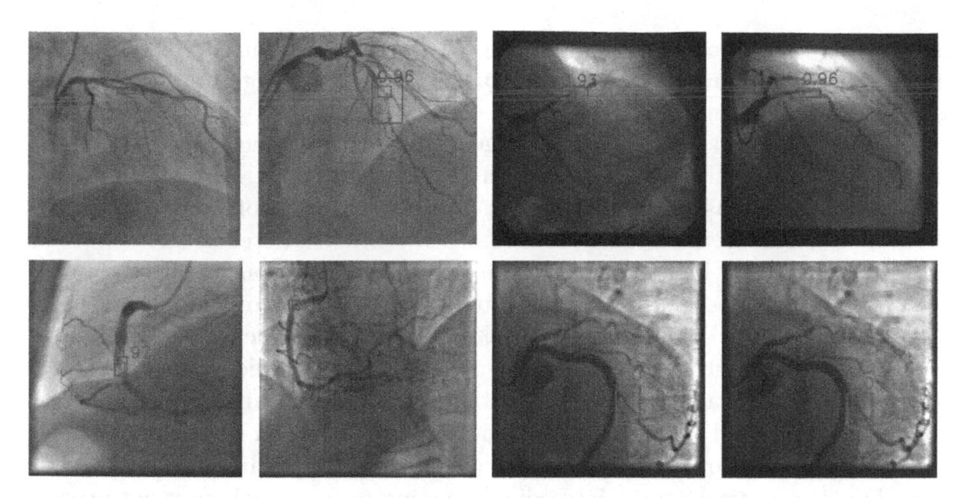

Fig. 4. Samples of stenosis predictions from Dataset 1 (first row) and Dataset 2 (second row) obtained with the proposed FL framework: ground truth is shown in green box, prediction with confidence score in red. (Color figure online)

of *Rec* and $F1$. The *Prec*, instead, could have not manifested a similar decline, as it is less susceptible to overfitting compared to the other metrics. Client 2 performance, then, showed stable trends, without improvements at increasing number of rounds. In addition to current ablation studies (**FL1**, **FL2**, **FL3**), we consider also to explore in the future the impact of uniform training epochs across communication rounds for both clients, to evaluate the effect of FedAvg aggregation in standard settings and its influence on convergence behaviour.

Figure 4 shows clients performance highlighting the difference between their datasets. Client 1 (first row), with its smaller and highly variable dataset, exhibited poorer performance, as visible from the presence of several missing predictions and less precise bounding box overlapping. In contrast, for client 2 (second row) stenoses are almost always recognized and localized accurately in the images. In Fig. 4 also two images belonging to the same patient, one from Dataset 1 and one from Dataset 2, are displayed as third and forth images of the row: for each patient CA procedure, during the dataset annotation process, at client 1 only significant frames were selected, whereas at client 2 sequential frames were all annotated, disregarding only the frames in which no stenosis was visible. This difference further accentuated datasets differences and the lower variability of Dataset 2 compared to Dataset 1.

The overall results showed the effectiveness of proposed FL framework in improving stenosis detection performance of client 1, by leveraging information provided by client 2. Sharing backbone weights allowed us to transfer knowledge from dataset 2, overcoming limits given by a small dataset size. On the other hand, client 2, already performing well locally, was not substantially impacted by the FL framework. However, sharing information with Dataset 1, with its smaller but highly variable nature, could have been helpful for client 2 in making its model more robust and more generalizable. In addition, the adoption of a partial model sharing approach though the aggregation of backbone weights only, as in [22], enhanced further data privacy protection, extracting only intrinsic and general features from CA images.

FL offers several potential advantages for stenosis detection in CA. First, it enables collaborative learning across multiple institutions, allowing the inclusion of diverse datasets and facilitating the development of more generalized models [18]. This is particularly crucial in the context of stenosis detection, as datasets can be very heterogeneous and exhibit wide variations in terms of imaging protocols, annotation procedures, patient populations, and equipment used. By involving multiple institutions and datasets, multicentric studies could provide a more comprehensive and representative view of clinical practice. Moreover, FL offers privacy-preserving capabilities: by performing training process locally on individual clients data, FL mitigates the need for data sharing while still enabling collaboration and knowledge sharing among different institutions [18]. This is a critical aspect in medical imaging research, where strict privacy regulations, ownership, regulatory compliance, and ethical considerations come into play [12]. The significance of multicentric studies and privacy-preserving approaches is further emphasized by the growing interest and attention from organizations such as the European Commission. A document to promote ethical principles in DL model design and deployment, called Ethics Guidelines for Trustworthy AI[1], was published in 2018 from the European Commission, which also encourages the use of FL in DL development providing useful information about FL impact on data protection[2].

Overall, our study highlights the importance of considering data heterogeneity and privacy concerns in the development of stenosis detection models and even though further research is needed to optimize the FL process and include multiple institutions for a wider representation of data heterogeneity, it opens the way for an efficient clinicians support to stenosis detection form CA, ultimately leading to improved patients clinical outcomes.

4 Conclusion

In this study, we explored the use of FL for stenosis detection in CA. Training with data from different institutions is particularly relevant in this context, where datasets exhibit wide variations in imaging protocols, annotation procedures, patient populations, and equipment used, in addition to the intrinsic CA imaging challenges. Our FL framework, by sharing the Faster R-CNN backbone weights, improved stenosis detection accuracy for client 1 achieving an increase in $Prec$, Rec and $F1$ of +3.76%, +17.21% and +10.80% respectively, while client 2, which already achieved high stenosis detection ability training the model locally, did not benefit significantly from a FL framework. We hope our study may pave the way for future studies on privacy-preserving computer-assisted algorithms for CAD diagnosis.

[1] https://ec.europa.eu/futurium/en/ai-alliance-consultation.1.html.

[2] https://edps.europa.eu/press-publications/publications/techsonar/federated-learning_en.

References

1. Cong, C., Kato, Y., Vasconcellos, H.D., Lima, J., Venkatesh, B.: Automated stenosis detection and classification in X-ray angiography using deep neural network. In: 2019 IEEE International Conference on Bioinformatics and Biomedicine, pp. 1301–1308. IEEE (2019)
2. Danilov, V.V., et al.: Real-time coronary artery stenosis detection based on modern neural networks. Sci. Rep. **11**(1), 1–13 (2021)
3. Fiorentino, M.C., Villani, F.P., Di Cosmo, M., Frontoni, E., Moccia, S.: A review on deep-learning algorithms for fetal ultrasound-image analysis. Med. Image Anal. **83**, 102629 (2022)
4. Han, T., et al.: Coronary artery stenosis detection via proposal-shifted spatial-temporal transformer in X-ray angiography. Comput. Biol. Med. **153**, 106546 (2023)
5. Lawton, J.S., et al.: 2021 ACC/AHA/SCAI guideline for coronary artery revascularization: executive summary: a report of the American College of Cardiology/American Heart Association Joint Committee on Clinical Practice Guidelines. Circulation **145**(3), e4–e17 (2022)
6. Li, X., Jiang, M., Zhang, X., Kamp, M., Dou, Q.: FedBN: federated learning on non-IID features via local batch normalization. In: International Conference on Learning Representations (2021)
7. Li, Y., Xie, S., Chen, X., Dollar, P., He, K., Girshick, R.: Benchmarking detection transfer learning with vision transformers. arXiv preprint arXiv:2111.11429 (2021)
8. Lin, T.-Y., et al.: Microsoft COCO: common objects in context. In: Fleet, D., Pajdla, T., Schiele, B., Tuytelaars, T. (eds.) ECCV 2014. LNCS, vol. 8693, pp. 740–755. Springer, Cham (2014). https://doi.org/10.1007/978-3-319-10602-1_48
9. Lu, M.Y., et al.: Federated learning for computational pathology on gigapixel whole slide images. Med. Image Anal. **76**, 102298 (2022)
10. McMahan, B., Moore, E., Ramage, D., Hampson, S., y Arcas, B.A.: Communication-efficient learning of deep networks from decentralized data. In: Artificial Intelligence and Statistics, pp. 1273–1282. PMLR (2017)
11. Moon, J.H., Cha, W.C., Chung, M.J., Lee, K.S., Cho, B.H., Choi, J.H., et al.: Automatic stenosis recognition from coronary angiography using convolutional neural networks. Comput. Methods Programs Biomed. **198**, 105819 (2021)
12. Myrzashova, R., Alsamhi, S.H., Shvetsov, A.V., Hawbani, A., Wei, X.: Blockchain meets federated learning in healthcare: a systematic review with challenges and opportunities. IEEE Internet Things J. **10**(16), 14418–14437 (2023)
13. Nazir, S., Kaleem, M.: Federated learning for medical image analysis with deep neural networks. Diagnostics **13**(9), 1532 (2023)
14. Ovalle-Magallanes, E., Avina-Cervantes, J.G., Cruz-Aceves, I., Ruiz-Pinales, J.: Hybrid classical-quantum convolutional neural network for stenosis detection in X-ray coronary angiography. Expert Syst. Appl. **189**, 116112 (2022)
15. Pang, K., Ai, D., Fang, H., Fan, J., Song, H., Yang, J.: Stenosis-DetNet: sequence consistency-based stenosis detection for X-ray coronary angiography. Comput. Med. Imaging Graph. **89**, 101900 (2021)
16. Ren, S., He, K., Girshick, R.B., Sun, J.: Faster R-CNN: towards real-time object detection with Region Proposal Networks. IEEE Trans. Pattern Anal. Mach. Intell. **39**, 1137–1149 (2015)
17. Saraste, A., Knuuti, J.: ESC 2019 guidelines for the diagnosis and management of chronic coronary syndromes: recommendations for cardiovascular imaging. Herz **45**(5), 409 (2020)

18. Sheller, M.J., et al.: Federated learning in medicine: facilitating multi-institutional collaborations without sharing patient data. Sci. Rep. **10**(1), 1–12 (2020)
19. Wang, Y., Shi, Q., Chang, T.H.: Batch normalization damages federated learning on non-IID data: analysis and remedy. In: ICASSP 2023–2023 IEEE International Conference on Acoustics, Speech and Signal Processing (ICASSP), pp. 1–5. IEEE (2023)
20. Wu, W., Zhang, J., Xie, H., Zhao, Y., Zhang, S., Gu, L.: Automatic detection of coronary artery stenosis by convolutional neural network with temporal constraint. Comput. Biol. Med. **118**, 103657 (2020)
21. Xiao, J., et al.: CateNorm: categorical normalization for robust medical image segmentation. In: Kamnitsas, K., et al. (eds.) MICCAI Workshop on Domain Adaptation and Representation Transfer. LNCS, vol. 13542, pp. 129–146. Springer, Cham (2022). https://doi.org/10.1007/978-3-031-16852-9_13
22. Yang, Q., Zhang, J., Hao, W., Spell, G.P., Carin, L.: Flop: Federated learning on medical datasets using partial networks. In: Proceedings of the 27th ACM SIGKDD Conference on Knowledge Discovery & Data Mining, pp. 3845–3853 (2021)
23. Zhang, D., Yang, G., Zhao, S., Zhang, Y., Zhang, H., Li, S.: Direct quantification for coronary artery stenosis using multiview learning. In: Shen, D., et al. (eds.) Medical Image Computing and Computer Assisted Intervention-MICCAI 2019: 22nd International Conference, Shenzhen, China, 13–17 October 2019, Proceedings, Part II 22, pp. 449–457. Springer, Cham (2019). https://doi.org/10.1007/978-3-030-32245-8_50
24. Zhao, C., et al.: Automatic extraction and stenosis evaluation of coronary arteries in invasive coronary angiograms. Comput. Biol. Med. **136**, 104667 (2021)

Benchmarking Federated Learning Frameworks for Medical Imaging Tasks

Samuele Fonio(✉)

University of Turin, Turin, Italy
samuele.fonio@unito.com

Abstract. This paper presents a comprehensive benchmarking study of various Federated Learning (FL) frameworks applied to the task of Medical Image Classification. The research specifically addresses the often neglected and complex aspects of scalability and usability in off-the-shelf FL frameworks. Through experimental validation using real case deployments, we provide empirical evidence of the performance and practical relevance of open source FL frameworks. Our findings contribute valuable insights for anyone interested in deploying a FL system, with a particular focus on the healthcare domain-an increasingly attractive field for FL applications.

Keywords: Federated Learning · Medical Image Classification · Scalability · Usability · FL Frameworks · Benchmark · Real Case Deployment · Cross Silo

1 Introduction

Federated Learning (FL) [18] has emerged as a crucial area of research in the field of Machine Learning (ML) in response to growing concerns surrounding data privacy [2,15]. This is especially relevant in the healthcare domain, where data is typically managed by hospitals and medical centers that must adhere to ethical and legal regulations, such as the General Data Protection Regulation (GDPR). Consequently, alternative approaches are necessary to address these data restrictions.

In this context, FL offers a valuable solution by enabling diverse data stakeholders to collaboratively train ML algorithms, overcoming the challenge of decentralized datasets. The core concept of FL involves training ML algorithms by aggregating clients' models without sharing the underlying data. A central server (referred to as Centralized FL) receives the local models and broadcasts the aggregated model at each iteration. The strength of FL lies in its ability

This work receives EuroHPC-JU funding under grant no. 101034126, with support from the Horizon2020 programme (the European PILOT) and by the Spoke 1 "FutureHPC & BigData" of ICSC - Centro Nazionale di Ricerca in High-Performance Computing, Big Data and Quantum Computing, funded by European Union - NextGenerationEU.

G. L. Foresti et al. (Eds.): ICIAP 2023 Workshops, LNCS 14366, pp. 223–232, 2024.
https://doi.org/10.1007/978-3-031-51026-7_20

to ensure data privacy, which aligns with the requirements of the healthcare domain. Moreover, FL proves highly effective for Deep Neural Networks (DNNs), particularly when models need to adapt to complex, non-linear patterns found in images or text. In fact, such DNNs often demand large quantities of data, making it challenging to aggregate multiple sources due to privacy concerns. FL facilitates the creation of shared models without compromising the privacy of local datasets, thus addressing this limitation.

However, deploying an FL system in a real-case scenario is not straightforward. Many Federated Learning Frameworks (FLF) are available and open-source, but they differ in many aspects: communication protocol, security, FL tools available, customization, and many others. The two main characteristics explored in this study are *scalability* and *usability*. At the best of our knowledge, literature lacks of works that compare FLFs regarding these two aspects. This gap hinders the research towards high-performing FLFs.

In this work, scalability refers to how computational time varies as the number of clients grows for a fixed problem size. As the number of clients grows, the time to complete the task is supposed to decrease, since the data volume for each party is smaller (*strong* scalability). On the other hand a growing number of clients usually brings more communication costs, which impacts on the performances of the FLF. Scalability is indeed regarded as an important future direction [3,14,26] for the design of FLFs.

Usability in the context of Federated Learning (FL) refers to the convenience and ease of deploying an FL system. However, a systematic literature review conducted by Witt et al. [26] highlights a significant limitation in the existing research. Among the 34 reviewed papers, only a small fraction (11 out of 34) considered a non-iid (non-independent and identically distributed) setting, while the majority focused on experiments with MNIST or CIFAR-10 datasets for classification tasks. This narrow focus suggests that FL frameworks may be optimized for specific datasets, making it challenging to adapt them to new datasets. It is essential to address this issue to ensure that FL frameworks can be effectively applied to a wide range of real-world scenarios. For this reason, the customization of FLF is a key aspect, and in the design of the architecture is often taken into account. We aim to provide valuable insights into the adaptability of FL frameworks, shedding light on this crucial usability concern.

To summarize, the contribution of this work is threefold, we:

1. study the scalability of FLFs;
2. provide insights about the usability of FLFs;
3. conduct experiments by deploying multiple FLFs in a realistic environment for the task of Medical Image Classification.

These contributions collectively enhance our understanding of FLFs, addressing the critical aspects of scalability, usability, and practical application in the healthcare domain.

2 Related Works

There are already different benchmarks and surveys for the application of FL to the healthcare domain [14,17,20], but they usually concentrate on the Decentralized Federated Learning. In our case, we deal with Centralized Federated Learning, which uses a trusted server to deal with the clients. At the best of our knowledge, there aren't works treating the scalability of the FLFs, so we extend the insights suggested by the cited literature review providing experimental results on scalability for the centralized case.

For the specific task of Image Classification there are many studies available regarding FL approaches in general [1,6,7,24] and for specific tasks: Brain tumor segmentation [16], Prediction of SARS-COV2 from Chest X-Ray [10], multi-disease X-Ray classification [18], Breast density classification [22]. Our work does not focus on performances of different FL algorithms, but we enrich the performance evaluation with results on scalability.

For a real-case deployment, there are many possible choice of FLF: OpenFL [21], NVFlare [23], FedML [12], FedScope [27], Flower [4], SecureBoost [8], Substra [11] and in particular for the healthcare domain [9]. In this work we compare only some of them: OpenFL, NVFlare, FedML and Flower. In future works extending this list is a key point to find similarities and differences that impacts on the communication cost.

3 Experiments

In this section we are going to present the experiments which are the core contribution of the proposed study. The task we choose is Image Classification on MedMNIST [28], and in particular on the organAMNIST dataset [5]. It consists of 58,850 images (MNIST-like, 28×28, grayscale) labeled with 11 classes, split into 34.581 for training, 6.491 for validation and 17.778 for test (available to all the clients). The transformations used for data augmentation are normalization, random flip and random rotation. The real case deployment was experimented on a cluster with 10 nodes (each node provided with a Tesla T4 GPU) and a frontend node (with no GPU). The frontend node was used as aggregator and/or administrator (for the frameworks that required it), and other nodes were used as clients of the FL system. This experimental setting mimics a realistic scenario where all clients and the aggregator are on different machines that can only communicate via network requests.

The FL algorithm used is FedAvg [18], with an iid split of data among the clients. As backbone network we chose ResNet18 [13] trained from scratch using the Adam optimizer with an initial learning rate set to 0.0001. A total of 100 FL rounds were performed, with 1 epoch of local training performed by each client using a batch size of 64. The communication protocol used is gRPC [25].

Table 1. Execution times with different frameworks and different numbers of clients.

Number of clients	Frameworks			
	OpenFL	NVFlare	Flower	FedML
2	01:11:24	00:55:43	00:51:08	00:45:38
4	01:07:55	00:40:27	00:38:52	00:36:48
6	01:21:22	00:35:30	00:34:30	00:33:44
8	01:40:13	00:34:18	00:33:07	00:34:15
10	02:12:43	00:33:30	00:29:27	00:28:33

The results obtained are reported in Table 1 and displayed in Fig. 1. There are plenty of architectural details that impact on the performance and usability of a FLF. We are going to discuss about them highlighting the similarities and differences among FLFs that may impact on the performances in terms of scalability and usability.

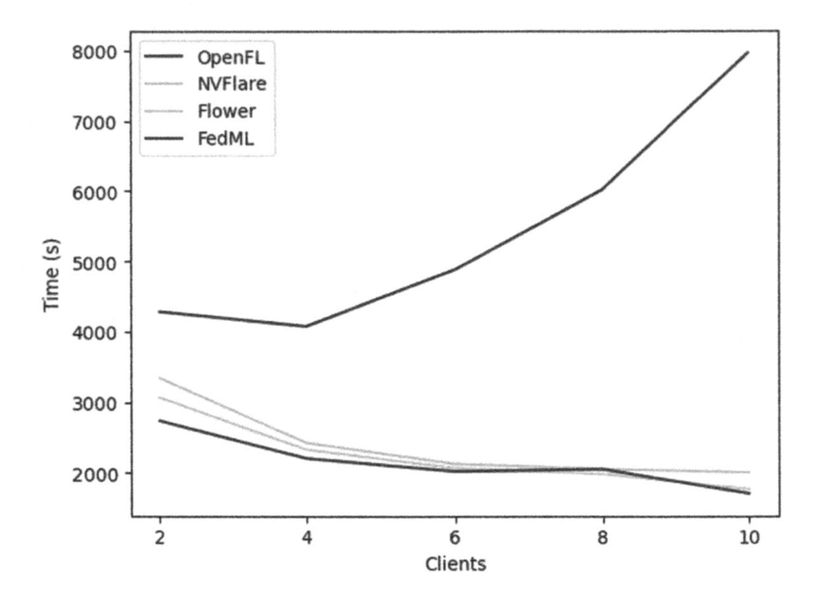

Fig. 1. Execution times for different number of clients.

As we can see, OpenFL does not scale efficiently after 4 clients. On the contrary, the computational time increases when the number of clients goes beyond 4. As highlighted by [19], changing the code at a very low level provided some improvements in the time performances, but the scaling behavior remains inefficient with a growing number of clients.

Because of these low performances we decided to deepen the investigation for OpenFL.

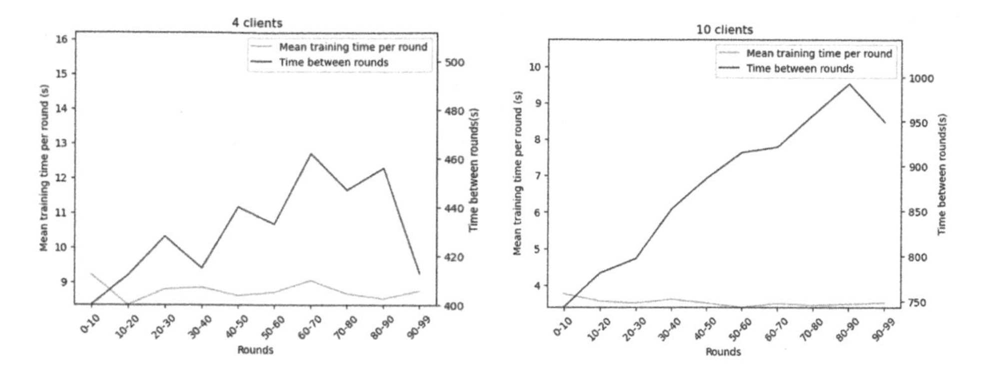

Fig. 2. Execution time with OpenFL for completing windows of 10 rounds along the experiment (blue line) and average training time for each client in each window (red line). (Color figure online)

In Fig. 2 we plot the execution time for completing 10 rounds along the experiment and the average training time in the examined window of rounds. It's clear that the training time is almost constant, showing that the problem stands in the communication cost. In fact, we can see that the time needed for completing the first 10 rounds is way less than the time needed for completing the last ones. This behavior was investigated for 4 and 10 clients. This latter scenario shows a linear trend, which clearly indicates the presence of communication overhead. The structure of OpenFL is clear and simple, using gRPC for connecting aggregators and collaborators and transport layer security (TLS) for network connection. A task based programming interface is used, focusing on the whole workflow design rather than the single client customization. We will see in the following that a similar approach is used by NVFlare, but with different results. As a consequence, the detected communication overhead must be investigated properly to see what is the reason of the low performances. An effort has already been done by [19], but more studies are needed to avoid this behavior in the future FLFs.

On the contrary, Flower shows a very good scaling behavior. In fact, execution time decreases with a growing number of clients, highlighting a good implementation for the communication system. With these results, we confirm the good scaling behavior of this FLF, which has been presented in [4] as one of the goal in its design. Comparing Figs. 3 and 2 we can see that the execution time every ten rounds does not increases, as expected by an efficient scaling. From an architectural perspective, this is probably due to the Virtual Client Engine: a tool that virtualizes the Flower client in order to maximize the utilization of its hardware capacities. This decision helps to address the resource consumption, which is the bottleneck when large-scale experiments are conducted. In addition, they developed the *Strategy* module for describing the FL workflow chosen, which makes the customization of the experiments straightforward. From this analysis, we may advocate that having resource-aware agents improves the scalability of a FLF.

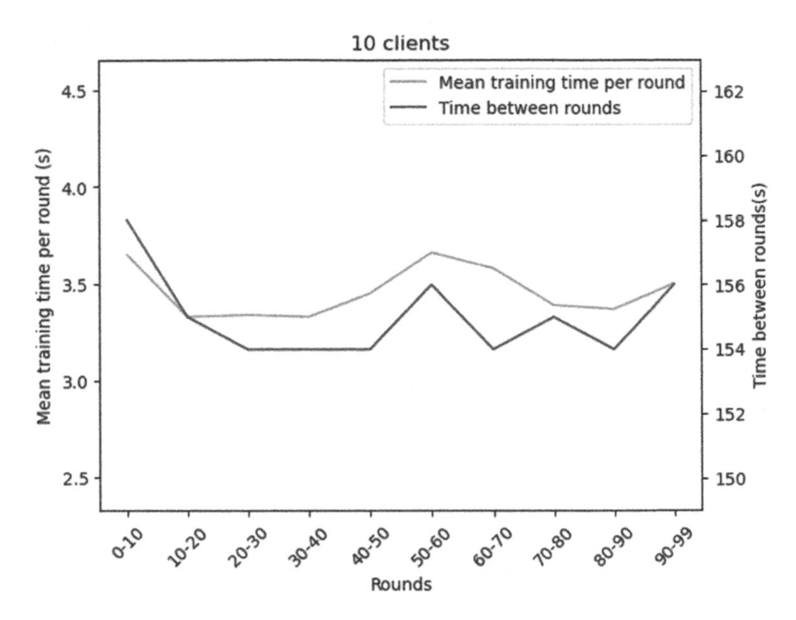

Fig. 3. Execution time with Flower for completing windows of 10 rounds along the experiment (blue line) and average training time for each client in each window (red line). (Color figure online)

In a similar manner, FedML relies on a worker-oriented architecture, avoiding the description of the entire training procedure. In order to do so they introduced the *WorkerManager* class, which utilizes an API system to manage the communication, instead of using a training procedure-oriented programming. In particular FedML-API and FedML-core are the main innovative modules of this framework. The first module is responsible to provide the customization of the algorithms, making the implementation of new FL scenarios straightforward. The second module separates the training engine and the communication system, enabling the customization of the whole procedure at many levels. This architectural choice makes the framework flexible and robust, providing good results for what concerns the scalability. In addition, it is the only one among the FLFs presented that provides different communication backends: MPI, mqtt, tRPC, gRPC. The analysis of FedML enforces the idea that a worker-oriented architecture is useful to make the FLF scalable. The use of APIs and the separation between communication and training may be considered as a winning strategy for the design of FLFs both in terms of scalability and usability.

We conclude with NVFlare. This FLF presents a very good scaling behavior. Their architectural focus is on the controller-worker interaction rather than a worker-oriented structure. Similarly to FedML, they implemented an API controller interface, which supports the typical controller-client interaction making the configuration of the workflow very practical. However, the central concept of collaboration stands in the notion of *task*, similar to OpenFL. On the contrary, they use a *Shareable* object to store a different information (like the model weights) and API at an architectural level. As highlighted with FedML, this seems to be a key aspect to take into account for developing scalable and flexible FLFs, and it may shed light on the different performances with respect to OpenFL.

4 Results

With these experiments we provided empirical evidence of the performance of open source FLFs for what concerns the strong scalability, proposing a comparison that is new in the literature. In addition, all along the experiments we provided insights about the design of the FLFs which impacts on scalability and usability and may help in developing and deploying FL systems in a real world scenario.

In particular we can recognized two possible patterns in designing FLFs: *client-oriented* programming and *training procedure-oriented* programming. The former is developed by Flower and FedML in different fashions, obtaining in both cases good results of scalability. The *training procedure-oriented* results effective for the customization of the workflow, but some architectural choices may impact heavily on the scalability of the framework, as highlighted by the difference in performances between NVFlare and OpenFL. A more detailed study is needed to understand what hinders the performances of OpenFL.

For what concerns usability, the implementation of an API system makes the framework very functional, and its customization straightforward. However, the same result can be obtained using particular abstractions like Flower does with the *Strategy* module. In addition, the sharp separation between communication system and training procedure developed by FedML results to be effective both for scalability and usability.

5 Conclusions

In conclusion, we have presented a study that compares different Federated Learning Frameworks (FLFs) when accomplishing the task of Medical Image Classification. The task, which may initially appear simple, has been challenging in multiple aspects. In the end our contribution is threefold:

i) we have provided insights about the usability of open-source FLFs. We examined their implementations and discussed key aspects that make a FLF flexible.

ii) We have tested the scalability of FLFs, which is a crucial aspect of their future development, through detailed experiments. In addition, we highlighted possible key features for designing scalable FLFs.
iii) We conducted a real-case deployment, which makes this study useful also from a practical perspective.

To conclude, scalability remains a critical focus for further advancements in FLFs. Our research provides empirical results on the performances of some of the main open-source FLFs available, with an additional focus on the usability that gives a practical impact to this work. This proposed study represents a starting point in an unexplored area and has the potential to provide valuable insights leading to FLFs improvement.

6 Future Works

This work represents a preliminary research for deepening our knowledge of FLFs. In particular the empirical results are useful when it comes to choosing which FLF suits the best for the task needed. However there are many aspects that we did not touch and may have relevance in the future.

First of all, a bigger number of clients should be considered, as much as a detailed study of the architectures both from empirical and theoretical points of view. In fact, recording the time needed for every computational step may shed lights on the positive and negative design choices of an FLF. If this is matched with a theoretical treatment of the FLF, then the study would bring important advancements to FLFs' development.

Furthermore, we have considered only the *strong* scalability, which provides an analysis when the amount of operations needed decreases with an increasing number of clients. On the other hand the *weak* scalability provides results when the amount of work is constant and the number of clients increases. Treating both strong and weak scalability would result in a more complete evaluation of the FLFs.

To conclude, a broader selection of FLFs would bring more comparisons between architectural choices in designing FLFs enabling a broader view on the possible directions towards more performative FLFs.

References

1. Adnan, M., Kalra, S., Cresswell, J.C., Taylor, G.W., Tizhoosh, H.R.: Federated learning and differential privacy for medical image analysis. Sci. Rep. **12**(1), 1953 (2022)
2. Al-Rubaie, M., Chang, J.M.: Privacy-preserving machine learning: threats and solutions. IEEE Secur. Privacy **17**(2), 49–58 (2019)
3. Beltrán, E.T.M., et al.: Decentralized federated learning: fundamentals, state-of-the-art, frameworks, trends, and challenges. arXiv preprint arXiv:2211.08413 (2022)

4. Beutel, D.J., et al.: Flower: a friendly federated learning research framework. arXiv preprint arXiv:2007.14390 (2020)
5. Bilic, P., et al.: The liver tumor segmentation benchmark (LiTS). Med. Image Anal. **84**, 102680 (2023)
6. Casella, B., Esposito, R., Cavazzoni, C., Aldinucci, M.: Benchmarking FedAvg and FedCurv for image classification tasks. In: Anisetti, M., Bonifati, A., Bena, N., Ardagna, C.A., Malerba, D. (eds.) Proceedings of the 1st Italian Conference on Big Data and Data Science (itaDATA 2022), Milan, Italy, September 20-21, 2022. CEUR Workshop Proceedings, vol. 3340, pp. 99–110. CEUR-WS.org (2022). https://ceur-ws.org/Vol-3340/paper40.pdf
7. Casella, B., Esposito, R., Sciarappa, A., Cavazzoni, C., Aldinucci, M.: Experimenting with normalization layers in federated learning on non-IID scenarios. arXiv preprint arXiv:2303.10630 (2023)
8. Cheng, K., et al.: SecureBoost: a lossless federated learning framework. IEEE Intell. Syst. **36**(6), 87–98 (2021)
9. Cremonesi, F., et al.: Fed-BioMed: open, transparent and trusted federated learning for real-world healthcare applications. arXiv preprint arXiv:2304.12012 (2023)
10. Flores, M., et al.: Federated learning used for predicting outcomes in SARS-COV-2 patients. Research Square (2021)
11. Galtier, M.N., Marini, C.: Substra: a framework for privacy-preserving, traceable and collaborative machine learning. arXiv preprint arXiv:1910.11567 (2019)
12. He, C., et al.: FedML: a research library and benchmark for federated machine learning. arXiv preprint arXiv:2007.13518 (2020)
13. He, K., Zhang, X., Ren, S., Sun, J.: Deep residual learning for image recognition. In: Proceedings of the IEEE Conference on Computer Vision and Pattern Recognition, pp. 770–778 (2016)
14. Joshi, M., Pal, A., Sankarasubbu, M.: Federated learning for healthcare domain-pipeline, applications and challenges. ACM Trans. Comput. Healthcare **3**(4), 1–36 (2022)
15. Li, T., Sahu, A.K., Talwalkar, A., Smith, V.: Federated learning: challenges, methods, and future directions. IEEE Signal Process. Mag. **37**(3), 50–60 (2020)
16. Li, W., et al.: Privacy-preserving federated brain tumour segmentation. In: Suk, H.-I., Liu, M., Yan, P., Lian, C. (eds.) MLMI 2019. LNCS, vol. 11861, pp. 133–141. Springer, Cham (2019). https://doi.org/10.1007/978-3-030-32692-0_16
17. Lian, Z., et al.: DEEP-FEL: decentralized, efficient and privacy-enhanced federated edge learning for healthcare cyber physical systems. IEEE Trans. Netw. Sci. Eng. **9**(5), 3558–3569 (2022)
18. McMahan, B., Moore, E., Ramage, D., Hampson, S., y Arcas, B.A.: Communication-efficient learning of deep networks from decentralized data. In: Artificial Intelligence and Statistics, pp. 1273–1282. PMLR (2017)
19. Mittone, G., Riviera, W., Colonnelli, I., Birke, R., Aldinucci, M.: Model-agnostic federated learning. In: Cano, J., Dikaiakos, M.D., Papadopoulos, G.A., Pericás, M., Sakellariou, R. (eds.) Euro-Par 2023: Parallel Processing. Euro-Par 2023. LNCS, vol. 14100, pp. 383–396. Springer, Cham (2023). https://doi.org/10.48550/arXiv.2303.04906, https://doi.org/10.1007/978-3-031-39698-4_26
20. Mothukuri, V., Parizi, R.M., Pouriyeh, S., Huang, Y., Dehghantanha, A., Srivastava, G.: A survey on security and privacy of federated learning. Futur. Gener. Comput. Syst. **115**, 619–640 (2021)
21. Reina, G.A., et al.: OpenFL: an open-source framework for federated learning. arXiv preprint arXiv:2105.06413 (2021)

22. Roth, H.R., et al.: Federated learning for breast density classification: a real-world implementation. In: Albarqouni, S., et al. (eds.) Domain Adaptation and Representation Transfer, and Distributed and Collaborative Learning: Second MICCAI Workshop, DART 2020, and First MICCAI Workshop, DCL 2020, Held in Conjunction with MICCAI 2020, Lima, Peru, 4–8 October 2020, Proceedings 2. LNCS, vol. 12444, pp. 181–191. Springer, Cham (2020). https://doi.org/10.1007/978-3-030-60548-3_18
23. Roth, H.R., et al.: NVIDIA FLARE: federated learning from simulation to real-world. arXiv preprint arXiv:2210.13291 (2022)
24. Sheller, M.J., Reina, G.A., Edwards, B., Martin, J., Bakas, S.: Multi-institutional deep learning modeling without sharing patient data: a feasibility study on brain tumor segmentation. In: Crimi, A., Bakas, S., Kuijf, H., Keyvan, F., Reyes, M., van Walsum, T. (eds.) Brainlesion: Glioma, Multiple Sclerosis, Stroke and Traumatic Brain Injuries: 4th International Workshop, BrainLes 2018, Held in Conjunction with MICCAI 2018, Granada, Spain, 16 September 2018, Revised Selected Papers, Part I 4, vol. 11383, pp. 92–104. Springer, Cham (2019). https://doi.org/10.1007/978-3-030-11723-8_9
25. Wang, X., Zhao, H., Zhu, J.: GRPC: a communication cooperation mechanism in distributed systems. ACM SIGOPS Oper. Syst. Rev. **27**(3), 75–86 (1993)
26. Witt, L., Heyer, M., Toyoda, K., Samek, W., Li, D.: Decentral and incentivized federated learning frameworks: a systematic literature review. IEEE Internet Things J. **10**(4), 3642–3663 (2022)
27. Xie, Y., et al.: FederatedScope: a flexible federated learning platform for heterogeneity. arXiv preprint arXiv:2204.05011 (2022)
28. Yang, J., et al.: MedMNIST v2-a large-scale lightweight benchmark for 2D and 3D biomedical image classification. Sci. Data **10**(1), 41 (2023)

Artificial Intelligence for Digital Humanities (AI4DH)

Examining the Robustness of an Ensemble Learning Model for Credibility Based Fake News Detection

Amit Neil Ramkissoon(✉), Kris Manohar, and Wayne Goodridge

Department of Computing and Information Technology, The University of the West Indies at St. Augustine, St. Augustine, Trinidad and Tobago
amit.ramkissoon@my.uwi.edu, {kris.manohar, wayne.goodridge}@sta.uwi.edu

Abstract. Ensemble learning is a technique of combining multiple base machine learning models and using the blended results as the final classification output. Such models provide a unique perspective on the classification results as it produces a more comprehensive and encompassing output. As such ensemble learning techniques are widely used for classification in today. Hence it is important that any ensemble learning model be robust and resilient to any type of data and not just applicable to one dataset. This research investigates and evaluates the robustness and the resilience of the proposed Legitimacy ensemble learning model. This ensemble learning model was previously proposed for Credibility Based Fake News Detection. This research evaluates Legitimacy's performance with a variety of datasets. In the first scenario, the Legitimacy ensemble learning model is evaluated with 3 different binary classification datasets for training and testing purposes, respectively. In the second scenario, the Legitimacy model is assessed where one dataset is used for training whilst another dataset is used for testing. In the final scenario the Legitimacy ensemble learning model is evaluated against a multiclass dataset for multiclass classification. The results of all the above tests are assimilated and evaluated. The results suggest that the Legitimacy ensemble learning model performs well in all three scenarios giving AUC values all equal to or greater than 0.500. As such it can be concluded that the Legitimacy model is a robust and resilient ensemble learning technique and can be employed for the task of classification with any dataset.

Keywords: Ensemble Learning · Fake News Detection · Legitimacy · Resilient · Robust

1 Introduction

Nowadays, there is a significant amount of popular false information referred to as "fake news." This type of news intentionally presents untrue information and can be verified as such. A study showed that between 2014 and 2016, 171,365 stories from 60 fake news websites were discovered, indicating the prevalence of this phenomenon. When fake

G. L. Foresti et al. (Eds.): ICIAP 2023 Workshops, LNCS 14366, pp. 235–246, 2024.
https://doi.org/10.1007/978-3-031-51026-7_21

news spreads via social media and other online platforms, it undermines the efforts made to prevent the spread of misinformation. This allows for false narratives, misinformation, and unfounded conspiracies to proliferate, creating a harmful environment. Therefore, detecting fake news on social media is crucial in reducing these negative effects and benefiting the public and the news industry.

Detecting and predicting fake news is a difficult task, particularly on social media and other online platforms where users often do not verify the news they consume. Therefore, there is a need for automated methods to validate the credibility of news, as noted in [18]. Detecting fake news on social media is particularly challenging, as fake news is deliberately written to mislead users and can mimic real news in various styles. As a result, detecting fake news has become a popular research topic and is receiving significant attention from the research community, as highlighted in [16].

One way to predict fake news is by analysing its features to look for any connections between them. Fake news detection can be categorized into four perspectives: knowledge-based, which fact-checks the content of the news, style-based, which looks for similarities in writing style with both genuine and fake news, propagation-based, which examines how the news spreads, and credibility-based, which investigates the credibility of the publisher and spreaders. These categories were identified in a study cited as [20].

Credibility-based fake news detection relies on the reputation of the news publisher, as mentioned previously. To detect fake news based on credibility, data about the publisher's online behaviour and news-related information is analysed. For instance, if an untrustworthy publisher shares a message that is then forwarded by unreliable network clients, it is less likely to be credible and more likely to be fake news compared to news posted by authoritative and credible users. These points were also made in reference to study [20].

The author, editor, and publisher of a news document are closely tied to the news content and context, as mentioned in [6]. The news context is reflected in the social engagement of the news article on social media platforms, which represents how the news spreads over time and the group of users that interact with it. Therefore, social-based features such as the number of followers, friends count, registration age, number of authored posts or tweets, related social groups, demographic information, user stance, and average credibility scores can be extracted for users, groups, and postings.

In order to perform credibility-based fake news detection, various features of users must be examined to determine how they relate to the credibility of the news, as explained in [14]. It is suggested that non-human accounts like social bots or cyborgs are involved in the creation and propagation of fake news. Therefore, analysing user profiles and user-based features of those who have engaged with the news on social media can be helpful in detecting fake news.

This study employs an ensemble learning approach with credibility-based features to determine whether news is genuine or fake based on the publisher's credibility. Credibility-based fake news detection is utilised in this research to maintain a sense of objectivity and ensure that a change in the type of data does not skew the results. Machine learning (ML) is used to detect patterns in the data, which aids in making predictions. The ML field is a highly successful discipline that is proficient in handling various learning tasks, as stated in [9]. The primary tasks that ML typically deals with are classification and regression. However, this work solely focuses on classification, as it identifies patterns between the features of fake news and the authenticity of the news.

This paper presents an analysis of the Legitimacy ensemble learning model. This model has been proposed by [13] and is used for the task of Credibility based Fake News Detection. This ensemble model consists of two underlying techniques namely, a Two Class Boosted Decision Tree and a Two Class Neural Network. These two individual models are ensembled in a pseudo-mixture-of-experts manner with Two Class Logistic Regression as the gating model. Though the Legitimacy model is proposed in [13], this research furthers the results of that paper by attempting to analyse the Legitimacy model to establish the robustness of the Legitimacy ensemble learning model. The model is employed with four different datasets from different distributions involving both two class and multiclass classifications. The Legitimacy model is also tested using a combination of datasets with one dataset being used for training whilst the other is used for testing. The results of each of these classifications is analysed and thoroughly interrogated.

This paper proposes the following:

1. To present an understanding of the current datasets used for Fake News Detection and their limitations.
2. To analyse the Legitimacy ensemble model using a variety of datasets
3. To present and evaluate initial results of the model for credibility based fake news detection using both two class and multiclass classification datasets.

The remainder of this paper is structured as follows. Section 2 presents the related work in this field whilst Sect. 3 presents an understanding of the machine learning algorithm. Experiments are conducted in Sect. 4 using the algorithm and the results of these are discussed in this section. Section 5 concludes the paper.

2 Related Works

Fake news detection is a binary classification problem. We compared the Legitimacy across a variety of dataset and classification problems. This section summarizes these related works, approaches and their classification problems and datasets.

2.1 The Liar Dataset

In 2017 fake news datasets lacked sufficient data to train machine learning models for automatic fake news detection. This issue motivated the creation of a dataset with 12836 short statements. Wang [17] carefully selected and fact-checked each short statement's subject, context/venue, speaker, state, party and prior history. Thus, his fake news dataset accurately represents fake news statements for a decade. This improved representation facilitated automatic fake news detection using logistic regression, support vector machine, bi-directional long short-term network and convolutional neural network classifiers. Wang [17] showed that combining text and metadata features (e.g., subject, speaker, context, party etc.) with a hybrid-convolutional neural network yielded the best performance. Wang's [17] implementation and dataset are available at http://aclweb.org/anthology/W18-5513.

2.2 The FakeNewsNet Dataset

Shu et al.'s [15] main contribution was the creation of the FakeNewsNet dataset. This dataset consists of 602,659 news articles sampled from PolitiFact and GossipCop websites. The staff of these fact-checking websites are diligent and transparent in assessing news content. Hence Shu et al. [15] utilized these websites as the ground truths for the FakeNewsNet dataset. They also proposed that the social context surrounding a news article is an essential feature for fake news detection. The FakeNewsNet dataset represents this context as related engagements on social media platforms. It also stores changes to news content or its social context as dynamic information. This information adds a temporal aspect to the problem of fake news detection. Shu et al.'s [15] implementation and dataset are available at https://arxiv.org/abs/1809.01286.

2.3 The Fake and Real News Dataset

In 2017 Ahmed et al. [2] tackled the opinion spam detection problem. Opinion spam consists of fake reviews and content. Ahmed et al.'s [2] proposed scheme extracted n-gram frequency profiles from the training data. It utilized the top p terms as features to represent fake and real opinions. Using these features across two datasets, they compared the performance of a stochastic gradient descent, support vector machine, linear support vector machine, k-nearest neighbour, logistic regression and decision trees classifiers. Their proposed scheme achieved 90% and 87% accuracy for fake reviews and news datasets, respectively. Ahmed et al.'s [2, 3] dataset and implementation are available at https://www.kaggle.com/datasets/clmentbisaillon/fake-and-real-news-dataset.

2.4 Spawned Dataset

The dataset is inspired by that proposed by [19] and contains 17,551 generated news records with 17,055 fake and 496 genuine messages. The dataset is a generated and synthetic one, that has been produced by the execution of the simulation environment. The experiments are validated on only this one dataset as there are no other that contains the features set required for the Veracity architecture. The features generated include: the text, eyewitness, label, source, date/time, language, listed count, location, statuses count, followers count, favourites count, time zone, user language, friends count, screen name, credibility score, text similarity and eyewitness score. The dataset is described in [12].

3 Methods

This study examines the effectiveness of an ensemble learning model for identifying false news using models from Microsoft Azure Machine Learning Studio (classic). AzureML is a user-friendly tool that allows you to develop and deploy predictive analytics solutions using drag-and-drop functionality [8]. It is a platform that integrates data science, predictive analytics, cloud resources, and data. AzureML publishes models created in it as web services for use in custom applications or business intelligence tools. This ensemble model includes various classification models, as detailed in [8].

3.1 Two-Class Boosted Decision Tree (BDT)

The Two-Class Boosted Decision Tree model was offered by Microsoft Azure Machine Learning Studio (classic). This study employs the Two Class Boosted Decision Tree, based on the findings presented in [11]. According to [11], the experiments conducted and results obtained indicate that the Two Class Boosted Decision Tree performed optimally. Thus, it can be concluded that, based on the dataset used, the Two Class Boosted Decision Tree is the most suitable method for detecting and predicting fake news with regards to credibility.

A Boosted Decision Tree, as outlined in [8], is an ensemble learning technique where subsequent trees rectify the errors made by previous trees. The final prediction is made based on the combined results of all trees in the ensemble. Boosted Decision Trees are known for their high performance in various machine learning tasks and are relatively straightforward to use when properly configured. However, they consume a significant amount of memory and the current implementation requires all data to be stored in memory, which can limit their ability to process extremely large datasets in comparison to linear learning methods.

Boosting can be implemented using two algorithms. The first method, the AdaBoost algorithm, creates an ensemble by emphasizing instances that were previously misclassified. The focus is determined by assigning weights to instances in the training set, with all instances having the same weight in the first iteration. In subsequent iterations, the weights of misclassified instances increase, while the weights of correctly classified instances decrease. When making predictions using the ensemble, weights are also assigned to individual base learners based on their predictive performance. The ensemble created using AdaBoost with decision stumps as weak learners serves as a baseline for comparing prediction metrics with scalable GBDT systems.

The second boosting method is the gradient-descent based formulation known as gradient boosting machines (GBMs), according to [5]. GBMs construct base learners by iteratively reweighting misclassified observations, but they differ from AdaBoost in that they determine weights by operating on the negative partial derivatives of the loss function at each training observation. GBMs using decision trees as base learners are labelled as gradient boosted decision classifiers (GBDCs) and serve as a baseline for comparison with the performance metrics of scalable GBDT systems, such as XGBoost, LightGBM, and CatBoost. The Two Class Boosted Decision Tree implements the decision tree algorithm and boosts the tree using the GBM methodology.

The GBM equation can be seen in (1).

$$(\rho_t, \theta_t) = arg \min_{\rho,\theta} \sum_{i=1}^{N} -g_t(x_i) + \rho h(x_i, \theta) \quad (1)$$

To summarize, the complete form of the gradient boosting algorithm was formulated by Friedman. The exact form of the derived algorithm with all the corresponding formulas will heavily depend on the design choices of $\psi(y, f)$ and $h(x, \theta)$.

3.2 Two Class Neural Network

The paper [8] states that a neural network is made up of interconnected layers, with inputs being the first layer and the output layer connected to it through a weighted graph of edges and nodes that does not form a cycle.

In summary, a neural network is composed of multiple interconnected layers, with inputs as the first layer and outputs as the last layer, connected by an acyclic graph of weighted edges and nodes. The number of hidden layers can vary and can range from one to many depending on the complexity of the task. Recent research has shown that deep neural networks with many hidden layers can be highly effective in complex tasks such as image or speech recognition. These successive hidden layers are used to model increasing levels of semantic depth.

The weights of the edges are determined by a training process, during which the network is exposed to examples of inputs and the desired outputs. The weights are then adjusted so that the difference between the actual output and the desired output is minimized. The process of training the neural network is typically done using an optimization algorithm, such as stochastic gradient descent (SGD), which iteratively updates the weights of the edges to minimize a loss function that measures the difference between the actual and desired outputs. After the training process, the neural network can then be used to make predictions on new data.

The activation function is used to introduce non-linearity in the output and is important for capturing complex relationships between inputs and outputs. Common activation functions include sigmoid, tanh, ReLU and its variants. The output from the activation function is then passed to the next layer, and the process is repeated until the output layer is reached. The final output of the network is compared with the target output, and the difference is used to adjust the weights and biases in the network, through a process called backpropagation, so as to minimize the prediction error. This process of adjusting weights and biases is repeated until the network has learned to predict the target output accurately, based on the input data. The softmax function can be defined as:

$$y_i = \frac{e^{x_i}}{\sum_{j=1}^{C} e^{x_j}} \tag{2}$$

The ANN seeks to replicate the brain's abilities to learn, generalize, and process information. The human brain has the ability to learn from experience and adjust its responses to new inputs. Similarly, an ANN can be trained on a set of inputs and their corresponding outputs to learn a mapping between inputs and outputs, and then make predictions for new inputs. The ANN can be trained using various optimization algorithms to minimize the error between its predictions and the actual outputs. Thus, ANNs have found applications in various fields such as computer vision, speech recognition, natural language processing, and robotics, among others.

According to [1], the layers in a Neural Network (NN) are independent of each other, meaning that a specific layer can have an arbitrary number of nodes, including a bias node. The bias node is a constant value that is always set to one and its purpose is to provide a constant value for the network node, similar to the offset in linear regression. The bias value allows for the adjustment of the activation function either to the right or to the left, which is crucial for the success of the training of the ANN. When the NN is used as a classifier, the input and output nodes will match the input features and output classes. On the other hand, when the NN is used as a function approximation, it generally has an input and an output node, with the number of hidden nodes being greater than the number of input nodes.

3.3 Mixture of Experts

According to [7], the authors explain the original mixture of experts (ME) regression and classification models. In this architecture, a group of experts and a gate work together to solve a nonlinear supervised learning problem by dividing the input space into multiple nested regions for classification. The gate makes a smooth division of the entire input space, and the experts learn simple parameterized functions within these divided regions. The parameters for these functions in both the gate and the experts can be learned using the Expectation-Maximization (EM) algorithm.

3.4 Two Class Logistic Regression

According to [8], Logistic Regression is a popular statistical method used for various problem-solving and is a type of supervised learning. In this method, a dataset that includes the desired outcome must be provided for training the model.

According to [8], Logistic regression is a widely recognized statistical approach that is used to model the probability of an event happening. This method is commonly used for classification problems and predicts the likelihood of a particular outcome by fitting the data to a logistic function.

According to [10], linear models are composed of one or multiple independent variables that describe a relationship to a dependent response variable. In machine learning, this is known as supervised learning, where input features are mapped to a target variable that is to be predicted, such as financial, biological, or sociological data, when the labels are known. One of the most commonly used linear statistical models for discriminant analysis is logistic regression.

According to [10], logistic regression is a simple and interoperable linear statistical model that is commonly used for binary classification problems. Despite its simplicity, logistic regression can outperform other complex nonlinear models in certain situations. However, when the response variable is drawn from a small sample size, logistic regression models may not perform well and other learning algorithms should be considered. The aim of the study is to specifically focus on the examination of logistic regression as a linear model for binary classification problems.

By combining the above individual learning models the Legitimacy ensemble model presented in this paper is built. Legitimacy combines the functionality of a Two Class Boosted Decision Tree and a Two Class Neural Network. The ensemble model is designed as seen in Fig. 1.

In the study, the features in the dataset are divided into two groups: demographic features and social behaviour features. The ensemble model is then constructed based on these two separate groups with the goal of improving the performance of each model by using a specific subgroup. The standard data processing steps are then followed,

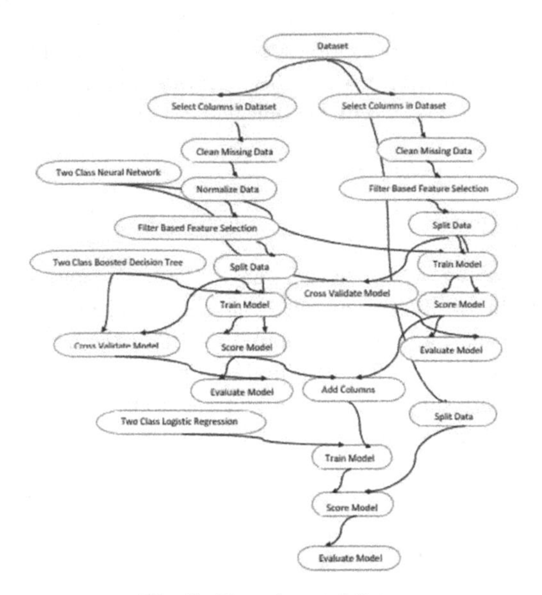

Fig. 1. Experimental Setup

including cleaning and normalizing the data. The data is split into a 65% training and 35% testing set, with the training set being 65% and the testing set being 35%. The Two Class Boosted Decision Tree algorithm is applied to the social behaviour data features due to its exceptional performance with this subgroup.

The process continues by applying the Two Class Neural Network model to the demographic features due to its demonstrated proficiency in this feature group. Following standard machine learning and ensemble learning procedures, each model is trained on its respective subgroup of features. Both models are then evaluated and scored independently. These scores are combined with the scores from the testing dataset to form a larger scored dataset, which is then used to train the logistic regression model. The logistic regression uses both models' outputs to train itself.

As stated in [4] Logistic Regression can be defined using the formula found in (3).

Let Y denote the binary response variable of interest and X_1, ..., X_p the random variables considered as explaining variables, termed features in this paper. The logistic regression model links the conditional $P(Y = 1|X_1, ..., X_p)toX_1, ..., X_p$ through

$$P(Y = 1|X_1, \ldots, X_p) = \frac{\exp\left(\beta_0 + \beta_1 X_1 + \cdots + \beta_p X_p\right)}{1 + \exp(\beta_0 + \beta_1 X_1 + \cdots + \beta_p X_p)}, \tag{3}$$

where β_0, β_0, ..., β_p are regression coefficients, which are estimated by maximum-likelihood from the considered dataset. The probability that $Y = 1$ for a new instance is then estimated by replacing the X's by their estimated counterparts and the 's by their realizations for the considered new instance in (3).

The Logistic Regression model acts as the decision maker in the pseudo mixture-of-experts approach. It uses the combined dataset to train itself by making a weighted combination of the scored datasets from both the Two Class Boosted Decision Tree and the Two Class Neural Network models. The features in the merged dataset serve as the inputs (X) and the target variable (Y), which is the binary classification of the news as true or false, is the predicted outcome. The Logistic Regression model determines the appropriate percentage of each data feature to use in each situation during training. After being trained, the model is scored and evaluated, and the result from this scoring is used as the final prediction value of the model.

4 Experimental Results

We conducted several experiments to investigate the robustness of Legitimacy's ensemble learning strategy. For the related works we used the online implementations provided by the authors. For all we used an 80-20 split with 5-fold validation. Except for the implementation provided by the authors of the related works, all other experiments were implemented and executed in AzureML.

We design several experiments to test Legitimacy's performance with a variety of combinations of datasets. First, we used the same dataset for both training and testing purposes. Then we varied the test and training datasets independently. That is, one dataset was used for training, whilst another (independent) dataset was used for testing. Since we are attempting to demonstrate the generalizability of Legitimacy, we needed to measure the similarity between the train and test datasets. The datasets used were the FakeNewsNet dataset [15], the Liar dataset [17], the fake and real news dataset [7] and the generated dataset [12]. For these experiments, we utilized the Kolmogorov-Smirnov (K-S) test to measure the similarity between datasets. The closer the distributions, the better the expected performance.

4.1 Experiments Where the Train and Test Set are the Same

The Legitimacy ensemble learning model was tested on the Fake and real news dataset proposed by [7] (referred to as "Fake"), the FakeNewsNet dataset proposed by [15] (referred to as "FNN") and the Liar dataset proposed by [17] (referred to as Liar). The results from these experiments are summarized in Table 1. In all test cases, Legitimacy improved the performance of the respective related work. These experiments indicate that the proposed model can detect fake content across several domains.

Our model was also tested with muti-model classification. The Liar dataset contains 6 labels "pants-fire", "false", "barely-true", "half-true", "mostly-true" and "true". Legitimacy achieved an accuracy of 0.774 which improved Liar's accuracy of 0.277. Both experiments utilized text and contextual information for classification. Hence, we conclude that Legitimacy is also acceptable from multi-class fake news detection.

4.2 Experiments Where the Train and Test Set are Different

The robustness of the Legitimacy ensemble model has been thoroughly evaluated through various experiments. Each experiment utilizes different (independent) datasets for training and testing. This procedure was repeated for each combination of the three datasets utilized in this research. Table 2 summarizes each scenario's result.

Table 1. Comparison of ensemble learning model's performance to the related works across various datasets

Model	F1 (%)	AUC	Ref
ATD2017SP	36.4	0.501	[7]
Legitimacy (with the Fake dataset)	70.4	0.965	[7]
SMWLL2020BG	45.6	0.502	[15]
Legitimacy (with the FNN dataset)	86.7	0.504	[15]

Table 2. Comparison of ensemble model's performance with different combinations of train and test datasets

Train dataset	Test dataset	K-S	F1 (%)	AUC
Fake	Fake	1	70.4	0.965
Spawned	Fake	0.697	0	0.500
FNN	Fake	0.299	0	0.500
FNN	FNN	1	86.7	0.504
Spawned	FNN	0.697	0	0.500
Fake	FNN	0	44.8	0.500
Spawned	Spawned	1	0	0.708
FNN	Spawned	0.697	0	0.500
Fake	Spawned	0.697	10.2	0.500

As the K-S score increases the performance of Legitimacy increases. When the train and test samples are likely to be drawn from the same distributions (i.e., K-S = 1) the proposed model improves the performance of the respective related works. In contrast, when the train and test samples are unlikely to be drawn from the same distributions (i.e., K-S $<$ 1) the model converges to a random guess. Under these circumstances, the optimal choice is a random guess because train and test datasets do not share any meaningful patterns that can be exploited for classification. The experimental results in Table 2 indicate that the ensemble model's behaviour is consistent regardless of the patterns in the training data. Hence, we conclude that Legitimacy is a good candidate for any binary classification problem.

5 Conclusion

In this paper we analyse the robustness of the Legitimacy ensemble model [13] with various state-of-the-art fake news datasets. Legitimacy utilizes a pseudo-mixture-of-experts methodology where it combines a Two-Class Boosted Decision Tree and a Two-Class

Neural Network models for fake news detection. The experimental results indicate that Legitimacy improves the performance recent fake news detection algorithms proposed by [15] and [7] using their datasets. It also improves the performance of [17] for multi-class fake news detection problems. We also conducted several experiments to establish the consistency of the ensemble model. Specifically, when the train and test distributions are dissimilar (i.e., they have low k-s score) the proposed model converges to a random guess which is optimal under these circumstances. However, when the train and test set are similar (i.e., the k-s score is equal to 1) the proposed model improves the performance of the respective related works. Hence, we conclude that Legitimacy is a robust model that can be used for a wide variety of binary and multi-class fake news detection use cases.

References

1. Abiodun, O.I., Jantan, A., Omolara, A.E., Dada, K.V., Mohamed, N.A., Arshad, H.: State-of-the-art in artificial neural network applications: a survey. Heliyon **4**(11), e00938 (2018)
2. Ahmed, H., Traore, I., Saad, S.: Detection of online fake news using N-gram analysis and machine learning techniques. In: Traore, I., Woungang, I., Awad, A. (eds.) ISDDC 2017. LNCS, vol. 10618, pp. 127–138. Springer, Cham (2017). https://doi.org/10.1007/978-3-319-69155-8_9
3. Ahmed, H., Traore, I., Saad, S.: Detecting opinion spams and fake news using text classification. J. Secur. Privacy **1**(1) (2018). Wiley
4. Couronné, R., Probst, P., Boulesteix, A.-L.: Random forest versus logistic regression: a large-scale benchmark experiment. BMC Bioinform. **19**(1), 1–14 (2018)
5. Dev, V.A., Eden, M.R.: Formation lithology classification using scalable gradient boosted decision trees. Comput. Chem. Eng. **128**, 392–404 (2019)
6. Elhadad, M.K., Li, K.F., Gebali, F.: Fake news detection on social media: a systematic survey. In: 2019 IEEE Pacific Rim Conference on Communications, Computers and Signal Processing (PACRIM), pp. 1–8. IEEE (2019)
7. Hakak, S., et al.: An ensemble machine learning approach through effective feature extraction to classify fake news. Futur. Gener. Comput. Syst. **117**, 47–58 (2021)
8. Martens, J.: Machine Learning Studio (Classic) Documentation - Azure." Machine Learning Studio (classic) documentation - Azure|Microsoft Docs. https://docs.microsoft.com/en-us/azure/machine-learning/studio/. Accessed 22 Apr 2020
9. Kazllarof, V., Karlos, S., Kotsiantis, S.: Active learning Rotation Forest for multiclass classification. Comput. Intell. **35**(4), 891–918 (2019)
10. Kirasich, K., Smith, T., Sadler, B.: Random forest vs logistic regression: binary classification for heterogeneous datasets. SMU Data Sci. Rev. **1**(3), 9 (2018)
11. Ramkissoon, A.N., Mohammed, S.: An experimental evaluation of data classification models for credibility based fake news detection. In: 2020 International Conference on Data Mining Workshops (ICDMW), pp. 93–100. IEEE (2020)
12. Ramkissoon, A.N., Goodridge, W.: Detecting fake news in MANET messaging using an ensemble based computational social system." In: Mazzeo, P.L., Frontoni, E., Sclaroff, S., Distante, C. (eds.) Image Analysis and Processing. ICIAP 2022 Workshops: ICIAP International Workshops, Lecce, Italy, 23–27 May 2022, Revised Selected Papers, Part II, pp. 278–289. Springer, Cham (2022). https://doi.org/10.1007/978-3-031-13324-4_24
13. Ramkissoon, A.N., Goodridge, W.: Legitimacy: an ensemble learning model for credibility based fake news detection. In: 2021 International Conference on Data Mining Workshops (ICDMW), pp. 254–261. IEEE (2021)

14. Shu, K., Sliva, A., Wang, S., Tang, J., Liu, H.: Fake news detection on social media: a data mining perspective. ACM SIGKDD Explor. Newsl. **19**(1), 22–36 (2017). https://doi.org/10.1145/3137597.3137600
15. Shu, K., Mahudeswaran, D., Wang, S., Lee, D., Liu, H.: FakeNewsNet: a data repository with news content, social context and dynamic information for studying fake news on social media. arXiv preprint arXiv:1809.01286 (2019)
16. Shu, K., Wang, S., Liu, H.: Understanding user profiles on social media for fake news detection. In: 2018 IEEE Conference on Multimedia Information Processing and Retrieval (MIPR), pp. 430–435. IEEE (2018)
17. Wang, W.Y.: "liar, liar pants on fire": a new benchmark dataset for fake news detection. arXiv preprint arXiv:1705.00648 (2017)
18. Khan, Y., Junaed, T.I.K., Iqbal, A., Afroz, S.: A benchmark study on machine learning methods for fake news detection. arXiv preprint arXiv:1905.04749 (2019)
19. Zahra, K., Imran, M., Ostermann, F.O.: Automatic identification of eyewitness messages on twitter during disasters. Inf. Process. Manage. **57**(1), 102107 (2020)
20. Zhou, X., Zafarani, R.: Fake news: a survey of research, detection methods, and opportunities. arXiv preprint arXiv:1812.00315 (2018)

Prompt Me a Dataset: An Investigation of Text-Image Prompting for Historical Image Dataset Creation Using Foundation Models

Hassan El-Hajj[1,2(✉)] and Matteo Valleriani[1,2,3,4]

[1] Max Planck Institute for the History of Science, Boltzmannstr. 22, 14195 Berlin, Germany
{hhajj,valleriani}@mpiwg-berlin.mpg.de
[2] BIFOLD – Berlin Institute for the Foundations of Learning and Data, 10587 Berlin, Germany
[3] The Cohn Institute for the History and Philosophy of Science and Ideas, Faculty of Humanities, Tel-Aviv University, 6997801 Tel-Aviv, Israel
[4] Institute of History and Philosophy of Science, Technology, and Literature, Faculty I, Technische Universität Berlin, Straße des 17. Juni 135, 10623 Berlin, Germany

Abstract. In this paper, we present a pipeline for image extraction from historical documents using foundation models, and evaluate text-image prompts and their effectiveness on humanities datasets of varying levels of complexity. The motivation for this approach stems from the high interest of historians in visual elements printed alongside historical texts on the one hand, and from the relative lack of well-annotated datasets within the humanities when compared to other domains. We propose a sequential approach that relies on GroundDINO and Meta's Segment-Anything-Model (SAM) to retrieve a significant portion of visual data from historical documents that can then be used for downstream development tasks and dataset creation, as well as evaluate the effect of different linguistic prompts on the resulting detections.

Keywords: SAM · GroundingDINO · Digital Humanities · Dataset Creation · Historical Documents · Text Prompts

1 Introduction

Technological advancements of the last decades have led to major digitization efforts focused on historical documents, such as the Google Book Search (GBS) and Open Content Alliance (OCA) [12]. This rapid growth of digitized historical documents has paved the way for computational historical document analysis, allowing researchers to comb through large number of documents and test hypotheses at scale.

With the advent of neural networks, new methods of image analysis and information extraction came to light. However, these methods are data intensive,

G. L. Foresti et al. (Eds.): ICIAP 2023 Workshops, LNCS 14366, pp. 247–257, 2024.
https://doi.org/10.1007/978-3-031-51026-7_22

and require a large amount of annotated and curated datasets in order to be trained. The lacuna of such datasets pushed the digital humanities community towards collecting and publishing annotated and curated datasets to facilitate the training of state-of-the-art models. However, given the heterogeneous nature of historical data, and the high degree of inter –and intra– domain variability, such datasets often cover very specific historical topics and domains, with limited generalization possibilities.

In this paper, we propose a pipeline for information extraction from historical documents using image foundation models to support the work of historians. We discuss the current state of research for information extraction pipelines within the humanities in Sect. 2. In Sect. 3, we discuss our current pipeline as well as the current experience within the Max Planck Institute for the History of Science, evaluate it on three datasets in Sect. 4, and conclude with an overview of possible extensions to the proposed pipeline in Sect. 5.

2 Current State of the Research

Information Extraction (IE), including image (e.g., visual elements) extraction, from historical documents is playing an increasingly important role in formulating historical hypotheses [23], allowing researchers to tap into a large pool of information that would have been impossible to assemble without computational methods. Meanwhile, there have been many advances in text processing with regard to both printed and handwritten sources [7,9,10]. In this paper, we tackle the well-addressed technical problem of image extraction from historical documents while relying on foundation models and text prompts.

Current approaches to extract images from historical texts can be divided into two main groups: Segmentation and Object Detection approaches. Segmentation approaches often rely on FCN architectures such as U-Net [20] or Mask-RCNN [11] to generate masks of the desired image region. One of these approaches is the one proposed by [16] to extract images from a wide range of historical texts using a modified U-Net. Another similar approach is proposed by [7] to segment text lines in handwritten historical documents. Numerous further approaches have treated information extraction and, more specifically, image extraction as an object detection problem and tackled it with models such as EfficientDet [22], for instance in [8], where the authors extracted images from a corpus of Scottish Chapbooks. [4] used instead YOLOv5 [19] to extract different classes of visual elements from a large corpus of early modern books.

While the above-mentioned approaches are far from representing a comprehensive review of the current state of image extraction from historical documents, they highlight general trends within the community. Despite their differences, these approaches share a fundamental feature, namely that they were all trained on carefully annotated datasets. This is notable because, in contrast to other industry domains, *annotated* data within the humanities remains relatively scarce due to numerous reason including a lack of expertise (compared to the difficulty of defining classes within heterogeneous data, as well as ambiguous data interpretations to name a few). Many of the approaches discussed above provide their own datasets, such as the Synthetic *SynDoc* dataset presented in [16], the Chapbook dataset presented by [8], as well as the S-VED [5] presented by [4,5]. The amount of humanities and historical document datasets is continuously growing with numerous datasets covering different aspects of these fields [17]; this is also manifested by the growing number of datasets published on Hugging-Face's BigLAM: Big-Science, Libraries, Archives and Museums group [1].

Despite the consistently growing number of datasets, the high level of heterogeneity of historical documents means that many of these datasets cover a small, often *niche-like*, group of target documents (e.g., Fig. 1). This essentially means that the image extraction models – whether segmentation or object detection – often perform very well on in-domain data, but suffer from high performance degradation on out-of-domain data. This is clearly shown in the results presented in [4], where the performance of the YOLOv5 model trained on S-VED [5] (a dataset containing visual elements from early modern books on astronomy) degrades on out-of-domain datasets, such Mandragore [2], a dataset consisting of diverse manuscripts from the Bibliothèque National de France (BnF), or RASM [18], a dataset of historical Arabic manuscripts.

Fig. 1. Images of diverse types and styles. (Left to Right) A diagram from the S-VED dataset [4]. A colored image from the IlluHistDoc dataset [16]. An image from the Chapbook dataset [8]. A miniature from the HORAE dataset [3]

While these target-specific models are dependent on the presence of well-curated domain datasets, new foundation models are being developed, trained on immensely large datasets and able to perform **zero-shot** inference, which means that these models are able to perform well on out-of-domain images without requiring extra training.

3 Pipeline

While stand-alone models perform excellently on in-domain data, we aim to leverage *image* foundation models to help humanities researchers extract visual elements from their datasets as an end goal and, more importantly, quickly generate image datasets from broad data sources without requiring in-domain training.

This pipeline relies heavily on prompting these foundation models to achieve the best possible image extraction results without the use of domain-specific data. In this case, we chain a GroundingDINO [15] model with a Segment-Anything-Model (SAM) [13] to create a pipeline to generate visual element region masks from historical manuscripts (see Fig. 2).

GroundingDINO is a model that relies on a Transformer-based end-to-end object detection DINO (DETR with Improved DeNoising Anchor Boxes for End-to-End Object Detection) [24], and fuses it with a Text-Encoder in order to detect objects based on human language input on open-domain data, achieving good zero-shot results on the COCO dataset [14]. GroundingDINO takes an image-text input pair, and returns a bounding box that corresponds to the image region that in turns semantically corresponds to its textual counterpart. These bounding boxes are then passed on as data prompts to SAM [13] in order to segment the desired semantically relevant object.

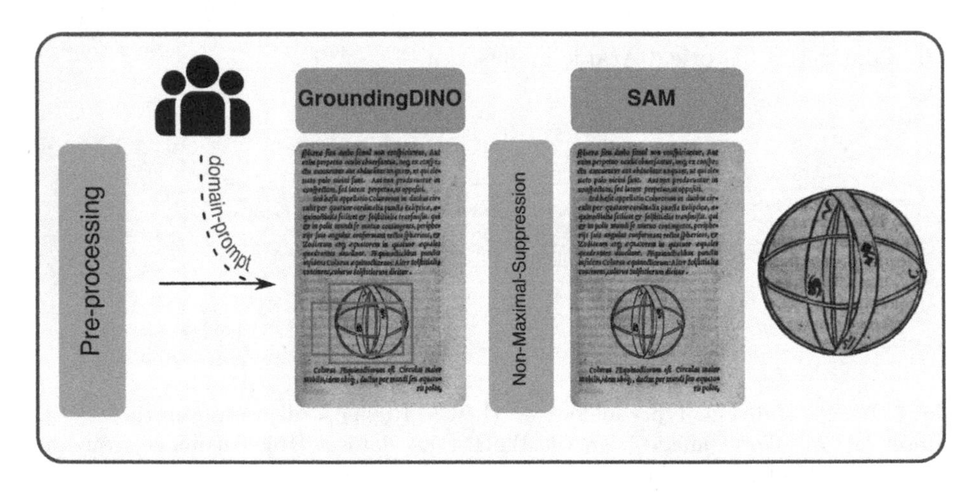

Fig. 2. Workflow from out-of-domain data entry on the left towards data extraction as bounding box with GroundingDINO [24] and as masks with SAM [13]

One obvious downside of such models is that despite the fact that they are trained on very large datasets (e.g., SA-1B dataset released by Meta contains 11 million images with 1 billion segmentation masks [13]), they often overlook humanities or historically oriented data, excluding data classes needed for manuscript and historical document information extraction. One of the major causes of the current status concerning data is the relatively low number of annotated manuscript and historical text data, as well as the difficulty in retrieving the domain knowledge required to annotate these images (e.g., think of the difference between different classes of images within the same manuscript or printed book). This situation makes it difficult to accommodate or create such data using Mechanical Turk[1] workers with little to no historical domain knowledge. To circumvent these shortcomings, we propose utilizing targeted domain-aware prompts that can hone in on the desired objects, and fine-tuning GroundingDINO as part of future developments, as discussed in Sect. 5.

The proposed pipeline is composed of three blocks: A pre-processing block, an object detection block relying on Prompt engineering GroundingDINO, and a finer segmentation block relying on SAM. The pre-processing block resizes each image to a standard size of 1000×1000 px, and includes an autocontrast step with a 2nd and 98th percentile cutoff. These images are then passed on to the GroundingDINO module with engineered prompts. These prompts are designed in a way to inject domain knowledge while remaining general enough so that the Feature Enhancer block of the GroundingDINO model is able to fuse text and image features in an efficient way to return reliable results. Examples of these engineered prompts are shown in Sect. 4.

With the multiple prompt classes, multiple bounding box detections are expected. We thus add a Non-Maximal Suppression module that operates on the selected prompt group classes to ensure that each object is detected once. The cleaned results, i.e., the bounding boxes, are then passed on as box-prompts to the SAM block to return clean segmentation masks of the desired regions.

We acknowledge that this pipeline relies on two very large models and might not be efficient to run in production. However, we believe that this approach can drastically increase the amount of data at the disposal of humanities researchers, and allows them to create large datasets using language prompts. For production scenarios or domain-specific requirements, the proposed approach can be used for an initial data collection phase in preparation for the training of bespoke object detector or segmentation models.

4 Text-Image Prompt Evaluation

We conduct a preliminary evaluation of the pipeline above on subsets of the S-VED [5], Chapbook [8], and the HORAE datasets [3] and report the preliminary results below. These datasets are object detection datasets in historical documents dating from the 15–17th, 17–19th, and the 14th–16th centuries respectively. The S-VED dataset contains four semantically different classes, the most abundant being Content Illustrations which covers visual elements within the body of the text and intend to enrich it. Other classes include Initials, which represent often decorated letters (or drop cap) at the beginning of chapters and paragraphs, Decorations which represents small decorative elements on pages, and Printer's Marks, which represent the emblem of the printer(s) who produced the book in question [4]. The Chapbook dataset consists of a single image class representing every image within a text page [8] while the pages of the HORAE dataset have the most detailed annotation scheme [3]. These cover Miniatures,

[1] Mechanical Turk is an Amazon based marketplace platform where organizations can hire workers, often for relatively low wage, to conduct some low-level work. This service is often used to annotate images and create large datasetes.

which are illustrations embedded in the text, Decorations which are elements often placed along the page borders, as well as different types of Initials, such as simple initials (initials differing from the main body of the text in ink and size), decorated initials (initials with purely ornamental decoration style), and historiated initials (initials whose decoration depicts an iconographic element such as a scene or a character) [3].

Beyond the difference in classes, these three datasets represent different types of content which contain different styles of visual elements. The S-VED dataset derives from the Sphere Corpus[2], which contains scientific books on geocentric astronomy used in pedagogical settings; the Chapbooks were booklets containing popular content ranging from literature, poems, religious texts, and riddles; and the HORAE dataset contains pages from the books of hours, which were a type of handwritten prayer book owned and widely circulated in the late middle ages. The difference between these types of primary sources is naturally reflected in the types of images they contain, with the S-VED containing a large number of orbital diagrams and geometric drawings, the Chapbooks dataset containing a wide range of daily life drawings featuring humans, animals, and in some cases abstract and stylized figures, and the HORAE containing a large amount of decorative elements places around the textual area of the page, as well as a lot of religious illustrations (Fig. 1).

In our attempt to evaluate the pipeline on the two models, we set the text and image thresholds to 0.35 within the GroundingDINO parameters and perform non-maximal suppression on the output boxes. We also cast all classes of the S-VED into a single visual element class in order to obtain a comparable result between the three chosen datasets. We evaluate the Average Precision (AP) [6] of different language prompts in order to examine their effect on the model's ability to extract the needed information on such as out-of-domain data.

The first language prompt that we applied uses simple language prompts (i.e., single words) to try to extract the visual elements from both datasets. In this case, the prompt is constituted of the single word **{figure}**, which resulted in very good semantically meaningful results on the S-VED dataset, but appears to perform poorly on both the Chapbook and the HORAE dataset (see Table 1).

We investigate the detection and segmentation results from our pipeline in an effort to improve our prompts and retrieve a larger amount of visual elements. In the S-VED, the error sources were manifold. The first consists of missing small visual elements placed in the marginalia; the second concerns missing abstract geometric shapes that the model did not deem to be a fit for the given textual prompts. However, the highest contributor to the relatively modest AP score reported in Table 1 is the difference between the bounding boxes that our pipeline considered to be representing a figure, and the bounding boxes created by the annotators of the S-VED, which is highly abstract. A simple example is the presence of three different drawings on an S-VED representing semantically related topics, and thus annotated as a single image by the S-VED annotators. However, relying solely on the image and the text prompt, our pipeline returns

[2] https://sphaera.mpiwg-berlin.mpg.de/.

multiple smaller bounding boxes with low Intersection-over-Union scores, leading to False Positive results (see Fig. 3).

In the case of the Chapbook and HORAE datasets, the main cause of this error was a semantic mismatch between our prompt and the desired outcome. The suggested prompt of **{figure}** has led the GroundingDINO module to return bounding boxes of human *figures* within the visual elements, instead of the desired output of a figure in the literal sense (see Fig. 3). Thus the low score was the result of the identification and segmentation of parts of the visual elements showing a human figure, diverging from the annotated ground truth data. This simple example proved interesting and highlights the multifaceted meaning that a single word prompt could have, and how it could affect the results (see Fig. 3).

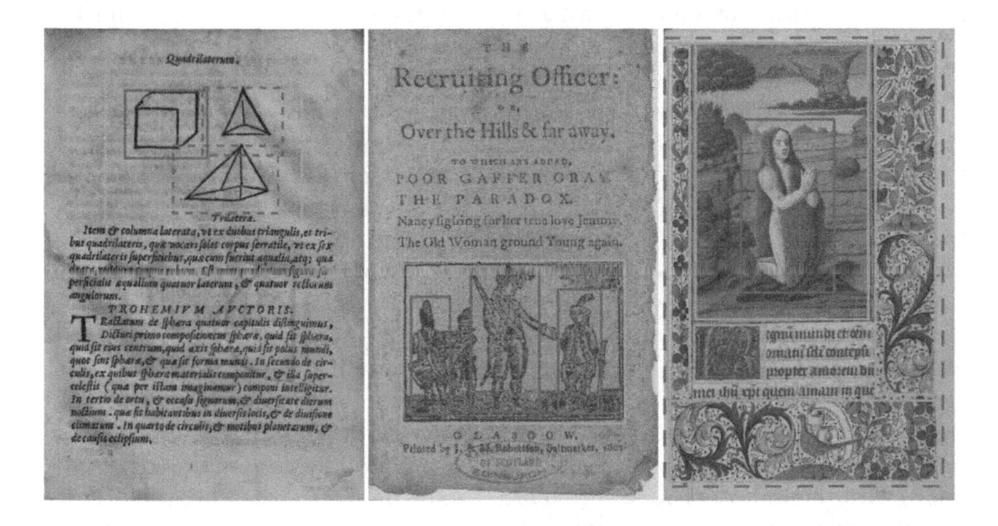

Fig. 3. (Left) A page from the S-VED dataset showing multiple detected regions. The solid red line represents the results obtained with the prompt **{figure}**, the dashed red lines represent the two extra boxes detected with the prompt **{figure - diagram - geometry - sketch}**, while the green box represent the ground truth data. (Center) A Chapbook page with multiple region predictions from a **{figure}** prompt in red, and a region prediction given a prompt **{image - square - rectangle - photo}**, in green corresponding to the ground truth. (Right) A HORAE page with ground truth bounding boxes in green showing a miniature, initial, and three decoration boxes around the page. The solid red box shows the prediction from a **{figure}** prompts, while the dashed red line shows the prediction from a **{floral - rectangle - flower - decorative - abstract}** prompt. (Color figure online)

To inject more domain-knowledge into our prompts, we provide dataset-tailored prompts. For S-VED, we provide a textual prompt that better describes the content of the majority of its visual elements: **{figure - diagram - geometry - sketch}**. In the Chapbook dataset, we focus on identifying the complete visual element, which is often square or rectangular in shape, thus the prompt in this case is **{image - square - rectangle - photo}**, while in the HORAE dataset, where we aim to detect miniatures, we provide a prompt that aims to describe their content based on an our observations, which in this case is **{figure - lanscape - scene - square}**. We re-evaluate the pipeline with the aforementioned linguistic prompts, and notice a small increase in performance for the S-VED, largely due to the detection of some previously missed geometric shapes, and a large increase in performance on the Chapbook and HORAE datasets due to the fact that the new textual prompt aligns with the ground truth annotation scheme.

In order to better probe the limits of text-image pairings on a very specific dataset such as historical documents, we attempt to differentiate between the different classes of the S-VED and HORAE dataset. In the case of the S-VED, we focus our attention on differentiating between the Content Illustration and Initial classes, the most abundant classes in the S-VED. In the HORAE dataset we focus on differentiating between the Initial, Decoration, and Miniature classes. In the first case, We prompt the following in order to retrieve the S-VED Initials class **{dropcap - decorated letter - large letter}** and the following for the Content Illustration class **{figure - diagram - circle - planets}**. To differentiate between the HORAE classes, we utilize the same S-VED prompt for the Initials class, and use the following **{floral - rectangle - flower - decorative - abstract}** and **{scene - landscape - square}** for the Decoration and Miniatures classes respectively. However in the above cases, we see that we have possibly reached the limit of the pre-trained model's text-image understanding, which is likely hindered by our efforts to differentiate the classes using very generalized terms. In the S-VED case, this is noticeable for example when an Initial such as "O" or "D" is classified as Content Illustration with high confidence due to its circular characteristic. In the HORAE examples, we encountered the same issues with the Initials class; but also faced some problems detecting the decorative elements according to the annotation scheme which divides the decorative elements according to their orientation in the manuscript pages. This meant that while our pipeline often recognizes decorative elements in the page, the detection box does not recognize the distinct decorative elements as per the annotation scheme, resulting in poor performance (see Fig. 3 for a clear comparison between the annotation scheme and the detected areas). Such mishaps ultimately resulted in almost random class detections (AP scores of 0.1, 0.08, and 0.12 for the S-VED Initials, HORAE Initials, and Decorative elements respectively), and proved to be an inefficient avenue.

The results presented above are, despite their limitations, very promising, especially for researchers aiming to collect large image datasets from archival material at scale. It is clear that such models soon hit their limits when it comes to differentiating between image classes that might be of interest for historical research (e.g., Initials, Content Illustrations, and Decorations). However, the possibility of quickly collecting thousands of images from historical documents using descriptive language remains enticing, and will increase the efficiency of

Table 1. Average Precision score for object detection on a subset of S-VED, Chapbook, and HORAE datasets

Dataset	Prompt	AP
S-VED	{figure}	0.42
S-VED	{figure - diagram - geometry - sketch}	**0.51**
Chapbook	{figure}	0.19
Chapbook	{image - square - rectangle - photo}	**0.82**
HORAE	{figure}	0.15
HORAE	{figure - lanscape - scene - square }	**0.74**

data collection, which is often a major barrier to applying ML algorithms in the frame of historical research. In this case, the scholars could invest human power on fine-tuning the retrieved data and creating well-curated sub-classes, which can then power the training of an in-domain model.

4.1 A Note on the Environment

As we are living in a climate-critical era, it is imperative that we take environmentally conscious choices when dealing with computational data at scale. In this case, we acknowledge that the use of both GroundingDINO and SAM comes at a high computational cost. Although these models have zero-shot capabilities, which means we do not need to spend energy on training them, a single inference across this pipeline takes ca. 40 times longer (on CPU) than a single inference using models such as YOLOv8. Thus, we highly recommend using such a pipeline for preliminary data collection followed by training a specific-domain model that can then perform inferences at scale.

5 Conclusion

In this paper, we explored the fast emerging field of multi-modal models and investigated its suitability for the digital humanist. The results of our investigation using the proposed pipeline show great potential from a technical aspect. We believe that this potential will lead to the generation of larger humanities datasets in the near future, but also to a larger interest and engagement from humanities scholars in computational approaches. This, we believe, is largely due to the linguistic interaction between the scholars and the machine, which is becoming one of the most human-computer interaction modes. This paper builds on the "multimodal turn in the Digital Humanities" [21]. This language interaction also forces us, as digital humanists, to reconsider object and class definitions, and reformulate them in a more computer-suited linguistic approach, which can often be very challenging, and often lead to new definitions and hypotheses.

The work on this pipeline is part of an ongoing infrastructure project at the Max Planck Institute for the History of Science that aims to collect large amounts of visual content from heterogeneous historical documents. We used the pre-trained GroundingDINO as our object extractor in this paper; however, in the medium term we also plan to fine-tune this model on humanities-specific datasets in order to allow specific linguistic prompts to match the desired image region. In the long run, we plan to slowly build an application with a simple GUI around this pipeline to allow humanists with minimal computer science knowledge to extract such information from historical documents.

Funding

This work was supported by the German Ministry for Education and Research as BIFOLD – Berlin Institute for the Foundations of Learning and Data (grant 01IS18037A) and the Max Plank Institute for the History of Science.

Code Availability

The code for this pipeline is available here: https://github.com/hassanhajj910/prompt-me-a-dataset.

Acknowledgments. We would like the thank Lindy Divarci and Luis Melendrez Zehfuss for the English proofreading.

References

1. Biglam: Bigscience libraries, archives and museums (2023). https://huggingface.co/biglam
2. Bibliothèque Nationale de France. Èchantillon segmenté d'enluminures de mandragore (2019). https://api.bnf.fr/mandragore-echantillon-segmente-2019
3. Boillet, M., Bonhomme, M.L., Stutzmann, D., Kermorvant, C.: Horae: an annotated dataset of books of hours. In: Proceedings of the 5th International Workshop on Historical Document Imaging and Processing (HIP 2019), pp. 7–12. Association for Computing Machinery, New York (2019). https://doi.org/10.1145/3352631.3352633
4. Büttner, J., Martinetz, J., El-Hajj, H., Valleriani, M.: Cordeep and the sacrobosco dataset: Detection of visual elements in historical documents. J. Imaging **8**(10) (2022). https://doi.org/10.3390/jimaging8100285
5. Büttner, J., Martinetz, J., El-Hajj, H., Valleriani, M.: Sacrobosco visual element dataset (s-ved) (2022). https://doi.org/10.5281/zenodo.7142456
6. Cartucho, J., Ventura, R., Veloso, M.: Robust object recognition through symbiotic deep learning in mobile robots. In: 2018 IEEE/RSJ International Conference on Intelligent Robots and Systems (IROS), pp. 2336–2341 (2018)
7. Droby, A., Kurar Barakat, B., Alaasam, R., Madi, B., Rabaev, I., El-Sana, J.: Text line extraction in historical documents using mask r-cnn. Signals **3**(3), 535–549 (2022). https://doi.org/10.3390/signals3030032

8. Dutta, A., Bergel, G., Zisserman, A.: Visual analysis of chapbooks printed in Scotland. In: The 6th International Workshop on Historical Document Imaging and Processing (HIP 2021), pp. 67–72. Association for Computing Machinery, New York (2021). https://doi.org/10.1145/3476887.3476893
9. Fischer, A., Liwicki, M., Ingold, R.: Handwritten Historical Document Analysis, Recognition, and Retrieval - State of the Art and Future Trends. World Scientific (2020). https://doi.org/10.1142/11353
10. Gaur, S., Sonkar, S., Roy, P.P.: Generation of synthetic training data for handwritten Indic script recognition. In: 2015 13th International Conference on Document Analysis and Recognition (ICDAR), pp. 491–495 (2015). https://doi.org/10.1109/ICDAR.2015.7333810
11. He, K., Gkioxari, G., Dollár, P., Girshick, R.: Mask r-cnn. In: 2017 IEEE International Conference on Computer Vision (ICCV), pp. 2980–2988 (2017). https://doi.org/10.1109/ICCV.2017.322
12. Jones, E.: Large-scale book digitization in historical context: outlines of a comparison. In: Proceedings of the 2011 IConference (iConference 2011), pp. 829–830. Association for Computing Machinery, New York (2011). https://doi.org/10.1145/1940761.1940925
13. Kirillov, A., et al.: Segment anything (2023)
14. Lin, T., et al.: Microsoft COCO: common objects in context. arXiv preprint arXiv:1405.0312 (2014)
15. Liu, S., et al.: Grounding dino: marrying dino with grounded pre-training for open-set object detection (2023)
16. Monnier, T., Aubry, M.: docExtractor: an off-the-shelf historical document element extraction. In: ICFHR (2020)
17. Nikolaidou, K., Seuret, M., Mokayed, H., Liwicki, M.: A survey of historical document image datasets (2022). https://doi.org/10.48550/arxiv.2203.08504
18. Pattern Recognition and Image Analysis Research Lab. University of Salford, Manchester: RASM 2019 Dataset (2019). https://www.primaresearch.org/RASM2019/resources
19. Redmon, J., Divvala, S.K., Girshick, R.B., Farhadi, A.: You only look once: Unified, real-time object detection. arXiv preprint arXiv:1506.02640 (2015)
20. Ronneberger, O., Fischer, P., Brox, T.: U-net: convolutional networks for biomedical image segmentation. arXiv preprint arXiv:1505.04597 (2015)
21. Smits, T., Wevers, M.: A multimodal turn in digital humanities. Using contrastive machine learning models to explore, enrich, and analyze digital visual historical collections. Digital Scholarship in the Humanities, fqad008 (2023). https://doi.org/10.1093/llc/fqad008
22. Tan, M., Pang, R., Le, Q.V.: Efficientdet: scalable and efficient object detection. In: 2020 IEEE/CVF Conference on Computer Vision and Pattern Recognition (CVPR), pp. 10778–10787. IEEE Computer Society, Los Alamitos (2020). https://doi.org/10.1109/CVPR42600.2020.01079
23. Valleriani, M., Vogl, M., el Hajj, H., Pham, K.: The network of early modern printers and its impact on the evolution of scientific knowledge: automatic detection of awareness relationships. Histories 2(4), 466–503 (2022). https://doi.org/10.3390/histories2040033
24. Zhang, H., et al.: Dino: Detr with improved denoising anchor boxes for end-to-end object detection (2022)

Artificial Intelligence in Art Generation: An Open Issue

Giuseppe Mazzola(✉), Marco Carapezza, Antonio Chella, and Diego Mantoan

Università degli Studi di Palermo, Palermo, Italy
{giuseppe.mazzola,marco.carapezza,antonio.chella, diego.mantoan}@unipa.it

Abstract. This paper aims to give a contribution to one of the most discussed issues in recent times, in both scientific and art communities: the use of Artificial Intelligence (AI) based tools for creating artworks. As the issue is strongly multidisciplinary, we structured the paper as a debate between experts in several fields (computer science, art history, philosophy) to listen to their specific points of view on the topic. The first part of the paper is focused on the relationship between the artists and the use of AI techniques. Furthermore, we organized an art exhibition with images created by an AI-based tools, to also collect people's feedbacks. We submitted to the viewers a questionnaire and their answers are reported in the experimental section. This, the second part is more focused on the visitors' perspective and about their perception on the use of these tools.

Keywords: Artificial Intelligence · AI Art · Computational Creativity · Fine Art

1 Introduction

The intersection of Artificial Intelligence (AI) and the world of art has given rise to a fascinating collaboration, leading to both awe and debate among artists, critics, and enthusiasts. AI-powered tools, leveraging the power of deep learning algorithms and neural networks, have emerged as capable creators, pushing the boundaries of artistic expression and challenging traditional notions of authorship. These tools can analyze vast amounts of data, learn patterns, and generate new artistic works across various mediums. From algorithms generating paintings, that rival those of renowned artists, to composing intricate musical compositions and even writing thought-provoking poetry, a new era of artistic production has dawned, one that blurs the lines between human ingenuity and machine creativity.

The implications of AI in art are profound. On one hand, it opens unprecedented possibilities for artists to explore new aesthetic landscapes, pushing the boundaries of their own imagination. AI tools can serve as a wellspring of inspiration, providing artists with novel ideas and alternative perspectives. By analyzing vast repositories of existing artworks, AI algorithms can identify trends, themes, and styles, enabling artists to build upon existing traditions or break away from them entirely. This collaboration between

G. L. Foresti et al. (Eds.): ICIAP 2023 Workshops, LNCS 14366, pp. 258–269, 2024.
https://doi.org/10.1007/978-3-031-51026-7_23

human creativity and machine intelligence has the potential to unlock new dimensions of artistic expression that were previously unexplored.

However, the rise of AI-generated art also raises complex questions and challenges. Critics argue that AI-produced artworks lack the emotional depth, intentionality, and authenticity that are intrinsic to human creations. They question the role of the artist as a conscious creator and argue that AI tools are merely replicating patterns without true understanding or personal experience. Concerns about copyright, ownership, and the commercialization of AI-generated artworks have surfaced, but they are out of the scope of this paper. These debates highlight the need for thoughtful discourse and a reevaluation of our definitions and perceptions of art and creativity in the digital age.

This paper investigates the dynamic relationship between AI-based tools and the generation of artworks, exploring the implications, possibilities, and controversies, according to the points of view of both experts in different academic fields and the viewers of the final artistic results. The aim is to give some contributions to the debate, and not a conclusive answer, as the issue is still open and too complex.

The paper is structured as follows: Sect. 2 presents the State of the Art, focusing on the technological aspects behind the AI tools for generating artworks; in Sect. 3, we report the points of view of three academic experts (a philosopher, an art historian and a computer scientist) on the relationship between art and AI; in Sect. 4, we discuss the results of some questionnaires that have been administered to the visitors of an exhibition of artistic pictures generated by AI tools; a conclusive section ends the paper.

2 State of the Art

Even though the use of Artificial Intelligence algorithms for artistic applications has been explored in the past [1, 2] the recent media overexposure of the newest AI-based tools opened new scenarios and perspective in the interaction between these techniques and the arts. The generation of artworks through algorithms and machine learning models has attracted interest both in the artistic and scientific communities. This state of the art will explore the major developments and challenges related to the intersection of AI and art generation [3], with a particular focus on Computer Science aspects.

In recent years, artificial neural networks have been widely employed for art generation. The most common approaches involve the use of generative neural networks, such as Generative Adversarial Networks (GANs) [4], which have been applied to generate artistic images [5, 6]. GANs enable the creation of new artwork by learning from the features of a reference artistic dataset.

The deep learning models DALL-E [7] and DALL-E 2 [8], developed by OpenAI [9], opened the way to the newest generation of text-to-image tools, that have been released by the major world ICT companies (Microsoft, Adobe, etc.). The user can type a text(prompt) and the system returns one or more images depicting the meaning of the text. Nowadays, these have become commonly used tools for those who work in the field of graphics and illustration.

Another significant area in AI-driven art generation is style transfer models. These models allow the application of the style of one artwork to another image, producing remarkable results [10, 11]. Implementing such models requires a sound understanding of Convolutional Neural Networks (CNNs) and their ability to extract stylistic features from artworks.

The evaluation and analysis of generated art pose critical challenges. Objective metrics for assessing the quality and creativity of AI-generated artwork are still being developed. Research efforts have been directed towards devising methods to measure aesthetic appeal, novelty, and emotional impact of AI-generated art [5, 12, 13]. These assessments often involve the collaboration between computer scientists, artists, and domain experts. The intersection of AI and art generation also raises important ethical and legal considerations. Issues such as copyright, authorship, and the impact of AI on the artistic landscape are subjects of debate. Research in this area aims to address questions related to ownership, attribution, and the boundaries between human creativity and AI-generated art [14].

3 The Experts' Point of View

To better investigate on the topic, we organized an exhibition of artistic images generated by an AI-based tool (more details in Sect. 4) and, contextually, a panel discussion, involving experts in different fields. The goal was to discuss about several questions, among which:

- Will artificial intelligences be able to replace artists?
- Can art, an activity that has always been exclusive to humans, exist without the emotional component present in the creative process?
- Who is the artist? The developers who created the AI? The AI itself? The vast amount of collective knowledge used for training? The person who selected the prompts and the generated images, among thousands?
- Can AIs simply be considered tools, like a camera or a graphics tablet?

In the next subsections, we will report the contributions of the speakers who were invited to the panel.

3.1 The Philosopher's Point of View

A long-standing tradition dating back to the Renaissance considers the artist as the creator of artworks and, whether they relied on external contributions or not (in the Renaissance, the main contribution came from the artisan shops, and from the religious and literary traditions), the artist remained the creator of the work. This prestigious tradition permeates common sense and intersects with a profound intuition that can be formulated as follows: if an artifact exists, someone must have made it.

What happens to this intuition in the face of phenomena such as artistic creations (and not only) generated by Artificial Intelligence? On the one hand, there is no doubt that the image is generated by a code, and therefore the code should be responsible for the creation, even though it clashes with our intuition of "creation". On the other hand, the process of generating images through AI involves someone selecting texts from which the code will produce the images, and then selecting the most interesting images according to their subjective taste. Equally important, someone decides to organize an exhibition or an installation with these works. We are in a situation where the creation of the image is fragmented into moments that refer to different figures. Moreover, since Duchamp [15], we know that creativity in art is not necessarily tied to the production of an object but can be found in the way it is experienced or presented for enjoyment.

However, the problem with AI-generated images is quite different. In these cases, we are not discussing operations as those of Duchamp, typical of conceptual art, where an existing object is repurposed or is given a new meaning to it. Here, in the purest artistic tradition, we are dealing with objects, traditionally artistic like images, that are created and only make sense within artistic practices. And the notion of the image (as well as the texts generated by AI) leads us to question the notion of authorship.

This is not a recent issue; it has been discussed for years. Artists like Miltos Manetas [16] have been using computers to create art since the late 1990s and now also work with AI. Authorship is not even a question for artists like Mario Klingermann, for instance, a German artist who coined a term that applies to AI-generated works: "neurophotos," photos produced by neural networks [17]. Another case that sparked much discussion, particularly due to its practical implications, is the portrait of Edmond Bellamy (2018), an artwork produced by an AI and sold for around 400,000 euros. This work was created by the French art collective Obvious [18], or at least that's how it appears in the texts and art catalogs. However, it's worth noting that the work is signed with a fragment of the programming code rather than the name of an artist, despite being credited to the collective Obvious. This artwork became the subject of a legally intriguing dispute between the Obvious collective, who claimed their creative act, and Robert Barrat, who created the code and wanted it to be open source. He argued that his code couldn't be used for commercial purposes. Barrat wasn't after the money; he wanted to assert that the artist was the code itself, not the human who used it. As wrote the linguist Emily Bender [19]: we have learned to build machines that mindlessly generate images, but we haven't learned to stop imaging the mind behind them.

At this moment, how to consider works generated by AI is an open problem, but it cannot be resolved by staying within the notion of traditional creation. We must undertake a significant interdisciplinary theoretical effort to radically rethink the notion of artistic creation, not as the result of individual "genius". But probably more as the result of collective activity. The creative applications of AI force us to think about this too.

3.2 The Art Historian's Point of View

What art is or who an artist is are questions that often go unasked in the art world, as the definition of what constitutes art or who qualifies as an artist is often taken for granted. Within the art system, there are specific roles and a chain of actors involved in the production of meaning, as proposed by the institutional theory of philosopher George Dickie [20]. Within this context, practically anything can be legitimized. However, it has been technological innovations such as Gutenberg's printing press in the early modern period or the disruptions caused by photography in the 19th century that have posed existential questions to art in the Western world. Similarly, when it comes to artificial intelligence applied to creativity, these disruptive technical advancements in image creation require a multidisciplinary perspective to address the problem at hand, using a more interesting, in-depth, and theoretical approach [21]. This perspective involves questioning a software's ability, based on a Gestalt perspective [22], to create new forms or something that exists, not necessarily in material form, such as code. It becomes evident that provocations like the recent cases of artworks created in collaboration with AI software - the winning photograph in a category of the 2023 Sony World Photography Award presented and later withdrawn by German artist Boris Eldagsen [23], or the image created by Jason M. Allen with Midjourney, winner at the 2022 Colorado State Fair [24] - only serve as provocations that fail to address the core issue. In this regard, it is useful to reflect on three terms: intention, innovation, and unconscious.

Intention is the first problem that artists must confront. One of the definitions, albeit somewhat elusive, of what constitutes art is that if there is artistic intentionality, then it is art. This was exactly the subject of contention in the debate highlighted by champions of Art Brut like Jean Dubuffet and Jean Fautrier [25], who focused on art produced by individuals whom rational Western society did not attribute intentional capacity to, such as children and people with mental illnesses. This naturally becomes a problem when it comes to AI. Can we recognize any form of intentionality in AI? Can it serve as a criterion to understand the kind of intentionality neural networks can express?

The second term, innovation, is important to question whether it is necessary from an artistic standpoint. Intentional capacity can be an element of creativity or innovation as it conveys character, style, and a point of view. In this regard, can AI express its own style, albeit reworked or patched together? However, the term innovation is slippery when it comes to art, at least from a historical perspective. It has become a cliché that art must possess intentionality and be innovative compared to what preceded it. On the other hand, this is a very recent view, linked to a mindset that emerged from industrialization, where there must always be evolution, progress, innovation, some form of growth in every human endeavor [26].

The last term to explore is the unconscious. If the surrealists had access to these tools, they would have produced images like those we are discussing about. In fact, the juxtaposition between rationality and the unconscious, that the surrealists managed to create through their automatic practices, provides a good example of what can be achieved when approaching the creative use of artificial neural networks [27]. Particularly when exploring the use of automatism techniques that excluded reason as much as possible to channel what the unconscious suggested, the surrealists produced works that formally resemble the recombination of images that neural networks are capable of. In a sense, due to the structure of these neural networks, which draw from an enormous amount of data, AI merely replicates what all humans do, including artists, in producing something new.

Following this line of thought, what AI helps us do is look at the images on which our contemporary society is based, as it draws from the collective images produced by us as a society, as a world, as humanity. AI is based on data created by humans, so the extraction of this information, made possible through collaboration with AI, speaks to the unconscious of our collective.

3.3 The Computer Scientist's Point of View

Art has always been linked to new technologies. Throughout art history, artists have consistently embraced new technologies and innovations, using them for artistic purposes. Consider the piano, which revolutionized music consumption, or the invention of the MP3. Even in the realm of music, both the consumption and generation of music have been greatly influenced by new technologies.

Today, the "new" technology is Artificial Intelligence (AI). Just as in previous centuries, art is drawing inspiration from AI and utilizing tools like ChatGPT, DALL-E, and others that generate art from text to create new concepts. The question is: what are these modern systems based on? These systems rely on neural networks, mathematical systems loosely inspired by the functioning of the brain. Instead of being programmed with algorithms, they learn from vast amounts of data. Huge quantities of data are collected, and with social networks, we ourselves contribute to creating this data, which is then used to train the neural networks. The neural networks compress and recombine this information in an interesting, sometimes very intelligent, way, but they do not create anything truly new. I do not expect AI to bring about a paradigm shift. Great artists have created new paradigms. We have AI-generated songs that closely resemble those of The Beatles or Mozart, combining existing elements. But AI is not The Beatles; it does not bring about that paradigm shift in the artistic realm like Mozart and The Beatles did. The new Beatles will not be an AI.

Similarly, in the realm of visual arts, the interesting aspect could be allowing artists to use these tools, along with other technological instruments, to create new ideas themselves. These tools can serve as stimuli, akin to new color palettes or new systems for generating artistic content. The role of researchers is to provide artists with these tools to create new artistic content. I have doubts that computer-generated content can be considered art itself. It can be imitations and recombination of existing elements but lacks that spark of novelty inherent in how these systems are constructed.

Given that these systems will continually propose new neural networks with increasing computational power, we can expect greater precision in generating texts and artistic content. But this type of learning is heavily tied to symbolic aspects. It lacks corporeality. It lacks the emotions and sensations that an artist experiences, which these systems, as currently designed, can never truly feel. We may have increasingly precise systems, but I doubt that the spark can emerge from them. Even artists draw from their biographies, but they are closely tied to sensations. A machine can simulate emotions, even for facilitating interactions with people, but it does not genuinely feel them [28].

The issue of ownership of artistic content is also a non-trivial problem, not limited to art alone but applicable to all AI-based systems. For example, in the case of autonomous vehicles, who is responsible in the event of an accident? The engineers who designed it? The owner of the company that built it? These are problems that our society is beginning to grapple with, not only in the artistic domain. Who is responsible for an artificial intelligence? Obviously, there is a broad chain of responsibility.

I see these experiments as new artistic systems because they provide new inputs and new expressive forms for artists, especially when used in situations where art could not be created without these new technologies.

4 Experimental Results

One of the goals of this paper is to collect feedbacks from people, from different points of view, and to compare and analyze them. With this aim, we organized an exhibition, entitled "I.A. – Io, Artista?" (A.I. – I, Artist?) of pictures generated by an AI-based tool. The title is inspired by that of the book of Asimov (I Robot), but the question mark is added to highlight the question of the authorship of the artworks. The pictures were exhibited during the month of March at the University of Palermo, and in May at the cultural event "Settimana delle Culture" in Palermo (see Fig. 1), giving us the opportunity to reach different types of audience.

We also organized two different talks, involving experts from different academic fields. The first panel was focused on the figure of the artist (see Sect. 3), the second one on the user perception of the risks and the opportunities that AI-based tools may present in the field of Arts, and not only. During the second talk, we administered a questionnaire to the audience, to collect their feedbacks about the exhibition and the topic of the discussion. Results are presented and discussed in Sect. 4.2.

4.1 The Art Exhibition

The art installation consisted of 45 images and texts, selected among 3000 images and 200 texts, generated by using the website craiyon.com [29], which was based on the first version of the DALL-E [7] algorithm. We decided to use this tool for several reasons: it was free to use; there were no limitations on the number of images to be generated; there were no restrictions on the words that could be used in the prompts (no censorship); with respect to the other tools (in November 2022), it gave more interesting results, from an artistic perspective (the goal was to create artistic images and not photos), while the others were better designed for photorealistic images (see Fig. 2). In future we plan to compare different tools for other applications.

Each image has been created by typing a prompt onto the search bar, generating different results, eventually re-generating them more times, and selecting the most interesting one for each text. Both the processes of selecting the texts and the "better" images were subjective and guided by Giuseppe Mazzola, while the exhibitions were curated together with Diego Mantoan.

The prompts span over different issues: portraits, current events, memes, but also abstract concepts and feelings (see Fig. 2). The generated digital images had a resolution of 1024×1024. 21 of the artworks were printed on 15×15 cm square forex panel (due to the resolution, a larger print format was not recommended). As well, the texts, that were part and parcel of the installation, were printed as 15×15 cm cardboards. The other 24 were added as square tiles of three 4×4 mosaic, and printed as a forex panel (40×40 cm, each square 10x10cm), consisting in 8 images and 8 captions each (see Fig. 2), grouped by content (politics, current events, portraits).

Fig. 1. The installations of the two exhibitions. At the Department of Human Sciences of the University of Palermo (on the left, part of the installation), and at the International Center of Photography of Palermo for the "Settimana delle Culture" cultural event (on the right). The second installation was made with a subset of the images.

Fig. 2. Some of the images of the exhibitions: on the right some single panels, with the texts used to generate them. On the left one of the mosaics.

4.2 Users' Feedbacks

During the second talk, organized as part of the cultural event "Settimana delle Culture", which was held in May 2023 in Palermo, we administered a questionnaire to the participants, asking them to answer the following questions:

1. Age range (0–18; 19–30; 31–50; 50–65; over 65)
2. Educational background (Elementary School diploma; Middle School diploma; High School diploma; Degree; PhD)
3. On a scale from 1 to 5 (1-not at all, 5-very much), how would you describe yourself as passionate about new technologies?
4. On a scale from 1 to 5 (1-not at all, 5-very much), how would you describe yourself as passionate about art?

5. On a scale from 1 to 5 (1-not at all, 5-very much), how realistic, meaning similar to those created by a human, would you consider the images in the exhibition?

Table 1. Questionnaire results

Question	1	2	3	4	5
3. Passionate about new technologies	3%	13%	28%	34%	22%
4. Passionate about art	0%	0%	16%	25%	59%
5. Realistic images	13%	6%	41%	28%	13%
6. Interesting images	3%	16%	22%	44%	16%
7. AI as a threat	6%	19%	38%	25%	13%
8. AI as an opportunity	3%	6%	31%	41%	19%
9. Use in the future	9%	28%	22%	22%	19%

6. On a scale from 1 to 5 (1-not at all, 5-very much), how artistically interesting would you consider the images in the exhibition?
7. On a scale from 1 to 5 (1-not at all, 5-very much), how much of a threat do you think AI-based applications could pose?
8. On a scale from 1 to 5 (1-not at all, 5-very much), how much do you believe AI-based applications could represent an opportunity?
9. On a scale from 1 to 5 (1-not at all, 5-very much), how often do you imagine using AI-based applications in the near future?

The first four questions were used to frame people's background, in terms of age, education and interests. The latter five questions were related to the topics covered during the debates. We collected 32 questionnaires, whose results are exposed and commented below.

Regarding the age range, none of the interviewed participants was under 18, 16% between 19 and 30, 31% between 31 and 50, 13% between 51 and 65 and 41% over 65. More than a half of the participants was over 50, and this is an interesting aspect to take into consideration, knowing that this is only a limited and partially biased set of testers.

Regarding the educational background, 19% had the High School diploma, 75% the degree and 6% a PhD. As expected for a cultural event, the education level of the audience was very high. The results about the other questions are shown in Table 1.

In terms of their interests, the participants were moderately interested in new technologies, while highly passionate about art, as expected, due to the context.

They found the images of the exhibition quite realistic but interesting. No trends in the answers for the questions 7 and 9, while most of the interviewed people thought that AI-based tools may represent an opportunity, not only in the fields of arts.

We decided also to analyze the results according to the people's background, then we split the questionnaires in different groups: according to the age of the participants (0–50, over 50) and to their interest in new technologies (≤3, >3). In the first case, we did not notice any relevant differences in the answers between the two categories. The second case was more interesting, and the results are reported in Fig. 3.

According to the people who are more interested in new technologies, the images of the exhibition are more realistic and more interesting with respect to the opinions of the other participants. Furthermore, they think that these tools could represent more a threat and more an opportunity, and they imagine they will use them more often in the future. So, analyzing the results, those who are more curious about the technological innovations are, at the same time, more fascinated and more frightened by the new AI tools.

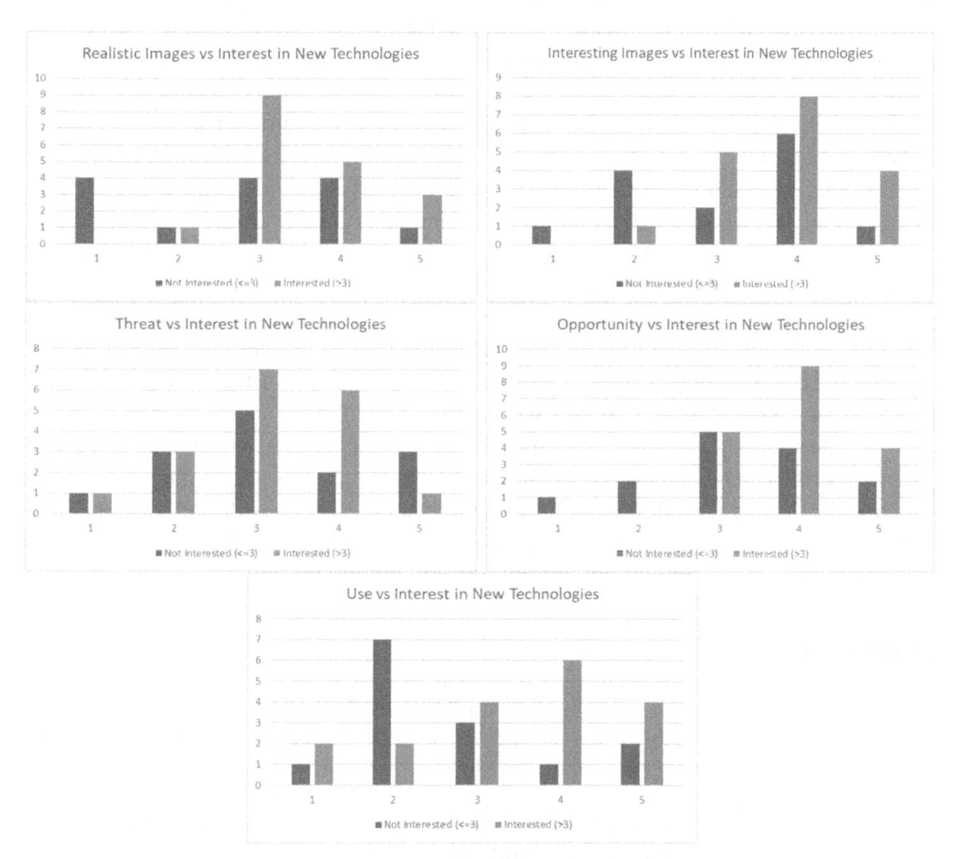

Fig. 3. Answers to the questions 5–9 of the questionnaires, for people who are less interested in new technologies (in red) and people that are more interested (in blue). (Color figure online)

5 Conclusions

The interaction between new technologies and the world of arts is, by nature, a multidisciplinary field. In this paper, we aimed to contribute to the ongoing discussion about the use of AI-based tools as instruments for creative and artistic applications. We presented the perspectives of experts from various fields and the feedbacks from visitors to an exhibition of artistic images generated by AI, without having the presumption to give exhaustive answers to the open questions on the issue.

We analyzed the problem of authorship in AI-generated artworks and the relationship between the artist and these tools. Additionally, we explored how these new instruments are perceived by their future users.

On one hand, the use of AI-based tools can be seen as an opportunity for art, as they enable the exploration of new creative approaches and the production of original and innovative artworks. For instance, AI can be used to automatically generate images, sounds, and texts, combine different art forms in novel and surprising ways, or assist artists as inspiration in creating works with greater efficiency and precision.

On the other hand, the use of AI-based tools can also pose a threat to art, as it may lead to a standardization of creativity and a loss of control by the artist over their own work. For example, AI could be used to replicate existing artistic styles without introducing anything new, or produce works completely autonomously, without any interaction from the human artist. Not to mention the question of deepfake images and videos, because it is outside the scope of the paper.

Moreover, this research focuses exclusively on a singular tool, giving rise to concerns regarding its replicability. The subjective element in the image selection process and the potential for bias introduced by the demographic composition of the exhibition further complicate the interpretability of the experimental outcomes.

In conclusion, according to the perspective of the authors of this paper, AI-generated systems for arts can be considered as new tools, like a new type of graphics tablet or a new model of camera. And, like all tools, they can be used for different purposes and in different ways, some more ethically acceptable than others, depending on the user intentions.

References

1. Kugel, P.: Artificial intelligence and visual art. Leonardo **14**(2), 137–139 (1981)
2. Santos, I., Castro, L., Rodriguez-Fernandez, N., Torrente-Patino, A., Carballal, A.: Artificial neural networks and deep learning in the visual arts: a review. Neural Comput. Appl. **33**, 121–157 (2021)
3. Cetinic, E., She, J.: Understanding and creating art with AI: review and outlook. ACM Trans. Multim. Comput. Commun. Appl. (TOMM) **18**(2), 1–22 (2022)
4. Goodfellow, I., et al.: Generative adversarial networks. Commun. ACM **63**(11), 139–144 (2020)
5. Elgammal, A., et al.: Generating 'Art' by Learning about Styles and Deviating from Style Norms (2017). Medium.com
6. Tan, W.R., Chan, C.S., Aguirre, H.E., Tanaka, K.: ArtGAN: artwork synthesis with conditional categorical GANs. In: 2017 IEEE International Conference on Image Processing (ICIP), pp. 3760–3764. IEEE (2017)

7. Ramesh, A., et al.: Zero-shot text-to-image generation. In: International Conference on Machine Learning, pp. 8821–8831. PMLR (2021)
8. Ramesh, A., Dhariwal, P., Nichol, A., Chu, C., Chen, M.: Hierarchical text-conditional image generation with clip latents. arXiv preprint arXiv:2204.06125 (2022)
9. OpenAI. https://openai.com/
10. Gatys, L.A., Ecker, A.S., Bethge, M.: Image style transfer using convolutional neural networks. In: Proceedings of the IEEE Conference on Computer Vision and Pattern Recognition, pp. 2414–2423 (2016)
11. Hien, N.L.H., Van Huy, L., Van Hieu, N.: Artwork style transfer model using deep learning approach. Cybern. Phys. **10**, 127–137 (2021)
12. Hong, J.W., Curran, N.M.: Artificial intelligence, artists, and art: attitudes toward artwork produced by humans vs. artificial intelligence. ACM Trans. Multim. Comput. Commun. Appl. (TOMM), **15**(2s), 1–16 (2019)
13. Ragot, M., Martin, N., Cojean, S.: AI-generated vs. human artworks. A perception bias towards artificial intelligence? In: Extended Abstracts of the 2020 CHI Conference on Human Factors in Computing Systems, pp. 1–10 (2020)
14. Selbst, A.D., Boyd, D., Friedler, S.A., Venkatasubramanian, S., Vertesi, J.: Fairness and abstraction in sociotechnical systems. In: Proceedings of the Conference on Fairness, Accountability, and Transparency, pp. 59–68 (2019)
15. Buskirk, M., Nixon, M. (Eds.): The Duchamp Effect. MIT Press (1996)
16. Miltos Manetas. http://www.gamevideoart.org/miltos-manetas
17. Klingemann, M.: Photography through the eyes of a machine', interview within the framework of the 2017 EyeEm Photography Festival, 15–17 September (2017). https://www.eyeem.com/blog/marioklingemann-ai-art. Accessed 18 Aug 2022
18. Obvious Collective. www.obvious-art.com
19. Bender E.M.: You are not a parrott and chatbot is not a human. Intelligencer (2023)
20. Dickie, G.: Defining Art. Am. Philos. Quart. **6**(3), 253–256 (1969). http://www.jstor.org/stable/20009315
21. Fried, M., Welling, J.: Why photography matters as art as never before: Michael Fried in conversation with James Welling. Aperture **195**, 82–85 (2009). http://www.jstor.org/stable/24473437
22. Wertheimer, M.: Gestalt Theory (1938)
23. Boris Eldagsen Blog. https://www.eldagsen.com/sony-world-photography-awards-2023/
24. New York Times. https://www.nytimes.com/2022/09/02/technology/ai-artificial-intelligence-artists.html
25. Minturn, K.: Dubuffet, Lévi-Strauss, and the idea of art brut. RES: Anthropol. Aesthet. **46**(1), 247–258 (2004)
26. Mantoan, D.: Arte convenzionale – ovvero – perché non possono esistere artisti realmente anticonformisti. In: POST, vol. 4, pp. 102–109. Mimesis, Milano (2016)
27. Jenny, L., Trezise, T.: From Breton to Dali: the adventures of automatism. October **51**, 105–114. https://doi.org/10.2307/778893
28. Chella, A., Manzotti, R.: Jazz and machine consciousness: towards a new turing test. In: Revisiting Turing and His Test: Comprehensiveness, Qualia, and the Real World (AISB/IACAP Symposium, Alan Turing Year 2012), pp. 49–53 (2012)
29. Craiyon. https://www.craiyon.com/

A Deep Learning Approach for Painting Retrieval Based on Genre Similarity

Tess Masclef(✉), Mihaela Scuturici, Benjamin Bertin, Vincent Barrellon, Vasile-Marian Scuturici, and Serge Miguet

Univ Lyon, Univ Lyon 2, CNRS, INSA Lyon, UCBL, LIRIS, UMR5205, Lyon, France
tess.masclef@univ-lyon2.fr

Abstract. As digitized paintings continue to grow in popularity and become more prevalent on online collection platforms, it becomes necessary to develop new image processing algorithms to effectively manage the paintings stored in databases. Image retrieval has historically been a challenging field within digital image processing, as it requires scanning large databases for images that are similar to a given query image. The notion of similarity itself, varies according to user's perception. The performance of image retrieval is heavily influenced by the feature representations and similarity measures used. Recently, Deep Learning has made significant strides, and deep features derived from this technology have become widely used due to their demonstrated ability to generalize well. In this paper, a fine-tune Convolutional Neural Network for the artistic genres recognition is employed to extract deep and high-level features from paintings. These features are then used to measure the similarity between a given query image and the images stored in the database, using an Approximate Nearest Neighbours algorithm to get a real time result. Our experimental results indicate this approach leads to a significant improvement in the performance of content-based image retrieval for the task of genre retrieval in paintings.

Keywords: Art classification · Deep Learning · Content-Based Image Retrieval · Approximate Nearest Neighbours Algorithm

1 Introduction

Our work is part of the Augmented Artwork Analysis (AAA) project which aims to produce a tool for assisted interpretation of artistic images. This tool has to adapt to a specific museum, by allowing the study of different aspects and levels of organization of an artwork being observed in situ and by cross-linking it with an open access corpus of images to highlight genealogies and to stimulate intertextual dialogues. The goal is to help guides visits, pedagogical activities and scientific research. Our focus is on searching by image similarity and identifying the closest neighbours of the artwork in real-time. In recent years, the increasing digitisation of artworks has led to the creation of datasets from museums and

G. L. Foresti et al. (Eds.): ICIAP 2023 Workshops, LNCS 14366, pp. 270–281, 2024.
https://doi.org/10.1007/978-3-031-51026-7_24

private collections that are available online, like the WikiArt collection[1], which provides expert metadata such as artist, date, artistic genre, style, etc. Louvre[2], Tate[3], Metropolitan[4] and many other museums have on-line galleries. Thus, projects such as Henri Focillon's [6], which tends to bring out the genealogy of forms, unaccomplished due to lack of data, are becoming technically feasible.

The genealogy of forms in painting is the study of the evolution of artistic elements, techniques and styles and their transmission from one artist or period to another. It involves tracing the development and transformation of visual elements, such as composition and the use of colour, in painting. By analysing these links, art historians can understand how artists have built on and responded to the work of their predecessors, contributing to the diversity and constant evolution of painting throughout history.

The literature offers many methods to enable such a study, notably in the exploration of painting based on images, based on textual descriptions or by combining images and textual descriptions. In the context of this project, we are mainly interested in the exploration of paintings based on images and the search for similarities between them. Content-based image retrieval is one of the methods used to obtain artworks similar to the one being explored.

Content-based image retrieval (CBIR) is a technique for searching images on the basis of their visual characteristics. Traditionally, these features are obtained using conventional extraction techniques (HSV histogram, local binary pattern, etc.) [13]. However, we propose a deep learning approach to identify visually similar images. Deep learning has successfully addressed the challenges posed by the semantic gap, which refers to the disparity between the low-level features extracted from images and the high-level semantic meanings perceived by humans. Using deep neural networks, we can extract complex features encompassing texture, colour and composition, exceeding the capabilities of conventional extraction methods [11,19]. These feature vectors can be fed into a query system, represented as a nearest neighbour graph, which suggests similar images. To guide our unsupervised search, we integrate a classification system that matches artworks by artist, genre and style.

In order to allow a supervised search, neural networks have indeed shown their efficacy for artwork processing, mainly for the recognition of a school, an era or a style (art movements) [3,16,23]. In most cases, these tasks are done in parallel. For the classification of artistic styles with high complexity due to high intra-class variation and low inter-class variation, self-supervised methods are also used [10]. Algorithms for the recognition of different painting techniques such as the clustered multiple kernel learning algorithm that extracts features from three aspects (colour, texture and spatial arrangement) for the recognition of oil paintings can be used for this task [15]. Comparative studies of different pre-

[1] http://www.wikiart.org.
[2] https://collections.louvre.fr/.
[3] https://www.tate.org.uk/search?type=artwork.
[4] https://www.metmuseum.org/art/the-collection.

trained neural networks on ImageNet, ResNet50, ResNet101 [7], and DenseNet [8] showed better performance [12] for artist classification than VGG16 [22].

VGG Face have also shown their efficiency in the recognition of characters in a painting using domain transfer for data augmentation [17]. Siamese networks are used for the recognition of visual links with the help of experts in the annotation of clusters of similar paintings. These networks are more efficient when they are pre-trained [1,20]. A CNN approach to estimate the pose of characters in a painting has shown that the pose criterion can be effective in finding visual links between artworks [9].

Several approaches exist to perform a nearest neighbour search. The most classical one is to use an exact algorithm that from a chosen distance calculates each of the distances between the features of the query image and those of the dataset. Another approach is to use an Approximate Nearest Neighbour algorithm which has the advantage of being faster with relatively low error using tree forests, Voronoi diagrams [2], or other partitioning methods [14].

In this paper, we present an architecture composed of a neural network that has the task of classifying paintings by genre and that is used as a feature extractor combined with an Approximate Nearest Neighbour algorithm allowing a real-time search of similar images, with a quality close to an exact algorithm.

We obtain a more refined search after fine-tuning the model. Indeed, the search for similar neighbours focus more on the type of requested painting. Furthermore, optimising parameters for the Approximate Nearest Neighbour algorithm shows that we can obtain a search close to the exact one with a considerably shorter time than with an exhaustive approach with statistical guarantees for the quality of the results.

2 Methodology and Experiments

In this section, we provide the architecture of the Convolutional Neural Network used. We briefly present the dataset and the functionality of the approximate nearest neighbour algorithm.

2.1 Convolutional Neural Network

We use the Resnet50 network as our feature extractor. After testing several pre-trained networks, ResNet50 is the one we found to be the most visually relevant. ResNet50 is a 50-layer Convolutional Neural Network (48 convolutional layers, one MaxPool layer, and one average pool layer).

To fine-tune ResNet50, we keep the convolutional layers (the encoder gives a vector of 2048 components as output) and use an average pooling followed by a fully connected layer with an output space size of 512 and a second fully connected layer using the activation Softmax for the classification. The idea behind the addition of the first fully connected layer permits to reduce the dimension of the feature vector.

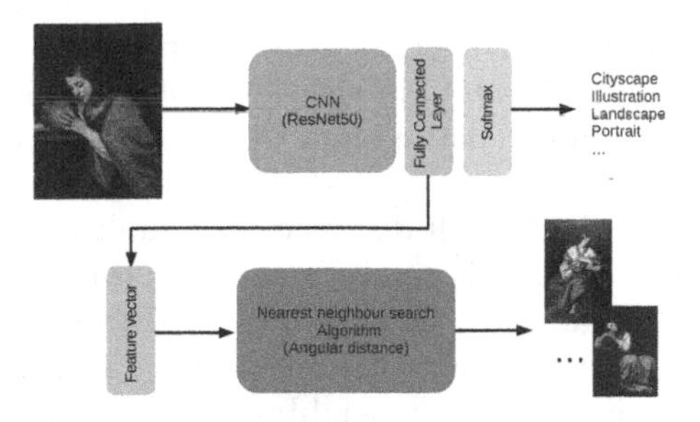

Fig. 1. Schematic representation of our similarity-based image search tool

2.2 Dataset

We use two datasets: the base ResNet50 is pre-trained on ImageNet [5] and the network is fine-tuned on WikiArt for genre recognition.

Among various subsets of the ImageNet dataset, the most used is "ImageNet Large Scale Visual Recognition Challenge (ILSVRC) 2012–2017 image classification and localization dataset". A database composed of about 1.5 million annotated images and grouping 1,000 object categories.

WikiArt from WikiArt.org is a visual art encyclopedia which contains 81,444 artworks (paintings) covering the periods between the 15*th* and 20*th* centuries [18] (for our application, we are interested in artworks dating from before the beginning of the 20*th* century). They are labelled based on artist (129), genre (abstract painting, cityscape, genre painting, illustration, landscape, nude painting, portrait, religious painting, sketch and study and still life) and style (a total of 27 styles like romanticism, baroque, impressionism, cubism, realism, etc.). In this work, we are going to train the model to perform genre-based classification. In the WikiArt dataset, because the classes are unbalanced we chose to randomly sample of around 2500 images per class for fine-tuning, a total of 27,613 images.

2.3 Nearest Neighbour Algorithm and Similarity Measure

To expect real time performances, we have chosen to use an Approximate Nearest Neighbour algorithm. In fact, the computation time of the distance between the features of the query image and those of the 81,444 images of the dataset is long, knowing that these features are in a 512 dimensional space. The choice is ANNOY (Approximate Nearest Neighbour Oh Yeah)[5], developed in 2015 for the Spotify platform by Erik Bernhardsson, this algorithm has the advantage of being efficient in high dimensional spaces. ANNOY is selected due to its exceptional search efficiency and resilience in handling various datasets, along with its straightforward customization of hyperparameters. Another advantage of ANNOY is that it can achieve a superior balance between search performance and index size/construction time in comparison to proximity graph-based methods. This is because ANNOY allows for a reduction in the number of trees

[5] https://github.com/spotify/annoy.

without compromising the search performance significantly [14]. ANNOY uses a tree forest to organise the vectors in the data space[6]. The forest is traversed in order to obtain a set of candidate points from which the closest to the query point is returned.

The next step is to calculate all distances and rank the points. Finally, the nodes are sorted by distance, so it can return the k-Nearest Neighbours (kNN). The distance used is the angular distance, obtained from the cosine similarity measure, between the feature vector of the query image and all the feature vectors of the dataset, to obtain the nearest neighbours of the query. This distance is defined as:

$$Cosine\ similarity = 1 - \frac{q.p}{||q||^2||p||^2}$$

$$Angular\ distance = \sqrt{2 * Cosine\ similarity}$$

where q is the feature vector of the request image and p, the feature vector of the image in the dataset.

The assembly of these two algorithms is shown in Fig. 1.

2.4 Experiments

In a first step, we fine-tune ResNet50 pre-trained on ImageNet, with the WikiArt dataset, using Adam as an optimizer, with a learning rate of 10^{-5}. For the loss function, we chose the categorical cross entropy [24] and train on 50 epochs. Since we have chosen to downsample our data to rebalance the classes, we increase the data by making small rotations (of the order of 10 degrees). For the repartition of the data, we consider 80% for the training and 20% for the test. Once the model is pre-trained for genre recognition in paintings, it is used as a feature extractor by removing the Softmax layer. For each image, we obtain a signature formed of a feature vector of size 512.

Each of these vectors is associated and indexed to an image. They are then given to the Approximate Nearest Neighbour algorithm, whose parameters are 25 for the number of trees in the forest, *angular distance* for the distance and 1,000 for the number of nearest neighbours searched, parameters we are seeking to optimise in Sect. 3.3.

In order to optimise these parameters, we search for a compromise between the execution time and the desired quality.

3 Results

In this section, we will compare the performance of our approach with state-of-the-art methods. We show that fine-tuning our model allows us to guide our search according to genre recognition. Indeed, even if we know the genre of most of the images, some experts do not agree. This could explain why some genres get confused, as shown on Fig. 2.

[6] https://erikbern.com/2015/10/01/nearest-neighbors-and-vector-models-part-2-how-to-search-in-high-dimensional-spaces.html.

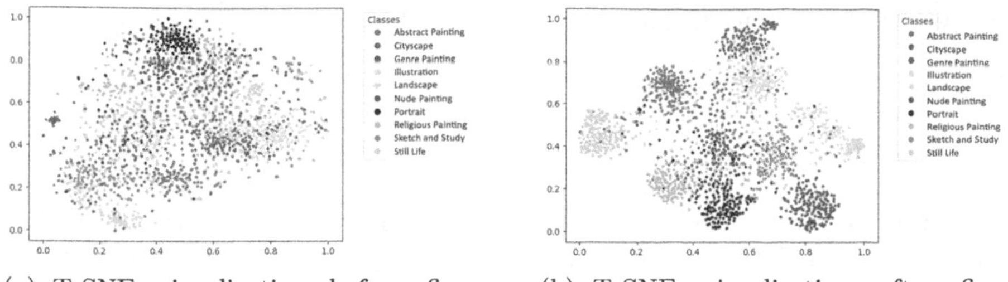

(a) T-SNE visualisation before fine-tuning

(b) T-SNE visualisation after fine-tuning

Fig. 2. T-SNE visualisation before and after fine-tuning ResNet50 for artistic genre recognition

Genre recognition helps to list and classify paintings for quicker access to artworks of the same genre. Artistic genre recognition ensures that query results give priority to artworks of the same genre, whereas without it, results may not match the desired genre preference. The aim is not to classify but to guide the search, by maintaining visual similarity. This is why the use of neural networks is useful for extracting visual information from images in addition to genre recognition. We also show that by optimising the parameters of the Approximate Nearest Neighbour algorithm, we can obtain results close to the exact algorithm while keeping a reasonable execution time.

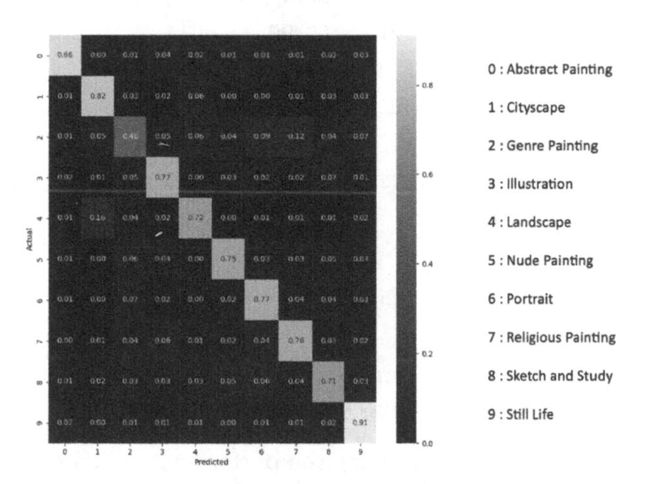

Fig. 3. Confusion matrix on WikiArt test data

3.1 Classifier Performance

Our first experiment evaluates the performance of our model in genre classification. Figure 3 represents the confusion matrix of the 10 genre classes of the WikiArt database.

The results show that the network performs relatively better on some classes than others because of their distinct visual appearance, such as still life (a 91% recall), abstract painting(a 86% recall) and cityscape (a 82% recall). In contrast, the network perfoms the worst for the genre painting class, mainly due to the fact that this class includes paintings featuring scene of anecdotal or familiar character (the illustration of scenes from everyday life, where the characters are anonymous human types) and can be confused with other genres such as religious painting (12%) or portraits (9%), as can be seen in the Fig. 3 representing the confusion matrix for the WikiArt dataset for each class.

The model obtains an accuracy of 75.12% with a standard deviation of 0.07 (with 97.27% for top 3 classes) on the validation data. In the Table 1, each model is trained on a number of images from WikiArt and with a class number equal to 10. It can be seen that despite a smaller amount of data due to class rebalancing, we obtain an accuracy close to that obtained with a larger amount of data.

Table 1. Genre recognition: a State-of-the-art

	genre	
References	Samples	Acc (%)
Tan et al. [21]	66,993	74.14
Saleh et al. [18]	63,691	60.28
Cetinic et al. [4]	86,087	77.6
Zhong et al. [26]	28,760	76.27
Zhao et al. [25]	64,993	78.03
Ours	27,613	75.12

3.2 Comparison of CBIR Performance Before and After Fine-Tuning with Specific Domain Knowledge

Our next experiment aims at comparing the performance of content-based image search before fine-tuning and after fine-tuning. Before fine-tuning, only the convolutional layers of ResNet50 are used for feature extraction, i.e. vectors of dimensions 2048. We make this comparison for all the test data on all genres and then on each genre treated separately. The results are presented in Table 2.

We can observe a clear improvement in performance after fine-tuning. Indeed, among the 1,000 nearest neighbours found, the percentage of images belonging to the same class as the query image is higher after fine-tuning. The average

Table 2. Percentage of images of the same genre as the query image in the nearest neighbours list before and after fine-tuning

	With 1,000 neighbours	
	Before fine-tuning	After fine-tuning
Genre	31.04%	38.09%
Abstract Painting	5.49%	19.24%
Cityscape	8.55 %	12.29%
Genre Painting	14.15%	14.62%
Illustration	2.03%	3.08%
Landscape	18.44%	29.77%
Nude Painting	1.60%	2.66%
Portrait	23.50%	23.82%
Religious Painting	9.47%	14.50%
Sketch and Study	4.32%	4.66%
Still Life	2.47%	6.67%

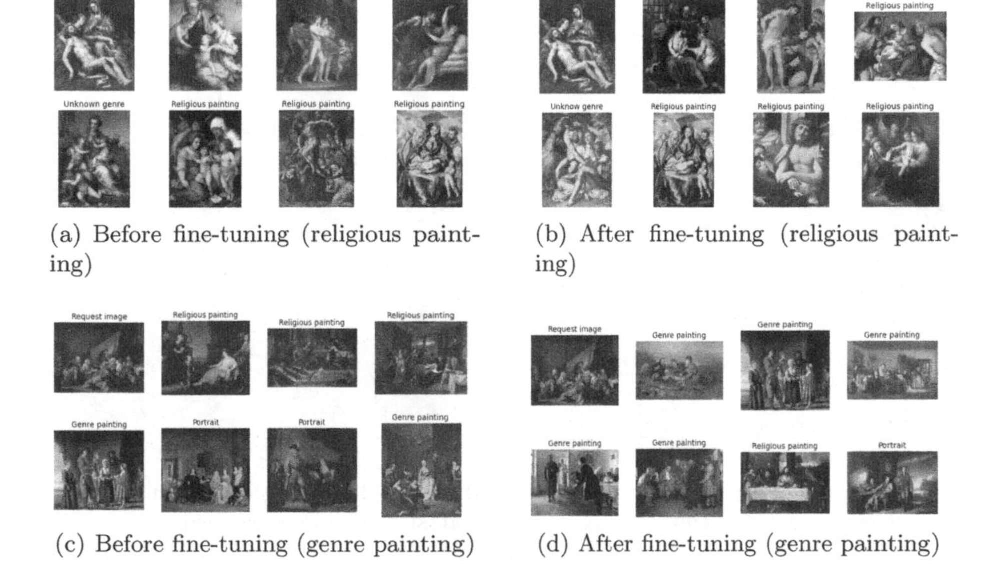

(a) Before fine-tuning (religious painting)

(b) After fine-tuning (religious painting)

(c) Before fine-tuning (genre painting)

(d) After fine-tuning (genre painting)

Fig. 4. CBIR performance on an example before and after fine-tuning on a religious painting and on a genre painting

improvement is 7%. Figures 4(a) and 4(b) show the query image belonging to religious painting genre with the seven most similar images. After the fine-tuning of ResNet50, we can see that the model fine-tuned for genre recognition focuses more on genre attributes. Figures 4(c) and 4(d) show that this is also revealing in the case of a genre painting. Among the classes most dependent on fine-tuning (with the best recalls), there is a clear improvement, such as an increase of 14% for abstract painting, 4% for still life painting, or 10% for landscapes. Despite some incorrect class images, the CBIR tool can still find similar paintings in terms of contents by including some elements (such as colour, texture, etc.) and similar compositions.

3.3 Parameters Optimization of the Approximate Nearest Neighbour Algorithm

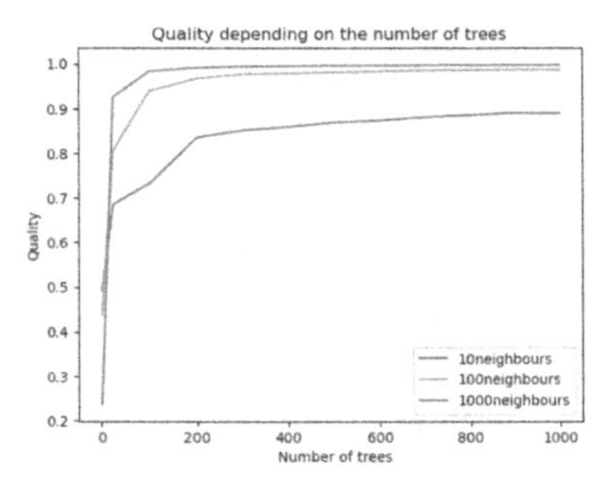

(a) Quality depending on the number of trees

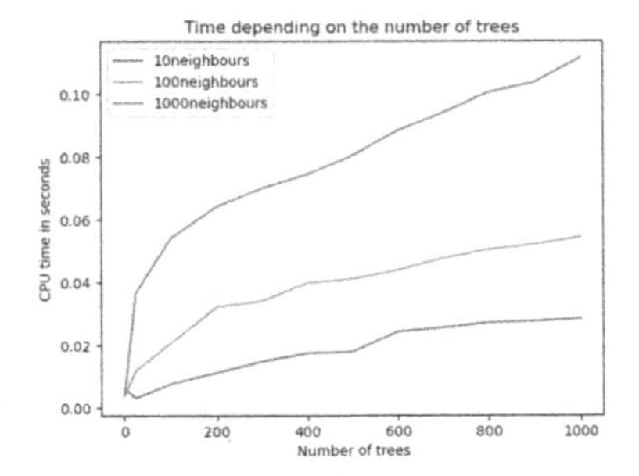

(b) Time depending on the number of trees

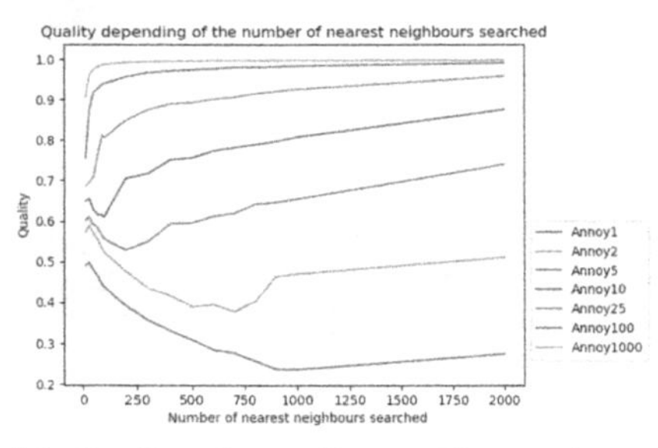

(c) Quality depending on the number of nearest neighbours searched

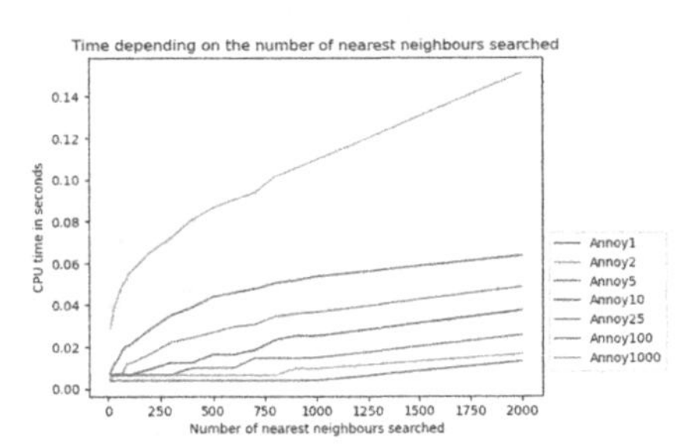

(d) Time depending on the number of nearest neighbours searched

Fig. 5. Optimisation of the number of trees and of the number of nearest nieghbours

We use an Approximate Nearest Neighbour algorithm, mainly for its speed of execution. We are interested in optimising its parameters, with the objective of getting the best possible list of kNN while keeping low execution times.

The parameters to be taken into account are the number of trees, the number of nearest neighbours searched and the execution time.

Let $(i_1, ..., i_n)$ be the image database sorted by increasing distance with a request image q. The list of k nearest neighbours of q is $(i_1, i_2...i_k)$, and the list of k-approximate nearest neighbours of q is $(i_{n_1}, i_{n_2}, ...i_{n_k})$, with n_i being an increasing integer function from $[1...k]$ to $[1...n]$. We define the quality $\mathcal{Q}$ of the request as:

$$\mathcal{Q} = \frac{k}{n_k}$$

Figure 5 shows that as soon as the number of trees is larger than 5, the quality increases with the number of trees and with the number of neighbours, and rapidly tends to 1. This means that the list of approximate nearest neighbours approaches the exact one. However, asking for a high number of neighbours or using a high number of trees requires a higher computation time. For our application, we suggest to fix the number of nearest neighbours and to make the number of trees vary to have the best possible quality, while keeping our time constraints under a acceptable threshold.

Suppose we want the 250 nearest neighbours with an execution time of less than 0.05 s. And taking into account the experiments done on Fig. 5. Figure 5(d) shows that we can choose a number of trees between 1 and 100. However, to obtain a result with a quality higher than 0.9 then the choice is made on a number of trees equal to 100, as shown on Fig. 5(c). Indeed, we obtain a quality of about 0.95 in about 0.03 s in CPU time on a MacBook Pro with an Apple $M1$ Max chip, 10 core and 64 Go RAM, which is much powerful than the final target device of our project, a tablet.

3.4 Introducing SimArt: A Web Application for Efficiently Searching Similar Artworks

To demonstrate the effectiveness of our approach, we developed SimArt[7], a web application for searching similar artworks. This application enables to explore wide and diverse art collections, spanning across different styles and eras, based on a nearest neighbour, image-based search engine. The request image can be a member of the dataset or can be uploaded by user.

We used two types of ANNOY indexes to enable efficient nearest neighbour search: one with features obtained from a pre-trained ResNet50 and the other with features obtained from a fine-tuned ResNet50 for genre recognition. The indexes were constructed using a 100 nodes tree structure to allow for efficient approximate nearest neighbour search.

We believe that SimArt can be a valuable resource for art enthusiasts and professionals alike, allowing for exploration and discovery of artworks across various styles and eras.

4 Discussion and Conclusions

In this work, we presented a content-based image retrieval tool using Convolutional Neural Networks for feature extraction to measure similarity between

[7] https://simart.datavalor.com.

images taking into account the genre of the paintings. We fine-tuned ResNet50 for genre recognition with an accuracy of 75%. We applied the features extracted from genre recognition model to increase the proximity of retrieved images belonging to the same genre as the request image. To obtain a real-time search, we used an Approximate Nearest Neighbour algorithm. Indeed, we were able to obtain the results with a factor of 1000 times faster (i.e. 0.04 s in CPU time instead of more than 30 s, for 81,444 images). These results allow us to project how to guide a search for nearest neighbours, taking into account for example the style or the artist. The performance of classification by genre can still be improved, which would allow to better distinguish similar images by genre. This can be done by enriching the data from other collections or by using attention modules or using another pre-trained model. We plan to use style and artist recognition to offer several search dimensions, while having the possibility to do a multi-criteria search (e.g. by selecting style and genre), by doing multi-label image recognition with Graph Convolutional Networks. We also plan to refine the search using Siamese networks by proposing groups of images of the same and different classes, in order to observe whether this improves the search. We would also like to refine the search by including the detection and location of objects representing strong symbolism in the history of art.

Acknowledgement. This work was funded by french national research agency with grant ANR-20-CE38-0017.

References

1. Alaasam, R., Kurar, B., El-Sana, J.: Layout analysis on challenging historical Arabic manuscripts using siamese network. In: 2019 International Conference on Document Analysis and Recognition (ICDAR), pp. 738–742. IEEE (2019)
2. Aurenhammer, F.: Voronoi diagrams-a survey of a fundamental geometric data structure. ACM Comput. Surv. (CSUR) **23**(3), 345–405 (1991)
3. Castellano, G., Vessio, G.: Deep learning approaches to pattern extraction and recognition in paintings and drawings: an overview. Neural Comput. Appl. **33**(19), 12263–12282 (2021)
4. Cetinic, E., Lipic, T., Grgic, S.: Fine-tuning convolutional neural networks for fine art classification. Expert Syst. Appl. **114**, 107–118 (2018)
5. Deng, J., Dong, W., Socher, R., Li, L.J., Li, K., Fei-Fei, L.: Imagenet: a large-scale hierarchical image database. In: 2009 IEEE Conference on Computer Vision and Pattern Recognition, pp. 248–255. IEEE (2009)
6. Focillon, H.: Vie des formes. Librairie Ernest Leroux Paris (1934)
7. He, K., Zhang, X., Ren, S., Sun, J.: Deep residual learning for image recognition. In: Proceedings of the IEEE Conference on Computer Vision and Pattern Recognition, pp. 770–778 (2016)
8. Huang, G., Liu, Z., Van Der Maaten, L., Weinberger, K.Q.: Densely connected convolutional networks. In: Proceedings of the IEEE Conference on Computer Vision and Pattern Recognition, pp. 4700–4708 (2017)
9. Jenicek, T., Chum, O.: Linking art through human poses. In: 2019 International Conference on Document Analysis and Recognition (ICDAR), pp. 1338–1345. IEEE (2019)

10. Joshi, A., Agrawal, A., Nair, S.: Art style classification with self-trained ensemble of autoencoding transformations. arXiv preprint arXiv:2012.03377 (2020)
11. Kapoor, R., Sharma, D., Gulati, T.: State of the art content based image retrieval techniques using deep learning: a survey. Multim. Tools Appl. **80**(19), 29561–29583 (2021)
12. Kelek, M.O., Calik, N., Yildirim, T.: Painter classification over the novel art painting data set via the latest deep neural networks. Procedia Comput. Sci. **154**, 369–376 (2019)
13. Latif, A., et al.: Content-based image retrieval and feature extraction: a comprehensive review. Math. Prob. Eng. **2019** (2019)
14. Li, W., Zhang, Y., Sun, Y., Wang, W., Li, M., Zhang, W., Lin, X.: Approximate nearest neighbor search on high dimensional data-experiments, analyses, and improvement. IEEE Trans. Knowl. Data Eng. **32**(8), 1475–1488 (2019)
15. Liao, Z., Gao, L., Zhou, T., Fan, X., Zhang, Y., Wu, J.: An oil painters recognition method based on cluster multiple kernel learning algorithm. IEEE Access **7**, 26842–26854 (2019)
16. Liu, S., Yang, J., Agaian, S.S., Yuan, C.: Novel features for art movement classification of portrait paintings. Image Vis. Comput. **108**, 104121 (2021)
17. Madhu, P., Kosti, R., Mührenberg, L., Bell, P., Maier, A., Christlein, V.: Recognizing characters in art history using deep learning. In: Proceedings of the 1st Workshop on Structuring and Understanding of Multimedia heritAge Contents, pp. 15–22 (2019)
18. Saleh, B., Elgammal, A.: Large-scale classification of fine-art paintings: learning the right metric on the right feature. arXiv preprint arXiv:1505.00855 (2015)
19. Saritha, R.R., Paul, V., Kumar, P.G.: Content based image retrieval using deep learning process. Clust. Comput. **22**, 4187–4200 (2019)
20. Seguin, B., Striolo, C., diLenardo, I., Kaplan, F.: Visual link retrieval in a database of paintings. In: Hua, G., Jégou, H. (eds.) ECCV 2016. LNCS, vol. 9913, pp. 753–767. Springer, Cham (2016). https://doi.org/10.1007/978-3-319-46604-0_52
21. Tan, W.R., Chan, C.S., Aguirre, H.E., Tanaka, K.: Ceci n'est pas une pipe: a deep convolutional network for fine-art paintings classification. In: 2016 IEEE International Conference on Image Processing (ICIP), pp. 3703–3707. IEEE (2016)
22. Tan, W.S., Chin, W.Y., Lim, K.Y.: Content-based image retrieval for painting style with convolutional neural network. J. Inst. Eng. Malaysia **82**(3) (2021)
23. Yang, Z.: Classification of picture art style based on vggnet. J. Phys. Conf. Ser. **1774**, 012043 (2021). IOP Publishing
24. Zhang, Z., Sabuncu, M.: Generalized cross entropy loss for training deep neural networks with noisy labels. Adv. Neural Inf. Process. Syst. **31** (2018)
25. Zhao, W., Zhou, D., Qiu, X., Jiang, W.: Compare the performance of the models in art classification. PLoS ONE **16**(3), e0248414 (2021)
26. Zhong, S.H., Huang, X., Xiao, Z.: Fine-art painting classification via two-channel dual path networks. Int. J. Mach. Learn. Cybern. **11**, 137–152 (2020)

GeomEthics: Ethical Considerations About Using Artificial Intelligence in Geomatics

Ermanno Petrocchi[1], Simona Tiribelli[1], Marina Paolanti[1(✉)], Benedetta Giovanola[1], Emanuele Frontoni[1], and Roberto Pierdicca[2]

[1] Department of Political Sciences, Communication and International Relations, University of Macerata, Via Don Minzoni 22/A, 62100 Macerata, Italy
{e.petrocchi,simona.tiribelli,marina.paolanti,benedetta.giovanola, emanuele.frontoni}@unimc.it

[2] Dipartimento di Ingegneria Civile Edile e dell'Architettura (DICEA), Università Politecnica delle Marche, 63100 Ancona, Italy
r.pierdicca@staff.univpm.it

Abstract. Artificial intelligence (AI) has made significant advancements in the field of geomatics, revolutionizing the way geospatial data is processed, analyzed, and interpreted. While these advancements have brought numerous benefits, they also raise ethical risks that must be carefully considered. The improvement of AI in geomatics has introduced ethical considerations such as data privacy, algorithmic bias, transparency, accountability, and the responsible use of AI technology. As AI algorithms process and analyze vast amounts of geospatial data, concerns regarding data privacy and security become paramount. Geospatial data often contains sensitive information, and the use of AI requires robust measures to protect individual privacy and prevent unauthorized access or misuse of data. This paper examines the ethical implications of the use of AI in geomatics and proposes the concept of GeomEthics as a framework for analyzing these ethical considerations. It explores the technical aspects of AI in geomatics and highlights the ethical principles of fairness, privacy, bias, accountability, and transparency. By coining the term GeomEthics, the paper emphasizes the importance of addressing these ethical concerns. It proposes the development of ethical guidelines and best practices for the responsible integration of AI in geomatics and discusses future research directions in the field. This paper contributes to a comprehensive understanding of the ethical implications of AI in geomatics and provides insights for ensuring the responsible and beneficial use of AI technologies in the geospatial domain. By addressing these ethical challenges, the field of geomatics can harness the benefits of AI while mitigating its potential risks and ensuring that geospatial analysis and decision-making processes are conducted ethically and responsibly.

Keywords: GeomEthics · GeoAI · Geomatics · Ethics · Artificial Intelligence

G. L. Foresti et al. (Eds.): ICIAP 2023 Workshops, LNCS 14366, pp. 282–293, 2024.
https://doi.org/10.1007/978-3-031-51026-7_25

1 Introduction

Digital technologies and artificial intelligence (AI) have been marking a revolution that has both transformed society and determined how people relate to the world. These technologies play a decisive role in shaping people's preferences and habits, satisfying their interests, monitoring their health, and so on [34]. Nowadays, every aspect of society has essentially turned into a collection of data and such a societal shift toward a data-driven environment has altered how individuals perceive themselves, relate to others, and make decisions [12]. At the same time, digital technologies and AI offer new possibilities for more efficient multilevel management of society. Techniques such as machine learning (ML) and deep learning (DL) have revolutionized data processing and analysis; consequently, they have become applicable across different sectors, enabling their widespread use and impact. This paper focuses on one field in which AI algorithms are gaining ground and becoming more widely used, that of geomatics[1]. Geomatics is the discipline that deals with the automated management of 2D and 3D information relating to the territory. Through a multidisciplinary and integrated approach, geomatics collects, archives, and models accurately in digital format spatial-geographical data acquired with different methods [16]. These include RGB and RGB-D cameras, infrared cameras, laser scanners, and satellite imagery. These acquisition methods allow obtaining geospatial and spatiotemporal data types, such as images (RGB, multispectral and thermal) and 3D point clouds [38]. This type of data analyzed and processed through ML and DL techniques is applied in many fields of geomatics for numerous tasks, including classification, prediction, segmentation, or clustering [21]. The interconnection of AI and geomatics determines the possibility of obtaining an infinite number of services much more quickly than in the past. Today, for example, people have easy access to topographic maps of cities (satellite imagery) or traffic information updates (RGB-D cameras) [26]. Or in the field of cultural heritage (CH), through laser-scanning technology, it is possible to digitally document monuments and archaeological sites [39]. The multi-spectral data, instead, make it possible to classify the vegetation of entire rural areas. And so on for various other fields of application. Finding application in many sectors, geospatial data play a fundamental role in public and private decision-making processes on multiple levels. For example, through geo-referenced information, companies can define marketing strategies by identifying areas for new stores while examining socio-demographic data or the presence of any competitors. Thanks to this data it becomes also possible to optimize logistics and transport, or track shipments more easily and quickly. In the sector of public interventions, instead, geospatial information is useful for the design and implementation of territorial planning actions, such as for the identification and management of agricultural and rural areas. Moreover, such information is particularly useful for monitoring and preventing environmental disasters and for the categorization of risk. Therefore, geospatial data has a significant socio-political valence and impact, as they represent the basis

[1] https://www.iso.org/committee/54904.html.

of certain choices rather than others. However, while this kind of data informs and shapes decision-making processes, surprisingly, specific attention and work to the ethics of data and the ethical consequences of AI in geomatics have been poorly developed thus far [13,14,31].

In order to comprehensively examine the ethical implications arising from the use of AI in geomatics, in this paper we coin the term GeomEthics. GeomEthics refers to the dedicated analysis and study of the ethical considerations and challenges that arise at the intersection of AI and geomatics. By proposing this concept, we aim to emphasize the importance of addressing ethical issues in the application of AI within the geospatial domain. GeomEthics provides a framework for exploring and addressing concerns related to fairness, privacy, bias, accountability, transparency, and the responsible use of AI technology in geomatics. Through the lens of GeomEthics, we can foster a deeper understanding of the ethical implications and promote the development of ethical guidelines and best practices to ensure the responsible and beneficial integration of AI in geomatics. The paper, having a preliminary purpose, is mainly focused on three ethical principles, namely, fairness, local territorial identity, and geo-privacy. The choice of these principles follows the belief that they represent topics of primary importance for GeomEtichs and are explicitly ascribable to the guidelines for the responsible use and development of AI. Indeed, given the pervasiveness and transformative power of AI, many guidelines of this type have been drawn up, both by public authorities and by the private high-tech sector. The UNESCO recommendations on AI ethics (2021) represent one of the most important documents in defining the values that must guide the responsible use and development of AI[2]. As highlighted by Jobin et al. [23], these guidelines converge in the definition of some agreed-upon ethics principles: transparency, fairness, nonmaleficence, responsibility, accountability, freedom, autonomy, privacy, safety, sustainability, trust, dignity, and solidarity (see also Fjeld et al. [11]). In light of this, we chose to focus on fairness, privacy, and territorial identity, which can be traced back to the principle of autonomy.

This paper makes several significant contributions to the field: i) it introduces the concept of GeomEthics as an ethical framework for analyzing the ethical implications of AI in geomatics; ii) it highlights the importance of addressing ethical concerns such as fairness, privacy, bias, accountability, and transparency in the context of geospatial applications of AI; iii) building upon the concept of GeomEthics, the paper proposes the development of ethical guidelines and best practices for the responsible use of AI in geomatics; iv) by outlining these guidelines, the paper offers practical recommendations to guide practitioners, policymakers, and researchers in ensuring the ethical integration of AI technologies in the field; v) it identifies emerging areas such as the integration of AI in

[2] Moreover, several committees of AI experts have been commissioned to draft guidelines for the correct use of AI. Examples of these committees are the High-Level Expert Group on Artificial Intelligence of the European Commission (HLEGAI) and that of the OECD. At a private level, Google has released its own indications for the ethical use of AI.

data acquisition processes, the trustworthy generation of geospatial data, and the interpretability and explainability of AI-driven geospatial analysis. These insights pave the way for further investigations and advancements in the field. These contributions enhance our understanding of the ethical implications of AI in geomatics and provide valuable insights for the responsible development and deployment of AI technologies in the field.

The paper is organized as follows. Following this introduction, Sect. 2 provides an in-depth examination of the application of AI in geomatics, covering technical aspects. In Sect. 3, we delve into the intricate relationship between AI ethics and geomatics. Specifically, this section focuses on the ethical principles of fairness, territorial identity, and geo-privacy, which are closely intertwined with the ethical guidelines of AI and hold great significance at the intersection of ethics and geomatics. Finally, Sect. 4 presents the conclusions drawn from our study and explores future research directions in this field.

2 The Use of Artificial Intelligence in Geomatics

GeoAI, also known as Geospatial Artificial Intelligence, is an emerging field that combines the power of AI with geospatial data and techniques [38]. It encompasses the application of AI algorithms and methodologies to geospatial problems, enabling advanced analysis, interpretation, and decision-making in the context of spatial data. GeoAI leverages the capabilities of AI, including machine learning, deep learning, computer vision, and natural language processing, to extract meaningful insights from geospatial datasets. These datasets can include satellite imagery, aerial photographs, LiDAR data, digital maps, GPS coordinates, and other location-based information. One of the key advantages of GeoAI is its ability to process and analyze vast amounts of geospatial data quickly and efficiently. Through automated feature extraction, pattern recognition, and classification algorithms, GeoAI can identify and map objects, land cover types, urban development patterns, transportation networks, and more. GeoAI finds applications across various domains, including urban planning, environmental monitoring, disaster management, agriculture, transportation, and public health. For example, in urban planning, GeoAI can analyze satellite imagery and demographic data to identify areas at risk of urban heat islands, helping policymakers develop strategies to mitigate the effects of heat stress on urban populations. In environmental monitoring, GeoAI can analyze remote sensing data to detect changes in land cover, monitor deforestation, and assess the health of ecosystems. It can also support precision agriculture by analyzing crop health, soil moisture levels, and weather patterns to optimize irrigation, fertilization, and pest control strategies. Furthermore, GeoAI plays a crucial role in disaster management by analyzing satellite imagery and social media data to identify affected areas, assess damage, and facilitate rescue and relief efforts. It can also aid in infrastructure planning by analyzing transportation patterns, predicting traffic congestion, and optimizing routing algorithms. However, as stated in the Introduction, the application of GeoAI also raises ethical considerations. Issues such as privacy, data bias, algorithmic transparency, and the responsible use of AI in decision-making processes become paramount when dealing with sensitive geospatial information. It is essential to ensure that GeoAI systems are developed and deployed in a manner that upholds ethical standards and respects societal values. Currently, AI is not integrated into the data acquisition process, which is still carried out by operators. This aspect remains true until automated acquisition systems (e.g., robots) are equipped with "intelligence" even in the initial phase of data detection. Within the three aforementioned areas and considering the numerous applications of geomatics, we can categorize the contribution of AI as follows: in the data processing phase, AI can intervene to reduce "machine" errors, thereby mitigating the risk introduced by instrumental tolerances and minimizing uncertainty in the system. This is particularly relevant in terrestrial photogrammetry, which enables the creation of realistic 3D models but is still prone to errors. The first ethical consideration arising from this process is as follows: *Do the processed data faithfully reproduce a cultural asset or any other entity?* and importantly, *how much deviation from reality is acceptable before it becomes an ethical concern regarding its reproducibility?* A second aspect, primarily related to object reconstruction (2D or 3D), involves the use of generative methods. Techniques like NeRF and GAN can generate data that is not controlled by humans but rather derived from processed data. How much does this generated data deviate from reality? Are there any guidelines or criteria to determine the truthfulness of such data? (Here, the concept of trustworthiness may be relevant, encompassing ethical and legal aspects). Lastly, in the interpretation of data, which is currently crucial in geomatics and extends beyond terrestrial sensing to satellite systems, there has been a shift toward intelligent systems. While segmentation and classification methods were traditionally rule-based and well-known within the community, today, the interpretation of satellite maps, for instance, relies on intelligent systems that extract semantic information from geospatial data. This aspect should be linked to the discussion below on Geo legality, a topic of significant importance, as highlighted by the open data movement and the entire FOSS (Free and Open-Source Software) community, particularly those involved in GIS.

3 Ethics of Artificial Intelligence in Geomatics

In 1998 American vice president Al Gore expressed all his confidence in the technological progress related to the exploitation of geospatial data [2]. His concern at the time was that limiting the collection of geospatial data by government and private companies represented only a possible slowdown in the development of society [17]. Al Gore, therefore, understood the importance of geospatial data well before the beginning of the big data era. As said, despite this enormous potential, today the integration of ML and DL with geomatics presents some ethical issues related to those principles to be safeguarded for the correct use of AI. In the first part of this section, we focus on the principle of fairness in geospatial data. In the second part, we delve into the concept of territorial identity. In the third part, it is analysed the principle of geoprivacy.

3.1 Geospatial Data Fairness

Over the last few decades, there has been a remarkable advance in AI systems, in particular ML and DL algorithms. However, algorithmic models have often turned out to be unfair in the results they generate, instead of being, as expected, accurate and fair [15]. Consequently, it is now widely acknowledged that biases can be concealed within algorithms. Research has revealed that predictive algorithms often yield unfavourable outcomes disproportionately affecting disadvantaged minority groups compared to others [5]. Additionally, even if algorithmic predictions exhibit equal accuracy across different groups, the nature of their errors varies: algorithms tend more strongly to wrong pessimism when dealing with members of disadvantaged groups but more strongly to wrong optimism when dealing with members of advantaged groups [29]. In light of these findings, the concept of "algorithmic fairness" has been coined, referring to the fairness required to assess algorithm processes and outcomes [25]. For instance, algorithmic fairness mandates that the scores generated by an algorithm should be equally accurate for legally protected groups, such as blacks and whites or men and women. Moreover, algorithmic fairness necessitates that the algorithm produces an equivalent percentage of false positives or false negatives for each of the relevant groups [18]. For this reason, the UNESCO recommendations on AI ethics mentioned in the introduction emphasize the importance of AI actors exerting all reasonable efforts to minimize and avoid perpetuating discriminatory or biased applications. Even in geospatial data, unfortunately, some biases are still present and discrimination therefore persists. Also in the field of geomatics, avoiding the presence of bias represents one of the major technical difficulties. Let's consider disaster risk management (DRM). It helps prevent disasters and mitigate their impact. In this sector, geomatics is very useful in mapping territorial areas and consequently in the identification and classification of disaster risk [24]. Obviously, AI algorithms have become increasingly popular in this discipline as well, as they improve the quality of DRM models [9]. However, there is a discrepancy in the amount of geospatial data between middle-poor and rich countries. Two factors play an important role in this gap. On the one hand, the absence of technological devices in poor countries makes the amount of geospatial data smaller than in countries where technologies are widespread[3]. On the other hand, the lack is greater for rural or non 'priority' areas, which are those areas where disasters have not already been witnessed [19][4]. To clarify the impact of the lack of geospatial data in poor countries, it is useful to recall this example. If we look at AI algorithm areal imagery of some African cities characterized by the presence of slums and compare them with imagery of the same city but produced by members of the local community, the bias is soon

[3] Additionally, by looking at mobile phone transcripts of call records it is possible to estimate population sizes pre and post a disaster. However, it is important to note that this method may underestimate vulnerable populations that do not have access to cell phones [37].

[4] In this sense, OpenStreetMap helped to reduce the gap between rich and medium-poor countries. Despite this, broad areas still remain uncovered.

evident [14]. Slums are characterized by very small buildings, very close to each other. In these cases, the algorithm happens not to be able to recognize such buildings as settlements and therefore does not reproduce them when mapping the city [35]. Hence, this indicates the presence of an algorithm bias towards the most informal and poorest areas of the city. As with non-geospatial data, the AI algorithm favours the most advantaged population groups and disadvantages the most disadvantaged. Using AI in this case provides inaccurate information. However, it must be considered that for disaster prevention, areal imagery still plays a very important role in assisting the choices of public decision-makers. To this end, it is essential to have the most accurate images possible of cities, especially those of the most fragile parts correctly reported. After all, due to overcrowding and the precarious conditions of the houses, slums represent precisely those areas at the highest risk and where disasters are often more harmful. Nevertheless, biases present in geospatial data are very difficult to detect. The example shows that here there is really a need both to compare the images provided by the AI with those provided by the community and to understand the local context as characterized by slums. Only in this way, it is possible to detect the geospatial bias of the algorithm towards informal settlements.

3.2 Local Identity

The areal imagery example reported in the previous section highlights how the maps produced by the AI algorithm through satellite data are in contrast with those integrated by community members. This example, in addition to spotting the bias favouring the most advantaged populations, also brings to light the importance of the point of view of local communities' members. One of the ethical principles present in the guidelines with which to integrate AI is autonomy, which can be defined as the people' right to decide for themselves [31]. However, this interpretation of autonomy has an exclusively Western matrix [42]. In most African communities, indeed, the concept of autonomy assumes a communitarian meaning and personal autonomy is declined in relational autonomy [30]. In these communities, the value of autonomy cannot be understood as intrinsically valuable, as it happens in the Western world, but must be related to what it brings both to the individual and to the community at the same time [22]. Hence, the concept of autonomy does not imply unrestricted free will. Rather, it encompasses the manner in which a person exercises that freedom, specifically how individual's decisions are influenced by internalized communal principles and beliefs. Considering this value difference, it appears that a principle that should guide AI towards a more respectful use for each individual can instead end up harming entire communities. Disagreement in how values are understood creates conflict between global guidelines and local realities [8]. Therefore, integrating AI algorithms with local points of view is not only useful to avoid the algorithm outcome being less biased, as explained above for the slum images. This would also be essential to respect the various territorial identities, different and sometimes opposite to the Western world [31]. In addition, overhead imagery is not distributed equally across all world countries. For poorer countries and less densely populated areas, such as deserts, there is little or no geospatial data available. This gap in the collection of geographical data produces a further injustice as for poorer parts of the world there is a lack of ML models which continue to be set up in the areas for which more data is available. This fortifies the need to build geographically and demographically diverse machine learning teams in order to ensure good quality model development processes. Thus, a deep understanding of the local context appears crucial to examine both which values should be considered as priorities and to identify which social groups require greater control in the algorithm [31]. To design an algorithm that respects the principle of territorial identity, a collaboration of experts from various local contexts might help to improve the process of AI design [13].

3.3 Geo-Privacy

Geoprivacy can be defined as the entitlement to control the disclosure of personal location data to external entities, including the ability to decide the extent, conditions, and timing of such information sharing [1,10,41]. The analysis of geoprivacy is divided into two parts. The first is related to the dangers of sharing one's coordinates and the possible identification in cases of data breach. The second recalls the previous section and analyses the different understandings of privacy in the world.

Geoprivacy and Geomatics. The spread of smartphones and the success of social networks made it possible to collect geospatial data on people as never before. In today's digital world, indeed, users themselves publicly share their coordinates and accept apps that have access to their location. Location-based social networks (LBSN) have gone mainstream. However, even georeferencing operations are not free from producing distortions and errors. Errors related to data privacy are relevant as such personal data could allow people to be identified. Privacy concerns have then increased as location information is closely tied to personal identity [41]. In fact, characteristics such as income, race, gender, education are all inferable from location. In order to avoid individual identification, obfuscation and geomasking techniques have been developed, which consist in masking the true geographical location of individuals, while preserving the geographical characteristics of the data itself.

While novel privacy-preserving techniques are in place today, phenomena such as data breaches or data leaks or the voluntary malicious use or selling to third-party of geospatial data can lead to individuals' privacy violations. This may happen especially when the surveillance/geo monitoring and data collection concern small or medium size places and such data are processed in combination with other more identifying data streams. In these situations, geomatics methods could provide a way to bypass individual identity protection techniques. Indeed, as demonstrated by Li et al. [28] through thermal images it is now possible to determine age and gender of a person. In this sense, thermal images in a certain way result even more accurate than visual images as they are not affected by illumination variations, indoor and outdoor lighting conditions, poses and disguise [4]. Although less accurate, it is worth noting that another form of remote sensing such as spectral images can also allow for individual recognition [3]. Hence, these techniques represent a tool that can assist in personal identification or tracking. Therefore, if combined even partially with geo-referenced data, they can pose a serious threat to personal privacy. These considerations highlight that the integration of AI in geomatics represents great opportunities but on the other hand it also hides many issues that need to be addressed thoroughly.

Geo-Privacy and Culture. As with local territorial identity, to understand what people mean by privacy it is necessary to look at their cultural background. Rzeszewski and Luczys [40] highlighted the difference in users' perspectives on location-based services. Some think that these services will change the world, while others believe that these systems have already changed our understanding of the concept of location. The impact of different cultural backgrounds on people's decisions cannot be ignored [27]. As described above, different cultures have different ways of understanding the community value. Western societies are more individualistic and have different levels of trust in the legal system. Both of these factors, individualism and trust, present positive correlations with privacy concerns. The most individualistic societies are those that place the greatest value on wealth and do not consider social interconnectedness and emotions important [20]. These societies, recognizing an economic value in the collection of personal data, present a high level of privacy concerns. On the other hand, in societies where there are stringent legal privacy regulations, there are high levels of trust in the legal system. Concerns about privacy decrease as this type of trust increases [32][5]. For this paper, it is important to highlight that all these privacy considerations are equally valid for geoprivacy as well [44]. In addition, considering the universal reach of UNESCO recommendations for the responsible use of AI [6], it is also useful to look at the understanding of privacy in East Asian countries. As well highlighted by Zhang and McKenzie [44], traditional East Asian cultures place great importance on harmony, courtesy and human relationships based on trust. These core values imply that the concept of privacy is considered foreign and requires time to be accepted [36,43]. Moreover, other factors, such as collectivism, a strong idea of altruism but also social conditions such as housing overcrowding favour the concept of 'group privacy', in which private matters can be shared rather than remain personal [7,43]. These considerations show once again how the attempt to safeguard universal principles without considering local diversity could be harmful to entire communities.

[5] A shift in privacy attitudes is observed across Europe which confirms these correlations. Western countries, such as France, and northern ones, such as Poland and Estonia, show more concern for individual responsibility. Conversely, more socially based countries in the south, such as Greece and Spain, and those in Eastern Europe, show greater trust in their government and regulations [33].

4 Conclusions and Future Works

This paper has examined the ethical implications of using AI in geomatics and proposed the concept of GeomEthics as a framework for addressing these considerations. The advancements in AI have brought significant benefits to geomatics, including improved data processing, image analysis, and decision support systems. However, ethical risks such as data privacy, algorithmic bias, transparency, and accountability have emerged. To address these ethical risks, the development of ethical guidelines and standards specific to AI in geomatics is crucial. These guidelines should promote privacy protection, fairness, transparency, and responsible AI use. Ongoing monitoring and evaluation of AI systems are necessary to detect and mitigate unintended negative consequences and biases. Future works in this field should focus on exploring human-AI collaboration in geomatics and investigating the role of GeoAI in sustainable development. Understanding how humans and AI can effectively work together and leveraging AI for sustainable urban planning, resource management, and climate change adaptation will be instrumental in achieving positive societal impact. By addressing these future research directions, the field of AI in geomatics can advance ethical practices, overcome challenges, and maximize the benefits of AI while upholding values such as fairness, transparency, and privacy.

References

1. AbdelMalik, P., Boulos, M.N.K., Jones, R.: The perceived impact of location privacy: a web-based survey of public health perspectives and requirements in the UK and Canada. BMC Public Health **8**(1), 1–9 (2008)
2. Al, G.: The digital earth: understanding our planet in the 21st century (1998). http://www.digitalearth.gov/VP19980131.html
3. Allen, D.W.: An overview of spectral imaging of human skin toward face recognition. In: Face Recognition Across the Imaging Spectrum, pp. 1–19 (2016)
4. Bhowmik, M.K., et al.: Thermal infrared face recognition-a biometric identification technique for robust security system. Rev. Refinem. New Ideas Face Recognit. **7**, 113–138 (2011)
5. Burrell, J.: How the machine 'thinks': understanding opacity in machine learning algorithms. Big Data Soc. **3**(1), 2053951715622512 (2016)
6. Camara, A.: International council of museums (ICOM): code of ethics. In: Encyclopedia of Global Archaeology, pp. 5868–5872. Springer, Cham (2020). https://doi.org/10.1007/978-3-030-30018-0_1049
7. Capurro, R.: Privacy. An intercultural perspective. Ethics Inf. Technol. **7**, 37–47 (2005)
8. Carman, M., Rosman, B.: Applying a principle of explicability to AI research in Africa: should we do it? Ethics Inf. Technol. **23**(2), 107–117 (2021)
9. Deparday, V., Gevaert, C.M., Molinario, G., Soden, R., Balog-Way, S.: Machine learning for disaster risk management (2019)
10. Duckham, M., Kulik, L.: A formal model of obfuscation and negotiation for location privacy. In: Gellersen, H.-W., Want, R., Schmidt, A. (eds.) Pervasive Computing, pp. 152–170. Springer, Heidelberg (2005). https://doi.org/10.1007/11428572_10

11. Fjeld, J., Achten, N., Hilligoss, H., Nagy, A., Srikumar, M.: Principled artificial intelligence: mapping consensus in ethical and rights-based approaches to principles for AI. In: Berkman Klein Center Research Publication (2020-1) (2020)
12. Floridi, L.: The Fourth Revolution: How the Infosphere is Reshaping Human Reality. OUP Oxford (2014)
13. Gevaert, C.M., Carman, M., Rosman, B., Georgiadou, Y., Soden, R.: Fairness and accountability of AI in disaster risk management: opportunities and challenges. Patterns **2**(11) (2021)
14. Gevaert, C.M.: Finding biases in geospatial datasets in the global south-are we missing vulnerable populations? In: 41st EARSeL Symposium 2022: Earth Observation for Environmental Monitoring (2022)
15. Giovanola, B., Tiribelli, S.: Weapons of moral construction? on the value of fairness in algorithmic decision-making. Ethics Inf. Technol. **24**(1), 3 (2022)
16. Gomarasca, M.A.: Basics of geomatics. Appl. Geomat. **2**, 137–146 (2010)
17. Goodchild, M.F., et al.: Next-generation digital earth. Proc. Natl. Acad. Sci. **109**(28), 11088–11094 (2012)
18. Hellman, D.: Measuring algorithmic fairness. Virginia Law Rev. **106**(4), 811–866 (2020)
19. Herfort, B., Lautenbach, S., Porto de Albuquerque, J., Anderson, J., Zipf, A.: The evolution of humanitarian mapping within the openstreetmap community. Sci. Rep. **11**(1), 3037 (2021)
20. Hoftede, G., Hofstede, G.J., Minkov, M.: Cultures and Organizations: Software of the Mind: Intercultural Cooperation and Its Importance for Survival. McGraw-Hill (2010)
21. Hong, D., Yokoya, N., Xia, G.S., Chanussot, J., Zhu, X.X.: X-modalnet: a semi-supervised deep cross-modal network for classification of remote sensing data. ISPRS J. Photogram. Remote. Sens. **167**, 12–23 (2020)
22. Ikuenobe, P.: African communal basis for autonomy and life choices. Dev. World Bioeth. **18**(3), 212–221 (2018)
23. Jobin, A., Ienca, M., Vayena, E.: The global landscape of AI ethics guidelines. Nat. Mach. Intell. **1**, 389–399 (2019)
24. Kemper, H., Kemper, G.: Sensor fusion, GIS and AI technologies for disaster management. Int. Arch. Photogram. Remote Sens. Spat. Inf. Sci. **43**, 1677–1683 (2020)
25. Kleinberg, J., Ludwig, J., Mullainathan, S., Sunstein, C.R.: Discrimination in the age of algorithms. J. Legal Anal. **10**, 113–174 (2018)
26. Krizhevsky, A., Sutskever, I., Hinton, G.E.: Imagenet classification with deep convolutional neural networks. Adv. Neural Inf. Process. Syst. **25** (2012)
27. Kummer, T.F., Recker, J., Bick, M.: Technology-induced anxiety: manifestations, cultural influences, and its effect on the adoption of sensor-based technology in German and Australian hospitals. Inf. Manag. **54**(1), 73–89 (2017)
28. Li, P., Dai, P., Cao, D., Liu, B., Lu, Y.: Non-intrusive comfort sensing: detecting age and gender from infrared images for personal thermal comfort. Build. Environ. **219**, 109256 (2022)
29. Lippert-Rasmussen, K.: Using (un) fair algorithms in an unjust world. In: Res Publica, pp. 1–20 (2022)
30. Mhlambi, S., Tiribelli, S.: Decolonizing AI ethics: relational autonomy as a means to counter AI harms. Topoi, pp. 1–14 (2023)
31. Micheli, M., et al.: Ai ethics and data governance in the geospatial domain of digital earth. Big Data Soc. **9**(2), 20539517221138770 (2022)
32. Milberg, S.J., Smith, H.J., Burke, S.J.: Information privacy: corporate management and national regulation. Organ. Sci. **11**(1), 35–57 (2000)

33. Miltgen, C.L., Peyrat-Guillard, D.: Cultural and generational influences on privacy concerns: a qualitative study in seven European countries. Eur. J. Inf. Syst. **23**(2), 103–125 (2014)
34. Mittelstadt, B.D., Allo, P., Taddeo, M., Wachter, S., Floridi, L.: The ethics of algorithms: mapping the debate. Big Data Soc. **3**(2), 2053951716679679 (2016)
35. Najmi, A., Gevaert, C.M., Kohli, D., Kuffer, M., Pratomo, J.: Integrating remote sensing and street view imagery for mapping slums. ISPRS Int. J. Geo Inf. **11**(12), 631 (2022)
36. Nakada, M., Tamura, T.: Japanese conceptions of privacy: an intercultural perspective. Ethics Inf. Technol. **7**, 27–36 (2005)
37. Pestre, G., Letouzé, E., Zagheni, E.: The abcde of big data: assessing biases in call-detail records for development estimates. World Bank Econ. Rev. **34**(Supplement_1), S89–S97 (2020)
38. Pierdicca, R., Paolanti, M.: Geoai: a review of artificial intelligence approaches for the interpretation of complex geomatics data. Geosci. Instrum. Methods Data Syst. **11**(1), 195–218 (2022)
39. Pierdicca, R., et al.: Point cloud semantic segmentation using a deep learning framework for cultural heritage. Remote Sens. **12**(6), 1005 (2020)
40. Rzeszewski, M., Luczys, P.: Care, indifference and anxiety-attitudes toward location data in everyday life. ISPRS Int. J. Geo Inf. **7**(10), 383 (2018)
41. Seidl, D.E., Paulus, G., Jankowski, P., Regenfelder, M.: Spatial obfuscation methods for privacy protection of household-level data. Appl. Geogr. **63**, 253–263 (2015)
42. Tiribelli, S.: Inequalities and artificial intelligence. In: Filosofia Morale/Moral Philosophy, vol. 3, pp. 187–198. Mimesis Edizioni (2023)
43. Yao-Huai, L.: Privacy and data privacy issues in contemporary china. Ethics Inf. Technol. **7**, 7–15 (2005)
44. Zhang, H., McKenzie, G.: Rehumanize geoprivacy: from disclosure control to human perception. GeoJournal **88**(1), 189–208 (2023)

Fine Art Pattern Extraction and Recognition (FAPER)

Enhancing Preservation and Restoration of Open Reel Audio Tapes Through Computer Vision

Alessandro Russo[1(✉)], Matteo Spanio[2], and Sergio Canazza[1]

[1] Centro di Sonologia Computazionale (CSC), Department of Information Engineering (DEI), University of Padua (UniPd), Via Giovanni Gradenigo 6b, 35131 Padua, PD, Italy
alessandro.russo@dei.unipd.it, sergio.canazza@unipd.it
[2] Audio Innova s.r.l., Viale della Navigazione Interna 51/a, 35129 Padua, PD, Italy
http://csc.dei.unipd.it , https://www.audioinnova.com/

Abstract. Analog audio documents inevitably face degradation over time, posing a challenge for preserving their audio content and ensuring the integrity of the recordings. Analog document preservation is one of the main research topics of interest of the Centro di Sonologia Computazionale (CSC) of the Department of Information Engineering of the University of Padua, which over the years developed and implemented a methodology for preservation that includes, among other things, the video recording of the digitization process of the open-reel tapes for documenting irregularities on the top of their surface. Together with the corpus of digitized high-quality audio recordings, this led to the creation of an internal archive of video documents. This paper presents a software application that leverages computer vision techniques to automatically detect Irregularities on open-reel audio tapes, analyzing the video documents produced during the digitization interventions. The software employs a frame-by-frame analysis to automatically identify and highlight points of interest that may indicate tape damages, splices, and other Irregularities. The software uses Generalized Hough Transform and SURF algorithms to locate regions of interest within the tape. The proposed software is also part of the MPAI/IEEE-CAE ARP standard developed by Audio Innova s.r.l., spin-off of the CSC, and it may offer a robust and efficient solution for analyzing open-reel audio tapes, supporting archivists and musicologists in their activities.

Keywords: open reel audio tapes · irregularities detection · computer vision · preservation · restoration

1 Introduction

During the last years, many archives are facing the need to preserve their heritage from degradation. Digitization plays a central role in the preservation of

Supported by Audio Innova s.r.l. and University of Padua.

G. L. Foresti et al. (Eds.): ICIAP 2023 Workshops, LNCS 14366, pp. 297–308, 2024.
https://doi.org/10.1007/978-3-031-51026-7_26

analog archives by offering a powerful solution to mitigate the challenges associated with physical storage and deterioration by converting analog materials, such as photographs, documents, and audio recordings, into digital formats. In order to guarantee the correct preservation of the original documents and minimize the loss of information, digitization processes cannot be carried out with the so-called "massive approach", but they require a scientific approach, such as the methodology developed at the *Centro di Sonologia Computazionale* (CSC) for the preservation of audio documents. The *Centro di Sonologia Computazionale* (CSC) of the Department of Information Engineering at Padua University [8,9] is a multidisciplinary laboratory actively engaged in music production, audio research, education, and dissemination. A significant area of research at CSC is the digitization and preservation of cultural heritage, especially audio documents, such as speech and music documents [6,7,12,15,17]. In dealing with audio documents, the recorded audio signal plays a central role, being the main point of interest, but digitization processes cannot be focused on the recorded audio only. In fact, the methodology developed at CSC gives great importance to metadata gathering and to the completion of the preservation copy, as well as including contextual information, such as photos and video files. In particular, video documentation allows for capturing additional information beyond the audio itself. At the CSC Lab, the need of including a video-documentation while preserving open-reel audio tapes emerged for the first time while working on the development of preservation strategies for computer music archives, such as the ones belonging to Italian composers Luciano Berio and Luigi Nono. In fact, due to the lack of a standard music score, composers often acted directly onto the surface of the tapes writing notes, markers, and signs to identify the various part of the artwork, and also to give indications to the technicians on stage when performing live tracks in which open-reel tapes were used to add sequences and other musical contents to the solo part. Furthermore, the video documentation of the entire A/D migration may help in identifying mechanical-related issues (e.g. deformations of the tape that may cause its misalignment with the reading head), splices, and other irregularities on the surface of the tape, such as dirt and loss of magnetic paste. Video documentation helps in preserving this kind of information, which would otherwise be lost. For these reasons, video documentation is a fundamental part of the preservation methodology adopted by the CSC for the digitization of audio documents, especially within preservation interventions carried out on computer music archives. Together with the Preservation Audio File (.wav, high-res, 24 bit, 96 kHz), the Preservation Copy includes the Preservation Audio Visual File, the video of the digitization process with a low-resolution (typically 16 bit, 48 kHz) audio track recorded for syncing purposes. Nevertheless, considering the large amount of digitized documents, it may be really time-consuming for archivists to watch several hours of video files for identifying irregularities. Artificial intelligence is a powerful tool that may help the work of archivists and other people involved in preservation interventions, by automatically detecting the points of interest on the surface of the tape. During the years of activity, the CSC preserved and digitized numerous audio archives, working on fonds of different genres, such as classical, ethno and computer music, oral sources, linguistic studies etc. [14]. For several digitized fonds, together with the large amount of audio files, a related corpus of video documents is available. The video corpus recorded and archived at the CSC lab was used to carry out several research projects and develop computer-vision-based tools for automated video analysis. The outcome of this research was the implementation of the Video Analyser module, part of the recently approved MPAI/IEEE-CAE ARP standard [11]. The following sections will describe the MPAI/IEEE-CAE ARP use case (Sect. 2), the features and functioning of the Video Analyser (Sect. 3), and further implementations of the developed software and conclusions (Sect. 4).

2 MPAI IEEE-CAE ARP

Audio Innova srl, spin-off of the CSC, is a founding member of MPAI-CAE. MPAI is an international unaffiliated non-profit organization, active in developing standards for AI-based data coding. MPAI-CAE's aims to improve the user experience for audio-related applications such as entertainment, communication, teleconferencing, gaming, post-production and restoration [4]. As MPAI member, Audio Innova developed the Audio Restoring Preservation (ARP) use case, part of the MPAI-CAE standard. The MPAI/IEEE-CAE ARP standard provides precise software references to the preservation work. The CAE-ARP technical specifications adopt the preservation methodology developed at CSC, incorporating computer tools based on artificial intelligence.

In this use case, automated AI is used to extract information from the digitized audio/video of open reel tapes. The outputs of the ARP are the preservation master (digital copy for long-term preservation), and the access copy. Access copies are generally in a compressed format and, if necessary, restored for correct playback of the digitized audio. The usage of AI allows the automatic detection of Irregularities on the surface of the tape, improving precision and speed in selecting and extracting data for its primary storage and use. The MPAI-CAE international standard was approved in May 2022 and was later adopted by the IEEE Standards Association as 3302-2022 in December of the same year [1]. The technical architecture of the standard includes five modules that target and process different digital inputs: the *Audio Analyser*, the *Video Analyser*, the *Tape Irregularity Classifier*, the *Tape Audio Restoration*, and the *Packager*. The next section will explain in detail the functioning of the *Video Analyser*, the module used for the analysis of the video of the tape for automatically detecting irregularities.

3 Video Analyser

In the MPAI-CAE ARP use case the modules directly involved in the analysis of video input are the Video Analyser, whose purpose is to identify the areas of the tape that present specific points of interest, storing them as *Irregularity*

Images, and the Tape Irregularity Classifier which takes care of classifying the Irregularity Images based on their content.

For the entire duration of the tape, the video analysis concerns the portion of the tape below the reading head, also including other typical components of the open-reel recorder (see Fig. 2). In particular, video files contain a recording of open-reel tapes played on a Studer A810 (see Fig. 1).

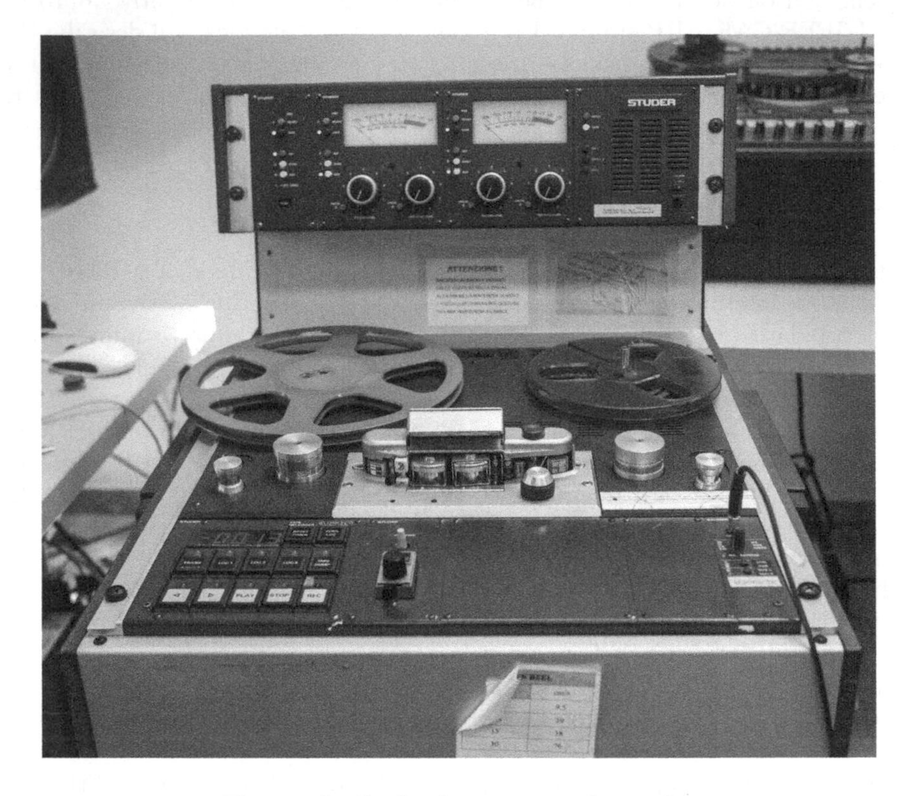

Fig. 1. The Studer A810 open-reel recorder

Since the CSC began to video documenting the entire digitization process in 2013, due to issues related to available equipment, most of the archived videos have a PAL resolution (720 × 576), at 25 interlaced fps. This choice doesn't affect the detection process but must be taken into account during the image classification stage.

The Video Analyser employs a frame-by-frame analysis approach to identify and capture frames exhibiting significant differences when compared to their preceding ones. In the subsequent step (Tape Irregularity Classifier), the focus is placed on these distinctive frames to detect and highlight potential anomalies that may indicate tape damages, scratches, splices, or other Irregularities. This is accomplished through the utilization of a carefully calibrated classifier module based on a convolutional neural network. The implementation of the Irregularity detection process underwent multiple iterations. After the refinement process, the following algorithm was devised as follow:

1. Identify the Regions of Interest.
2. Traverse through the video, comparing consecutive pairs of frames.
3. If the count of dissimilar pixels between a pair surpasses a predetermined threshold, an Irregularity is detected.

3.1 ROI Detection

Preliminary studies explored the possibilities of background subtraction algorithms [10]. This method utilized previously gathered information to separate new elements from recurring ones. However, this approach yielded numerous false positives, capturing insignificant variations in brightness, reel movement, and other undesired artifacts. Furthermore, there were inevitable performance issues due to the long operation time associated with the BackgroundSubtractorKNN tool. Unfortunately, improving the execution time was not feasible due to limitations imposed by the OpenCV algorithm implementation [5].

To address these challenges, a reevaluation of the Irregularities' characterization was conducted, with a focus on scene framing. It was observed that all anomalies shared a lack of vertical movement and typically consisted of small clusters of points related to the frame size.

Given that Irregularities exhibited only horizontal movement, the approach was to concentrate on variations in a specific part of the frame. The most effective solution was to shift the focus away from the moving elements of the scene (i.e., the Irregularities) and select areas of interest based on stationary elements instead. In this case, the capstan and the reading head, which are the components closest to the tape, remained stationary within the frame and served as reference points for automatically identifying the pixel regions where Irregularities appear (see Fig. 2).

The capstan is a rotating shaft used to move the tape through the mechanisms and magnetic heads (for erasing, recording, and playback) of the tape recorder. During playback, the tape passes through the capstan and a rubber wheel called the pinch roller. The pinch roller presses the tape against the capstan, providing the necessary friction for the tape to continue moving. Typically, the pinch roller is located after the magnetic heads in the direction of the tape's movement (on the right side of the video in this case).

To detect Irregularities, it is sufficient to examine the pixels below the reading head. Since the tape only moves horizontally, the anomalies, flowing from left to right, inevitably enter this area. The capstan can also be used for the same purpose, but in this case, it is useful for different events of interest: the start and the end of the tape. When the playback ends, the pinch roller releases the tape by moving away from the capstan, and sometimes this causes the tape to come out of the reading head's slot. The movement of the pinch roller is clearly visible in all the videos, and for this reason, it was chosen as the reference point to detect the moment when the tape reaches its end. A similar situation occurs

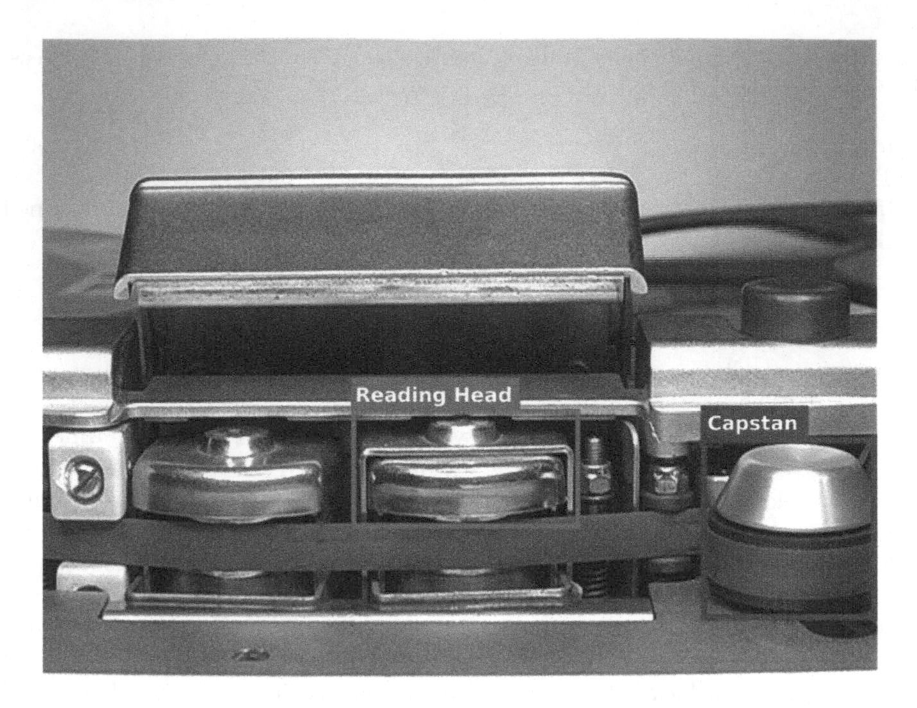

Fig. 2. Fixed elements of interest in the frame.

at the beginning of the video when the tape is not being played and the capstan is in its "rest" position (see Fig. 3 for a visual representation of the capstan and pinch roller positioning). By focusing only on the area below the capstan, it is possible to detect when it moves to release the tension from the tape.

To identify stationary elements within the scene, well-known algorithms capable of finding patterns within an image were employed: the Generalized Hough Transform [2] and SURF (Speeded Up Robust Features) [3]. Their coordinates within the image could be determined by providing these algorithms with the image of the capstan and the reading head. The coordinates of the underlying areas were defined through empirical methods. Figure 4 shows the identified areas below the capstan and the reading head.

The actual implementation of this step involves examining the middle frame of the video, which should represent a normal situation regarding the framing or the presence of an Irregularity. During this step, the position of ROIs is determined by searching for template images in the frame using the aforementioned algorithms.

3.2 Detection of Irregularities

Once the areas of interest within the scene have been established, an activation function is employed to determine the presence of Irregularities. The number of differing pixels in each pair can be calculated By utilizing pairs of consecutive

Fig. 3. Tape scrolling mechanism of the Studer A810

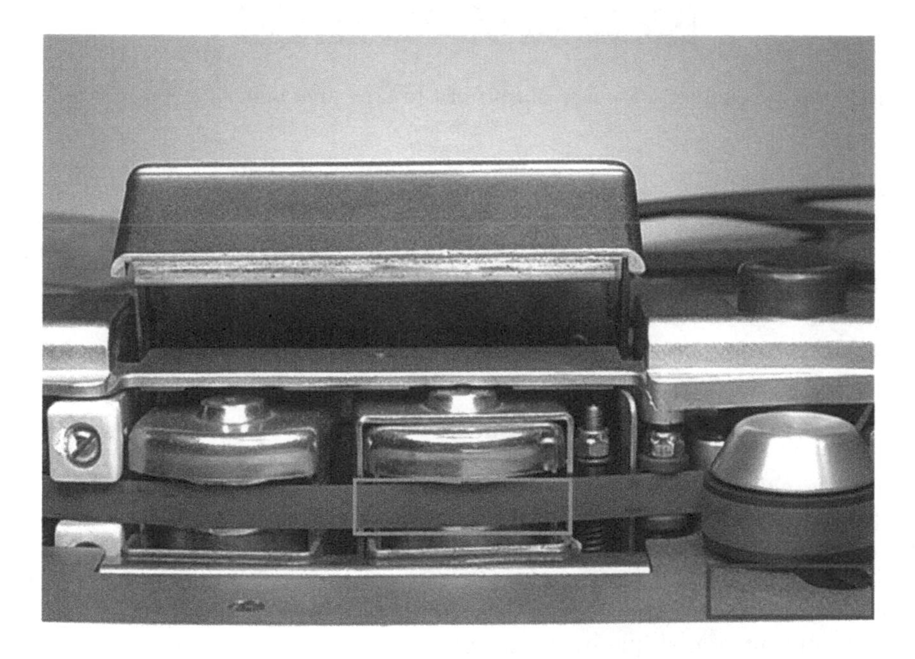

Fig. 4. ROIs under the capstan and the reading head.

frames. Focusing on the identified ROIs, it has been observed that approximately 80% of the pixels consistently exhibit variations in terms of colors and shadows. Therefore, it has been established that when the quantity of differing pixels exceeds the set threshold, a significant difference between the two frames has been detected. To quantify the differing pixels between two images, a new image is generated with white pixels representing the matching ones, and black ones for indicating their differences. The generation of the comparison image can be described as follows:

$$\mathbf{D}(i,j) = \begin{cases} 255 & \text{if } \mathbf{C}_{\text{red}}(i,j) - \mathbf{P}_{\text{red}}(i,j) = 0 \quad \wedge \\ & \quad \mathbf{C}_{\text{green}}(i,j) - \mathbf{P}_{\text{green}}(i,j) = 0 \quad \wedge \\ & \quad \mathbf{C}_{\text{blue}}(i,j) - \mathbf{P}_{\text{blue}}(i,j) = 0 \\ 0 & \text{otherwise} \end{cases}$$

where $i = 1, \ldots, n$ and n is the number of rows in the matrix, $j = 1, \ldots, m$ and m is the number of columns in the matrix, matrix $\mathbf{D}$ is the difference frame, $\mathbf{C}_{\text{red}}$, $\mathbf{C}_{\text{green}}$, and $\mathbf{C}_{\text{blue}}$ are the current frame matrices for the red, green, and blue color channels respectively, and $\mathbf{P}_{\text{red}}$, $\mathbf{P}_{\text{green}}$, and $\mathbf{P}_{\text{blue}}$ are the previous frame matrices for the red, green, and blue color channels respectively.

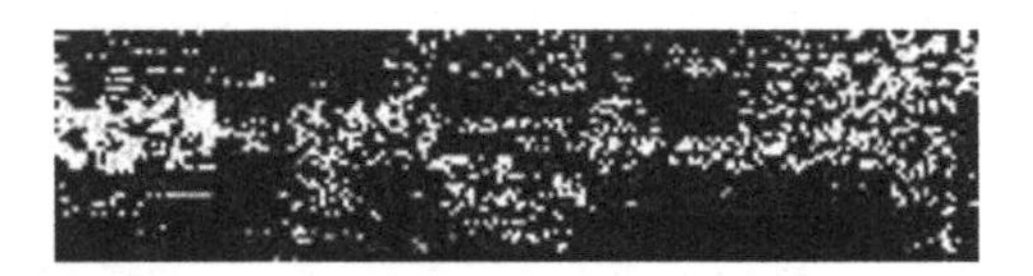

Fig. 5. Actual comparison image output of the tape area under the reading head.

As shown in Fig. 5, the generated comparison image provides a visual representation of the differences between consecutive frames in the tape area under the reading head. The white regions indicate areas where the frames match, while the black regions represent variations between the frames. This image serves as a valuable tool for identifying Irregularities and assessing the extent of their presence.

3.3 Resolution of Identified Issues

Mitigating False Positives. In order to ensure that all Irregularities within an input tape are accurately identified, it is fundamental to test the accuracy of the software. The testing of the developed software showed good results, but an occasional detection of false positives also occurred. If an Irregularity extends for several centimeters on the length of the tape, it may be detected several times, with the output of multiple Irregularity images.

To address this issue, considering that the region of interest (ROI) in correspondence of the tape's reading head is 3 cm wide, the analyzed tapes run at a speed of 7.5 or 15 in. per second (ips), and the videos were recorded at 25 fps, it was measured that, depending on the speed, discarding two or three subsequent frames upon detecting an Irregularity is sufficient to avoid duplicating the same Irregularity (although false positives may still occur in the case of exceptionally long Irregularities). This approach significantly reduces the number of images saved as Irregularities by the software.

Handling Interlacing. Another issue related to the tape format concerns the storage of Irregularity images. Since the majority of the video files archived at CSC was recorded long before the design and development of the software, the videos were recorded in PAL format at 25 fps, interlaced. Interlacing is a notable drawback of the input files, since each frame is divided into even and odd lines, resulting in misalignment between the two fields, particularly noticeable in areas with motion (Fig. 6).

Fig. 6. Interlaced picture: notice the misalignment between odd and even lines

In the Irregularity detection phase, dealing with interlaced images does not introduce particular problems. Since the comparison occurs on a pixel-by-pixel basis, the analyzed frames in pairs are interlaced in the same manner, resulting in consistent misalignment of even and odd pixel rows in subsequent frames. However, interlacing can adversely affect the classification phase, which employs a convolutional neural network. In fact, this type of neural network reduces the number of pixels as it progresses deeper into the network, approximating the colors of neighboring pixels to identify characteristic patterns within an image.

Unfortunately, interlacing alters the colors of neighboring pixels and significantly impairs the classifier's performance.

Various techniques to reduce the effects of interlacing have been explored and tested. One widely used technique in dedicated playback software involves separating the semi-quads within the analyzed frame. This process generates two images: one containing the even lines and the other containing the odd lines. By utilizing only one of these semi-quads, half of the original frame's information is lost, but the other half image gives more reliable results for this specific use case. Thus, it has been chosen to save only the semi-quad containing the odd lines, resulting in images with a resolution of 720×228 pixels. This approach mitigated the misalignment caused by interlacing and provided a more accurate representation of the tape's content.

3.4 Irregularities Classification

Following the image acquisition process, a CNN based on the GoogLeNet network [16] is employed to classify the acquired images. The CNN has undergone training on specific classes, namely splices, brands, shadows and end of tape. While additional types of Irregularities do exist, their inclusion in the classifier's training set was not feasible due to their significantly lower frequency of recurrence, especially compared to the aforementioned classes. Such inclusion would have led to an imbalanced dataset, undermining the effectiveness of the classification model. Therefore, the number of classes was reduced, ensuring a more balanced input for the classifier [13].

Separate analyses were conducted on two different datasets: the first one with tapes played at 7.5 ips and the other for tapes played at 15 ips. Both classifiers exhibited an accuracy of 95% during the validation phase, a noteworthy achievement. This high accuracy was further confirmed during the subsequent testing phase, which involved evaluating the performance of the classifiers on frames extracted from the dataset prior to the network's training. This comprehensive evaluation solidifies the robustness and reliability of the developed system.

4 Conclusions

The work carried out at CSC is an example of the application of modern technologies to the field of cultural heritage, especially for preserving analog audio documents. Given the recent emphasis on digitizing audiovisual archives, the number of digitized documents is constantly increasing, and archivists and musicologists often find themselves studying and dealing with collections consisting of a large number of units.

For this reason, automated tools for detecting irregularities on the surface of the tape, such as the described Video Analyser, can be of great assistance in simplifying the work of musicologists and scholars, especially when combined with other tools, such as the Audio Analyser, which is also part of the developed MPAI/CAE ARP standard. The MPAI/CAE ARP standard was recognized by the committee of the world contest *Cannes Neurons 2023* (https://www.worldaicannes.com/en/neurons) at the World Artificial Intelligence Cannes Festival (https://www.worldaicannes.com/en/what-is-the-waicf) that awarded Audio Innova s.r.l. in the "Best Creative AI" category.

Future developments envision the implementation of the video dataset, training the model for recognition on an increasing number of videos. As said in Sect. 3, most of the videos in the archive are in PAL resolution, with 25 interlaced fps. This represents one of the main limitations of the first version of the software, and for this reason, authors are conducting tests on video with a FHD resolution (1920×1080), increasing the number of frames from 25 to 50 fps, in order to obtain video snapshots with a reduced motion blur and a higher resolution. In fact, thanks to high-definition images, it could be possible to recognize more detailed shapes in the Irregularity detection phase, allowing to consider the arrangement of different pixel and not only their quantity. Future implementations will target also the model of the classifier, to better discriminate the various classes of Irregularities.

Acknowledgements. We thank Prof. Marina Bosi (Stanford University, Ca, US) and Dr. Leonardo Chiariglione (President and Chairman of the Board of MPAI) for useful discussions, exceptional support and valuable and constructive suggestions during the planning and development of ARP IEEE 3302-2022 Standard.

This work is partially supported by SYCURI Project, funded by University of Padua, in the Program "World Class Research Infrastructure".

References

1. IEEE Standard Adoption of Moving Picture, Audio and Data Coding by Artificial Intelligence (MPAI) Technical Specification Context-Based Audio Enhanced (CAE) version 1.4. IEEE Std 3302-2022, pp. 1–94 (2023). https://doi.org/10.1109/IEEESTD.2023.10112597
2. Ballard, D.: Generalizing the Hough transform to detect arbitrary shapes. Pattern Recogn. **13**(2), 111–122 (1981). https://doi.org/10.1016/0031-3203(81)90009-1. https://www.sciencedirect.com/science/article/pii/0031320381900091
3. Bay, H., Ess, A., Tuytelaars, T., Van Gool, L.: Speeded-up robust features (surf). Comput. Vis. Image Underst. **110**(3), 346–359 (2008). Similarity Matching in Computer Vision and Multimedia. https://doi.org/10.1016/j.cviu.2007.09.014. https://www.sciencedirect.com/science/article/pii/S1077314207001555
4. Bosi, M., Pretto, N., Guarise, M., Canazza, S.: Sound and music computing using AI: designing a standard. In: Proceedings of the 18th Sound and Music Computing Conference 2021, SMC 2021, Virtual Conference (2021)
5. Bradski, G.: The OpenCV library. Dr. Dobb's J. Softw. Tools **120**, 122–125 (2000)
6. Bressan, F., Bertani, R., Furlan, C., Simionato, F., Canazza, S.: An ATR-FTIR and ESEM study on magnetic tapes for the assessment of the degradation of historical audio recordings. J. Cult. Herit. **18**, 313–320 (2016). https://doi.org/10.1016/j.culher.2015.09.004. http://www.elsevier.com
7. Canazza, S.: The digital curation of ethnic music audio archives: from preservation to restoration. Int. J. Digit. Libr. **12**(2–3), 121–135 (2012)

8. Canazza, S., De Poli, G.: Four decades of music research, creation, and education at Padua's Centro di Sonologia Computazionale. Comput. Music. J. **43**(4), 58–80 (2020)
9. Canazza, S., De Poli, G., Vidolin, A.: Gesture, music and computer: the Centro di Sonologia Computazionale at Padova University, a 50-year history. Sensors **22**(9) (2022). https://doi.org/10.3390/s22093465. https://www.mdpi.com/1424-8220/22/9/3465
10. Fantozzi, C., Bressan, F., Pretto, N.: Tape music archives: from preservation to access. Int. J. Digit. Libr. **18** (2017). https://doi.org/10.1007/s00799-017-0208-8
11. MPAI: Technical Specification MPAI Context-based Audio Enhancement (MPAI-CAE) (2022)
12. Pretto, N., Fantozzi, C., Micheloni, E., Burini, V., Canazza, S.: Computing methodologies supporting the preservation of electroacoustic music from analog magnetic tape. Comput. Music. J. **42**(4), 59–74 (2019)
13. Pretto, N., Fantozzi, C., Micheloni, E., Burini, V., Canazza, S.: Computing methodologies supporting the preservation of electroacoustic music from analog magnetic tape. Comput. Music J. **42**(4), 59–74 (12 2018). https://doi.org/10.1162/comj_a_00487
14. Pretto, N., Russo, A., Bressan, F., Burini, V., Rodà, A., Canazza, S.: Active preservation of analogue audio documents: a summary of the last seven years of digitization at CSC. In: Proceedings of the 17th Sound and Music Computing Conference 2020, SMC 2020, Torino, Italy (2020)
15. Salvati, D., Canazza, S.: Incident signal power comparison for localization of concurrent multiple acoustic sources. Sci. World J. **2014**, 1–13 (2014)
16. Szegedy, C., et al.: Going deeper with convolutions. CoRR abs/1409.4842 (2014). http://arxiv.org/abs/1409.4842
17. Verde, S., Pretto, N., Milani, S., Canazza, S.: Stay true to the sound of history: philology, phylogenetics and information engineering in musicology. Appl. Sci. **8**(2), 226 (2018)

Exploring the Synergy Between Vision-Language Pretraining and ChatGPT for Artwork Captioning: A Preliminary Study

Giovanna Castellano, Nicola Fanelli(✉), Raffaele Scaringi, and Gennaro Vessio

Department of Computer Science, University of Bari Aldo Moro, Bari, Italy
{giovanna.castellano,raffaele.scaringi,gennaro.vessio}@uniba.it,
n.fanelli10@studenti.uniba.it

Abstract. While AI techniques have enabled automated analysis and interpretation of visual content, generating meaningful captions for artworks presents unique challenges. These include understanding artistic intent, historical context, and complex visual elements. Despite recent developments in multi-modal techniques, there are still gaps in generating complete and accurate captions. This paper contributes by introducing a new dataset for artwork captioning generated using prompt engineering techniques and ChatGPT. We refined the captions with CLIPScore to filter out noise; then, we fine-tuned GIT-Base, resulting in visually accurate captions that surpass the ground truth. Enrichment of descriptions with predicted metadata improves their informativeness. Artwork captioning has implications for art appreciation, inclusivity, education, and cultural exchange, particularly for people with visual impairments or limited knowledge of art.

Keywords: ChatGPT · Computer vision · Cultural heritage · Deep learning · Digital humanities · Image captioning

1 Introduction

Artwork captioning refers to generating concise and informative text descriptions that capture the essence of an artwork, its visual elements, and underlying concepts [7]. This emerging field has significant potential not only to enrich art appreciation but also to promote inclusivity, education, and cultural exchange, particularly for people with visual impairments or limited artistic knowledge.

However, generating rich and semantically meaningful captions for artworks poses unique challenges [6]. Artistic expression often transcends literal representation, incorporating abstract concepts, emotions, and symbolic meanings. Capturing and effectively communicating these elements in textual descriptions requires a deep understanding of artistic intent, cultural references, and historical context. Moreover, the complexity of visual elements within artworks, including

G. L. Foresti et al. (Eds.): ICIAP 2023 Workshops, LNCS 14366, pp. 309–321, 2024.
https://doi.org/10.1007/978-3-031-51026-7_27

colors, textures, and spatial relationships, adds a layer of difficulty in generating comprehensive captions.

Multi-modal techniques for artwork captioning have gained attention in recent years, with studies exploring different approaches. For example, *neural style transfer* has been employed to transform images into paintings, creating a large-scale dataset with image-text pairs [20]. Iconographic captions and visual question answering in cultural heritage have also been explored [3,7]. However, there are still gaps in developing more comprehensive and accurate captioning techniques, addressing the limitations of dataset design, and improving the model's knowledge and understanding of artworks.

In this paper, we aim to contribute to this research by introducing a new dataset designed explicitly for automatically generating artwork captions. Using prompt-engineering techniques, we exploited ChatGPT [21] to generate visual descriptions of artworks based on title and artist information. Although these descriptions focus on the content of the artwork and contain conceptually rich elements, there is a significant presence of noise caused by ChatGPT hallucinations, particularly for lesser-known artworks. We found that CLIPScore [15] is an effective indicator of caption noise and used it to filter out poor examples and assign weights to the remaining descriptions. We then fine-tuned a vision-language pre-trained (VLP) model, the Generative Image-to-text Transformer (GIT) in its base version [31], resulting in a new framework for creating artwork captions that generate visually accurate captions that are superior to the ground truth. In addition, we enriched the visual descriptions with predicted metadata using a multi-task classification model based on the Vision Transformer (ViT) architecture [11], improving their informativeness.

The rest of the paper is organized as follows. Section 2 reviews related literature. Section 3 presents the data used in this study. Section 4 describes the proposed methodology. Section 5 presents our experimental evaluation. Section 6 concludes the paper and discusses future directions for our research.

2 Related Work

Since their introduction in neural machine translation, Transformers [30] have found several applications in the domain of image captioning [9,12,14]. In particular, using Transformers has facilitated the emergence of vision-language pre-training as a powerful approach to cross-modal learning by exploiting large-scale models and datasets. VLP models are commonly pre-trained on extensive collections of unlabeled or weakly labeled multimodal data, using pretraining objectives to develop a holistic understanding of vision and language. Subsequently, these models can be fine-tuned on various downstream tasks.

However, while much work in automatic image captioning has been done in the general domain of natural images, very few studies have tackled this task in the more challenging fine arts domain, arguably one of the most problematic domains in which to perform this task, both because of its complexity and the absence of rich task-specific datasets [26]. Initially, this research focused primarily on image-text and text-image retrieval to exploit the synergy between textual and visual content to improve the effectiveness and accuracy of the search. In a seminal paper, Garcia et al. [13] presented *SemArt*, the first dataset of fine art images paired with corresponding artistic commentaries. They conducted several experiments using this dataset, paving the way for further exploration in the field. Another significant contribution came from Stefanini et al. [27], who introduced *Artpedia*, a dataset containing paired fine art images and annotated texts. These annotations categorize the text into "contextual" and "visual" sentences.

Over time, the application of multi-modal techniques has broadened, attracting the interest of researchers working on more complex tasks, such as artwork captioning. In [2], a description generation system based on *SemArt* was proposed, which uses an encoder-decoder model (ResNet-LSTM) to generate multi-topic artwork descriptions covering content, form, and context, using placeholders instead of named entities. A parallel process performs metadata classification and object detection on the artwork image, generating prompts for DrQA [8] and using retrieved documents to fill placeholders in the generated description. Other works [19,25] explored data-driven approaches for generating captions for ancient artworks. In another study, Lu et al. [20] employed *neural style transfer* to transform images from the MS COCO dataset into paintings, creating a large-scale image caption dataset with original MS COCO captions. Cetinic [7] explored iconographic captions using the Iconclass AI Test Set dataset, developing a VLP model to recognize iconographic elements from images of artworks. However, this dataset was not explicitly designed for captions, and ground-truth captions were generated through preprocessing steps applied to image labels. An alternative study conducted by Ruta et al. [23] presented a new dataset called *StyleBabel*, encompassing artworks from various genres. This study focused on artwork tagging and captioning. More recently, Ishikawa and Sugiura [16] approached artwork captioning from a different perspective, emphasizing the affective dimension of image captions. Recent work studied visual question answering in the cultural heritage domain, developing models using the VISCOUNTH dataset [3] or employing specific prompts with GPT-3, demonstrating model knowledge of specific and famous artworks [4].

In our research, we curated a dataset designed explicitly for the automatic generation of artwork captions. Using ChatGPT, we generated descriptions based on artists and titles of artworks from our *ArtGraph* Knowledge Graph [5]. Although the descriptions focused on the content of the artwork and included rich concepts, noise was present due to ChatGPT hallucinations, especially for lesser-known artworks. We used CLIPScore [15] to filter out poor examples and fine-tuned GIT-Base [31], a VLP model, resulting in a new captioning framework that generates visually accurate captions that overcome the ground truth. In addition, we enriched descriptions with predicted metadata using a ViT-based model [11], enabling the integration of other textual information.

3 Materials

Building a comprehensive dataset of richly annotated artwork captions poses significant challenges, requiring human effort and expertise. To overcome these difficulties, we used an innovative approach based on an artificially created ground truth derived from ChatGPT, the widely adopted chatbot. This approach not only streamlines the data collection process but also provides a unique opportunity to explore the intersection of ChatGPT and art curation.

As a starting point, we used *ArtGraph*, our recently released Knowledge Graph on art, built by scraping WikiArt and DBpedia [5]. It collects 116,475 artworks spanning 18 genres and 32 styles, and many other metadata that characterize them. However, despite incorporating semantic concepts through metadata, *ArtGraph* lacks textual descriptions necessary to train an artwork captioning model. As said, we created a synthetically generated ground truth using the popular ChatGPT to fill this gap. As shown in [4], GPT-3, which was trained on a large corpus of textual data related to several domains, including art, can produce good descriptions of artworks by exploiting the information it used during the training process. However, these capabilities do not prevent the model from generating erroneous or partially erroneous descriptions.

Similarly, we asked ChatGPT to generate text descriptions for each artwork in *ArtGraph* with the following prompt followed by a list of artworks with their titles and authors:

Write visual descriptions for the following artworks.
RULES:
- *Descriptions must be between 20 and 40 tokens in length.*
- *The content of each description should only refer to the subjects, their attributes, and the scenes depicted.*
- *Avoid repeating the author's name or the painting's title within the descriptions.*
- *Begin each description with the phrase 'The artwork depicts'.*
- *List the descriptions using numbers and maintain the order of the provided artworks.*
- *Descriptions must not include false information.*

The prompt rules were refined manually after experimenting with various configurations. It is worth noting that ChatGPT rarely complied with the token limits and the rule prohibiting providing false information. In any case, the generated descriptions include information about the subjects, their attributes, and the scene depicted, and occasionally include iconographic, formal, or emotional elements to enhance the overall appeal of the captions.

To evaluate the quality of an image-caption pair, we leveraged CLIP [22], specifically its associated CLIPScore [15]. CLIP is a deep learning model that maps images and texts into a shared embedding space. It was trained using contrastive loss over 400M image-text pairs from the Internet. For an image with visual CLIP embedding $\mathbf{v}$ and the corresponding generated caption with textual CLIP embedding $\mathbf{c}$, we computed the CLIPScore as:

$$CLIPScore(\mathbf{c}, \mathbf{v}) = \max(\cos(\mathbf{c}, \mathbf{v}), 0)$$

The scores are within the range $[0, 1]$, where higher scores indicate a higher level of semantic matching between the image and the caption. In practice, scores typically fall within the [0, 0.4] range. Hessel et al. [15] showed that CLIPScore highly correlates with human judgment on image captioning tasks. Unlike traditional metrics, it does not require reference captions, so we could use it to evaluate the quality of our ground truth examples and to automatically filter out bad image-caption pairs from our dataset, a technique that has already been used to create open image-text datasets such as LAION-400M [24], where the authors heuristically chose a value of 0.3 as a threshold for filtering out bad Internet-collected examples.

In line with the findings in [29], we employed the NLP augmentation technique known as *back-translation* to generate two additional captions for each artwork. This technique involves translating the original English caption into another language and then translating it back into English, resulting in slightly different captions. For our back-translation, we utilized the OPUS-MT translation models [28] for French and German. As a result, our final dataset comprises 116,475 *ArtGraph* images, each associated with three English captions.

4 Methods

Our framework comprises two models: a caption generator and a metadata classifier. Both models take as input the image of an artwork without any additional information, allowing our framework to be applied to any artwork whose only information is its visual appearance. When the digitized image of an artwork is fed into our framework, it is processed by the two models in parallel. Both models use ViT-B [11] as the image encoder. The caption generator is responsible for generating a visual description of the artwork and is trained using the captions generated synthetically by ChatGPT. Simultaneously, the metadata classifier predicts the artist, genre, style, tags, and media associated with the artwork and is trained using the *ArtGraph* connections of the artwork as supervised labels. The outputs of the metadata classifier are used to populate a predefined template, which is then combined with the visual caption generated by the first model. This results in a description of the artwork that highlights both information about the artwork and a visual description. Figure 1 shows the general outline of the proposed method; the functioning of the two core models is described below.

4.1 Caption Generation

For caption generation, our method involves fine-tuning GIT-Base [31]. The model uses an encoder-decoder architecture. The encoder is ViT-B/16, initialized with CLIP weights, while the decoder is a standard Transformer decoder. The entire encoder-decoder model is pre-trained on 10M image-text examples from MS COCO, SBU, Conceptual Captions (CC3M), and Visual Genome.

In this approach, the image information of an example is embedded in the caption input tokens through linear projections of the patch embeddings generated by the ViT encoder. At each time step, the decoder generates probability

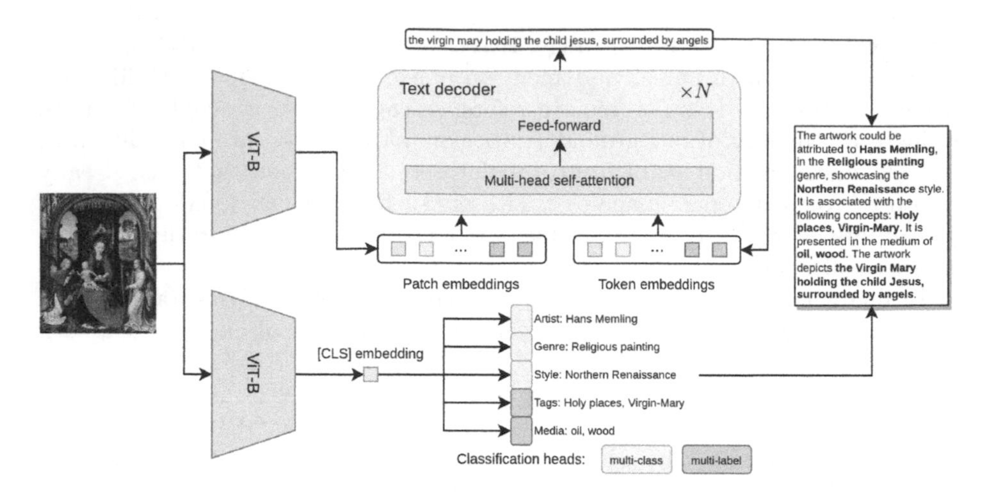

Fig. 1. Our framework includes a caption generator and a metadata classifier using ViT-B as an image encoder. The caption generator produces a visual description, trained on synthetic captions, while the metadata classifier predicts artist, genre, style, tags, and media based on *ArtGraph*. The combined results provide a complete description of the artwork.

distributions over the vocabulary associated with the BERT-Base [10] uncased tokenizer, considering the context of the image. These distributions are then used to compute the language modeling loss, which is employed to train the model. Specifically, for each example corresponding to the triple (I, T, c), where I is the image, $T = t_0, t_1, t_2, \ldots, t_{N+1}$ is the sequence of caption tokens, being t_0 the `[BOS]` token and t_{N+1} the `[EOS]` token, and c is the CLIPScore computed as the cosine similarity between the CLIP embeddings of I and T, we apply the weighted variant of the loss as follows:

$$\ell = w(c)\frac{1}{N+1}\sum_{i=1}^{N+1} CE(t_i, p(t_i|I, t_0, t_1, ..., t_{i-1}))$$

where CE is the cross-entropy loss and $w(c)$ is a linear function of c, which grows proportionally as c increases, used to weight the importance of an example in the computation of the loss, based on its quality, estimated by c. This approach was designed to produce superior performance compared to the unweighted loss, as shown by the CLIPScore results in Table 2.

In all experimental settings, we excluded the final period and the initial sub-string "*The artwork depicts*" from each caption, elements that ChatGPT included during ground truth caption generation. Additionally, we converted the text inputs to lowercase to align with the same (BERT-Base uncased) tokenizer employed by GIT. Furthermore, we imposed a maximum text length of 40 tokens. To restore case information, the output of the caption generator, which is in lowercase, is modeled using *truecasing* [18].

4.2 Metadata Classification

In parallel, we fine-tuned a ViT-B/16 model, pre-trained on ImageNet-21K, to perform multitask classification on artwork images. This involved using the connections in *ArtGraph* as our ground truth. Specifically, we performed multi-class classifications for artist, genre, and style for each artwork while employing multi-label classifications for tags and media. This information is intriguing to incorporate into a visual description of an artwork, as it references the artwork's context, form, content, and style. Therefore, we injected this information into the description using a predefined textual template in which it was embedded.

To fine-tune a single image encoder, we employed a multi-classification setup by adding a linear projection of the embedding corresponding to the `[CLS]` patch as the classification head for each task. Artists with fewer than 100 associated artworks were assigned the class *other*, while media and tags with fewer than 100 associated artworks were ignored. To counteract the problem of class imbalance in multi-class classifications, each class was associated with a weight inversely proportional to its frequency in the loss calculation. Each of the five classification tasks, whether multi-class or multi-label, was associated with its own cross-entropy loss $CE_i, i \in \{1, ..., 5\}$ during training. In traditional multi-task learning, these losses are typically aggregated by summing them with empirical weights, often determined by trial and error, which can be costly and time-consuming. To overcome this problem, we opted for a more efficient approach by allowing the model to learn task weights, using the uncertainty-based approach described in [17]. By taking advantage of this method, we avoided manually adjusting the weights, making the training process more streamlined and efficient. Specifically, our model was trained using the combined loss:

$$\ell = \sum_{i=1}^{5} \frac{CE_i}{\sigma_i^2} + \sum_{i=1}^{5} \log \sigma_i$$

where σ_i^2 values represent the task variances, which are used as weights to adjust the contribution of each task to the overall loss. The model learns these weights through backpropagation (particularly, we allowed the model to learn $\log(\sigma_i^2)$ for numerical stability).

It is essential to mention that for examples without associated tags or media in the dataset, the tags and media losses were ignored. This means the model was not penalized for predicting tags or media for instances lacking these annotations, either due to missing annotations or the removal of infrequent tags or media.

5 Experiments

5.1 Experimental Setting

For all experiments, we divided the entire dataset into training, validation, and test set using a 70/15/15 stratified split on the *genre* attribute to distribute the data variability equally among the three splits. All images were treated at a resolution of 224×224. We excluded examples with a CLIPScore less than 0.15 from the validation and test sets.

For caption generation, we fine-tuned several versions of GIT-Base, using a batch size of 64 (simulated using gradient accumulation). The learning rate was warmed up for the first 500 steps and followed a cosine decay to 0 for five epochs. The optimizer was AdamW with $\beta_1 = 0.9$ and $\beta_2 = 0.999$. Slight image augmentations were applied to the images (large-scale random crops, random horizontal flips and rotations, color jittering), with one epoch corresponding to three passes over the training samples, considering one of the three associated captions and an image variation as an example. All GIT-Base trainings were stopped after three epochs due to improvements of less than 1% in BLEU-1 on the validation set to save computational time.

Table 1 shows the training configurations for the tested caption generators. We conducted an ablation study to investigate the impact of different choices on the results. In addition to the basic version of GIT-Base, we trained three other variants: one without using instance weights (nw), one with frozen image encoder and word embeddings (fr), and finally, one with an increased CLIPScore threshold to exclude bad examples from the training set (gs). We also used the pre-trained version of GIT-Base without further fine-tuning on our dataset (nft) to establish an image captioning baseline for our work.

Table 1. Training configurations for caption generators (c corresponds to the instance's CLIPScore).

Model	Threshold	Train images	Learning rate	Instance weight	Encoder
GIT-Base	0.15	80, 127	4.5×10^{-7}	$8c - \frac{1}{5}$	Not frozen
GIT-Base-nw	0.15	80, 127	9.0×10^{-7}	No	Not frozen
GIT-Base-fr	0.15	80, 127	4.5×10^{-7}	$8c - \frac{1}{5}$	Frozen
GIT-Base-gs	0.25	47, 924	4.5×10^{-7}	$\frac{20}{3}c - \frac{2}{3}$	Not frozen

For metadata classification, we fine-tuned ViT-B/16 with five classification heads. We used a batch size of 32 (simulated using gradient accumulation), the AdamW optimizer with $\beta_1 = 0.9$ and $\beta_2 = 0.999$. The encoder's weights were frozen for the first five epochs, during which only the classification heads were trained with a learning rate of 10^{-3}. After that, we unfroze the Transformer and continued training with a lower learning rate of 5×10^{-5}. We applied regularization techniques to avoid overfitting, including gradient clipping to a maximum norm of 1 and dropout just before the classification heads, with a dropout probability of 0.3. We chose to keep the model with the highest average macro F1 score across the tasks on the validation set.

The models were trained on an NVIDIA Tesla P100 GPU. Our Python implementation used PyTorch and Hugging Face for fine-tuning the pre-trained models. For evaluation, we computed BLEU-N, SPICE, METEOR, ROUGE-L, CIDEr and CLIPScore for captioning, and accuracy and macro-averaged F1 score for metadata classification on the test set predictions.

5.2 Results

Table 2 shows the results obtained with our GIT-Base models on the entire test set. We compare the results of all versions of GIT-Base we tested to perform an ablation study of our artwork's visual captioning method. The table shows that the best results in traditional captioning metrics are obtained from GIT-Base without weighting the instances. The worst results are obtained by freezing the ViT-B encoder and word embeddings, suggesting that it is better to let the model modify the image and word representations according to our loss. We also show that in terms of image-text matching as measured by CLIPScore, each of our models outperforms the ground truth, with the best attempt obtained by CLIPScore weighting and with a higher threshold for the selection of training samples (GIT-Base-gs). When assessing the average CLIPScore on the test set, we discovered that our top configuration (GIT-Base-gs) achieved a slightly lower value than 0.28, which is still regarded as promising. It is worth noting that Schuhmann et al. [24] established a threshold of 0.3 as a significant benchmark for image-text alignment when creating their dataset. As expected, running GIT-Base-nft, i.e., the image captioner as is, results in poor performance in relation to both traditional metrics on our dataset and CLIPScore.

Table 2. Captioning results using greedy decoding (S: SPICE; B@N: BLEU-N; M: METEOR; RL: ROUGE-L; Cr: CIDEr; C: CLIPScore).

Model	S	B@1	B@2	B@3	B@4	M	RL	Cr	C
Ground truth	–	–	–	–	–	–	–	–	25.8
GIT-Base-nft	4.9	9.0	4.2	2.0	1.0	4.9	16.2	7.3	26.1
GIT-Base	10.0	35.3	20.7	12.4	7.6	12.0	**30.5**	31.8	26.9
GIT-Base-nw	**10.1**	**35.7**	**20.9**	**12.6**	**7.8**	**12.1**	**30.5**	**32.5**	26.6
GIT-Base-fr	9.0	35.1	20.1	11.7	7.0	11.5	29.7	27.1	26.1
GIT-Base-gs	9.9	33.9	19.8	11.7	7.2	11.6	30.0	30.1	**27.9**

Table 3. Classification results (Acc: Accuracy; F1: macro-averaged F1 score).

Model	Artist		Genre		Style		Tags	Media
	Acc	F1	Acc	F1	Acc	F1	F1	F1
ViT-B (multitask)	69.93%	58.63%	72.78%	65.94%	59.98%	57.41%	39.61%	53.55%

Regarding metadata classification (Table 3), our results align with the genre and style classification outcomes reported in [5], albeit employing a different approach for multi-task learning. Qualitative analysis revealed that artist classification is remarkably accurate for well-known artists. Additionally, we achieved favorable F1 scores for tags and media, considering that the model can identify tags not initially added to artworks by WikiArt annotators.

A qualitative evaluation (Fig. 2) demonstrates that our GIT-Base-gs can successfully identify well-known places, personalities, and different objects and activities in a wide range of artistic styles. However, the model still experiences hallucinations resulting from the noisy ground truth. For example, the model may identify objects that are not present. Additionally, the model can generate completely inaccurate captions when the content of the painting is too chaotic.

 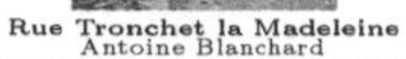 **Rue Tronchet la Madeleine** Antoine Blanchard	 **Harlequin** Pablo Picasso	 **The Lamentation** Gerard David	 **Masks** Jose Gutierrez Solana
Rue Tronchet la Madeleine, a busy street in Paris with its signature buildings. (0.28)	*A Harlequin wearing a hat, with his arms crossed.* (0.28)	*The Lamentation, with Mary holding the body of Jesus while other figures mourn around them.* (0.30)	*Three grotesque masks with exaggerated features and vivid colors, on a neutral background.* (0.21)
A street scene in front of the Capitol building in Washington D.C. (0.26)	*A woman with a guitar in her hair and a hat.* (0.24)	*Death by painting artist depicting the death of builder.* (0.25)	*A street scene in the city of Rome, Italy. The building in front is the house of the artist.* (0.24)
The artwork could be attributed to ***Antoine Blanchard****, in the* ***Cityscape*** *genre, showcasing the* ***Impressionism*** *style. It is associated with the following concepts:* ***houses-and-buildings****,* ***roads-and-vehicles****,* ***streets-and-squares****. It is presented in the medium of* ***canvas, oil****. The artwork depicts* ***a street scene in Paris, with people and buildings in the background****.* (0.26)	*The artwork could be attributed to* ***Pablo Picasso****, in the* ***Portrait*** *genre, showcasing the* ***Cubism*** *style. It is associated with the following concepts:* ***female-portraits****. It is presented in the medium of* ***canvas, oil****. The artwork depicts a* ***self-portrait of a man with a serious expression, wearing a hat and holding a guitar in his hand****.* (0.28)	*The artwork could be attributed to* ***Albrecht Durer****, in the* ***Religious painting*** *genre, showcasing the* ***Northern Renaissance*** *style. It is associated with the following concepts:* ***Christianity****,* ***Jesus-Christ****,* ***Mythology****,* ***saints-and-apostles****. It is presented in the medium of* ***oil****. The artwork depicts* ***the Virgin Mary and the child Jesus, surrounded by a group of people and trees, with a landscape in the background****.* (0.27)	*The artwork could be attributed to* ***Pablo Picasso****, in the* ***Genre painting*** *genre, showcasing the* ***Expressionism*** *style. It is presented in the medium of* ***canvas, oil****. The artwork depicts* ***a group of musicians playing music in a street with buildings in the background****.* (0.30)

Fig. 2. Captioning examples. We present captions from our ChatGPT-generated ground truth (first row), GIT-Base without fine-tuning using our dataset (second row), and our GIT-Base-gs, which includes predicted metadata aggregation. Alongside each caption, we provide the corresponding image's CLIPScore (for the last row, computed solely based on the visual caption).

6 Conclusion and Future Work

This paper introduced a new framework for artwork captioning. Our approach involves training a VLP model on the visual descriptions of artworks generated by ChatGPT. Moreover, it improves these descriptions by incorporating predicted metadata from *ArtGraph*, which provides valuable information about the artwork's style, form, and context. Our research shows that accurate captions can be generated by using instance filtering, loss weighting with CLIPScore, and leveraging the prior knowledge of a VLP model such as GIT-Base.

To further improve artwork captions, we can focus on several key areas. First, it is critical to improve the quality of the dataset used. Second, we can improve the accuracy of captions by using VLP models with more parameters, and pre-trained on a more extensive collection of images and texts. Another possibility for improvement is the incorporation of contextual information into captions. This can be achieved by integrating external knowledge through template-based or context-based approaches. Template-based approaches use predefined structures to include contextual information, as demonstrated in a previous paper [2], while context-based approaches directly integrate external knowledge during the caption generation process. Finally, adding emotional information to captions can significantly increase reader engagement [1].

Acknowledgment. The research of Raffaele Scaringi is funded by a Ph.D. fellowship within the framework of the Italian "D.M. n. 352, April 9, 2022" - under the National Recovery and Resilience Plan, Mission 4, Component 2, Investment 3.3 - Ph.D. Project "Automatic analysis of artistic heritage via Artificial Intelligence", co-supported by "Exprivia S.p.A." (CUP H91I22000410007).

References

1. Aslan, S., Castellano, G., Digeno, V., Migailo, G., Scaringi, R., Vessio, G.: Recognizing the emotions evoked by artworks through visual features and knowledge graph-embeddings. In: Mazzeo, P.L., Frontoni, E., Sclaroff, S., Distante, C. (eds.) Image Analysis and Processing. ICIAP 2022. LNCS, vol. 13373, pp. 129–140. Springer, Cham (2022). https://doi.org/10.1007/978-3-031-13321-3_12
2. Bai, Z., Nakashima, Y., Garcia, N.: Explain me the painting: multi-topic knowledgeable art description generation. In: Proceedings of the IEEE/CVF International Conference on Computer Vision, pp. 5422–5432 (2021)
3. Becattini, F., et al.: VISCOUNTH: a large-scale multilingual visual question answering dataset for cultural heritage. ACM Trans. Multim. Comput. Commun. Appl. (2023). Just Accepted
4. Bongini, P., Becattini, F., Del Bimbo, A.: Is GPT-3 all you need for visual question answering in cultural heritage? In: Karlinsky, L., Michaeli, T., Nishino, K. (eds.) ECCV 2022. LNCS, vol. 13801, pp. 268–281. Springer, Cham (2023). https://doi.org/10.1007/978-3-031-25056-9_18
5. Castellano, G., Digeno, V., Sansaro, G., Vessio, G.: Leveraging knowledge graphs and deep learning for automatic art analysis. Knowl.-Based Syst. **248**, 108859 (2022)
6. Castellano, G., Vessio, G.: Deep learning approaches to pattern extraction and recognition in paintings and drawings: an overview. Neural Comput. Appl. **33**(19), 12263–12282 (2021)
7. Cetinic, E.: Towards generating and evaluating iconographic image captions of artworks. J. Imaging **7**(8), 123 (2021)
8. Chen, D., Fisch, A., Weston, J., Bordes, A.: Reading Wikipedia to answer open-domain questions. arXiv preprint arXiv:1704.00051 (2017)
9. Cornia, M., Stefanini, M., Baraldi, L., Cucchiara, R.: Meshed-memory transformer for image captioning. In: Proceedings of the IEEE/CVF Conference on Computer Vision and Pattern Recognition, pp. 10578–10587 (2020)
10. Devlin, J., Chang, M.W., Lee, K., Toutanova, K.: BERT: pre-training of deep bidirectional transformers for language understanding. arXiv preprint arXiv:1810.04805 (2018)

11. Dosovitskiy, A., et al.: An image is worth 16x16 words: transformers for image recognition at scale. arXiv preprint arXiv:2010.11929 (2020)
12. Fang, Z., et al.: Injecting semantic concepts into end-to-end image captioning. In: Proceedings of the IEEE/CVF Conference on Computer Vision and Pattern Recognition, pp. 18009–18019 (2022)
13. Garcia, N., Vogiatzis, G.: How to read paintings: semantic art understanding with multi-modal retrieval. In: Proceedings of the European Conference on Computer Vision (ECCV) Workshops (2018)
14. Herdade, S., Kappeler, A., Boakye, K., Soares, J.: Image captioning: transforming objects into words. Adv. Neural Inf. Process. Syst. **32** (2019)
15. Hessel, J., Holtzman, A., Forbes, M., Bras, R.L., Choi, Y.: CLIPScore: a reference-free evaluation metric for image captioning. arXiv preprint arXiv:2104.08718 (2021)
16. Ishikawa, S., Sugiura, K.: Affective image captioning for visual artworks using emotion-based cross-attention mechanisms. IEEE Access **11**, 24527–24534 (2023)
17. Kendall, A., Gal, Y., Cipolla, R.: Multi-task learning using uncertainty to weigh losses for scene geometry and semantics. In: Proceedings of the IEEE Conference on Computer Vision and Pattern Recognition, pp. 7482–7491 (2018)
18. Lita, L.V., Ittycheriah, A., Roukos, S., Kambhatla, N.: tRuEcasIng. In: Proceedings of the 41st Annual Meeting of the Association for Computational Linguistics, pp. 152–159. Association for Computational Linguistics, Sapporo (2003)
19. Liu, F., Zhang, M., Zheng, B., Cui, S., Ma, W., Liu, Z.: Feature fusion via multi-target learning for ancient artwork captioning. Inf. Fusion **97**, 101811 (2023)
20. Lu, Y., Guo, C., Dai, X., Wang, F.Y.: Data-efficient image captioning of fine art paintings via virtual-real semantic alignment training. Neurocomputing **490**, 163–180 (2022)
21. OpenAI: ChatGPT (2023). https://openai.com, version 3.5
22. Radford, A., et al.: Learning transferable visual models from natural language supervision. In: International Conference on Machine Learning, pp. 8748–8763. PMLR (2021)
23. Ruta, D., et al.: StyleBabel: artistic style tagging and captioning. In: Avidan, S., Brostow, G., Cissé, M., Farinella, G.M., Hassner, T. (eds.) ECCV 2022. LNCS, vol. 13668, pp. 219–236. Springer, Cham (2022). https://doi.org/10.1007/978-3-031-20074-8_13
24. Schuhmann, C., et al.: LAION-400M: open dataset of CLIP-filtered 400 million image-text pairs. arXiv preprint arXiv:2111.02114 (2021)
25. Sheng, S., Moens, M.F.: Generating captions for images of ancient artworks. In: Proceedings of the 27th ACM International Conference on Multimedia (MM 2019), pp. 2478–2486. Association for Computing Machinery, New York (2019)
26. Sirisha, U., Chandana, B.S.: Semantic interdisciplinary evaluation of image captioning models. Cogent Eng. **9**(1), 2104333 (2022)
27. Stefanini, M., Cornia, M., Baraldi, L., Corsini, M., Cucchiara, R.: Artpedia: a new visual-semantic dataset with visual and contextual sentences in the artistic domain. In: Ricci, E., Rota Bulò, S., Snoek, C., Lanz, O., Messelodi, S., Sebe, N. (eds.) Image Analysis and Processing - ICIAP 2019. LNCS, vol. 11752, pp. 729–740. Springer, Cham (2019). https://doi.org/10.1007/978-3-030-30645-8_66
28. Tiedemann, J., Thottingal, S.: OPUS-MT - building open translation services for the world. In: Proceedings of the 22nd Annual Conferenec of the European Association for Machine Translation (EAMT), Lisbon (2020)
29. Turkerud, I.R., Mengshoel, O.J.: Image captioning using deep learning: text augmentation by paraphrasing via backtranslation. In: 2021 IEEE Symposium Series on Computational Intelligence (SSCI), pp. 01–10 (2021)

30. Vaswani, A., et al.: Attention is all you need. Adv. Neural Inf. Process. Syst. **30** (2017)
31. Wang, J., et al.: GIT: a generative image-to-text transformer for vision and language. arXiv preprint arXiv:2205.14100 (2022)

Progressive Keypoint Localization and Refinement in Image Matching

Fabio Bellavia[1(✉)], Luca Morelli[2,4], Carlo Colombo[3], and Fabio Remondino[2]

[1] University of Palermo, Palermo, Italy
fabio.bellavia@unipa.it
[2] Bruno Kessler Foundation (FBK), Trento, Italy
{lmorelli,remondino}@fbk.eu
[3] University of Florence, Florence, Italy
carlo.colombo@unifi.it
[4] University of Trento, Trento, Italy

Abstract. Image matching is the core of many computer vision applications for cultural heritage. The standard image matching pipeline detects keypoints at the beginning and freezes them until bundle adjustment, by which keypoints are allowed to move in order to improve the overall scene estimation. Recent deep image matching approaches do not follow this scheme, historically imposed by computational limits, and progressively refine the localization of the matches in a coarse-to-fine manner.

This paper investigates the use of traditional computer vision approaches based on template matching to update the keypoint position throughout the whole matching pipeline. In order to improve the accuracy of the template matching, the usage of the coarse-to-fine refinement is explored and a novel normalization strategy for the local keypoint patches is designed. Specifically, the proposed patch normalization assumes a local piece-wise planar approximation of the scene and warps the corresponding patches according to a "middle homography", so that, after normalization, patch distortion is roughly equally distributed within the two original patches.

The experimental comparison of the considered approaches, mainly focused on cultural heritage scenes but straightforwardly generalizable to other common scenarios, shows the strengths and limitations of each evaluated method. This analysis indicates promising and interesting results for the investigated approaches, which can effectively be deployed to design better image matching solutions.

Keywords: Image matching · Keypoint refinement · Cross correlation · Middle homography · Patch normalization · Pixel-Perfect SfM · Cultural Heritage

G. L. Foresti et al. (Eds.): ICIAP 2023 Workshops, LNCS 14366, pp. 322–334, 2024.
https://doi.org/10.1007/978-3-031-51026-7_28

1 Introduction

1.1 Image Matching Perspectives

Image matching plays a key role in computer vision [26] and photogrammetric applications designed for cultural heritage and archaeology [11]. Among these, Structure-from-Motion (SfM) is generally devised as a downstream task of image matching and its advancements are significantly contributing to document and digitally preserve archaeological artifacts and 3D art works [8]. In order to obtain high-quality digital models replicating the geometry and texture of the original objects with accurate details, image matching needs to register images so that the localization precision of corresponding matches is the highest possible [16].

1.2 Common Ground of Deep and Non-deep Image Matching

Thanks to the ever increasing availability of both data and computational resources, the rise of deep learning has led to impressive advancements in computer vision and its sub-fields, including image matching. State-of-the-Art (SotA) deep image matching includes sparse methods such as SuperGlue [24], Accurate and Lightweight Keypoint Detection and Descriptor Extraction (ALIKE) [28] and Accurate Shape and Localization Features (ASLFeat) [20], or semi-dense methods such as Local Feature Transformer (LoFTR) [25] and the more recent Dense Kernelized Feature Matching (DKM) [7]. The mentioned approaches are end-to-end architectures, whose main advantage with respect to pipelines composed by standalone, separate modules is to allow a global optimization and synchronization of the process. Nonetheless, current end-to-end image matching methods are the final results of the efforts made by the research community on each individual part of the matching pipeline, which can be summarized in terms of multiple reiterations of these steps: keypoint detection, patch normalization, feature description extraction and matching.

Keypoint Detection. Traditional keypoint detectors combines image derivatives to define corners and Difference-of-Gaussian (DoG) blobs, extracted by the popular handcrafted Harris [13] and Scale Invariant Feature Transform (SIFT) [19] detectors, respectively. Filters designed according to the above functions of the image derivatives are applied to the images and the peaks in the filter response maps obtained by Non-Maximum Suppression (NMS) provide the final keypoints. The Keypoint Network (Key.Net) [2] was the first to introduce the softmax operator, that enables differentiable NMS on the filter response maps, obtained from learned convolutional layers but also by explicitly including first and second order derivatives of the input image. Moreover, differentiable NMS is used to achieve sub-pixel precision in ALIKE [28] or analogously to refine the matches established by correlation by LoFTR [25]. The basic idea for the sub-pixel keypoint estimation is to interpolate the discrete response map around the local neighborhood of the peak so as to obtain the true maximum. Classic approaches use parabolic interpolations [27] or approximate the response map by its derivative as in the case of SIFT [19]. Deep sub-pixel estimation acts instead as a Gaussian process regression interpolation, explicitly employed in DKM [7].

Patch Normalization. Patch normalization warps the local neighborhood of the keypoints so that patches become roughly aligned in order to compare them. The main assumption in the non-deep approaches is that any general spatial or radiometric transformation can be locally approximated by a simpler one with less degrees of freedom. Normalization by the mean and standard deviation of the intensity values of the patch is generally sufficient to achieve good radiometric invariance [19]. Robust spatial patch normalization is instead more complex to obtain. In the case of SIFT, normalization assumes to work with patches related only by a similarity (scale and orientation) transformation [19], and experiences decreasing performances in the presence of more severe perspective distortions. Local affine normalization [21] better tolerates these scene configurations. SotA affine patch normalization is achieved by the deep Affine Network (AffNet) [23] so that similar patches are clustered together in the transformed space of the normalized patches. This is not achieved explicitly according to some patch characteristic, e.g. edge shapes, but implicitly using hard negative mining triplet loss introduced in the Hard Network (HardNet) [22] descriptor. ASLFeat extends deep patch normalization on dense maps by employing Deformable Convolutional Networks (DCNs) [15] which basically act in two steps: first DCNs look locally inside the patch, then according to the gathered information decide the shape of the convolution filter to use. While a better shape adaptability is guaranteed, this is still dependent from the local data. This dependency has been surpassed by transformers, successfully employed by SuperGlue and LoFTR, which extract relations between distant image areas.

Feature Description Extraction and Matching. Keypoint description extracts features able to compare the keypoint local patches. Ideally, in the case of perfectly registered patches and in absence of noise, the cross correlation of the normalized image patches would be the optimal choice. In real scenarios, robust handcrafted feature descriptors are generally based on histograms of the orientations of the image gradient, as for SIFT [19], candidate matches are established by Nearest Neighbor (NN) strategies [3], and final matches are obtained by robust correspondence filtering based on spatial constraints through RANdom SAmple Consensus (RANSAC) [9]. Spatial constraints include strong ones such as planarity and stereo epipolar geometry [14], or loose constraints such as the spatial neighborhood consistency used in Adaptive Locally-Affine Matching (AdaLAM) [6] and Delaunay Triangulation Matching (DTM) [3]. Since the original aim of deep architectures is to extract features, feature descriptors were the first components of the image matching pipeline to be successfully implemented by deep networks. HardNet is a deep SotA standalone feature descriptor which extracts features by processing the patch through successive convolutional layers. Conversely, effective keypoint matching was accomplished by deep learning only later. The first architecture to succeed was SuperGlue, which employs the Sinkhorn algorithm behaving as a differentiable NN matching and graph neural networks (of which the transformer can be regarded as a later and lightweight version) to infer and apply spatial constraints to the matches. Match similarity is measured by the correlation in the feature space of the corresponding patches.

Image Matching Pipeline Evolution. A last, essential characteristic that has contributed to the success of end-to-end deep image architectures is to be sought in the deep structure of the networks composed by a sequence of stacked layers. Even without explicitly designing the network to have a coarse-to-fine architecture as for LoFTR [25] and DKM [7], the deep structure allows to progressively and successively refine the matching process. On the one hand, this can be associated to multiple successive passes of a base matching pipeline. On the other hand, patch normalization and the effective keypoint matching can be thought of as the same image matching process at micro and macro levels, respectively: inside a patch, point-like features are extracted and matched according to spatial constraints to get their correspondences and to decide if the patches match; inside the whole image, patch-like features are extracted and matched according to spatial constraints to get their correspondences and to decide if the images match. Pixel-Perfect SfM [18] is a deep architecture which extends this idea to the whole SfM pipeline, as it takes keypoint tracks computed by image matching on multiple image pairs and refines them both before SfM and after on the basis of the SfM output together with the 3D coordinates of the keypoints.

1.3 Paper Contribution

The aim of this paper is to investigate how to improve match localization accuracy according to the aforementioned design concepts, yet avoiding the use of deep architectures. The idea is to let every step of the matching pipeline to be explicitly described in an algorithmic way. Such analysis of the process can contribute to implement optimized handcrafted matching pipelines and to understand better and improve deep matching architectures.

The main idea is to re-process image matches already extracted by the matching pipeline. For this objective, template matching approaches [10] which require a robust initial solution can be used since after the first pass raw matches have been roughly detected. Moreover, raw matches define a planar piece-wise approximation of the scene in the local neighborhood of the match, which is more general and adheres better to the actual warping than an affine transformation. Patch normalization is updated according to the outcomes of the previous pass in order to improve the template matching. Worth to note that upgrading patch local transformation was already shown to be effective at improving the estimated scene structure [1]. Sub-pixel registration of the patch is also considered in order to refine the matches. Finally, multiple passes of the match localization refinement are considered too.

The rest of the paper is organized as follows. The different base modules employed for the match refinement are described in detail in Sect. 2, while the experimental analysis is presented and discussed in Sect. 3. Conclusions and future work are provided in Sect. 4.

2 Match Refinement Base Modules

2.1 Normalized Cross Correlation (NCC) Matching

Given two images $I_1, I_2 : \mathbb{R}^2 \rightarrow \mathbb{R}$ and a coarse match $(\mathbf{x}, \mathbf{x}')$, where $\mathbf{x}, \mathbf{x}' \in \mathbb{R}^2$ are the corresponding keypoint coordinates in the two images, NCC for the patches centered on the given keypoints with radius r is [10]

$$\mathcal{C}^r_{\mathbf{x},\mathbf{x}'} = \sum_{\|\Delta\|_\infty \leq r} \frac{(I_1(\mathbf{x}+\Delta) - \mu_{I_{1\{\mathbf{x},r\}}})(I_2(\mathbf{x}'+\Delta) - \mu_{I_{2\{\mathbf{x}',r\}}})}{\sigma_{I_{1\{\mathbf{x},r\}}} \sigma_{I_{2\{\mathbf{x}',r\}}}} \tag{1}$$

with μ_S, σ_S indicating the mean and standard deviation over the set S, respectively, and

$$I_{i\{\mathbf{w},r\}} = \{I_i(\mathbf{w}+\Delta) : \| \Delta \|_\infty \leq r\} \tag{2}$$

representing the patch centered in $\mathbf{w}$ as a set. Assuming I_1 as reference, the keypoint on I_2 is updated by cross correlation as $\mathbf{x}' + \Delta^\star$, where the discrete offset $\Delta^\star$ maximizes the correlation between the two patches, i.e.

$$\Delta^\star = \underset{\|\Delta'\|_\infty \leq r}{\operatorname{argmax}}\, \mathcal{C}^r_{\mathbf{x},\mathbf{x}'+\Delta'} \tag{3}$$

Notice that NCC is invariant to local affine illumination changes, as the intensity values are normalized by the mean and standard deviation over the local patch windows. Moreover, NNC provides a response map by which to refine the keypoint by sub-pixel interpolation.

2.2 Adaptive Least Square (ALS) Correlation Matching

ALS correlation [12] performs an iterative affine registration between the two patches. The aim of the method is to estimate an affine patch warping $\mathrm{A} \in \mathbb{R}^{2\times3}$ and an affine transformation of the intensity values $\mathrm{L} \in \mathbb{R}^{1\times2}$ to register the two patches. Defining $\tilde{\mathbf{z}} = [\mathbf{z}\ 1]^\mathrm{T}$ as the normalized homogenous vector associated to $\mathbf{z}$, the registration error is given by

$$\mathcal{E}^{\theta,r}_{\mathbf{x},\mathbf{x}'} = \sum_{\|\Delta\|_\infty \leq r} f_k(\theta) = \sum_{\|\Delta\|_\infty \leq r} \| I_1(\mathbf{x}+\Delta) - \mathrm{L}\tilde{I_2}(\mathrm{A}\tilde{\mathbf{x}}' + \Delta) \|^2 \tag{4}$$

where $\theta = \{\mathrm{A}, \mathrm{L}\}$ indicates the transformation parameters and $\Delta = [i\ j]^\mathrm{T}$ such that $k = (i+r) + (2r+1)(j+r)$ is an univocal linear index for each pixel of the patch. Assuming I_1 as reference, the keypoint on I_2 is updated as $\mathrm{A}^\star\tilde{\mathbf{x}}'$, where $\mathrm{A}^\star$ minimized the patch error, i.e.

$$\mathrm{A}^\star = \underset{\mathrm{A}\in\mathbb{R}^{2\times3}}{\operatorname{argmin}}\, \mathcal{E}^{\theta,r}_{\mathbf{x},\mathbf{x}'} \tag{5}$$

The best parameter set $\theta^\star = \{\mathrm{A}^\star, \mathrm{L}^\star\}$ is found by non-linear least square minimization, which is basically the gradient descent employed in deep learning.

The initial configuration $\theta' = \{\mathrm{A}' = [\mathbf{I}\ \mathbf{0}], \mathrm{L}' = [1\ 0]\}$ assumes that the original patches are almost registered, and the errors $f_k(\theta)$ are approximated linearly by Taylor expansions as

$$f_k(\theta) = f_k(\theta' + \Delta_\theta) = f_k(\theta') + \frac{\partial f_k}{\partial \theta}\Delta_\theta \tag{6}$$

The minimal error solution is then equivalent to

$$\mathrm{F}(\theta') + \mathrm{J}_\theta \Delta_\theta = \mathbf{0} \tag{7}$$

where $\mathrm{F}(\theta') = \left[f_1(\theta') \ \cdots \ f_{(2r+1)^2}\right]^{\mathrm{T}}$ and $\mathrm{J}_\theta = \left[\frac{\partial f_1}{\partial \theta} \ \cdots \ \frac{\partial f_{(2r+1)^2}}{\partial \theta}\right]^{\mathrm{T}}$ is the Jacobian matrix obtained from the discrete derivatives so that

$$\varDelta_\theta = -\mathrm{J}_\theta^{+}\mathrm{F}(\theta') \tag{8}$$

and the parameters of the transformation get updated iteratively until convergence as $\theta' + \alpha\varDelta_\theta$, where $\alpha = 0.5$ is introduced to prevent that the solution diverges due to the forced linearization of the error function.

2.3 Fast Affine Template Matching (FAsT-Match)

FAsT-Match [17] patch error formulation is analogous to ALS correlation but the Sum-of-Absolute-Differences (SAD)

$$\mathcal{A}_{\mathbf{x},\mathbf{x}'}^{\theta,r} = \sum_{\|\varDelta\|_\infty \le r} |I_1(\mathbf{x} + \Delta) - \mathrm{L}\tilde{I}_2(\mathrm{A}\tilde{\mathbf{x}}' + \varDelta)| \tag{9}$$

is used instead of the Euclidean distance. Again, assuming I_1 as reference, the keypoint on I_2 is updated by $\mathrm{A}^\star\tilde{\mathbf{x}}'$, where $\mathrm{A}^\star$ minimizes the error between the two patches, i.e.

$$\mathrm{A}^\star = \operatorname*{argmin}_{\mathrm{A}\in\mathbb{R}^{2\times 3}} \mathcal{A}_{\mathbf{x},\mathbf{x}'}^{\theta,r} \tag{10}$$

Unlike ALS correlation, which assumes continuous functions, FAsT-Match exploits discretization by partitioning the space of the allowable transformations and employing a branch-and-bound strategy to efficiently explore the solution spaces and find the best transformation. The vertical sub-pixel offset derivation is similar. FasT-Match is computationally intensive so that in the evaluation to bound the running times default parameters were set to $\epsilon = 0.5$ and $\delta = 0.75$ respectively, and the allowable scale factor to 3 and the orientation range were limited by $\pm\frac{\pi}{3}$. These settings improve the running times with no accuracy loss.

2.4 Parabolic Sub-pixel Peak Interpolation

Parabolic interpolation refines NNC response map $D(u,v) = \mathcal{C}_{\mathbf{x},\mathbf{x}'+\varDelta'}^{r}$, where $\varDelta' = [u\ v]^{\mathrm{T}}$, as follows. Assume that $\mathbf{x}'$ has been updated as described in

Sect. 2.1, so that the patch is centered in the peak, i.e. $\Delta^\star = \mathbf{0}$. Then, the keypoint sub-pixel offset is computed as

$$\Delta p = \begin{bmatrix} -\frac{b}{2a} & -\frac{b'}{2a'} \end{bmatrix}^{\mathrm{T}} \tag{11}$$

The horizontal sub-pixel offset corresponds to the vertex x-coordinate of the parabola $ax^2 + bx + c = y$, interpolated from the 3 points

$$P_d = (d, \mathcal{C}^r_{\mathbf{x},\mathbf{x}'+[0\ d]^{\mathrm{T}}}) = (d, y_d), \qquad d \in \{-1, 0, 1\} \tag{12}$$

along the horizontal dimension of D, which leads to $a = \dfrac{y_1 - 2y_0 + y_{-1}}{2}$ and $b = \dfrac{y_1 - y_{-1}}{2}$. The vertical sub-pixel offset is computed analogously.

2.5 Taylor Approximation Sub-pixel Peak Interpolation

This adapts the SIFT detector sub-pixel precision method [19]. In this case, the second order Taylor expansion of the response map gives around the peak $\Delta^\star$

$$D(\Delta') = D(\Delta^\star + \Delta_l) = D(\Delta^\star) + \frac{\partial D^{\mathrm{T}}}{\partial \Delta'}\Delta_l + \frac{1}{2}\Delta_l^{\mathrm{T}} \mathrm{H}_{\Delta'} \Delta_l \tag{13}$$

where $\mathrm{H}_{\Delta'}$ is the Hessian matrix of D, computed by discrete derivatives. The maximum is achieved when the derivative of $D(\Delta')$ is zero, i.e. when

$$\frac{\partial D^{\mathrm{T}}}{\partial \Delta'} + \mathrm{H}_{\Delta'}\Delta_l = 0 \tag{14}$$

which implies that the requested sub-pixel correction offset is

$$\Delta_l = -\mathrm{H}^{-1}_{\Delta'} \frac{\partial D^{\mathrm{T}}}{\partial \Delta'} \tag{15}$$

Actually, in the original SIFT paper, the offset space is 3D, since the DoG filter operates also on scales.

2.6 Middle Homography (MiHo) Patch Normalization Updating

Patch normalization updating assumes that a set of matches $M = \{(\mathbf{x}, \mathbf{x}')\}$ has been obtained after the first matching pipeline pass. It also assumes that matches for the subset $P \subseteq M$ are related by a planar homography $\tilde{\mathbf{x}}' = \mathrm{H}^\star \tilde{\mathbf{x}}$, where $\mathrm{H}^\star \in \mathbb{R}^{3\times3}$ is non-singular, using the same conventions of [14]. The idea of MiHo is to find an associated pair of planar homographies $(\mathrm{H}, \mathrm{H}')$ so that

$$\tilde{\mathbf{m}} = \mathrm{H}\tilde{\mathbf{x}} \quad \wedge \quad \tilde{\mathbf{m}} = \mathrm{H}'\tilde{\mathbf{x}}', \qquad \forall (\mathbf{x}, \mathbf{x}') \in P \tag{16}$$

where $\mathbf{m} = \frac{\mathbf{x}+\mathbf{x}'}{2}$. As shown in Fig. 1a, this heuristic procedure inspired by [4] tends to distribute equally the distortion error over the two patches when these are normalized by H and H′, respectively. Since interpolation degrades with up-sampling, MiHo aims to provide a balance with down-sampling the patch at finer resolution and up-sampling the patch at the coarser resolution. Also, a planar homography is a better local approximation than an affine transformation.

The Direct Linear Transform (DLT) [14] is used to find H and H′. Actually, MiHo pairs estimation can be repeated on the warped keypoint pairs $(\mathrm{H}\tilde{\mathbf{x}}, \mathrm{H}'\tilde{\mathbf{x}}')$ to refine the solution, by concatenating all the successive homographies. It was experimentally observed that three iterations generally suffice.

Defining an inlier match $(\mathbf{x}, \mathbf{x}')$ for a generic H according to the threshold r by the maximum reprojection error

$$\mathcal{P}^{\mathrm{H},r}_{\mathbf{x},\mathbf{x}'} = \begin{cases} 1 \text{ if } \max(\|\tilde{\mathbf{x}}'-\mathrm{H}\tilde{\mathbf{x}}\|,\|\tilde{\mathbf{x}}-\mathrm{H}^{-1}\tilde{\mathbf{x}}'\|)\leq r \\ 0 \quad \text{otherwise} \end{cases} \tag{17}$$

an inlier for the MiHo pair $(\mathrm{H}, \mathrm{H}')$ is straightforwardly defined as the product

$$\mathcal{P}^{(\mathrm{H},\mathrm{H}'),r}_{\mathbf{x},\mathbf{x}'} = \mathcal{P}^{\mathrm{H},r}_{\mathbf{x},\mathbf{m}} \mathcal{P}^{\mathrm{H}',r}_{\mathbf{x}',\mathbf{m}} \tag{18}$$

In order to discover simultaneously both the approximated planes on the scene and their associated MiHo pairs, the whole process is embedded into the RANSAC framework. Starting from $M_0 = M$, RANSAC is used at iteration i to extract the i-th best MiHo pair $(\mathrm{H}_i, \mathrm{H}'_i)$ using threshold r, and strong inliers are removed for the next iteration according to a stricter threshold $\frac{r}{2}$, i.e.

$$M_{i+1} = M_i \setminus \left\{(\mathbf{x}, \mathbf{x}') : \ (\mathbf{x}, \mathbf{x}') \in M_i \wedge \mathcal{P}^{(\mathrm{H}_i,\mathrm{H}'_i),\frac{r}{2}}_{\mathbf{x},\mathbf{x}'}\right\} \tag{19}$$

until $M_i = \emptyset$ or last MiHo has only 4 inliers, i.e. the minimum model size. Finally, since more than one MiHo pair can satisfy a match, the MiHo pair $(\mathrm{H}_{\mathbf{x},\mathbf{x}'}, \mathrm{H}'_{\mathbf{x},\mathbf{x}'})$ assigned to a match $(\mathbf{x}, \mathbf{x}')$ is the one with the larger consensus set

$$(\mathrm{H}_{\mathbf{x},\mathbf{x}'}, \mathrm{H}'_{\mathbf{x},\mathbf{x}'}) = \operatorname*{argmax}_{\mathcal{P}^{(\mathrm{H}_i,\mathrm{H}'_i),r}_{\mathbf{x},\mathbf{x}'}=1} \left(\sum_{(\mathbf{x}_j,\mathbf{x}'_j)\in M} \mathcal{P}^{(\mathrm{H}_i,\mathrm{H}'_j),r}_{\mathbf{x}_j,\mathbf{x}'_j} \right) \tag{20}$$

In case no MiHo pair is compatible with a match, the identity matrix will be used for the corresponding patch normalization.

Figure 1b shows an example of rough planes associated to a same MiHo pair. Notice that the input images should not be roughly rotated by 180° in order for MiHo to work. This can be understood considering the case when the global transformation within the images is close to a reflection through a point. In this case, the mid-points $\mathbf{m}$ corresponding to a match $(\mathbf{x}, \mathbf{x}')$ tend to accumulate about the center of reflection, thus providing a configuration close to degeneracy.

3 Evaluation

As shown in Fig. 1c, the evaluation dataset considers 12 image pairs representing scenes of interest for cultural heritage on which 20 matches $(\mathbf{x}, \mathbf{x}')$ have been manually selected by expert users as ground-truth (GT). The images have a resolution of 20 MegaPixel (MP), and the keypoint accuracy for the selected matches at the original resolution is up to 1 px. By down-scaling the images with a factor of 5, images maintain a feasible testing resolution and matches get a sub-pixel accuracy. Bilinear interpolation [10] is used to warp patches for its efficiency. Code and data are freely available in the additional material[1].

The patch radius is set to $r = 15$ px for NCC, ALS correlation and FAsT-Match. GT keypoints $\mathbf{x}$ on I_1 are used as reference, while keypoints $\mathbf{x}'$ on I_2 are perturbed by adding a noise offset of $n = 1, \ldots, 11$ px in one or both directions at testing resolution. Specifically, for a given noise offset n, 4 noisy matches $(\mathbf{x}, \mathbf{y}'_n)$ are obtained where

$$\mathbf{y}'_n \in \begin{cases} \mathbf{x}' + \left\{ n \begin{bmatrix} \pm 1 \\ 0 \end{bmatrix}, n \begin{bmatrix} 0 \\ \pm 1 \end{bmatrix} \right\} & \text{if } n \text{ is odd} \\ \mathbf{x}' + \left\{ n \begin{bmatrix} \pm 1 \\ \pm 1 \end{bmatrix}, n \begin{bmatrix} \pm 1 \\ \mp 1 \end{bmatrix} \right\} & \text{if } n \text{ is even} \end{cases} \tag{21}$$

for a total of $20 \times 11 \times 4 = 880$ tested keypoint matches for each image.

In order to evaluate MiHo patch normalization update, initial matches were estimated using SotA matching pipelines to which noisy matches were added. To make RANSAC plane discovery unrelated from the GT matches, matches within $2r$ of GT matches were removed before including the noisy matches, see Fig. 1b. The employed pipelines are Hz$^+$ [5] and Key.Net+AffNet+ HardNet+AdaLAM [2], both with and without upright constraints.

Table 1 shows the average keypoint shift error on the whole dataset for different noise offsets, ordered by their magnitude. Lighter bars indicate the error percentage with respect to the noise offset magnitude when less than 100%, darker bars when greater than 100%. NCC is used with no sub-pixel refinement, ⊞ indicates the base run with $r = 15$ px and ⊡⊞ a two-step coarse-to-fine run. Specifically, in the latter case in the first step the keypoint is coarsely refined at half testing resolution with $r = 7$ px, and in the next step the updated keypoint is refined again at full resolution with $r = 15$ px. MiHo initial matches have been estimated with Hz$^+$. Detailed results are reported in the additional material.

ALS correlation increases the accuracy only when the noise offset magnitude is limited, due to the fact that the image approximation by its derivatives is valid only in a small local neighborhood. NCC and FAst-Match absolute improvements generally do not depend on the noise. For NCC the absolute error is about 4, 2 px respectively without and with MiHo, for FAsT-Match this is 3 px. MiHo patch update remarkably helps NCC, roughly halving the error. On FAsT-Match MiHo improvements are lower, since the method itself uses affine adaptation. Nevertheless, the MiHo solution does better, which implies that planar homography approximation is better than the affine one. When ALS correlation decreases the error, MiHo normalization makes ALS behaves as FAsT-Match, since both approaches perform an affine warping. Coarse-to-fine two-step solutions ⊡⊞ degrade the localization with respect to the base approaches ⊡, besides

[1] https://drive.google.com/drive/folders/12jPMbU4doWoDRv57unBctjyxXUKhDHRF.
[2] https://kornia.github.io/

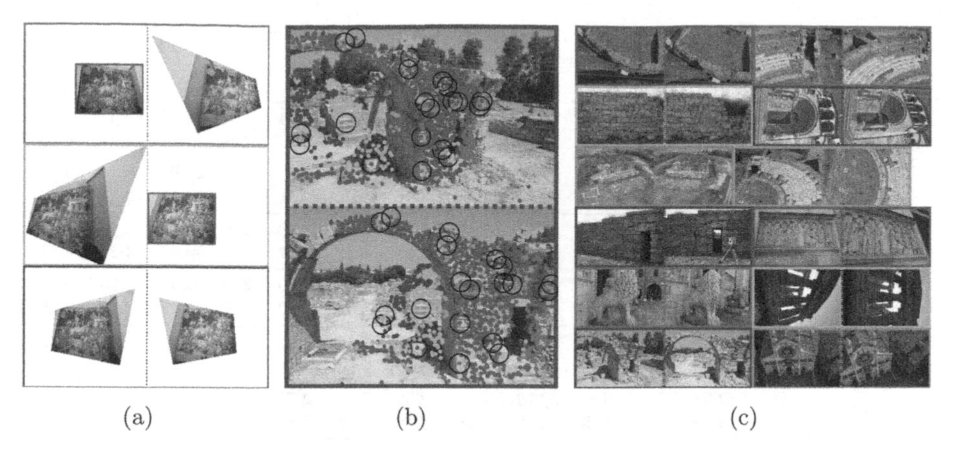

(a) (b) (c)

Fig. 1. (a) Differences between common homography warping and MiHo. I_1 and I_2 original and warped images are respectively on the left and right sides. In the top and middle rows the original image (framed in blue) is used as reference to apply the standard warping to the other image (sided), MiHo warped pairs are shown in the bottom row. (b) Corresponding clusters of raw planes associated to a same MiHo pair for an image pair. Matches are extracted by Hz^+, GT matches are highlighted by black circles. (c) Image pairs of the evaluation dataset (best viewed in color, for high resolution images see the additional material).

doubling the running times. Concerning running times, code was implemented in Matlab with no optimizations and was run on a Intel Core I9 10900K. The refinement of a single match takes 0.05 s for both ALS correlation and NCC, while it is close to 2 s for FAsT-Match. MiHo code is excluded from the current analysis. Clearly, multiple matches can be refined in parallel and code optimization could speed the computation.

According to this comparison, NCC with MiHo provides the best solution. Table 2 adds further experiments with NNC, by including parabolic and Taylor sup-pixel estimation. Moreover, a further two-step approach (indicated by ⊞⊠) is evaluated where the NCC solution is further refined by ALS correlation since the latter can better cope with small noise shifts. According to these results, both parabolic interpolation and ALS refinement provide modest incremental improvements, while Taylor sub-pixel offset generally degrades the base solution. Notice also that ALS refinement doubles the running times.

On average, no methods achieve a sub-pixel refinement, i.e. an error less than 1 px. Nevertheless, as reported by further analyses in the additional material, the situation is more articulated. Specifically, considering the sub-pixel accuracy in terms of percentages of keypoints with error less than 1 px after the refinement, ALS correlation achieves for noise offset magnitude less than 2 px values about 55%, 70% without and with MiHo, respectively. FAsT-Match percentage is stable around 40% in any case. For the base NCC, sub-pixel accuracy percentage is the same of FAsT-Match but increases to about 55%, 59% and 60% as MiHo, parabolic fitting and ALS correlation are incrementally included, respectively.

Table 1. Average keypoint shift error (px) on the whole dataset.

	Noise offset mag. (px)		1	$2\sqrt{2}$	3	5	$4\sqrt{2}$	7	$6\sqrt{2}$	9	11	$8\sqrt{2}$	$10\sqrt{2}$	avg.
ALS	no patch norm.	⊞	1.25	1.76	1.82	4.03	3.35	7.45	5.61	10.84	7.87	14.29	10.33	6.24
		⊡⊞	1.87	2.27	2.23	4.15	3.44	7.45	5.53	10.85	7.74	14.17	10.25	6.36
	MiHo	⊞	0.92	1.28	1.39	3.52	2.89	7.28	5.38	10.59	7.78	14.04	10.26	5.94
		⊡⊞	1.38	1.75	1.71	3.71	2.93	7.14	5.27	10.74	7.68	14.12	10.16	6.06
NCC	no patch norm.	⊞	3.62	3.64	3.64	3.57	3.66	3.70	3.69	3.76	3.78	3.81	3.78	3.70
		⊡⊞	4.04	4.00	4.06	4.08	4.09	4.25	4.22	4.37	4.28	4.57	4.38	4.21
	MiHo	⊞	1.75	1.92	1.79	2.06	2.00	2.14	2.07	2.20	2.19	2.44	2.25	2.07
		⊡⊞	1.93	2.09	1.99	2.23	2.12	2.40	2.30	2.40	2.40	2.72	2.50	2.28
FAsT-Match	no patch norm.	⊞	2.01	2.18	2.10	2.40	2.35	2.72	2.66	3.06	2.76	5.76	3.08	2.82
		⊡⊞	2.59	2.66	2.65	2.77	2.80	3.06	2.99	3.40	2.96	5.54	3.58	3.18
	MiHo	⊞	1.93	1.89	1.88	2.18	2.07	2.36	2.23	2.72	2.38	4.87	2.70	2.47
		⊡⊞	2.18	2.21	2.23	2.45	2.29	2.66	2.48	2.87	2.54	4.76	2.79	2.68

Table 2. Average keypoint shift error (px) of NCC sub-pixel on the whole dataset.

	Noise offset mag. (px)		1	$2\sqrt{2}$	3	5	$4\sqrt{2}$	7	$6\sqrt{2}$	9	11	$8\sqrt{2}$	$10\sqrt{2}$	avg.
none	no patch norm.	⊞	3.62	3.64	3.64	3.57	3.66	3.70	3.69	3.76	3.78	3.81	3.78	3.70
		⊞⊠	3.51	3.51	3.50	3.42	3.50	3.55	3.54	3.61	3.63	3.65	3.64	3.55
	MiHo	⊞	1.75	1.92	1.79	2.06	2.00	2.14	2.07	2.20	2.19	2.44	2.25	2.07
		⊞⊠	1.73	1.85	1.75	1.98	1.93	2.07	1.98	2.13	2.10	2.36	2.17	2.00
parabolic	no patch norm.	⊞	3.69	3.78	3.76	3.66	3.77	3.92	3.90	3.98	4.02	4.44	4.35	3.93
		⊞⊠	3.52	3.59	3.56	3.45	3.60	3.71	3.70	3.77	3.84	4.24	4.19	3.74
	MiHo	⊞	1.67	1.78	1.72	1.94	1.88	2.00	1.96	2.22	2.10	2.66	2.37	2.03
		⊞⊠	1.62	1.75	1.69	1.90	1.85	1.98	1.94	2.17	2.07	2.63	2.36	2.00
Taylor	no patch norm.	⊞	3.81	4.00	4.09	4.02	4.39	4.07	4.39	4.15	4.49	4.13	4.24	4.16
		⊞⊠	3.64	3.75	3.82	3.73	4.03	3.78	4.05	3.86	4.06	3.84	3.89	3.86
	MiHo	⊞	1.99	2.32	2.28	2.60	2.90	2.48	3.07	2.61	3.20	2.82	3.04	2.66
		⊞⊠	1.90	2.17	2.10	2.42	2.55	2.32	2.73	2.40	2.90	2.62	2.70	2.44

The results are in accordance with the previous observations, but also show that it is possible to achieve sub-pixel accuracy with the investigated approaches.

4 Conclusions and Future Works

This paper has presented a thorough comparative analysis of non-deep, conventional approaches to improve the localization accuracy of keypoint matching, focusing in particular on cultural heritage and archaeological scenes. The results suggest that patch normalization is crucial for improving the match localization and that simple NCC paired with parabolic fitting, and optionally ALS correlation, can provide promising results. Future works will focus on further analyses, incorporating the evaluated modules in practical applications, even between the pipeline steps. Moreover, extension to multi-view patches will be explored and comparisons with deep solutions will be carried out. MiHo results are also quite interesting and will be further investigated, also in the context of its applications to planar matching and benchmarking.

References

1. Barath, D.: On Making SIFT Features Affine Covariant. Int. J. Comput, Vis (2023)
2. Barroso-Laguna, A., Riba, E., Ponsa, D., Mikolajczyk, K.: Key. Net: keypoint detection by handcrafted and learned CNN filters. In: Proceedings of the International Conference on Computer Vision (ICCV) (2019)
3. Bellavia, F.: SIFT matching by context exposed. IEEE Trans. Pattern Anal. Mach. Intell. **45**(2), 2445–2457 (2023)
4. Bellavia, F., Colombo, C.: Estimating the best reference homography for planar mosaics from videos. In: Proceedings International Conference on Computer Vision Theory and Applications (VISAPP), pp. 512–519 (2015)
5. Bellavia, F., Mishkin, D.: HarrisZ$^+$: Harris corner selection for next-gen image matching pipelines. Pattern Recognit. Lett. **158**, 141–147 (2022)
6. Cavalli, L., Larsson, V., Oswald, M.R., Sattler, T., Pollefeys, M.: AdaLAM: revisiting handcrafted outlier detection. In: Proceedings of the European Conference on Computer Vision (ECCV) (2020)
7. Edstedt, J., Athanasiadis, I., Wadenbäck, M., Felsberg, M.: DKM: dense kernelized feature matching for geometry estimation. In: Proceedings of the IEEE Conference on Computer Vision and Pattern Recognition (CVPR) (2023)
8. Farella, E.M., Morelli, L., Grilli, E., Rigon, S., Remondino, F.: Handling critical aspects in massive photogrammetric digitalization of museum assets. Int. Arch. Photogram. Remote Sens. Spat. Inf. Sci. XLVI-2/W1-2022, 215–222 (2022)
9. Fischler, M., Bolles, R.: Random sample consensus: a paradigm for model fitting with applications to image analysis and automated cartography. Commun. ACM **24**(6), 381–395 (1981)
10. Gonzales, R., Woods, R.E.: Digital Image Processing. Pearson College Division, 4th edn. (2017)
11. Gruen, A., Remondino, F., Zhang, L.: Photogrammetric reconstruction of the Great Buddha of Bamiyan. Afghanistan. Photogramm. Rec. **19**(107), 177–199 (2004)
12. Gruen, A.W.: Adaptive least squares correlation: a powerful image matching technique. South Afr. J. Photogram. Remote Sens. and Cartogr. **14**(3), 175–187 (1985)
13. Harris, C., Stephens, M.: A combined corner and edge detector. In: Proceedings of the 4th Alvey Vision Conference, pp. 147–151 (1988)
14. Hartley, R.I., Zisserman, A.: Multiple View Geometry in Computer Vision. Cambridge University Press, 2st edn. (2000)
15. J. Dai, H.Q., Xiong, Y., Li, Y., Zhang, G., Wei, H.H.Y.: Deformable convolutional networks. In: Proceedings of the International Conference on Computer Vision (ICCV) (2017)
16. Karami, A., Menna, F., Remondino, F.: Combining photogrammetry and photometric stereo to achieve precise and complete 3D reconstruction. Sensors **22**(21), 8172 (2022)
17. Korman, S., Reichman, D., Tsur, G., Avidan, S.: Fast-Match: Fast affine template matching. In: Proceedings of the IEEE Conference on Computer Vision and Pattern Recognition (CVPR), pp. 1940–1947 (2013)

18. Lindenberger, P., Sarlin, P., Larsson, V., Pollefeys, M.: Pixel-perfect structure-from-motion with featuremetric refinement. In: Proceedings of the IEEE International Conference on Computer Vision (ICCV) (2021)
19. Lowe, D.: Distinctive image features from scale-invariant keypoints. Int. J. Comput. Vis. **60**(2), 91–110 (2004)
20. Luo, Z., et al.: ASLFeat: learning local features of accurate shape and localization. In: Proceedings of the IEEE Conference on Computer Vision and Pattern Recognition (CVPR) (2020)
21. Mikolajczyk, K., Schmid, C.: Scale and affine invariant interest point detectors. Int. J. Comput. Vis. **60**(1), 63–86 (2004)
22. Mishchuk, A., Mishkin, D., Radenovic, F., Matas, J.: Working hard to know your neighbor's margins: local descriptor learning loss. In: Proceedings of the Conference on Neural Information Processing Systems (NeurIPS) (2017)
23. Mishkin, D., Radenovic, F., Matas, J.: Repeatability is not enough: Learning affine regions via discriminability. In: Proceedings of the European Conference on Computer Vision (ECCV) (2018)
24. Sarlin, P.E., DeTone, D., Malisiewicz, T., Rabinovich, A.: SuperGlue: learning feature matching with graph neural networks. In: Proceedings of the IEEE Conference on Computer Vision and Pattern Recognition (CVPR) (2020)
25. Sun, J., Shen, Z., Wang, Y., Bao, H., Zhou, X.: LoFTR: detector-free local feature matching with transformers. In: Proceedings of the IEEE Conference on Computer Vision and Pattern Recognition (CVPR) (2021)
26. Szeliski, R.: Computer Vision: Algorithms and Applications. Springer-Verlag, 2nd edn. (2022)
27. Trucco, E., Verri, A.: Introductory Techniques for 3-D Computer Vision. Prentice Hall (1998)
28. Zhao, X., et al.: ALIKE: Accurate and lightweight keypoint detection and descriptor extraction. IEEE Trans. Multimed. early access (2022)

Toward a System of Visual Classification, Analysis and Recognition of Performance-Based Moving Images in the Artistic Field

Michael Castronuovo, Alessandro Fiordelmondo(✉), and Cosetta Saba

Department of Humanities and Cultural Heritage (DIUM), University of Udine (UNIUD), Udine, Italy
{michael.castronuovo,cosetta.saba}@uniud.it, fiordelmondo.alessandro@spes.uniud.it

Abstract. This paper proposes a research program focused on the design of a model for the recognition, analysis and classification of video art works and documentations based on their semiotic aspects and audiovisual content. Focusing on a corpus of art cinema, video art, and performance art, the theoretical framework involves bringing together semiotics, film studies, visual studies, and performance studies with the innovative technologies of computer vision and artificial intelligence. The aim is to analyze the performance aspect to interpret contextual references and cultural constructs recorded in artistic contexts, contributing to the classification and analysis of video art works with complex semiotic characteristics. Underlying the conceptual framework is the simultaneous use of a set of technologies, such as pose estimation, facial recognition, object recognition, motion analysis, audio analysis, and natural language processing, to improve recognition accuracy and create a large set of labeled audiovisual data. In addition, the authors propose a prototype application to explore the primary challenges of such a research project.

Keywords: Recognition system · Performance art · Video art · Visual studies · AI for audiovisual analysis

1 Introduction

The exponential growth of digitally native audiovisual goods and the digitization of analogue cultural heritage[1] require new epistemic methodologies, new

[1] Archival facilities in the GLAM (Galleries, Libraries, Archives and Museums) and MAB (Museums, Archives, Libraries) sectors are invested in the European Union's strategic program for digitization, preservation and online accessibility of cultural heritage, supported by the Plan for Recovery, which will be completed by 2030. https://commission.europa.eu/strategy-and-policy/priorities-2019-2024/europe-fit-digital-age/europes-digital-decade-digital-targets-2030_en (last accessed 17 August 2023).

G. L. Foresti et al. (Eds.): ICIAP 2023 Workshops, LNCS 14366, pp. 335–346, 2024.
https://doi.org/10.1007/978-3-031-51026-7_29

cataloging criteria and interoperable classification techniques that allow direct and immediate access to the visual content of works and to the records that archives preserve. The aim is to initiate a research program to design a model for the visual recognition, analysis and classification of moving images based on what they contain on the levels of expression (technical-linguistic aspects) and content.

The model involves the application of computer vision technologies and the development of convolutional neural networks (CNNs) for the visual recognition, analysis and classification of audiovisual databases. The visual aspects to be recognized and analyzed within images, image segments and image sequences concern the ways in which objects, subjects, actions, gestures are activated or are activated in performative contexts and their organization in relational form. The main field of investigation focuses on the audiovisual *corpora* of artists' cinema and video art at the intersection with performance art [9]. The latter was defined in the course of the 20th century and has become a cross-sectional practice to art making tout court. As argued below, this type of content will form the basis of the dataset to be developed. The main theoretical frameworks in the humanities pertain to semiotics, film studies, visual studies and performance studies.

Regarding computer science and the specific applications of artificial intelligence for the purpose of this research, computer vision and related innovative technologies will be the main development tools. In particular, the project proposes to adopt tools aimed at pose estimation, motion analysis, object and facial recognition.

Video-recording transforms performance into something else, as it makes it lose its inherent character of transience and ephemerality, while at the same time capturing and memorizing its event dimension: this makes it possible, for the purposes of classification and dataset construction, to study what happens in a performative action.

While the flagrancy of the event is lost, the processes of repetition, variation and improvisation/transformation through which the event is produced in the performative action can be analyzed. Moreover, in art-historical terms, through video recording there is the simultaneous operation of audiovisual and performative languages, which may manifest themselves in different forms or may confuse, hybridize, or interfere. On this basis, it is possible to define a distinction between "documentation" and "artwork": there is documentation when the performative work is not the video; the work, on the other hand, is given when the performance is also the video (videoperformance); in that case, nonetheless, the work is documenting itself (so to speak, intrinsically, it is "self-documenting") [20]. On the level of cultural transmission, the video recording (both as artwork and as document) can be used as a score for the reactivation of performance in "liveness" or technology-mediated mode.

2 Construction of the Research Object

The challenge that the research faces for the purpose of establishing a methodology for the classification of audiovisual corpora with a performative character is not to search for the moment in which the creative/inventional act finds expression. It concerns the recognition, analysis and interpretation of the theoretical and operational conditions by which a performative action finds execution [1,8,10,17].

The project takes its cue from a series of questions that generate a problematic field of research on the ontological level: How does the same gesture or action performed in an artistic context or a non-art context differ? Why in moving from one context to another do they change in sign and meaning? How does the influence of contexts manifest itself with respect to their institutional qualities (inside and outside museum protocols)? What are the cultural aspects - formal, expressive, and content - that come into play (codes, conventions, languages/metalanguages) in a performative action? The underlying exploratory hypothesis is that performativity posits the issue of defining the ontological status of the image with respect to what in the image is done and, more importantly, what the image in enunciating itself does [21].

Fig. 1. Marina Abramović, Ulay, *Work Relation*, 1978 – Video documentation: 28':45"; live performance: 180', Sala Polivalente, Centro Video Arte, Ferrara, October 1978 © Gallerie d'Arte Moderna e Contemporanea di Ferrara

The main audiovisual reference database is that of the University of Udine's La Camera Ottica laboratory, which records more than six thousand works and documentations (collections, fonds, A/D audiovisual archives of museum institutions). Data mining from audiovisuals is aimed at identifying a set of performative actions, localized in space and time, in a large body of image sequences to label those that occur with greater frequency. It involves extracting visual information by classifying dynamic sets of images based on what they contain and show in order to design technical and cultural algorithms for recognizing patterns of data whose salient aspect is performativity and automate their analytical search (Fig. 2).

Fig. 2. Billboards / Chantier Barbès-Rochechouart, 2000 © Pierre Huyghe Courtesy @

As noted above, in an artistic context, audiovisual devices - notably analog and digital electronic devices - have the ability to capture and record the flagrancy of what in "corporeality" (human and non-human), "action" and "gesture" (in image, voice/sound) is "living process", creative, immediate, dynamic, transformative, transient, ephemeral, relational.

However, it is a question of understanding how video recording makes visible the salient stages of the living process and how these are recognizable, analyzable, and inferable through a "computational eye" and a "modeled gaze" [2,3]. One wonders: how would such a computational eye be able to recognize, analyze, and interpret the relational performative actions and gestures that take place, for example, in a video work such as *Untitled (Human Mask)* (in Fig. 3) and 4 created by artist Pierre Huyghe?

In *Untitled (Human Mask)* the "agents" are not immediately identifiable; the first "agent" watches and is a drone; the second is watched and wears a female "Noh" mask.

They alone interact in an abandoned place (an empty restaurant located near the Fukushima Daiichi exclusion zone, immediately after the natural and technological disaster).

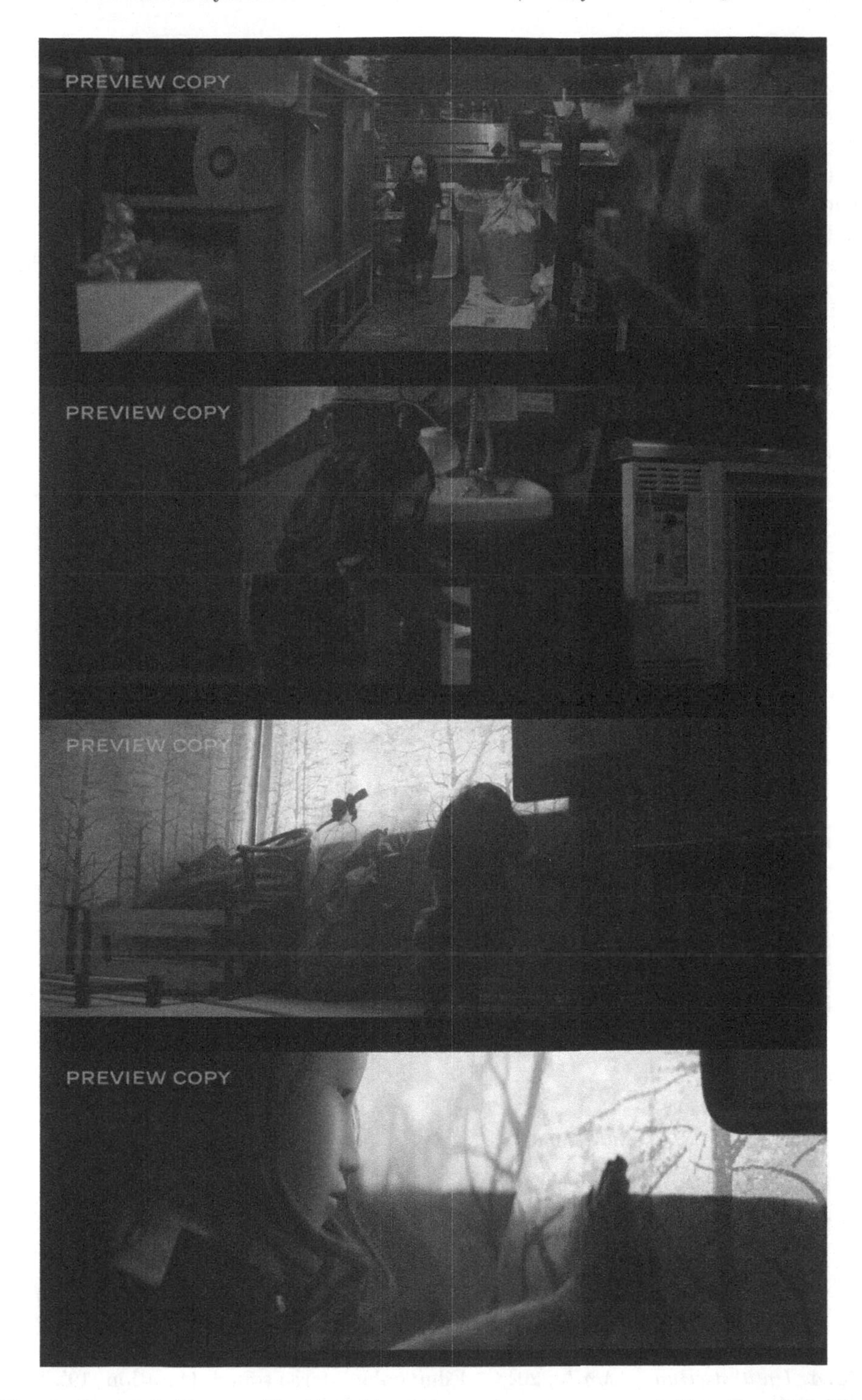

Fig. 3. *Untitled (Human Mask)*, 2014 – Film, color, stereo sound. Duration: 19', Film stills © Pierre Huyghe -1

The "agent" wearing the mask appears as an indistinct being, a young woman or a child. The other "agent" is a monkey trained to serve as a waitress in a sake house and is as the artist claims: "[...] an unconscious actor of human labor. [...] The monkey, left on its own, executes, like an automaton, the gestures it had been trained to do, in a pointless pattern of repetition and variation. Trapped inside a human represe ntation, the monkey has become its sole mediator" [12]

Based on these queries and examples - among other possible ones - the research project intends to define training sets of audiovisual sequences classified

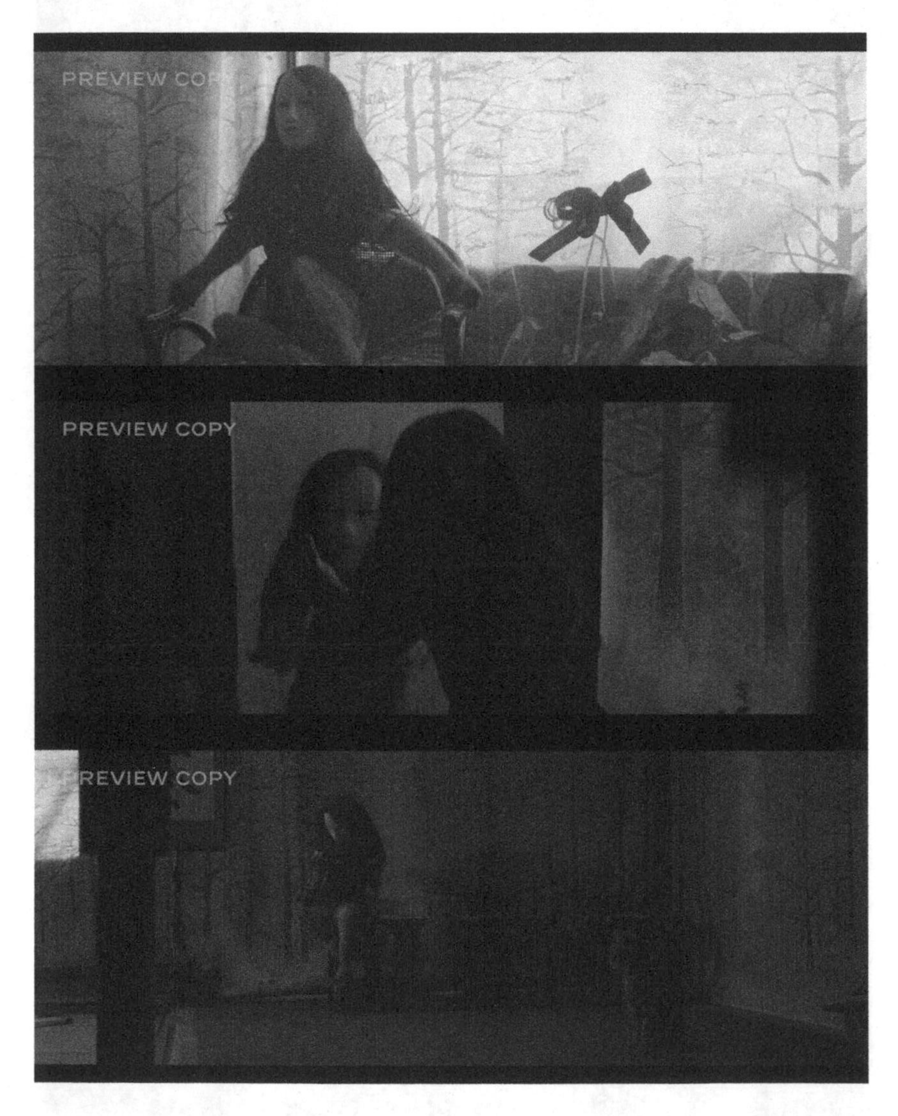

Fig. 4. *Untitled (Human Mask)*, 2014 – Film, color, stereo sound. Duration: 19', Film stills © Pierre Huyghe

through visual aspects referring to the history of images in the arts inherent in performative actions.

3 Conceptual Framework Proposal

The goal of elaborating a specific AI-based tool and datasets for the purposes of recognizing and analyzing the content of audiovisual works of performance-based video art must necessarily take into account its inherent nature. Considering that the performative action can be articulated on different levels (the act of the performer, the use of audiovisual language for performative and expressive purposes, and finally the fusion between these two declinations: the video performance), the project will first of all be structured by taking as reference a series of pre-existing tools useful for defining and distinguishing the relevant operations of recognition and analysis: catalographic codes that follow a principle of interoperability; one or more definitions of both audiovisual and performance language; existing taxonomies specific to video art.

Because of the complex ontological articulation of the performative act, the recognition and analysis of the works will have to be rooted in an effective distinction of the performance as such. As a reference, *Work Relation* of Marina Abramović and Ulay (in Fig. 1) can be taken as an example: the two artists' performance consists of repetitive actions of collecting and transporting elements present in the staged space. Bodies, movements, and props are invested with meaning by the very performative dimension of the work. The integration of AI technologies in video art works will therefore have to take into account the phenomenological distinction between a performative act and, for example, the staging or documentation of the work itself. It is, in essence, a matter of structuring the use of AI on the basis of a principle of traceability of a given set of gestures to a specific behavior or iconographic configuration, i.e., the performance.

In this sense, given the heterogeneous nature of such content, it will be necessary to integrate ad hoc datasets - stemming from labeling processes based on the semantic interpretation of spaces, actions, gestures, events, existents (bodies), objects ascribable to the dimension of artistic act and expression (video) and, possibly, related peritextual materials - with pre-existing ground truth datasets, already widely used for content recognition of audiovisual streams and for training dedicated deep learning models (e.g., ImageNet for image recognition, Kinetics for action recognition, AudioSet for sound information, and AVA for Atomic Visual Actions, i.e., human activities more broadly intended in comparison to single actions). For these purposes, it is evident that computational models of deep learning and machine learning must be multimodal, that is, they must be capable of combining visual and audio information to improve the accuracy of recognition. From a methodological standpoint, given the multifaceted and difficult-to-frame nature of some occurrences attributable to artistic (video) making, it would be effective to implement unsupervised learning processes that allow the model to learn from unlabeled data, in parallel with the adoption of the aforementioned supervised AI training. The tools and technologies belonging to the field of artificial intelligence to be adopted for the purposes of the project can be traced first to the field of computer vision and its potential applications (motion and gesture recognition, artist identification). Synthetically, computer vision algorithms can be used for the following operations:

- Pose estimation: computer vision algorithms can be employed to recognize poses and body movements in artistic performances, such as dance or theater and their video documentation. By analyzing video footage of the performance, the algorithm can track the poses and movements of performers, enabling detailed analysis of their gestures, choreographies and performances (Fig.5 reports an example of pose estimation) [4,15,22].
- Facial recognition: computer vision can be implemented in order to recognize facial expressions and emotions in artistic performances. By analyzing artists' facial expressions, the algorithms can determine the emotions the performer expresses in a given situation [11,16].
- Object recognition: recognizing objects in artistic performances, such as musical instruments, props or scenery through computer vision could be useful for the sake of analyzing the composition of the performance and how it contributes to the overall artistic experience [5,18,19].
- Movement analysis: computer vision algorithms can be implemented with the objective of analyzing the movement of objects and performers in artistic performances. This can be helpful for identifying iterative patterns and rhythms in the performance as well as for understanding its overall structure and inherent flow [13,23,24].

Along with computer vision, artificial intelligence technologies can be integrated to complement the operations described above. In particular:

- Audio analysis: audio analysis technologies can be used to recognize musical genres, instruments, and patterns in recorded musical performances. Additionally, they can be functional to sound pattern identification in those video artworks that make extensive use of audio as a means of artistic expression [6].
- Natural language processing: the technologies deriving from this AI field of studies (such as speech recognition and natural language understanding) can be used to analyze texts and dialogues in artistic performances, by highlighting some of their specific features (repetition, emphasis, etc.) [7,14].
- Neural networks: neural networks, such as RNNs (recurrent neural networks) and CNNs (convolutional neural networks), can be used synergistically with computer vision in order to develop a more effective interpretation of the information contained in the audiovisual streams to be analyzed. The typical properties of GANs (generative adversarial networks) can also be exploited, with the aim of optimizing AI training in data classification through competitive training between the generative data model and the discriminative model peculiar to this type of networks. For "composite" content such as audiovisual content, an effective technology to adopt as a complement could

be LSTM (long short-term memory) artificial neural networks [25], due to their adaptability to multimodal analysis models for image content processing and prediction.

The system of recognition and analysis of audiovisual content specifically pertaining to the field of video art here proposed, making use of the tools made available by artificial intelligence for the interpretation and identification of artistic performances, can be articulated along two directions: on the one hand, the extensive use of computer vision aided by technologies derived from the different declinations of artificial neural networks; on the other hand, the construction and training of dedicated datasets starting from the pre-existing models, with the objective of defining specific and pertinent semantic attributions aimed at labeling the distinctive elements of video art.

4 Objectives and Project Structure

Recognition in a large number of video recordings of performative actions in the artistic field can enable interpretation/inference of contextual/situational references, codes, conventions, and cultural constructs. Regarding these instances, it will be possible to analyze when, how, and why

- a performative action, the execution of which took place in an artistic space, differs from a similar action performed in a non-artistic space.
- the same action transported to different contexts, in different ways, changes its sign and transforms its sign function.
- the performance of the same action produces different realities and effects of reality.

Through the classificatory study of archives of moving images, epistemological, ontological, and gnoseological issues related to film studies, visual studies, and performance studies are addressed. The concatenations of these issues will emerge in the construction of datasets to define the following process:

1. definition of a dataset of sixty performance actions.
2. recognition and analysis of a large number of video recordings of performative actions performed in an artistic context (supervised operation). These recordings are collected and selected from a sample of hundreds of occurrences derived from video artworks digitized, and subsequently archived, by the La Camera Ottica Lab of the University of Udine.
3. recognition and analysis of a large number of video recordings of unintentionally performative actions in a non-art context (supervised operation).
4. data extraction from internal correlations at 2 and 3 (unsupervised operation/ Generative Adversarial Network/GAN).
5. comparison of data extracted from the internal correlations of 2 and 3 (unsupervised operations/ Generative Adversarial Network/GAN).
6. interpretation of the data extracted from the comparison of 2; classification of the data of 2 in visual form.

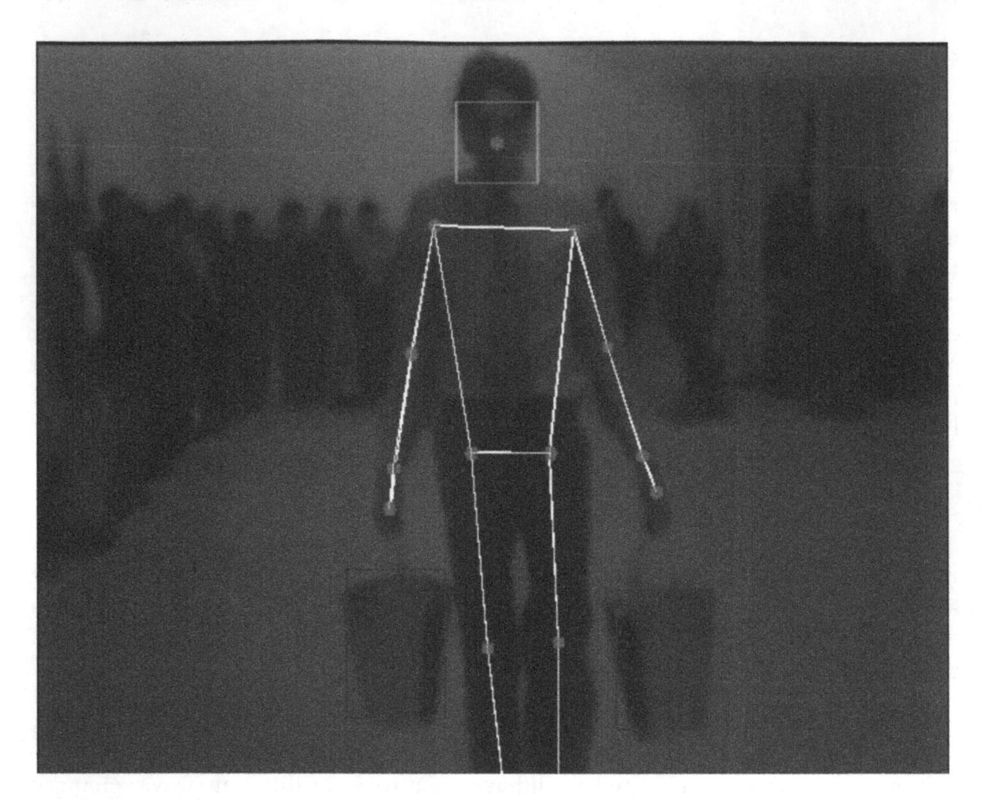

Fig. 5. A frame from the video documentation of *Work Relation* (in Fig. 1), for which a prototype action tracking system was tested. The system was built with OpenCV and the MediaPipe library for Python. MediaPipe is used to estimate the pose of the performers and detect performance objects (in this case two buckets for each performer). Before using MediaPipe the image was highly-processed, since both the quality of the video (very poor) and the camera movements made the pose and object estimations very inaccurate. These initial tests reveal central issues for the proposed research project.

Starting from this database, the huge amount of motion images that make up the performative audiovisual archives can be classified and analyzed, including the more complex ones that are growing exponentially in contemporary times and have complex semiotic features and are characterized by an equally complex semiosis process (that is, the process through which an activity or an expression becomes a sign)

Author contributions. The paper has been conceived, discussed and planned by all three authors. Michael Castronuovo has written Sects. 3-4, Alessandro Fiordelmondo planned and carried out the implementation of a prototype application, and Cosetta Saba has written Sects. 1-2.

References

1. Andrea, P., Antonio, S.: Teorie dell'immagine. il dibattito contemporaneo (2009)
2. Arcagni, S., et al.: L'occhio della macchina, vol. 705. Einaudi (2018)
3. Audry, S.: Art in the age of machine learning. Mit Press (2021)
4. Avola, D., Cinque, L., Fagioli, A., Foresti, G.L., Fragomeni, A., Pannone, D.: 3D hand pose and shape estimation from RGB images for keypoint-based hand gesture recognition. Pattern Recogn. **129**, 108762 (2022)
5. Bochkovskiy, A., Wang, C.Y., Liao, H.Y.M.: Yolov4: optimal speed and accuracy of object detection. arXiv preprint arXiv:2004.10934 (2020)
6. Bosi, M., Pretto, N., Guarise, M., Canazza, S.: Sound and music computing using AI: Designing a standard. In: Proceedings of the 18th Sound Music Computing Conference (SMC'21) (2021)
7. Falcon, A., Serra, G., Lanz, O.: Video question answering supported by a multi-task learning objective. Multimedia Tools and Applications pp. 1–28 (2023). https://doi.org/10.1007/s11042-023-14333-0
8. Fontanille, J.: Soma & séma. Figures du corps, Maisonneuve et Larose (2004)
9. Goldberg, R.: Performance now: Live art from the 21st Century. Thames and Hudson (2018)
10. Grespi, B.: Figure del corpo. Gesto e immagine in movimento (2019)
11. Hossain, M.S., Muhammad, G.: Emotion recognition using deep learning approach from audio-visual emotional big data. Inform. Fusion **49**, 69–78 (2019)
12. Huyghe, P., et al.: Pierre huyghe. (No Title) (1999)
13. Kazakos, E., Nagrani, A., Zisserman, A., Damen, D.: Epic-fusion: audio-visual temporal binding for egocentric action recognition. In: Proceedings of the IEEE/CVF International Conference on Computer Vision, pp. 5492–5501 (2019)
14. Khurana, D., Koli, A., Khatter, K., Singh, S.: Natural language processing: State of the art, current trends and challenges. Multimed. Tools Appl. **82**(3), 3713–3744 (2023)
15. Kim, J.W., Choi, J.Y., Ha, E.J., Choi, J.H.: Human pose estimation using mediapipe pose and optimization method based on a humanoid model. Appl. Sci. **13**(4), 2700 (2023)
16. Ko, B.C.: A brief review of facial emotion recognition based on visual information. Sensors **18**(2), 401 (2018)
17. Mitchell, W.J.: Pictorial turn. In: Visual Global Politics, pp. 230–232. Routledge (2018)
18. Redmon, J., Divvala, S., Girshick, R., Farhadi, A.: You only look once: Unified, real-time object detection. In: Proceedings of the IEEE Conference on Computer Vision and Pattern Recognition, pp. 779–788 (2016)
19. Redmon, J., Farhadi, A.: Yolov3: An incremental improvement. arXiv preprint arXiv:1804.02767 (2018)
20. Saba, C.: Per un supplemento d'indagine: la forza deterritorializzante del video. In: Valentini V., Saba C. (edited by), Medium senza medium. Amnesia e cannibalizzazione: il video dopo gli anni '90, pp. 79–127. Bulzoni (2015)
21. Saba, C.G.: Extended cinema: the performative power of cinema in installation practices. Cinéma & Cie **13**(1), 123–140 (2013)
22. Sun, K., Xiao, B., Liu, D., Wang, J.: Deep high-resolution representation learning for human pose estimation. In: Proceedings of the IEEE/CVF Conference on Computer Vision and Pattern Recognition, pp. 5693–5703 (2019)

23. Wang, L., et al.: Temporal segment networks: Towards good practices for deep action recognition. In: European Conference on Computer Vision pp. 20–36. Springer (2016)
24. Yao, G., Lei, T., Zhong, J.: A review of convolutional-neural-network-based action recognition. Pattern Recogn. Lett. **118**, 14–22 (2019)
25. Zamprogno, M., et al.: Video-based convolutional attention for person re-identification. In: Image Analysis and Processing-ICIAP 2019: 20th International Conference, Trento, Italy, September 9–13, 2019, Proceedings, Part I 20. pp. 3–14. Springer (2019)

CreatiChain: From Creation to Market

Enrico Maria Aldorasi[1], Remo Pareschi[2](✉), and Francesco Salzano[2]

[1] Department of Biosciences and Territory, University of Molise, Contrada Fonte Lappone, 86090 Pesche, Italy
[2] Stake Lab, University of Molise, Campobasso, Italy
remo.pareschi@unimol.it

Abstract. CreatiChain is a novel, integrated workflow management system that supports creating and monetizing AI-generated art. The system leverages grammars (Prompt Grammars) for semi-automated prompt generation, advanced AI algorithms for digital art creation, and blockchain technology for NFT minting and placement. The workflow begins with generating creative prompts using Prompt Grammars, offering creators a high level of customization. These prompts are fed into an AI-based art generation platform, producing unique digital art pieces. Once the art is created, the system automatically mints it into an NFT and places it on an NFT marketplace. CreatiChain streamlines access to AI art generation and NFT creation, offering a comprehensive solution for artists, designers, and digital creators to navigate the rapidly evolving digital art landscape.

Keywords: AI-generated Art · Prompt Generation · Blockchain · NFT Creation · Digital Art Workflow

1 Introduction

The intersection of artificial intelligence (AI), blockchain technology, and creativity has given rise to a new era in art and design. AI-generated art and Non-Fungible Tokens (NFTs) are transforming how creators produce and monetize their work. This paper introduces CreatiChain, a comprehensive workflow designed to streamline this process, making it more efficient and accessible to creators of various kinds, including artists, designers, and advertisers.

CreatiChain employs Prompt Grammars to generate prompts that guide the AI platform in creating art. This semi-automated approach allows creators to maintain a degree of control over the creative process while also benefiting from the efficiency and scalability of automation. The AI's capacity for rapid, large-scale image generation unlocks unprecedented opportunities for expansive creative exploration.

Once the creator is satisfied with the result, CreatiChain facilitates minting the artwork into an NFT and placing it on an NFT marketplace. Blockchain technology makes this process possible, providing a decentralized and secure mechanism for proving the ownership and uniqueness of digital assets.

G. L. Foresti et al. (Eds.): ICIAP 2023 Workshops, LNCS 14366, pp. 347–358, 2024.
https://doi.org/10.1007/978-3-031-51026-7_30

This paper is structured as follows. We first discuss related work, comparing CreatiChain with other systems and approaches in AI-generated art and NFTs. This will help to put CreatiChain in context and perspective of related efforts and to highlight its specific features and contributions. Then, we provide a background on AI-generated art, NFTs, and Prompt Grammars, which are the main concepts and technologies involved in CreatiChain. Next, we describe the overall architecture of CreatiChain, explaining its main components and functionalities. After that, we show the system in practice, exemplifying the from-prompt-to-nft end-to-end process through various cases. Finally, we conclude with a summary of the key points and discuss potential future directions for CreatiChain.

2 Related Work

CreatiChain is a system that uses Prompt Grammars to systematically generate prompts for different themes of visual content. Generative platforms then process these prompts to produce creative outputs, which human users can select, edit and mint as NFTs. This unique and comprehensive approach integrates the creation, editing, and minting processes into one workflow. However, CreatiChain is not the only system that explores the use of prompts for text-to-image generation tasks using generative AI platforms. In particular, Promptify is another system that uses large language models (LLMs) to generate diverse and creative prompts [2]. Promptify allows the user to interactively explore and refine the prompts based on the feedback from the generated images, thus enhancing the creative process. This technology relies on transformer models, a type of neural network architecture that uses self-attention mechanisms and has shown remarkable performance in various natural language processing tasks [13].

Promptify and CreatiChain, though methodologically distinct, can be viewed as complementary. Promptify excels in generating images from free text prompts, catering to an intuitive and exploratory creative process. On the other hand, CreatiChain's grammatical approach provides a structured way to generate prompts, ideal for maintaining coherence across prompts in domains like CryptoArt or systematically exploring new design concepts. It can also leverage open-access specialized lexicons related to products, art, and design. The potential for integration is significant. For example, CreatiChain's structured prompts could be enhanced with free text optimization from approaches like Promptify, combining the strengths of both systems. This could lead to a more comprehensive system for AI-generated art and design, broadening the scope of digital creation.

CryptoArt, a genre that uses blockchain technology to certify digital artwork's ownership and originality, often involves systematically generating related images, as seen in collections like CryptoPunks[1] and Bored Ape Yacht Club[2]. CreatiChain's Prompt Grammar approach aligns with CryptoArt principles, enabling systematic generation of prompts that maintain coherent features. Unlike traditional CryptoArt, which relies on static graphical traits, prompts

[1] https://cryptopunks.app/.
[2] https://boredapeyachtclub.com/.

generated via CreatiChain are interpreted by a generative AI platform, fostering creativity through unique and innovative outputs. This approach is particularly apt for creating large, coherent, and varied collections of digital artworks. These collections can be tokenized and licensed as unique digital assets, expanding opportunities for producing and distributing digital art.

Grammatical frameworks in architecture and industrial design, such as Shape Grammars [6] and CAD Grammars [10], have been used since the 19701970ss to automatically evolve shape patterns according to predefined rules. There is a crucial distinction between these frameworks and Prompt Grammars. Traditional grammars in design merge the provision of input with its processing, enabling an input form to be replaced by another derived from a fractal-like combination of input replications. While effective in fields like Computer-Aided Design, this approach lacks the creative contribution of generative platforms. In contrast, CreatiChain maintains a separation between input provision and processing. The system does generate prompts following predefined rules, yet generative platforms creatively interpret these prompts. This approach is especially effective in the exploratory phases of design projects, where many potential prototypes must be generated and evaluated.

Prompt Grammars were independently discovered in [12], where they guide LLMs in generating code in Domain Specific Languages (DSL) such as SQL and Python. Despite the difference in application domains, this approach shares with CreatiChain the common goal of harnessing the potential of AI to generate diverse and creative outputs guided by the structure and rules provided by Prompt Grammars. This shared methodology underscores the versatility and potential of Prompt Grammars in various fields of AI-assisted generation.

CreatiChain can be considered a specific instance of the MyBottega framework [8], which proposes an architecture for a comprehensive digital art production and distribution environment. MyBottega uses AI-based generative platforms and digital channels, such as blockchains and NFT marketplaces, to create and share digital art. It also envisions complex economic ties between galleries and artists, supporting a dynamic ecosystem for digital art. CreatiChain implements this framework by focusing on generating prompts for generative platforms and enabling the automated creation and placement of NFTs. CreatiChain simplifies the economic aspects, focusing on digital art's direct creation and distribution.

3 System Description

This section briefly overviews the background technology used in implementing CreatiChain, divided into two categories: system and applications.

3.1 System Stack

The MEAN stack[3], an acronym for MongoDB, Express.js, Angular, and Node.js, is a popular full-stack JavaScript solution that helps developers build fast,

[3] https://www.ibm.com/topics/mean-stack.

robust, and maintainable web applications. The MEAN stack is designed to provide a simple and fun starting point for cloud-native full-stack JavaScript applications. MEAN is a set of Open Source components that provide an end-to-end framework for building dynamic web applications, starting from the front end, progressing to the back end, and connecting to the database in the cloud.

Node.js is a JavaScript runtime built on Chrome's V8 JavaScript engine. It is used for building scalable network applications and can handle thousands of simultaneous connections with high throughput, which equates to high scalability. Node.js operates on a single thread, using non-blocking I/O calls, allowing it to support tens of thousands of concurrent connections without incurring the cost of thread context switching.

Express.js is a web application framework for Node.js that enhances its capabilities for building web and mobile applications. It simplifies the development process by offering features for managing routes, requests, and views. It also enables the creation of REST APIs that facilitate communication between the client and server-side applications.

Angular is a platform for building mobile and desktop web applications. It provides a way to build applications for any deployment target by reusing existing code. Using HTML as the template language, Angular 8 allows developers to build applications for web, mobile web, native mobile, and native desktop.

MongoDB is a source-available cross-platform document-oriented database program [4]. MongoDB is classified as a NoSQL database program and stores data as JSON-like documents.

3.2 Involved Applications

On the application side, We have Prompt Grammars, image generation platforms like DALL-E, MidJourney, and Stable Diffusion, and tokenization technology like blockchain and smart contracts.

Prompt Grammars, a variant of context-free grammars [7], are meant to describe images using specific words and rules. These grammars generate strings with placeholders, which users can replace with text to form a complete prompt for a generative platform. This assists users in crafting expressions that align with the style, concept, and vision of the image they aim to create. The simplicity and flexibility of context-free grammars are key advantages. Their declarative nature eliminates the need to specify how to parse or generate strings, a feature we leverage in our approach. While unsuitable for full-fledged natural language processing due to the complexity of the task, they are perfectly suited for controlled contexts, such as generating prompts for image creation. We will not delve into a formal definition of Prompt Grammars here; instead, we will illustrate how they operate in the following section, providing a practical understanding of their application in the context of CreatiChain.

Image generation platforms like DALL-E, MidJourney, and Stable Diffusion are AI models that generate images from textual descriptions. DALL-E [9], developed by OpenAI, is a platform that creates images from textual prompts. It is the only platform of this kind currently supported in CreatiChain, which, however, can be integrated with similar platforms like MidJourney[4] and Stable Diffusion[5].

Blockchain technology and smart contracts are the backbones of tokenization technology [5]. They allow the creation of NFTs (Non-Fungible Tokens), unique digital assets representing ownership of a unique item or piece of content [1]. NFTs are traded on dedicated portals such as OpenSea[6] and Rarible[7].

4 Architecture

CreatiChain is structured around a multi-tier architecture, decomposing complex functionalities into manageable sub-functionalities. This structure enhances clarity and ease of handling. The foundational model of CreatiChain's architecture adheres to a three-tier approach, as shown in Fig. 1, effectively partitioning the system into three primary modules: the presentation, application, and persistence layers. This division delineates roles and responsibilities within the system, enhancing its overall efficiency and maintainability.

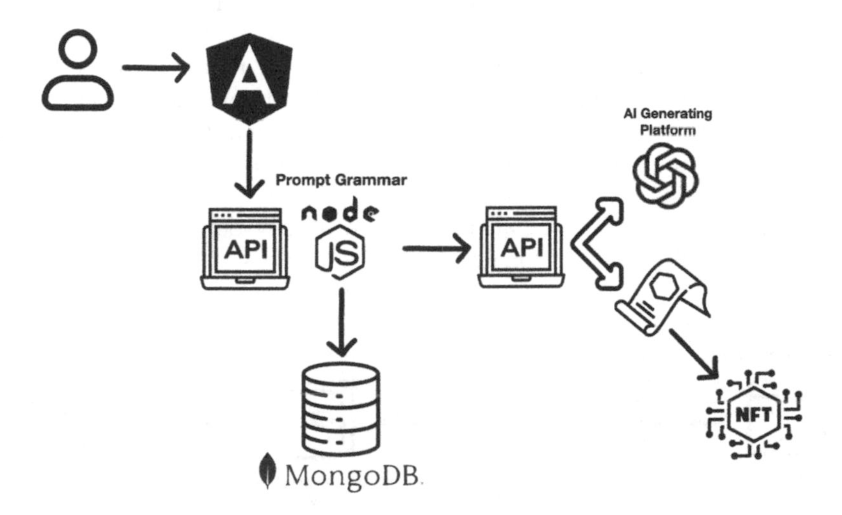

Fig. 1. Architecture of the framework

The first level, the presentation layer, manages the user interface and displays data to the end user. In our case, this layer has been implemented using the Angular framework. Angular provides a robust platform for developing dynamic and interactive user interfaces, allowing users to interact with the system seamlessly and intuitively.

---

[4] https://www.midjourney.com/.
[5] https://stablediffusionweb.com/.
[6] https://opensea.io/.
[7] https://rarible.com/.

The second level, the business logic layer, manages the application's rules. This layer implements the application's functionalities and interacts with the data persistence layer to retrieve and save necessary data. This layer has been implemented using Node.js, which is pivotal in orchestrating the system's business logic. Through REST APIs, it communicates with various applications to progressively realize the creative process. This includes prompt generation, AI-powered image creation based on the generated prompts, and the blockchain platform with associated smart contracts for creating NFTs. Node.js also acts as a persistence manager, providing control to the human creator over the generated images, which can be stored locally or in the cloud and manipulated before being returned to CreatiChain for the final step of creating the NFT.

The third level, the data persistence layer, manages data access and stores application information. This layer can be implemented using database technologies such as SQL, NoSQL, and graph databases. In our case, consistently with full leverage of the MEAN stack, the MongoDB NoSql DBMS was used. This layer maintains a dedicated DBMS at layer 3 for the grammars used for the prompt generation, ensuring the persistence of the system's data.

To ensure proper communication between the different layers of the application, well-defined communication mechanisms need to be implemented. For example, the presentation layer must be able to send requests to the business logic layer and receive responses using a well-defined communication protocol. Moreover, it is important to define a standard format for data transmitted between layers, to ensure the correct interpretation of data by all application components [3].

The human user can enter anytime the workflow depicted in Fig. 2, thus allowing for a flexible and personalized creative process.

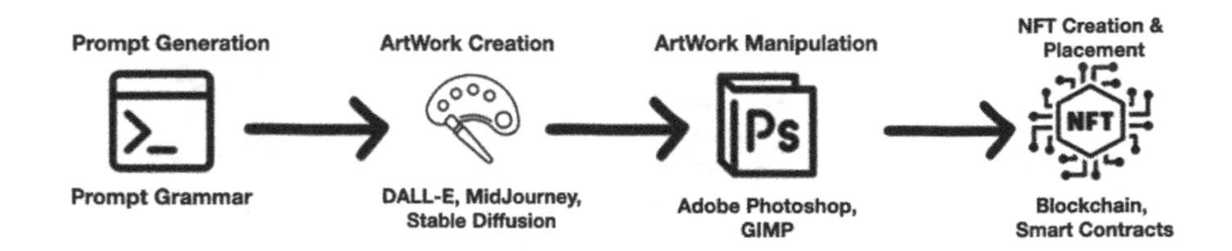

Fig. 2. CreatiChain Workflow

They can follow the complete process from prompt generation to image creation to NFT creation, or they can manually create one or more prompts to be given to the image generation platform. They can even skip this step entirely, bringing an independently generated digital image to the system and using CreatiChain only to create the NFT. This flexibility ensures that CreatiChain can cater to various creative needs and preferences, providing a comprehensive and adaptable tool for generating digital art.

5 Creation of Certified Digital Art Through Prompt Grammars and Generative AI

The developed grammars reflect specific usage contexts. We crafted three distinct grammars: Art, Car Restyling, and 3D Architecture. The Art grammar contains a wide range of terms for creating prompts for artworks, with rules that enable the creative and variable combination of artistic styles and techniques. The Car Restyling grammar aims to generate prompts that allow the creation of vintage car images that match the configuration of modern vehicles. The defined rules permit the combination of vintage car features with contemporary elements. Finally, the 3D Architecture grammar was devised for generating prompts for creating custom 3D architectural models, with rules that allow for the flexible and adaptable combination of architectural elements. Figures 3, 4, 5 show how CreatiChain grammar is used to guide the creator. Figures regarding prompt grammar depicted in this paper refer to the Art Prompt Grammar.

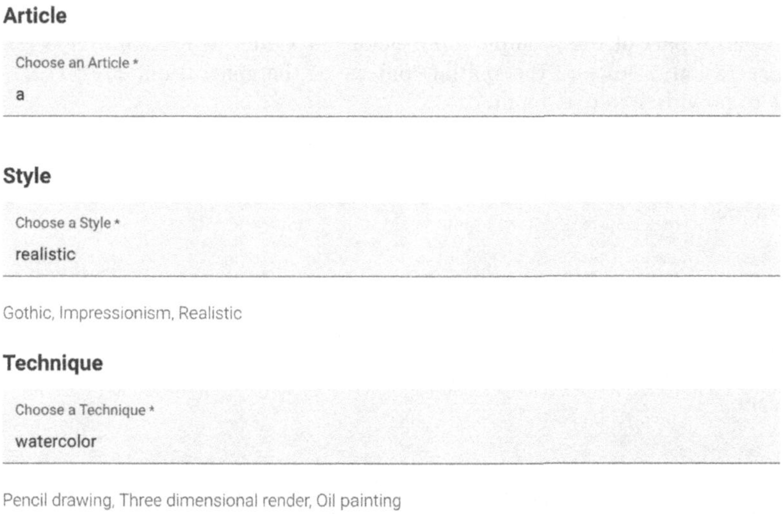

Article

Choose an Article *
a

Style

Choose a Style *
realistic

Gothic, Impressionism, Realistic

Technique

Choose a Technique *
watercolor

Pencil drawing, Three dimensional render, Oil painting

Fig. 3. Preamble of the prompt. *Article*: Represents the number of the subject to place in the image to generate. *Style*: Indicates the cultural trends with which depicting the image to generate associated with the subject, according to the chosen artistic movement. *Technique*: Represents the specific methods and approaches used to create the artwork, giving also the rules involved in manipulating materials and tools in order to achieve the artistic vision.

Subject

Choose a Subject *

of a woman

of a person, of a hand, of a child

Free Text

Write what you want *

reading a book

Location

Choose a Location *

in a forest

on a beach, in a city, in a forest

Fig. 4. Central part of the prompt. Input fields get values as follows: *Subject*: subject of the generation, *Location*: the spatial context of the generation, *Free Text*: Allows the user to provide free-text input.

Adverb

Choose an Adverb *

passionately

Passionately, Beautifully, Intensely

Verb

Choose an Verb *

painted

created, drawn, produced

Artist

Choose an Artist *

by Vincent Van Gogh

by Vincent van Gogh, by Claude Monet, by Salvador Dali

Fig. 5. Closing part of the prompt. Input fields get values as follows: *Verb* that describes the specific action performed by the chosen artist, *Adverb*: emotional, intellectual, or conceptual dimension to be expressed, *artist*: inspiring artist of the work with associated style and techniques.

The snippet below provides an example of the prompt grammar filled with the creation string supplied by the creator.

```
1  const artGrammar = {
2      prompt: ["preamble", "text", "closing"],
3      "preamble": ["items", "style", "technique"],
4      "text": ["subject", "freeText", "location"],
5      "closing": ["adverb", "verb", "artist"],
6      "items": ["a"],
7      "style": ["realistic"],
8      "technique": ["watercolor"],
9      "subject": ["of a woman"],
10     "freeText": ["reading a book"],
11     "location": ["in a forest"],
12     "adverb": ["passionately"],
13     "verb": ["painted"],
14     "artist": ["by Vincent van Gogh"]
15   };
```

Figures 6a and 6b depict two instances of digital art created by exploiting the workflow discussed in Sect. 4, generated based on the above defined Prompt Grammar. In detail, Fig. 6a is generated by using the prompt defined in the previous images.

(a) A realistic watercolor of a woman reading a book in a forest passionately painted by Vincent van Gogh

(b) A hyper-realistic 3D render of a full snapshot of the external body of a Ford Model A Deluxe Roadster restyled according to 2023 lines such that it looks like a Lamborghini car and bears the typical imprint of cars designed by Marcello Gandini

Fig. 6. Two instances of digital art created through the CreatiChain workflow

5.1 NFT Generation and Placement

After generating the image, the creator can post-process their digital art. CreatiChain allows them to mint a corresponding NFT and place it on an NFT marketplace when satisfied. One of the main requirements is to separate the minting process from the generation process. To achieve this, CreatiChain provides an API, which takes as input the following parameters:

- **chain**: The name of the blockchain on which to mint;
- **name**: The name of the NFT being generated;
- **description**: The description of the NFT being generated
- **mint_to_address**: The address of the wallet on which the NFT is to be associated with.

Thus, when the creator has finished their work, they can issue NFT minting request providing the data above. The request is handled by an API meeting the following requirements.

```
1  Define a web application:
2      Define a route "/mint-nft" that accepts POST requests:
3
4          When a POST request is received at "/mint-nft":
5              Extract the file and the parameters from the request
6
7              Prepare the request to send to NFTPort with the
8              extracted values
9
10             Add an authorization header with the API key to the
11             API request
12
13             Send a POST with the built request to the
14             "https://api.nftport.xyz/
15
16             Receive the response from the NFT Port API
17
18             Return the response from the NFT Port API
19
```

This pseudo-code creates a Web application with a single endpoint */mint-nft* that accepts a POST request. The request must include the file parameter containing the image file to be minted as an NFT and the other parameters (chain, name, description, mint_to_address). The */mint-nft* endpoint sends a POST request to the *NFTPort* API[8], passing the required headers and query parameters. The request sends the image file as part of the files parameter. *NFTPort* offers several APIs; the one used here mints an NFT in less than 5 min using smart contracts deployed on various blockchains, such as Ethereum,

[8] https://docs.nftport.xyz/reference/easy-minting-file-upload.

Polygon, and Solana. After minting, the NFT will appear in the specified mint_to_address wallet. After a few minutes, the minted NFT can also be seen on the NFT marketplace in the profile associated with the wallet.

6 Conclusion and Future Work

CreatiChain introduces a novel approach for generating structured prompts using Prompt Grammars, a type of context-free grammars. This method enables systematic exploration of new concepts and styles, which is particularly beneficial for CryptoArt and design projects. The potential of CreatiChain has been demonstrated through the successful generation of diverse prompts for the DALL-E 2 platform. The resulting digital artworks, tokenized and licensed as unique digital assets, showcase new avenues for producing and distributing digital art.

Future work will focus on automating the generation of Prompt Grammars. Currently, these grammars are manually crafted, but the goal is to employ Large Language Models (LLMs), such as GPT models, for grammar generation. This parallels using machine learning models for code generation from natural language specifications, as discussed in [11]. Both fields involve translating human-readable instructions into a structured format but also present unique challenges. Precision and correctness are crucial for code generation. In contrast, for grammar generation, defining efficient criteria to evaluate the validity of use and domain coverage of the grammars is a key challenge.

In conclusion, CreatiChain pioneers a novel approach to AI-driven digital art production by synergizing prompt-generating grammars with AI-based generative platforms. This enhances the creative process and, coupled with blockchain technology and NFTs, streamlines the distribution and monetization of AI-generated digital art. The prospective integration of machine learning techniques for automated grammar generation promises to further amplify the efficacy and reach of the CreatiChain workflow.

References

1. Bao, H., Roubaud, D.: Non-fungible token: a systematic review and research agenda. J. Risk Finan. Manag. **15**(5) (2022). https://doi.org/10.3390/jrfm15050215. https://www.mdpi.com/1911-8074/15/5/215
2. Brade, S., Wang, B., Sousa, M., Oore, S., Grossman, T.: Promptify: text-to-image generation through interactive prompt exploration with large language models. CoRR abs/2304.09337 (2023). https://doi.org/10.48550/arXiv.2304.09337
3. Evans, E., Evans, E.J.: Domain-Driven Design: Tackling Complexity in the Heart of Software. Addison-Wesley Professional, Boston (2004)
4. Fowler, M.: Patterns of Enterprise Application Architecture: Pattern Enterpr Applica Arch. Addison-Wesley, Boston (2012)
5. Garriga, M., Palma, S.D., Arias, M., Renzis, A.D., Pareschi, R., Tamburri, D.A.: Blockchain and cryptocurrencies: a classification and comparison of architecture drivers. Concurr. Comput. Pract. Exp. **33**(8) (2021). https://doi.org/10.1002/cpe.5992

6. Jowers, I., Earl, C.F., Stiny, G.: Shapes, structures and shape grammar implementation. Comput. Aided Des. **111**, 80–92 (2019). https://doi.org/10.1016/j.cad.2019.02.001
7. Jurafsky, D., Martin, J.H.: Speech and language processing: an introduction to natural language processing, computational linguistics, and speech recognition, 2nd edn. Prentice Hall series in artificial intelligence, Prentice Hall, Pearson Education International (2009). https://www.worldcat.org/oclc/315913020
8. Noviello, N., Pareschi, R.: Mybottega: An environment for the innovative production and distribution of digital art. In: Mazzeo, P.L., Frontoni, E., Sclaroff, S., Distante, C. (eds.) Image Analysis and Processing. ICIAP 2022 Workshops - ICIAP International Workshops, Lecce, Italy, 23–27 May 2022, Revised Selected Papers, Part I. Lecture Notes in Computer Science, vol. 13373, pp. 162–173. Springer, Heidelberg (2022). https://doi.org/10.1007/978-3-031-13321-3_15
9. Ramesh, A., Dhariwal, P., Nichol, A., Chu, C., Chen, M.: Hierarchical text-conditional image generation with CLIP latents. CoRR abs/2204.06125 (2022). https://doi.org/10.48550/arXiv.2204.06125
10. Rowe, P.D.G., Reed, C.: Cad grammars. In: Gero, J.S. (ed.) Design Computing and Cognition, pp. 503–520. Springer, Dordrecht (2006). https://doi.org/10.1007/978-1-4020-5131-9_26
11. Treude, C.: Navigating complexity in software engineering: a prototype for comparing gpt-n solutions. CoRR abs/2301.12169 (2023). https://doi.org/10.48550/arXiv.2301.12169
12. Wang, B., Wang, Z., Wang, X., Cao, Y., Saurous, R.A., Kim, Y.: Grammar prompting for domain-specific language generation with large language models. CoRR abs/2305.19234 (2023). https://doi.org/10.48550/arXiv.2305.19234
13. Yang, J., et al.: Harnessing the power of llms in practice: a survey on chatgpt and beyond. CoRR abs/2304.13712 (2023). https://doi.org/10.48550/arXiv.2304.13712

Towards Using Natural Images of Wood to Retrieve Painterly Depictions of the Wood of Christ's Cross

Johannes Schuiki[1(✉)], Miriam Landkammer[2], Michael Linortner[1], Isabella Nicka[2], and Andreas Uhl[1]

[1] Department of Artificial Intelligence and Human Interfaces, University of Salzburg, Jakob -Haringer -Straße 2, 5020 Salzburg, Austria
jschuiki@cs.sbg.ac.at

[2] Institute for Medieval and Early Modern Material Culture, University of Salzburg,Körnermarkt 13, 3500 Krems an der Donau, Austria

Abstract. A painting, much like written text, allows future viewers to draw conclusions about the time of its creation or its painter. Also, when looking at a whole corpus of images instead a single instance, trends in painting can be analyzed. One particular trend originating in the 14th century is the transfer of the visual impression of real world materials onto paintings. One object, which is often depicted in paintings around that time being made from wood, is Christ's cross. Scarce research has been done in the direction of automatically analyzing painterly depictions of the wooden cross of Christ. Hence, this study walks a step towards automatic annotation of wooden crosses in paintings by evaluating three publicly available databases containing natural images of wood for their applicability to use their images as queries to retrieve painterly depictions of the wood of the cross. Experimental results underline the demand for further investigations.

Keywords: painted material · depictions of wood · material detection

1 Introduction

Throughout the 14th century, we can observe new trends in painting all over the Latin West. One of the most important of these was the imitation of the material of objects and things from nature. This trend culminated in the 15th century in a new realism which - stimulated in particular by the achievements of Early Netherlandish painting - brought with it a new peak in the preoccupation with the intricate rendering of surface qualities. With regard to the representation of materials, the roots and course of such a major and sustained cultural innovation

This work has been partially supported by the Salzburg State Government project "How Material Came into the Picture: Exploring Cultural Innovations Interdisciplinarily with Artificial Intelligence (KIKI)."

G. L. Foresti et al. (Eds.): ICIAP 2023 Workshops, LNCS 14366, pp. 359–371, 2024.
https://doi.org/10.1007/978-3-031-51026-7_31

have not been the subject of research on a European scale. Nor are there detailed analyses for the different ways, surface qualities were rendered in art of the time for a lot of materials. One important reason for this is that hardly any annotation data to depicted material are available, making even the formation of a research corpus a very difficult task. Grouping similar ways of artistic approach to surface qualities would allow to study both the various narrative functions and the proximity of artistic concepts. To fill this research gap, the KIKI project (title: "How the material got into the image: Exploring cultural innovation with DH and AI") develops automated recognition of material in historical visual media, to enable in-depth analyses of the representation of materials and surface qualities in painting across different geographical areas and time spans, and to apply a distant viewing approach [1,16] that allows the recognition of underlying patterns and specific solutions in big image data.

In art history, the perspective of Materialikonologie - the analysis of the functions of the physical material of works of art - has been central to the study of medieval art since the late 1960 s (most eminent: [30]). More recently, the concept of Materiality opened up the view to broader phenomena such as material interactions [18], thereby also drawing the attention to materials evoked, imitated or fictionalised by artists in images. Research has been conducted on various rendered materials, like rock (i.a. [14,33]), marble (i.a. [17,23] or hair, skin and (semi-)transparent materials [19,20]. However, other materials, like rendered wood, have not been in the focus in a broader sense. The function of rendered materials in the narrative strategies of visual media is only touched upon by a few researchers, most notably in [7,14].

The art historical part of KIKI therefore sheds light on how artists reflected or fictionalised materials through imitation in painting, and how they made depicted materials productive for the pictorial narrative. The research corpus consists primarily of works from today's Austria as well as neighbouring regions in the database REALonline. One of the challenges faced by art historical researchers is to cope with large amounts of - in this case visual - data. Thus, computer-aided annotation of big corpora of data is desired. While there exists literature on object retrieval [5] object detection [6], face retrieval [4] and object segmentation [3] in paintings, very scarce literature exists on detection and segmentation of painted materials.

Zuijlen et al. [35] conducted a study regarding the perception of painted materials and their attributes. They used existing segmentation software on a rather small set of images to segment 15 material categories. Later, the same group [36] published a material database named materials in paintings (MIP) along with bounding boxes encapsulating objects from the same 15 material categories used in their earlier study. They utilized Amazon's Mechanical Turk for the exhaustive manual annotation work. By fine tuning an existing object detection network using the manually annotated materials, the number of bounding boxes was artificially increased.

In an approach to close these gaps in both research disciplines, the art historical and computer vision, this study focuses on the painted representation of wood, especially wooden planks used to depict Christ's cross. Although the MIP database would contain wood as a material category, the annotated bounding boxes often contain background but no segmentation mask, hence deliver insufficient information for our experiments. As a first step towards detection of cross wood in paintings, the experiments within this research aim to answer the question whether patches of natural images of actual could be used as a query to retrieve patches of painted cross wood from a larger pool of image patches.

The remainder of this paper is structured as follows: Sect. 2 motivates why analyzing the depictions of crosses, especially Christ's. cross, is of interest. Section 3 describes the experiments carried out in Sect. 4 along with a detailed description of the used databases. Finally, Sect. 5 summarizes the findings of the investigations.

2 Holy Wood: The Material of Christ's Cross

In the Golden Legend, compiled around 1260, a dilemma - the fact that the Son of God died in a particularly ignominious, humiliating way on the cross [11, ch. 1] - is resolved: Christ's redemptive death makes the cross an instrument of salvation, which is demonstrated by the reversal of characteristics of the wood into their opposite (from cheap to precious, from ignoble to sublime, from dark - and without any beauty - to light, from malodourous to fragrant, etc.) [10, p. 554]. The passage shows how densely the notion of the cross of Christ was intertwined with its material in the Middle Ages and how the wood of the cross served as a connecting point for different ideas and forms of metaphor. The significance of the material is evident from "wood" being a common synonym for Christ's cross in Christian literature. Also, cross relics have played a role in the Latin West especially since the High Middle Ages [32]. Unlike the body of Christ, particles of the cross were accessible as material remnants of Christ's passion. Veneration of the cross formed part of several liturgical feasts and included rituals such as the pouring of wine or water over the relic of the cross, which had healing powers when drunk [34, pp. 18–23]. The *lignum crucis* was believed to have "absorbed" Christ's power through his blood and the touch of his limbs, and the cross could become a substitute for Christ's body. In both contexts, it is relevant that wood was frequently characterised as a living material, with veins and "humours" (body fluids), which, like the human body, could fall ill and be injured [29, pp. 252–254].

A link between the cross particles and the historical cross was provided by the immensely popular Legend of the Cross: the "biography" of the wood of the cross, which was in turn firmly embedded in the liturgical year [2]. The Legend of the Cross establishes a link to Old Testament time by tracing the origin of the wood back to a branch of the Tree of Life in paradise. It also reports on events and miracles that happened long after Christ's sacrifice (i.a. the finding and miraculous identification of the True Cross by the Empress Helena in the fourth century).

These ideas about the wood of the cross, as well as the liturgical and devotional practices revolving around it, made painters dedicate special attention to Christ's cross concerning not only its construction but also its texture. Much research has been done on tree-like or vegetal depictions of the cross as a prospect of eternal life, and establishing the link to the tree in the Edenic garden and the Tree of Life in the Book of Revelation [9,15]. Depicted wood grains on crosses have however not been examined comparatively, which might partly result from the lack of data to systematically retrieve and compare all examples. The application of computer vision methods allows to examine the representation of the *lignum crucis* across a diverse corpus of images: In which pictorial contexts is the wood of the cross represented by the depiction of a grain? In which cases is there no texture, and where does the rendered materiality deviate decidedly from Christ's death-instrument (e.g. a golden cross)? A comparison of textures can provide an overview to the frequently depicted wood grains of Christ's cross, as well as outliers. On this basis further questions can be addressed, e.g. if a particularly abstract or unnaturally regular texture emphasizes the supernatural character of the wood, if there are textures that can be interpreted as references to other entities (e.g. Christ's body), if particularly explicit renderings of the wooden surface are due to the local context of use of the artworks (e.g. an existing cross relic) etc.

3 Experimental Setup

The conducted experiments aim to evaluate the similarity of painterly depictions of Christ's cross and the two thieves' crosses to natural images of actual wood. To do so, a simple query-based scheme is employed: Image patches containing real wood (query) are compared with image patches containing painted cross wood as well as patches containing other painted content (i.e. not cross wood). Since every comparison yields a similarity score, the closest N patches can be determined for every query. The main metric for the experiments in this study is the proportion of painted cross wood within the closest N patches.

Before comparison, images are cut into smaller, quadratic tiles (databases included in this study along with their corresponding tiling strategies are described in Sect. 3.1) and features are extracted from every tile using well-known texture descriptors. To gain a similarity measure between two tiles, the Euclidean distance is calculated on their corresponding feature vectors for a certain feature extraction method. Hence, a lower distance indicates a higher similarity between two patches. Experiments in this work are evaluated using three tiling sizes (64×64 px, 96×96 px, and 128×128 px).

In total, seven distinct techniques for feature extraction are employed, briefly introduced in the following. The first six of the algorithms were successfully used in classical texture classification, image tampering detection and paper identification [13]: Dense SIFT [21] (SIFT descriptors applied in a grid), Local Binary Pattern (LBP) [26], Weber Pattern (WP) [24], Local Phase Quantization (LPQ) [28], Rotation Invariant - LPQ [27] and Binarized Statistical Image Features (BSIF) [12]. While most of the extracted features are used without adaptation, the features from SIFT further undergo a dimensionality reduction using principle component analysis and soft-quantization utilizing a Gaussian mixture model, hereby minimizing the spatial-dependent information within the extracted features. In addition to that, a pre-trained ResNet50 [8] is employed, where the fully connected layer is skipped, resulting in an 2048 dimensional feature vector.

The experiments in Sect. 4 can be divided into four parts and the corresponding research questions defined as:

A) Is it generally feasible to employ image patches containing real wood as queries to successfully retrieve patches containing painted cross wood?
B) How does the behaviour change using patches of painted wood as query instead of real wood images?
C) Can image patches from other real objects (other than wood) be used to receive similar results as in experiment A?
D) Do variations in tiling size of query and test images have an effect on the results?

3.1 Databases & Tiling

Hereafter, the used databases and their corresponding tiling strategies are explained in detail. The number of resulting patches per database is presented in Table 1. Examples for each database can be seen in Fig. 1, Fig. 2 and Fig. 3.

Table 1. Number of available tiles per database and tile size.

	#Patches	Patch size [px]		
		64	96	128
Database	REALonline-Wood	1725	448	149
	REALonline-Non-Wood	1725	1725	1725
	Sharan14	3346	1255	535
	Elzaar21	6877	2955	1608
	Santos21	8598	3557	1364

REALonline: Unlike many other image databases of medieval and early modern visual cultural heritage, in REALonline [22,25] all image elements are recorded by name, category (i.e. clothing), colour, form and (when possible) material. The database contains more than 28.000 digital and digitised photographs and more than 22.500 in records on works of arts or their components. There are more than 1.2 million high-level annotations on image elements, manually recorded over almost 50 years by database editors trained in a discipline of medieval or early modern studies. For this research, we have selected 115 images based on the query of the existing annotation data. We also restricted the works in the dataset to painting and drawing techniques, and according to whether they were produced in the 14th or 15th century.

To locate Christ's cross and the two thieves' crosses, images were annotated manually as can be seen in Fig. 1-B. The annotation mask separates regions where tiles for the REALonline-Wood can be sampled (white areas) and background regions (black) where tiles for the REALonline-Non-Wood can be extracted. In case of the REALonline-Wood data, quadratic patches are cropped from a every annotated polygonal shape. This is accomplished by first applying morphological erosion with a squared structure element of size $S \times S$ to the annotated bitmask. S here is the size of the desired patch resolution. Doing so, every remaining logical 1 constitutes a center point of a patch that is guaranteed to be within the annotated polygon. The first tile is cropped around the left-most center point. Afterwards, the area is removed from the annotated bitmask. This procedure is repeated until no further valid patch is available. To construct the REALonline-Non-Wood database, 15 tiles per image are randomly cropped at locations which do not overlap with the annotated wooden areas.

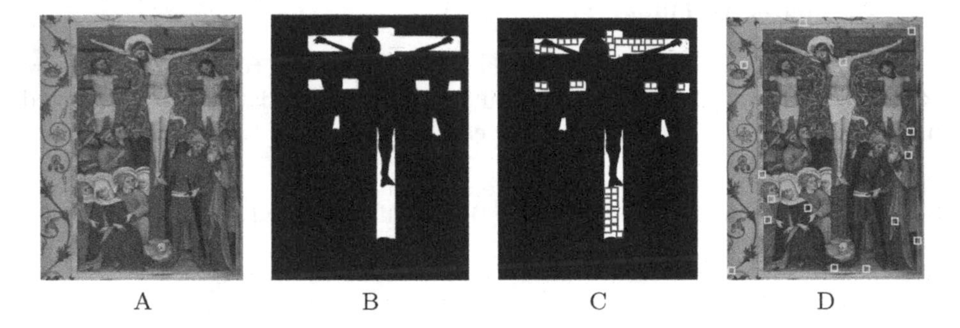

Fig. 1. Patch extraction pipeline for the REALonline data: A) example image from database, B) manually annotated mask of wooden crosses of Christ and the thieves, C) found patches of certain patch size within wooden areas, D) wooden patches (red) and 15 randomly chosen non-wooden patches (yellow). (Color figure online)

Flickr Material Database (Sharan14) [31]: The Flickr Material Database was constructed by [31] with the goal of capturing the natural range of material appearances. It comprises of 500 images from ten material categories. For the purpose of this work, only the images from the category *wood* are used. Since some wood images from this database also contain background, segmentation masks for the wooden areas are also available. In order to only draw samples stemming from the wooden areas, the same algorithm which is also used for tiling the REALonline-Wood data is employed. For experiment C (mentioned earlier this section), two other image categories (*stone* & *water*) from this database are used as query images.

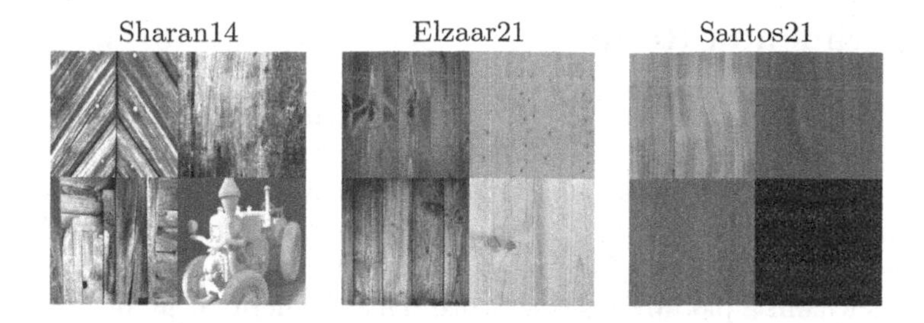

Fig. 2. Examplary depiction (before tiling) of images from the three query databases (resized to quadratic shape to fit the grid).

Plastic and Wood Pollution Dataset (Elzaar21) [1]: This dataset consists of 100 images from two classes. Similarly to the Flickr Material Database, only the class *wood* is used, resulting in 50 wood texture images of varying resolution. Tiling is accomplished by cropping adjacent patches without overlaps, starting on the upper left corner. Upon reaching the right image border, crops are continued underneath the just finished row of patches.

Fig. 3. Randomly picked tiles of size 128×128 px from the 5 wood databases. Note that for the six classical texture features, images were converted to grey-scale prior to the experiments. (Color figure online)

The Wood Database Images (Santos21) [2]: This database consists of images from the website https://www.wood-database.com/ acquired via web-scraping. Without modification, it comprises of 2105 images of varying resolution. For the purpose of automatically cropping patches from wooden areas, a subset of 1463 images has been selected by hand, only keeping images where solely wood is visible. Tiling is done in a similar manor as with the pollution dataset.

[1] https://www.kaggle.com/datasets/abdellahelzaar/plastic-and-wood-pollution-texture-dataset.

[2] https://www.kaggle.com/datasets/edhenrivi/the-wood-database-images.

4 Experimental Results

This section contains the results from the experiments motivated in Sect. 3. Although the REALonline-Non-Wood data set consists of 1725 patches for every tiling size (see Table 1), the number of patches are artificially reduced in a random fashion for the experiments where less patches from the REALonline-Wood are available in order for the pool of patches to be balanced. Note that data points within a plot always constitute the arithmetic mean over all patches from a set of query images.

Experiment A: Images from the databases containing real wood images are used as query to retrieve N closest samples from a pool of images combining all samples from REALonline-Wood and REALonline-Non-Wood . The proportion of retrieved wood patches within the N closest samples is analyzed and the results are depicted in Fig. 4.

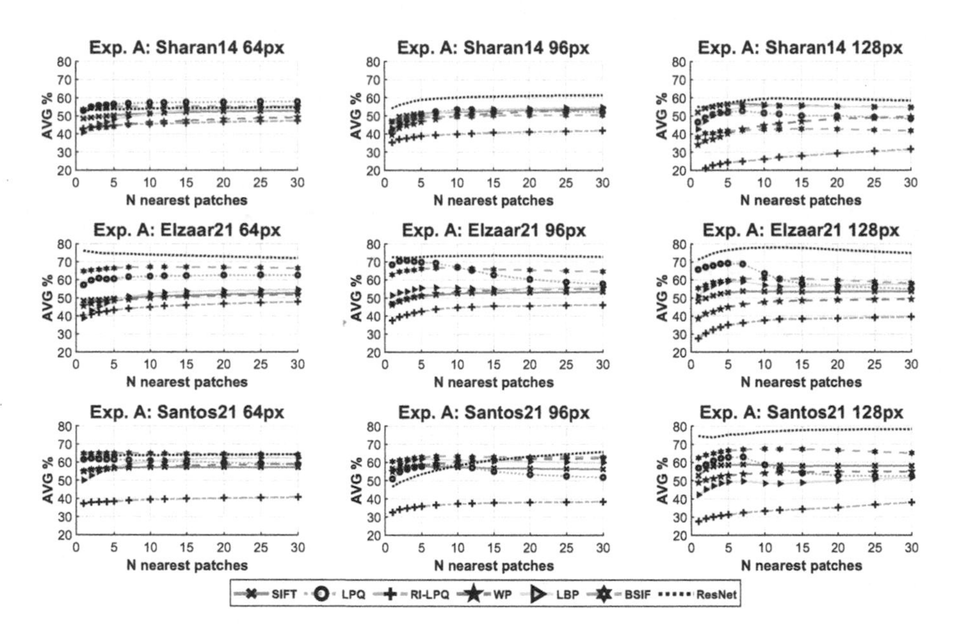

Fig. 4. Results experiment A.

Per row of plots, one particular real wood database was used for the query patches. Tiling size increases from left to right. LPQ and BSIF often perform better than SIFT and LBP. RI-LPQ is guessing at best (i.e. always near or below 50%). Query images from the Sharan14 database (first row) generally result in guessing (below 60% on average). The ResNet features often perform best. Using image patches from Elzaar21 or Santos21 as queries, retrieval performance can reach up to almost 80%.

Experiment B: Since results from the first experiment seem rather inconclusive, the second experiment uses a subset of the REALonline-Wood database as query images. In order to still include all tiles in the results, a 2-fold cross validation is employed. The plots (see Fig. 5) show better results for increasing tile size, however the ResNet and SIFT seems to be perform best for any size.

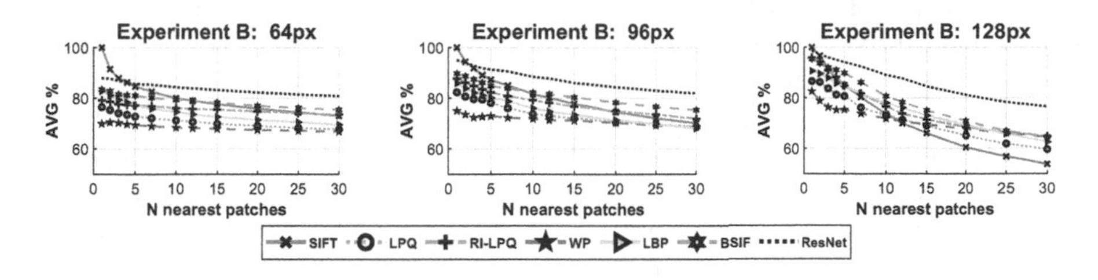

Fig. 5. Results experiment B. Using tiles containing painted wood as query.

Results indicate, at least for small values of N, tiles containing painted cross wood can indeed be used to find other tiles containing also painted cross wood rather than other painted contents. This experiment can be viewed as a sanity check whether the used feature descriptors are actually capable of capturing the structure within the tiles of painted cross wood.

Experiment C: This experiment is meant to evaluate the outcome when different natural images (other than wood) are used as queries. Two databases are tested. Results are depicted in Fig. 6. For Sharan14-stone (upper row) can be said that, on average, at best 50% of image tiles with highest similarity contain painted wood. Especially for the case 128 px, where results were best in experiment A, results tend to contain a higher amount of non-wooden content. Different results can be observed when looking at Sharan14-water (lower row), where the average paint wood hit-rate even reaches 70%.

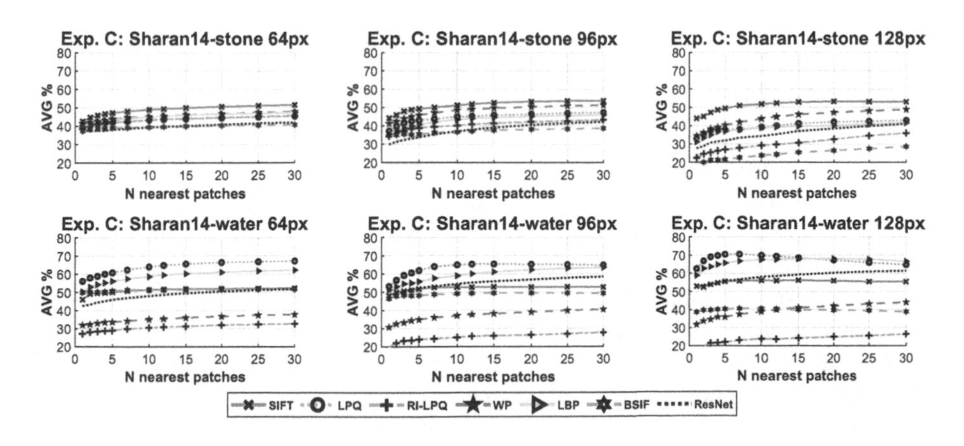

Fig. 6. Results experiment C. Using images containing stone and water as query.

Experiment D: This experiment is meant to investigate the effect of scaling. Tiles are cropped at a certain size and then scaled to another tile size. The experiment can be separated into two parts: First, only the query tiles are scaled and compared to un-scaled tiles from the REALonline databases (see Fig. 7-left). Second, tiles from query databases and REALonline databases are cropped with different tile sizes but then scaled to the same size in order to make them comparable (see Fig. 7-right). Note that for the experiment D, only the BSIF feature descriptor is used exemplary on the Sharan14 database. In order to compare the results to the results from experiment A, the pink dash-dotted line with hexagrams in the first row of Fig. 4 needs to be looked at. For better visualization, these data points are included in Fig. 7 as dashed lines. The colors correspond to one certain (initial) tile size for the painted tiles. The results from experiment A stagnate on around 55%, 50% and 42% for 64 px, 96 px and 128 px, respectively. A mostly similar behaviour can be observed for experiment D. Thus, it can be concluded that scaling entire databases does not improve the ability to retrieve painted wood.

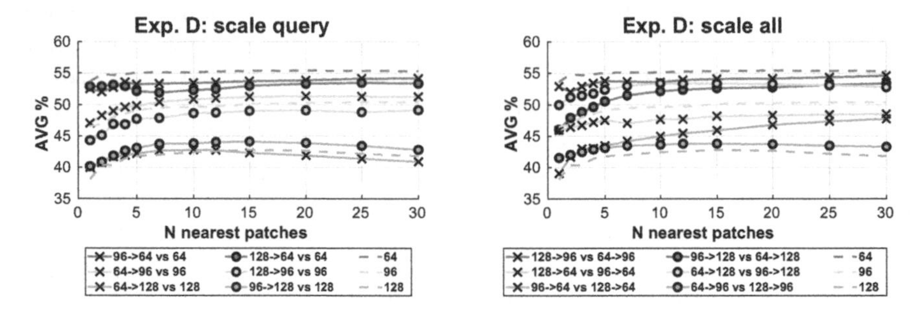

Fig. 7. Results experiment D. The legend shows the initial crop size and the final size after resizing. The first part (before vs) describes the cropping & scaling for the the query tiles, while the second part (after vs) belongs to the REALonline databases.

5 Conclusion

This study analyzed the usability of natural images containing real wood to filter image patches containing painted cross wood from a pool of images where also other painted motives are depicted. Several experiments were conducted using a three wood databases and varying patch sizes. It was found that experimental results often show a random characteristic and also the behaviour is very similar to using a different material as query. Experiments using patches of painted wood as query showed that the employed feature descriptors should be principally sufficient to capture the textural characteristics from the depicted wood. However, it appears the gap between the domains of natural wood and painted wood, at least for the used databases, is too high to deliver reliable results.

Future work will include an attempt to overcome the gap in domain by trying to model the domain shift using e.g. image to image (I2I) translation. Also paint to paint results hold the potential of improvement by resorting to more sophisticated learning based schemes. Furthermore, the annotation database will be extended. Once enough data is available, instance segmentation of crosses using state of the art segmentation tools will be done.

References

1. Arnold, T., Tilton, L.: Distant viewing: analyzing large visual corpora. Digital Scholarship in the Humanities 34(Supplement_1), i3–i16 (2019)
2. Baert, B.: A Heritage of Holy Wood: The Legend of The True Cross in Text and Image, Cultures, Beliefs and Traditions, vol. 22. Brill, Leiden (2004)
3. Cohen, N., Newman, Y., Shamir, A.: Semantic segmentation in art paintings. Comput. Graph. Forum **41**(2), 261–275 (2022)
4. Crowley, E., Parkhi, O., Zisserman, A.: Face painting: querying art with photos, pp. 1–13. British Machine Vision Association (2015)
5. Crowley, E., Zisserman, A.: The state of the art: Object retrieval in paintings using discriminative regions (01 2014)
6. Crowley, E., Zisserman, A.: The art of detection. vol. 9913, pp. 721–737 (10 2016)
7. Degler, A.: Parergon: Attribut. Material und Fragment in der Bildästhetik des Quattrocento. Verlag Wilhelm Fink, Paderborn (2015)
8. He, K., Zhang, X., Ren, S., Sun, J.: Deep residual learning for image recognition. In: Proceedings of the IEEE Conference on Computer Vision and Pattern Recognition, pp. 770–778 (2016)
9. Irvine, C.: The iconography of the cross as the green tree. In: Prickett, S. (ed.) The Edinburgh companion to the Bible and the arts, pp. 195–207. Edinburgh Univ. Press, Edinburgh (2014)
10. Jacobus de Voragine: The golden legend: Readings on the saints. Princeton University Press, Princeton, New Jersey, 2012 edition edn. (2012)
11. Jensen, R.M.: The Cross: History, Art, and Controversy. Harvard University Press, Cambridge, MA (2017)
12. Kannala, J., Rahtu, E.: Bsif: binarized statistical image features. In: Proceedings of the 21st International Conference on Pattern Recognition (ICPR2012), pp. 1363–1366 (2012)
13. Kauba, C., Debiasi, L., Schraml, R., Uhl, A.: Towards drug counterfeit detection using package paperboard classification. In: Advances in Multimedia Information Processing - Proceedings of the 17th Pacific-Rim Conference on Multimedia (PCM'16). Springer LNCS, vol. 9917, pp. 136–146. Xi'an, CHINA (2016)
14. Kim, D.Y.: Stonework and Crack in Giovanni Bellini's St. Francis in the Wilderness. In: Augart, I., Saß, M., Wenderholm, I. (eds.) Steinformen: Materialität, Qualität, Imitation, pp. 59–74. De Gruyter, Berlin, Boston (2019)
15. Kitzinger, B.E.: The Cross, the Gospels, and the Work of Art in the Carolingian Age. Cambridge University Press, New York (2019)
16. Klinke, H.: The Digital Transformation of Art History, chap. 3. Routledge (2020)
17. Kölle, J.: Versteinertes Blut. Heilig-Blut-Säulen in der flämischen Malerei um 1500. In: Augart, I., Saß, M., Wenderholm, I. (eds.) Steinformen: Materialität, Qualität, Imitation, pp. 267–278. De Gruyter, Berlin, Boston (2019)

18. Kumler, A.: Materials, Materia, "Materiality", chap. 4, pp. 95–117. John Wiley & Sons, Ltd (2019)
19. Lehman, Anne-Sophie; Bol, M.: Painting skin and water: towards a material iconography of translucent motifs in early netherlandish painting. In: Rogier van der Weyden in context, pp. 215–228 (2012)
20. Lehmann, A.S.: Small hairs: meaning and material of a multiple detail in the ghent altarpiece's Aadam and eve panels, pp. 104–116. Leuven University Press; Uitgeverij Peeters (2016)
21. Lowe, D.G.: Distinctive image features from scale-invariant keypoints. Int. J. Comput. Vision **60**(2), 91–110 (2004)
22. Matschinegg, I., Nicka, I., Hafner, C., Stettner, M., Zedlacher, S.: Daten neu verknoten: Die Verwendung einer Graphdatenbank für die Bilddatenbank REALonline, DARIAH-DE working papers, vol. 31. Niedersächsische Staats- und Universitätsbibliothek Göttingen, Göttingen (2019)
23. Mestemacher, I.: Marmor, Gold und Edelsteine. Ph.D. thesis, De Gruyter and Walter de Gruyter GmbH & Co. KG, Berlin/Boston (2021)
24. Muhammad, G.: Multi-scale local texture descriptor for image forgery detection. In: Industrial Technology (ICIT), 2013 IEEE International Conference, pp. 1146–1151 (2013)
25. Nicka, I.: Object Links in/zu Bildern mit REALonline analysieren. In: Object Links. Dinge in Beziehung. formate - Forschungen zur Materiellen Kultur, vol. 1, pp. 95–126. Institut für Realienkunde des Mittelalters und der frühen Neuzeit, Wien (2019)
26. Ojala, T., Pietikainen, M., Harwood, D.: Performance evaluation of texture measures with classification based on kullback discrimination of distributions. In: Proceedings of the 12th IAPR International Conference on Pattern Recognition. vol. 1, pp. 582–585 vol 1 (Oct 1994)
27. Ojansivu, V., Rahtu, E., Heikkila, J.: Rotation invariant local phase quantization for blur insensitive texture analysis. In: Pattern Recognition, 2008. ICPR 2008. 19th International Conference, pp. 1–4 (Dec 2008)
28. Ojansivu, V., Heikkilä, J.: Blur insensitive texture classification using local phase quantization. In: International Conference on Image and Signal Processing (2008)
29. Pastoureau, M.: Introduction à la symbolique médiévale du bois. In: Biget, J.L. (ed.) Le bois et la ville, pp. 251–264. Cahiers de Fontenay, Ecole Normale Supřieure de Fontenay Saint-Cloud, Fontenay-aux-Roses (1991)
30. Raff, T.: Die Sprache der Materialien. Zugl.: Augsburg, univ., veränd. habil.-schr., 1991, Dt. Kunstverl, München (1994)
31. Sharan, L., Rosenholtz, R., Adelson, E.H.: Accuracy and speed of material categorization in real-world images. J. Vision **14**(10) (2014)
32. Toussaint, G.: Kreuz und Knochen: Reliquien zur Zeit der Kreuzzüge. Reimer, Berlin (2011)
33. Tripps, J.: "Nimm Steine, rauh und nicht gereinigt." Cennino Cennini und das Gebirge als kunsttheoretisches Studienobjekt in der Malerei des Trecento. In: Fantasie und Handwerk. Cennino Cennini und die Tradition der toskanischen Malerei, pp. 109–120 (2008)
34. van Tongeren, L.: Ein heilsames Zeichen: die Liturgie des Kreuzes im Mittelalter. In: Heussler, C., Gensichen, S. (eds.) Das Kreuz, pp. 10–31. Regensburger Studien zur Kunstgeschichte, Schnell & Steiner, Regensburg (2013)

35. van Zuijlen, M., Pont, S., Wijntjes, M.: Painterly depiction of material properties. J. Vision **20**, 7 (07 2020)
36. van Zuijlen, M.J.P., Lin, H., Bala, K., Pont, S.C., Wijntjes, M.W.A.: Materials in paintings (mip): An interdisciplinary dataset for perception, art history, and computer vision. PLOS ONE **16**(8), 1–30 (08 2021)

Pattern Recognition for Cultural Heritage (PatReCH)

Feature Relevance in Classification of 3D Stone from Ancient Wall Structures

Giovanni Gallo[1](✉), Yaser Gholizade Atani[1], Roberto Leotta[1], Filippo Stanco[1], Francesca Buscemi[2], and Marianna Figuera[3]

[1] Department of Mathematics and Computer Science, University of Catania, Viale A. Doria 6, Catania, Italy
yaser.gholizade@phd.unict.it
[2] CNR, Institute of Heritage Sciences, Palazzo Ingrassia, via Biblioteca 4, Catania, Italy
[3] Department of Humanities, University of Catania, Piazza Dante, 32, Catania, Italy

Abstract. The increasing availability of quantitative data in archaeological studies has prompted the research of Machine Learning methods to support archaeologists in their analysis. This paper considers in particular the problem of automatic classification of 3D surface patches of "rubble stones" and "wedges" obtained from Prehistorical and Protohistorical walls in Crete. These data come from the W.A.L.(L) Project aimed to query 3D photogrammetric models of ancient architectonical structures in order to extract archaeologically significant features. The principal aim of this paper is to address the issue of a clear semantically correspondence between data analysis concepts and archaeology. Classification of stone patches has been performed with several Machine Learning methods, and then feature relevance has been computed for all the classifiers. The results show a good correspondence between the most relevant features of the classification and the qualitative features that human experts adopt typically to classify the wall facing stones.

Keywords: Machine Learning · Feature Importance · SHAP Analysis · Cultural Heritage · Archaeology

1 Introduction

Archaeologists classify artifacts integrating many knowledge sources. Training and professional experience allow them to integrate physical and geometrical observed data with historical frame, comparisons, and previous studies to formulate sound hypotheses. On the other side, the increasing availability of quantitative data, i.e. faithful 3D models, has prompted the research of computational methods to mimic the experts' skills. The use of quantitative data is not new to archaeology [5,16] nevertheless the recent advances in Machine Learning (ML) are suggesting new applications. In this context there are two major problems:

(a) Data set availability: annotated large data sets are expensive to create and validate;

G. L. Foresti et al. (Eds.): ICIAP 2023 Workshops, LNCS 14366, pp. 375–386, 2024.
https://doi.org/10.1007/978-3-031-51026-7_32

(b) Interpretation issues: ML models are frequently hard to interpret for the archaeologist. ML creates mathematical models that may combine in a non linear and indirect way the input data making difficult for the human expert to understand why these models reach their results. Lack of understanding, in turn, leads to a lack of trust slowing down the deployment of novel powerful techniques of ML to archaeology.

The W.A.L.(L) Project [19] addresses these issues by creating a semantically well structured data base of 3D models of the visible part of stones in Prehistorical and Protohistorical walls in Crete and by applying machine learning approaches to analyze such data. This project is aimed to create a large, semantically correct and well structured data base of 3D archaeological data and to experiment ML methods to extract information from these data. The project foresees the integration of the geometric information within a relational data base specifically focused on the management of ancient architecture [8].

F.B.

The W.A.L.(L) data base has been organized into many entities related to four different subject matters: localization, chronology, typology, and documentation. Two archives are the fulcrum of the whole conceptual system: "Wall facing elements" and "Masonries". Since the main purpose was the correct classification of the stones, the semantic and the vocabulary used in the "Stone types" archive has been examined in depth, taking into account the archaeological expertise involved into the project. This archive includes the label and the description of 19 types of stones, that has been distinguished essentially according to the archaeological parameters of the working degree and the size.

M.F.

The core of the W.A.L.(L) data is a collection of 3D meshes approximating the visible surface ("stone patches" in the following) of 1445 stones belonging to 12 unit walls from 4 archaeological sites in Crete (Phaistos, Ayia Triada, Sissi, Anavlochos). The stone patches have been manually segmented from the larger meshes of the wall units and expert archaeologists have labeled each stone patch through a semantic category. Stone patches are a special kind of 3D object of interest: they are very irregular open surface meshes.

Stone patches have been classified by archaeologists in several classes: Wedges, Rubble stones, one-face-worked stones, Rough extraction blocks, and so on (see Fig. 2 for more details). Among them the most common are Rubble stones and Wedges. To discriminate between Rubble stones and Wedges is a difficult issue. Archaeologists perform this task based on their knowledge and direct observation. This paper makes use of the project data to attempt an automatic discrimination between these two typologies.

3D object recognition and classification is a deeply investigated problem in computer vision and graphics. There is a huge literature addressing this problem (see the extensive reviews of [6,20]). The main approaches that have been proposed use in turn feature based representation. Among the methods based on feature representation, two type of methods are popular: using local features (e.g. SIFT) or using global features, i.e. topological structure, geometric shape invariants.

The principal aim of this paper, beyond the task of automatic classification of stone patches, is to investigate which features are more relevant for the discrimination and if they correspond with features that have a semantic relevance also for the experts of the archaeological domain. Because of this intention the classification approach that has been chosen is based on global shape features that can be easily visualized. To this purpose the raw geometrical data have been processed to obtain a tabular data set listing for each stone patch a set of global features. Six classification methods have been attempted on the tabular data: Logistic Regression [7], Support Vector Machine [7], Decision Tree [4], Random Forest [3], XGBoost [18], and LigthGBM [11].

As for Logistic Regression and Decision Tree, feature importance may be deduced from absolute value of coefficients in the regression or directly from the decision structure. On the other side, Support Vector Machine with RBF kernel, Random Forest, XGBoost, LightGBM are "black-box model", i.e., they do not provide a direct explanation of their internal reasoning. To explore the relevance of the data features a SHapley Additive exPlanations (SHAP) analysis has been performed in these cases [13]. The results show that the most relevant features for the model have a sound correspondence with qualitative features commonly taken into account by the archaeologists. This promising results suggest further lines of research to address the interpretation problem in this field.

G.G., Y.G., R.L., F.S.

The paper is structured as follows: Sect. 2 describes the data used for classification. Section 3 reports the experimental results obtained with the classifiers. Section 4 presents and discusses the obtained results with feature importance analysis on the previous models. Conclusions close the paper, summarize the results and point to further line of related research.

2 Stone Patches Data

3D reconstructions with triangular meshes of 12 wall units in 4 archaeological sites of Crete have been obtained by photogrammetry. The reconstruction has produced at first a dense mesh model that has been subsequently decimated to make computations feasible. The models used for the study are triangular meshes: mean area of triangles in the meshes is 0.1668 cm^2. Size of each model ranges from 83k triangles in the smallest wall unit, up to 1 million triangles in the largest wall unit. The 3D models will be openly accessed at the conclusion of the W.A.L.(L) project through a suitable search engine (completion of the project is estimated at the date of December 2023). Figure 1 shows a comprehensive view of the wall units that have been considered in this study.

Each wall unit model has been manually segmented by the archaeologists into stone patches. The stone patches represent only the visible surface of the whole

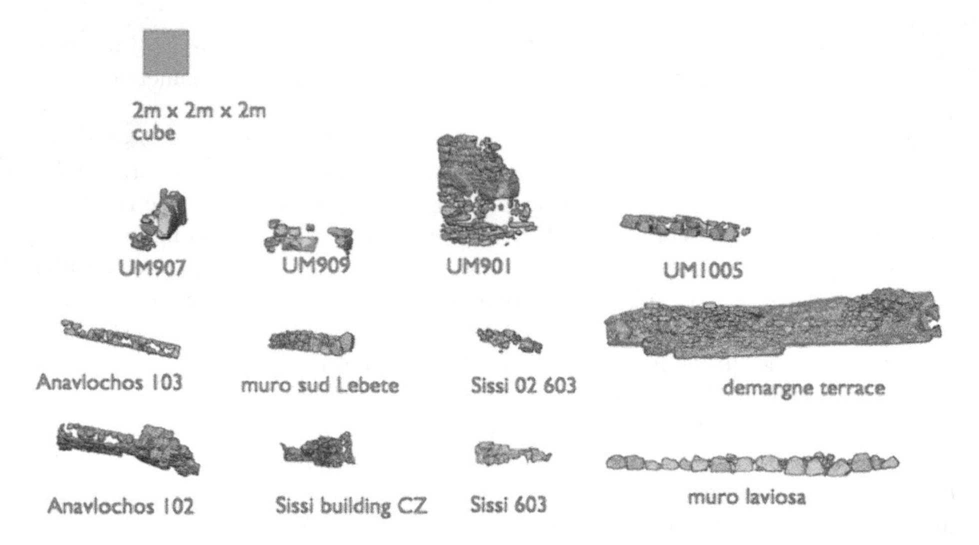

Fig. 1. Comprehensive view of the 12 wall units included in the present study.

stones. In other words they are not closed meshes, but open surfaces with very irregular geometric shapes. The number of stone patches from the models set ranges from 20 up to 364 and in total 1445 stone patches have been extracted. Archaeologists have classified these stone patches into 15 semantic categories, (see Fig. 2). Only the categories of "rubbles" and "wedges" are the object of the present study: a set of 906 stone patches (436 rubble stones and 470 wedges), see Figs. 3. More precisely:

(i) "*wedges*", are stones saturating the spaces between other larger stones;
(ii) "*rubble stones*", are unworked stones not bigger than 30 cm.

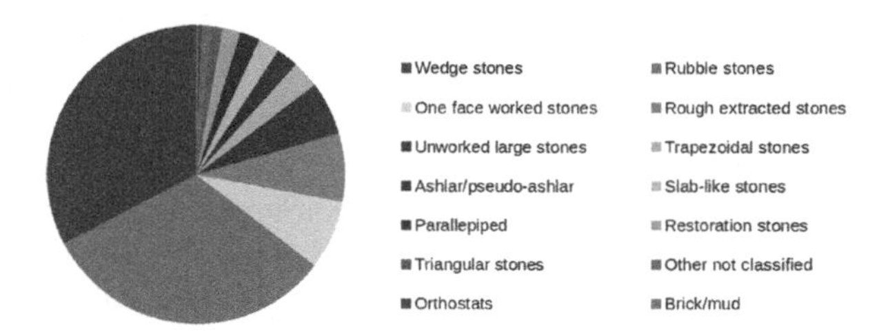

Fig. 2. Distribution of the stone categories of the complete W.A.L.(L) dataset.

Since their availability, low costs of supply and ease of use in the walls' set up, these two classes are the most recurrent in the not monumental Minoan and Geometric architecture. For this reason, they have been chosen for the present study. Other classes that can be classified according to their size (e.g. unworked

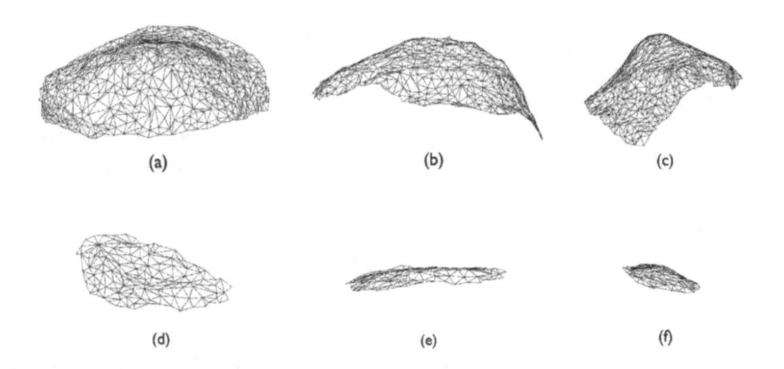

Fig. 3. (a),(b),(c) orthogonal views of a typical rubble stone patch; (d),(e),(f), orthogonal views of a typical wedge patch.

large stones) or geometric shape (e.g. ashlar) have been chosen for the study but are not part of the ML analysis reported in this paper.

F.B.

Computational approaches aimed to analyze and classify artifacts according to their shape are gaining momentum in archaeology [9,10]. Generally, these studies use 2D data (projection silhouettes of the artifacts) or 3D models that are closed meshes representation of the artifacts. Computational geometry has produced a vast literature of techniques and algorithms to deal with mesh shapes (for an overview, see [6,14,20]), nevertheless most of them are not well suited for fragmentary open surfaces as the stone patches of the present study are.

The study of stone shapes, on the other hand, has a great relevance in geo-engineering where the physical properties of stone aggregates are observed in conjunction with the irregularity of their shapes. Several shape indices have been proposed in this context. In [1] a survey of such indices is reported. For the present study the elongation and flakiness, indices proposed by Kong and Fouces [12], have been computed for the Minimum Oriented Bounding Box (MOBB) relative to each stone patch. Since these indices are computed on a MOBB relative only to a portion of the whole stone, they do not fully relate to the global stone shape but only to the visible part of it. Although this limitation makes these indices semantically less expressive, they have been demonstrated useful for automatic classification.

For each stone patch several geometrical information has been extracted as in the following list:

1. Number of faces (*integer*); since the spatial resolution is fixed, this index is related to the stone patch size.
2. Area (*float*); sum of the areas of the triangles in the mesh.
3. 3D coordinates of the stone patch centers (*float, three features*); these coordinates are referred to the median point of the unit wall.

4. Standard deviation of 3D coordinates of the stone patches (*float, three features*); these numbers are related to the spatial spread of the stone in the global reference of the unit wall.
5. Mean and standard deviation of the normal vectors of the triangles in the mesh (*float, six features*); these values provide information about the overall orientation of the patch.
6. Volume of the convex hull of the stone patches (*float*).
7. Minimum Oriented Bounding Box (MOBB) dimensions (*float, three features*); these values provide another estimate of the patch extension and orientation (see Fig. 4).
8. Azimuth and Elevation of the three axis of the MOBB (*float, two features*).
9. Elongation (*float*); it is defined according to [12]: if $a > b > c$ are the MOBB dimensions:

$$elongation(a,b,c) = 1 - \frac{b}{a} \tag{1}$$

10. Flakiness (*float*); same as above [12]: if $a > b > c$ are the MOBB dimensions:

$$flakiness(a,b,c) = 1 - \frac{c}{a} \tag{2}$$

11. Planarity (*float*); the mean distance of mesh vertices from the best approximately plane.
12. Mean and Variance of Gaussian Discrete Curvature (*float, two features*); these value are related to the roughness of the stone patches and it has been computed as in [2].
13. Position (*Categorical*); the substructure where stone belongs. For example: foundation, corner, elevation, etc.

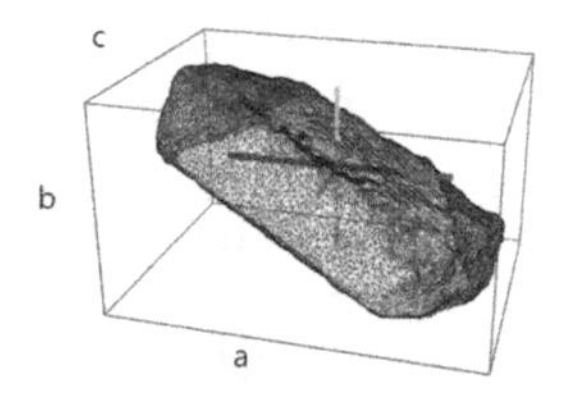

(a) Minimum Bounding Box of a stone patch.

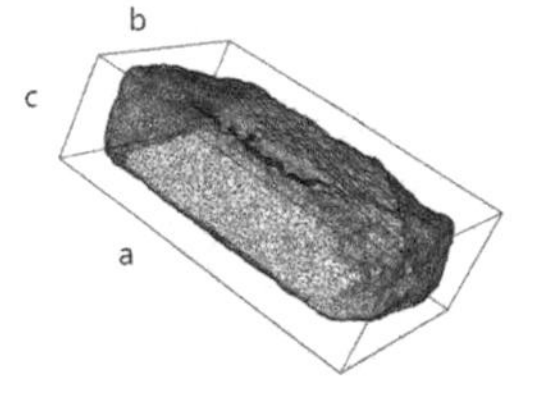

(b) Minimum Oriented Bounding Box of the same stone patch.

Fig. 4. In order to get more expressive geometrical information MOBB has been preferred as a classification feature.

It can be observed that features 2., 7., and 12. have a clear semantic correspondence to qualitative features taken into account by a human expert; on the other side the remaining features are more machinery. Notice that discrimination between wedges and rubble stones could take advantage their position inside the wall unit, i.e., wedges are typically found in between larger stones (see Fig. 5).

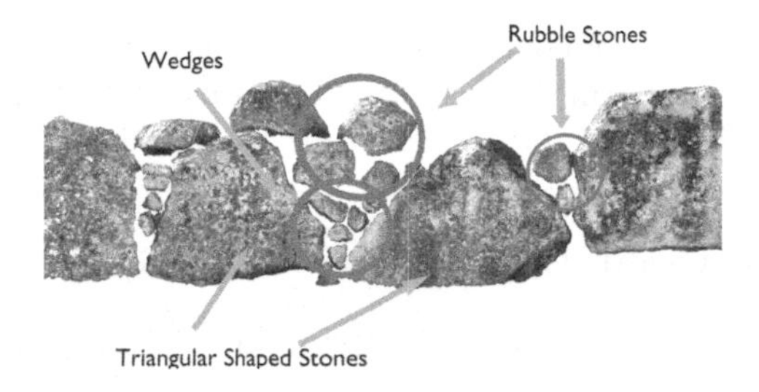

Fig. 5. A fragment of the wall unit "Muro Laviosa" (Ayia Triada) showing typical wedges and rubble stones.

In this study, for sake of simplification, this kind of information has not taken into account but it will be considered in future work. The data have been extracted into a .csv file that is available from the authors upon request.

G.G., Y.G., R.L., F.S.

3 Classification Results

The "rubble stones vs wedges" automatic classification is a challenging problem because for all the collected features the density distributions have large overlappings. Classification has been tried with Logistic Regression (LR), Support Vector Machine (SVM) with RBF kernel, Decision Trees (DT), Random Forest (RF), Extreme Gradient Boosting (XGBoost), and Light Gradient-Boosting Machine (LightGBM). The data were randomly divided into 80% and 20% slices to create respectively the training set and testing set respectively. To prevent overfitting the model the K-fold cross-validation with K=5 has been performed on the training set.

For each algorithm, classification attempts have been repeated 100 times where in each attempt the data was randomly splitted into training and test set. The mean accuracy over training data and test data for each algorithm are reported in Table 1. Random Forest, XGBoost, and LightGBM methods achieved the best accuracies.

Table 1. Mean, standard deviation and maximum of accuracy for each method.

	Training set		Testing set	
Method	Mean of Accuracy	Std of Accuracy	Mean of Accuracy	Std of Accuracy
DT	95.0 %	0.5	73.0 %	2.1
RF	**96.0** %	0.0	80.0 %	2.4
XGBoost	**96.0** %	0.2	**81.0** %	2.4
LightGBM	**96.0** %	0.1	80.0 %	2.3
SVM	73.0 %	0.8	73.0 %	3.1
LR	73.0 %	0.9	72.0 %	2.7

Table 2 reports the mean confusion matrix relative to these three algorithms and shows that they perform similarly. Each confusion matrix is the average of 100 confusion matrices which have been computed in classification attempts.

Table 2. Confusion matrix for the Random Forest, XGBoost, and LightGBM.

Model	RF		XGBoost		LightGBM	
True/Predicted	Rubble	Wedge	Rubble	Wedge	Rubble	Wedge
Rubble	38%	10 %	38%	10 %	38%	10 %
Wedge	10%	42%	10%	42%	10%	42%

G.G., Y.G., R.L., F.S.

4 Feature Importance

As stated above, the main aim of this study is to investigate the relationship among the most relevant features used by the machine learning classifiers and the features that are commonly used by the human expert. Logistic Regression and Decision Tree provide a direct way to evaluate feature importance. SVM, Random Forest, XGBoost, and LightGBM on the other side are "black-box" algorithms, i.e. the user does not have a direct way to know how they reach their conclusions. This lack of explanation may be an obstacle for the archaeologists to trust and adopt ML techniques in support to their works.

Feature importance for Logistic Regression can be deduced from the absolute value of coefficients in the regression formula. In the case of Decision Tree, the importance of a feature is computed as the normalized total reduction of the Gini impurity brought by that feature [17]. In the case of SVM, Random Forest, XGBoost and LightGBM the feature importance can be evaluated using SHAP method [15]. SHAP evaluates feature importance locally: each feature contribution is computed separately for each instance. In this way the non-linearity of the classification method is taken into account. One can observe that in the case of logistic regression and Decision Tree the feature importance is computed globally. In the case of Random Forest, XGBoost, and LightGBM the global feature importance is obtained by taking the average of the absolute SHAP values calculated for each item.

Although different techniques have to be adopted to estimate feature importance for the different classification algorithms, for sake of comparison the importance indices have been normalized for each method relatively to the maximum. Figure 6 provides a unique view of the feature importance for each method. It stacks the importance of each feature of the four methods in a unique bar.

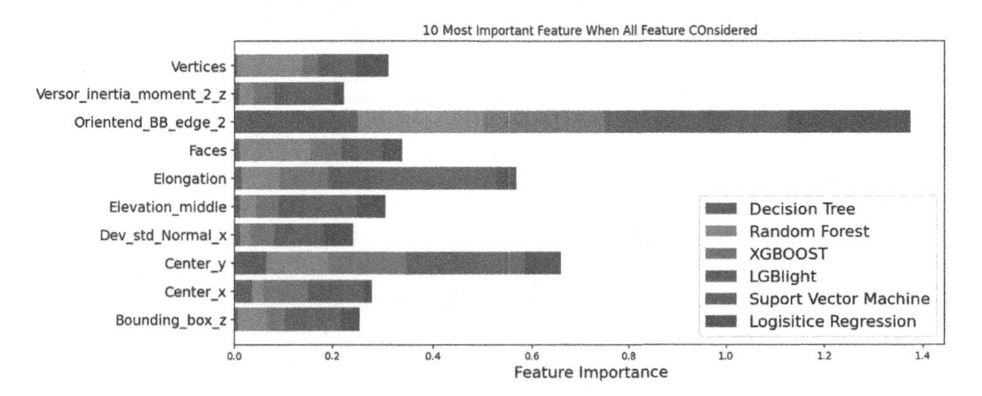

Fig. 6. Bar plot of feature importance by summing up the normalized feature score for each method.

A close examination of Fig. 6 suggests that the three more relevant features are:

1. $MOBB_2$, (middle value) of the MOBB dimensions.
2. *center_y*, i.e. the y-coordinate of the center of the stone patches.
3. elongation.

All the other features contribute with similar lesser weights.

For clear comparison between the role of each feature in the different classifiers Fig. 7 reports the importance of the top ten features. It can be observed that $MOBB_2$ is consistently the most important for all classifiers but SVM. On the other hand SVM relies on elongation more than the other methods.

Notice that $MOBB_2$ and elongation are related with the aspect ratio of the stones. In archaeologist's experience, indeed, this is perhaps one of the first evidence considered when assigning the stone type. This supports the claim that ML reasoning and expert evaluation mostly relay on the same features and encourages the adoption of automatic classifiers to support archaeology studies.

G.G., Y.G., R.L., F.S.

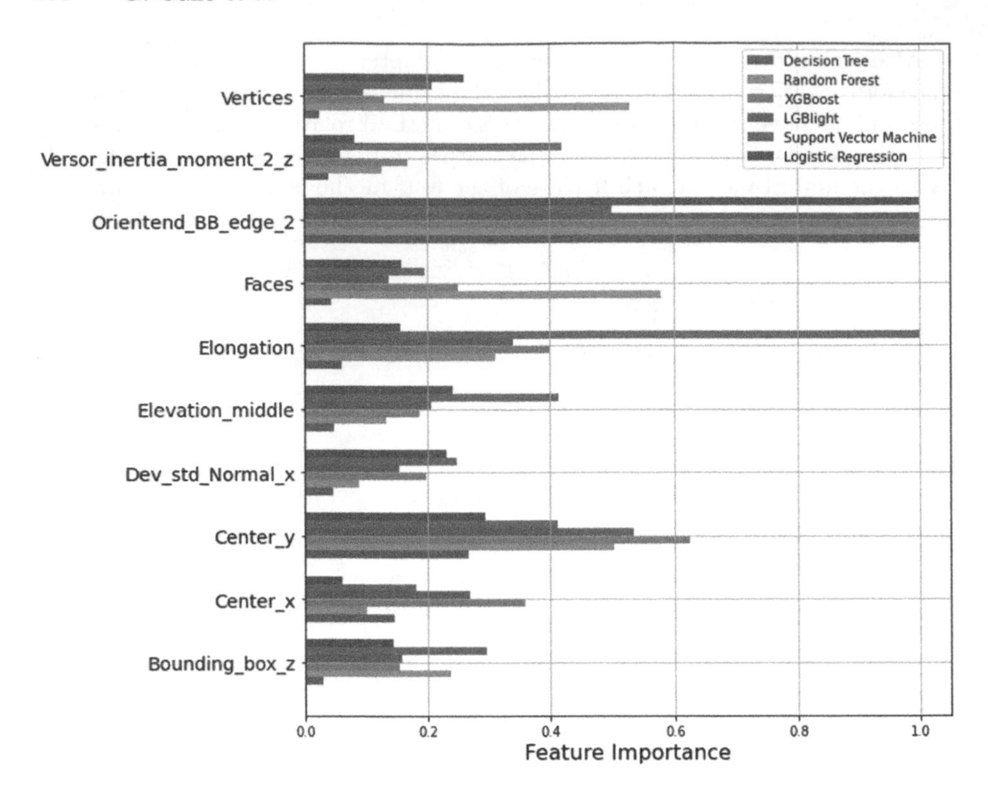

Fig. 7. Comparison feature importance of six methods for the ten most important features.

5 Conclusion

In this work, six machine learning algorithms have been tried to classify two categories of stones (Rubbles and Wedges) from the dataset of the W.A.L.(L) project. Classification has been repeated with different random splitting of the data into train and test data. Cross validation were also used in the model training process to prevent overfitting. The mean and standard deviation of accuracy for each methods show that the best result is obtained with the RF, XGBoost, and LightGBM.

Explanation is a crucial issue when automatic procedure are suggested to support studies in applicative domains, hence feature importance evaluation for the classifiers has been done. The analysis shows that all methods coherently give the same relevance to the features. Moreover, the most relevant features are in agreement with the qualitative traits that archaeologists adopt when assigning stone type.

Future work will take into consideration more stone categories as well as the spatial layout of the stones in the wall units.

References

1. Angelidakis, V., Nadimi, S., Utili, S.: Elongation, flatness and compactness indices to characterise particle form. Powder Technol. **396**, 689–695 (2022). https://doi.org/10.1016/j.powtec.2021.11.027
2. Borrelli, V., Cazals, F., Morvan, J.M.: On the angular defect of triangulations and the pointwise approximation of curvatures. Comput. Aided Geom. Des. **6**(20), 319–341 (2003). https://doi.org/10.1016/S0167-8396(03)00077-3
3. Breiman, L.: Random forests. Mach. Learn. **45**, 5–32 (2001). https://doi.org/10.1023/A:101093340432
4. Breiman, L.: Classification And Regression Trees, 1st edn. Routledge, New York (1984)
5. Caravale, A., Moscati, P.: La Bibliografia di Informatica Archeologica nella Cultura Digitale Degli Anni Novanta (30 anni di informatica archeologica - 1). All'Insegna del Giglio, Firenze (2021)
6. Carvalho, L.E., von Wangenheim, A.: 3D object recognition and classification: a systematic literature review. Pattern Anal. Appl. **22**, 1243–1292 (2019). https://doi.org/10.1007/s10044-019-00804-4
7. Dangeti, P.: Statistics for Machine Learning, 3rd edn. Packt Publishing Ltd., Birmingham (2017)
8. Gallo, G., Buscemi, F., Ferro, M., Figuera, M., Marco Riela, P.: Abstracting stone walls for visualization and analysis. In: Del Bimbo, A., et al. Pattern Recognition. ICPR International Workshops and Challenges. ICPR 2021, LNCS, vol. 12667, pp. 215–222. Springer, Cham (2021). https://doi.org/10.1007/978-3-030-68787-8_15
9. Grosman, L.: Reaching the point of no return: the computational revolution in archaeology. Annu. Rev. Anthropol. **1**(45), 129–145 (2016). https://doi.org/10.1146/annurev-anthro-102215-095946
10. Iovita, R.: Shape variation in Aterian tanged tools and the origins of projectile technology: a morphometric perspective on stone tool function. PLoS ONE **12**(6), 1–14 (2011). https://doi.org/10.1371/journal.pone.0029029
11. Ke, G., et al.: LightGBM: a highly efficient gradient boosting decision tree. In: 31st Conference on Neural Information Processing Systems (NIPS), Long Beach, CA, USA (2017)
12. Kong, D., Fonseca, J.: Quantification of the morphology of shelly carbonate sands using 3D images. Géotechnique **68**, 249–261 (2018). https://doi.org/10.1680/jgeot.16.P.278
13. Lundberg, S.M., Lee, Su-In.: A Unified Approach to Interpreting Model Predictions. In: Advances in Neural Information Processing Systems, Curran Associates Inc, vol. 30, pp. 1–10 (2017)
14. Mari, J.L., Hétroy-Wheeler, F., Subsol, G.: Geometric and Topological Mesh Feature Extraction for 3D Shape Analysis. Wiley-ISTE (2019)
15. Molnar, C.: Interpretable Machine Learning: A Guide for Making Black Box Models Explainable. Addison Wesley Longman Publishing Co., Inc., Boston (2020)
16. Moscati, P.: 30 Anni Di Archeologica e Calcolatori. Tra Memoria e Progettualita. All'Insegna del Giglio. CNR - Istituto di Scienze del Patrimonio Culturale, Firenze, Italy (2019)
17. Suthaharan, S.: Decision tree learning. In: Machine Learning Models and Algorithms for Big Data Classification. Integrated Series in Information Systems, vol. 36, Springer, Boston, MA. (2016). https://doi.org/10.1007/978-1-4899-7641-3-10

18. Tianqi, C., Carlos, G.: XGBoost: a scalable tree boosting system. In: Proceedings of the 22nd ACM SIGKDD International Conference on Knowledge Discovery and Data Mining, pp. 785–794 (2016). https://doi.org/10.1145/2939672.2939785
19. W.A.L. (L), Wall-facing Automatic images identification Laboratory. A quantitative analysis method for the study of ancient architecture, International Archaeological Joint Laboratories, Financed by the National Research Council(CNR), P.I. Francesca Buscemi, (2020–2021)
20. Wang, H., Zhang, J.: A survey of deep learning-based mesh processing. Commun. Math. Stat. **10**, 163–194 (2022). https://doi.org/10.1007/s40304-021-00246-7

Gamification in Cultural Heritage: When History Becomes SmART

Mario Casillo[1], Francesco Colace[2], Francesco Marongiu[2], Domenico Santaniello[1], and Carmine Valentino[2](✉)

[1] DISPAC – Dipartimento del Patrimonio Culturale, University of Salerno, Fisciano, Italy
{macasillo,dsantaniello}@unisa.it

[2] DIIn – Dipartimento di Ingegneria Industriale, University of Salerno, Via Giovanni Paolo II, 132, 84084 Fisciano, SA, Italy
{fcolace,fmarongiu,cvalentino}@unisa.it

Abstract. The term "Gamification" refers to the use of techniques and features typically found in the gaming industry, applied to contexts that extend beyond traditional entertainment video games. It has been observed that these techniques are particularly effective for learning information, as users are actively engaged. The field of cultural heritage can also benefit from the use of gamification techniques. Often, the dissemination of tangible and intangible cultural heritage poses certain challenges, especially from a communication standpoint. In this context, the use of gamification-based approaches can significantly enhance the ability to disseminate culture by placing users at the center of the visiting experience and fostering a stronger connection with the cultural heritage. By employing such approaches, it is possible to create immersive and memorable experiences that enhance users' learning capabilities. Innovative technologies like augmented reality and localization systems can provide valuable support in developing gamified visitor paths. A case study was proposed and developed within the Archaeological Park of Pompeii, and the experimental results were highly encouraging, particularly demonstrating greater benefits for younger age groups compared to a traditional visit.

Keywords: Gamification · Cultural Heritage · Augmented Reality

1 Introduction

Gamification is a concept that has gained increasing popularity in recent years. It is an innovative approach that applies game elements and mechanics to non-game contexts with the aim of engaging, motivating, and entertaining people. Gamification leverages humans' natural inclination for competition, gratification, and goal achievement, transforming everyday activities into more engaging and exciting experiences. By using rewards, points, leaderboards, challenges, and badges, gamification aims to create a game-like dynamic that encourages users to take specific actions or achieve certain results. Gamification has proven to be effective in fostering skill acquisition, customer loyalty, and performance improvement. It offers an interactive and rewarding experience

G. L. Foresti et al. (Eds.): ICIAP 2023 Workshops, LNCS 14366, pp. 387–397, 2024.
https://doi.org/10.1007/978-3-031-51026-7_33

that can positively influence behavior and participation. With its versatility and ability to actively engage individuals, gamification has emerged as a powerful strategy to motivate, entertain, and enhance the human experience across a wide range of contexts.

Gamification has proven to be a valuable resource for enhancing the visiting experience of large-scale archaeological sites and major museums, offering an innovative and exciting way to engage and entertain visitors of all ages. Through the application of game-like elements, gamification provides the opportunity to transform a static and passive visit into an interactive, engaging, and educational experience. Large-scale archaeological sites, such as ancient city sites or archaeological complexes, can often be complex and challenging to explore. Gamification offers a solution to address this challenge by providing interactive tools to guide visitors through the site and make their exploration more engaging. By utilizing mobile applications or augmented reality devices, visitors can participate in games, solve puzzles, or complete missions that lead them to the discovery of historical and cultural information. This approach makes the visiting experience more enjoyable, allowing visitors to immerse themselves in the atmosphere of the place and feel intimately connected to the history that underlies those sites.

Large museums can also benefit from the application of gamification to enhance the visitor experience. Often, museums house vast and complex collections that can be overwhelming for visitors. Gamification offers the opportunity to make the exploration of museum galleries more engaging and interactive. For example, visitors can participate in treasure hunts, quiz games, or interactive activities that encourage them to explore different sections of the museum in search of answers or clues. This not only stimulates visitors' interest but also allows them to deepen their knowledge in a fun and engaging way.

The use of innovative technologies, such as virtual reality or augmented reality, can further enrich the visitor experience. By utilizing these technologies, visitors can "travel through time" and immerse themselves in digital reconstructions of ancient sites or artworks. They can virtually explore inaccessible or destroyed spaces, engaging in a realistic and immersive experience. These technologies can also provide additional information, such as historical insights or details about art, through projections or interactive visualizations. To allow technological tools to support and enhance the user experience, information is needed to understand user preferences and the environment they are immersed in. By identifying user preferences, it is possible to provide a personalized gaming experience immersed in the world of cultural heritage, appropriately selecting Points of Interest. Understanding the user's environment, or the context [1] in which the experience takes place, is a focal point for understanding both the most convenient points to reach for the user and the environmental or crowd conditions that could limit the enjoyment of the visit for the user. By leveraging this information effectively, it is possible to develop a pleasant and seamless visitor path associated with the gaming experience. This will motivate users to successfully complete the objectives proposed by the game.

Therefore, for an effective gamification approach, it is necessary to identify the most suitable technological resources for the specific use case. Regarding the case under analysis, three main technologies have been identified that can provide added value compared to a traditional visit:

- **Localization systems**: By being able to track visitors in real-time within an archaeological park or museum, it is possible to offer users a much more precise and accurate visiting mode. Additionally, it allows for better visitor management by the facility itself, as it enables the precise monitoring of human traffic flow.
- **Augmented reality**: This innovative technological tool allows for the creation of immersive experiences for users in relation to their physical surroundings. These experiences are particularly intriguing as they place the user at the center of the visiting experience, creating a memorable encounter with what they have seen [2].
- **Recommendation systems**: Recommendation systems (RS) allow for the personalization of services provided to users. In the context of gamification, they contribute to making the experience more immersive and further motivate users to continue playing [3]. This is possible because RS can analyze and filter information to provide suggestions based on the preferences of the users they interact with. Based on the analyzed data, RS can be divided into three main categories: Content-Based, Collaborative Filtering, and Hybrid systems [4]. The first category aims to build user and service profiles to determine user-service affinity through measures of similarity. The second category utilizes interactions between users and services to make predictions about services that users have not yet encountered. Hybrid RS, on the other hand, combine the strengths of the first two categories to overcome the limitations of individual recommendation techniques.

The purpose of this work is to demonstrate how gamification techniques, combined with a robust technological foundation, can be a valuable aid in enhancing and promoting tangible and intangible cultural heritage. The study is structured around the presentation of the proposed approach at a theoretical level, along with a case study applied to the archaeological park of Pompeii.

2 Related Works

There are numerous examples in the literature that suggest how gamification techniques can be useful for the enhancement of cultural heritage. The creation of immersive virtual scenarios [5], for instance, can aid in widespread dissemination even towards cultures and backgrounds different from the original context.

Teaching methods for cultural heritage can also benefit from gamification techniques [6]. Approaches based on the concept of "challenge" enable a more effective dissemination of information, enhancing student learning, for example.

From a visitor experience perspective, there is certainly an improvement, particularly in terms of engagement [7].

In the field of cultural heritage, there is a vast literature on the use of Context-Aware Recommender Systems (CARS) integrated with contextual information [8]. The use of CARS enhances the cultural experience of users visiting a museum or an archaeological park [9, 10]. Specifically, the contextual information most utilized is related to the user's location, the crowd levels at Points of Interest (POIs), and, in the case of archaeological sites, the weather conditions [10, 11].

A critical phase of recommendation involves integrating context for generating suggestions. In the literature, there are three possible strategies:

- Contextual Pre-Filtering: Contextual information acts as a filter for the inputs of the recommendation phase, generating suggestions based on the evaluated context [8];
- Contextual Post-Filtering: The contextual filter allows for the selection of outputs from the recommendation phase, choosing the suggestions that are most suitable for the environmental conditions experienced by the user [8];
- Contextual Modeling: The contextual filter is integrated into the phase of generating contextual recommendations. In this case, new recommendation strategies are developed [8]. In the case of Content-Based approaches, it is necessary to incorporate context into the calculation of user similarity. For Collaborative Filtering, tensor factorization techniques can be employed, or alternatively, contextual information is integrated into matrix factorization-based methods [12, 13].

3 Proposed Approach

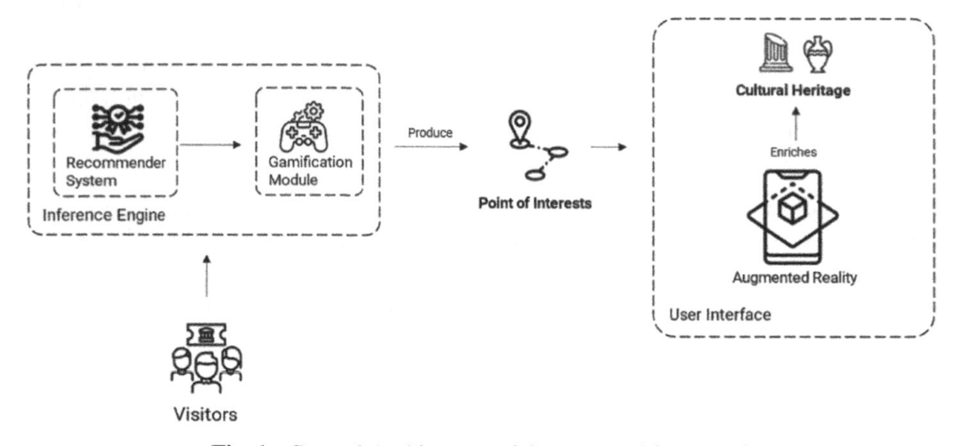

Fig. 1. General Architecture of the proposed framework

The architecture of the application primarily consists of two essential sections: the Inference Engine and the User Interface. The first section is responsible for creating a personalized path based on the specific needs of the visitor, utilizing contextual information from the location. The goal is to provide a tailored and targeted experience. On the other hand, the User Interface focuses on displaying the gamification elements generated in the previous phase (Fig. 1). This means creating an interactive and engaging environment for users that guides them through the personalized path created by the Inference Engine. The user interface's task is to present information and game elements in an intuitive and captivating way.

In the proposed architecture, augmented reality (AR) is extensively utilized to enhance the experience of accessing cultural heritage. AR technology enables the overlay of virtual elements onto the real world, thus enriching the visit and offering interactive and informative content. This innovative approach allows users to explore cultural heritage in a more engaging and immersive way, providing them with additional and stimulating information throughout the journey.

3.1 Inference Engine

This module is further divided into two distinct phases. First, the recommender system generates the most suitable Points of Interest (POIs) for the user. These POIs are then passed to a gamification module, which creates an additional presentation layer following a specific approach.

Recommender System The Recommender System aims to identify the most suitable Points of Interest (POIs) for the user, to determine a gamification path that can be of greater interest and motivation. To achieve this, it is necessary to identify the user's preferences through a questionnaire administered during the initial interaction with the game. In this phase, the user is asked to rate 10 POIs of the visited archaeological site on a scale ranging from 1 to 5. Concurrently, using sensors installed within the park, contextual information regarding weather conditions, POI crowding, and distance between the user and various Points of Interest can be collected. These pieces of information are processed by the Context-Aware Recommender System, utilizing Singular Value Decomposition [14] integrated with contextual biases [13] that adjust the recommendations based on the current context. Furthermore, to evaluate contextual conditions, the system considers a single "embedded" context in which contextual information is described through a tuple (weather, crowdedness, location).

In particular, the calculation of the recommendations $\hat{r}_{uic}$ is done as follows:

$$\hat{r}_{uic} = \overline{r}_c + b_{uc} + \tilde{b}_{ic} + p_u q_i^t \quad (1)$$

where:

- $\overline{r}_c$ is the average of known ratings in the embedded context c;
- b_{uc} is the bias associated with user u in the embedded context c;
- $\tilde{b}_{ic}$ is the bias associated with POI i in the context c;
- p_u and q_i describe the preferences of user u and the features of POI i.

The parameters related to the predictions generated by the system (b_{uc}, $\tilde{b}_{ic}$, p_u, q_i) are learned through the known contextual ratings using the technique of Stochastic Gradient Descent. The generated list of contextual recommendations serves as input for the Gamification Module.

Gamification Module The gamification module is responsible for creating a game environment in which users can participate and perform specific actions. Firstly, a type of gamification technique is selected, and the objectives to be achieved by the user or a group of users are defined. Utilizing the information provided by the previous module, the challenge is customized based on the user's personal preferences and available contextual information. The primary goal of this module is to offer the best possible experience to users by creating points of interest and a challenge that spans the entire journey. This means that the module focuses on structuring the game in a coherent and engaging manner, ensuring that users are motivated to actively participate and achieve the set objectives.

This process requires careful planning and meticulous design of the gaming experience. Points of interest along the journey are identified, such as specific areas within an archaeological site or artworks in a museum, which can capture the users' interest. These points of interest become an integral part of the challenge, providing users with specific objectives to achieve. Furthermore, the challenge is designed to adapt to the preferences and characteristics of the users. Contextual information, such as geographic location or personal preferences, is utilized to create a customized challenge that reflects the users' needs and tastes. In this way, the gaming experience becomes personalized and engaging, enhancing the level of involvement and entertainment for users of different age groups.

Overall, the gamification module focuses on creating an engaging and personalized gaming environment, providing users with a stimulating challenge that spans the entire journey. The goal is to offer the best possible experience to users by actively involving them and motivating them to achieve the set objectives.

3.2 User Interface

The user interface module plays a fundamental role in presenting all the results obtained from the previous modules to the end user. By using approaches typical of the gaming world and leveraging advanced technologies, the calculated paths from the Inference Engine are presented to the user in an engaging manner. Specifically, the use of augmented reality (AR) is particularly suited to this use case as it allows for the enrichment of the physical reality with additional information otherwise impossible to obtain. This technology enables the overlay of digital elements onto the real-world context, allowing users to view additional information, such as historical data, detailed descriptions, or virtual reconstructions, directly on the screen of their device. This contributes to creating an immersive and interactive experience that enhances user interest and engagement. Moreover, augmented reality seamlessly integrates with gamification mechanisms. Thanks to its ability to provide immediate and intuitive interaction, augmented reality allows users to interact with game elements in an easy and natural way, meeting their expectations. For example, users could be challenged to find hidden objects or solve puzzles in the real-world context using augmented reality as a supportive tool. This creates an engaging and enjoyable experience, stimulating active user participation.

The user interface fully leverages the potential of augmented reality to present the results obtained from the previous modules in an appealing and engaging manner. By using innovative technologies, users can enjoy an enriched experience where reality blends with the virtual element, creating an immersive and stimulating gaming environment. The integration of augmented reality with gamification mechanisms offers a comprehensive user experience, where interaction, gratification, and fun come together to provide an engaging and memorable cultural visit.

4 Case of Study

The proposed approach has been implemented in a real case study within the context of the Archaeological Park of Pompeii. A mobile platform video game has been developed with the aim of offering young users an innovative and engaging method to discover the archaeological park. The gamification technique employed is the treasure hunt. The basic idea is to guide visitors through the archaeological park using a map that showcases the points of interest along their personalized journey.

By using augmented reality, the points of interest take on a new dimension, transforming into hidden treasures that users must discover during their visit. Through careful storytelling, information is presented to users in a way that is tailored to their target audience, making it as understandable and assimilable as possible. This means that historical and cultural information is conveyed through engaging narratives, using a language and approach that are suitable for the cognitive and learning abilities of young users.

4.1 Internal Architecture

The game was internally developed using a range of open-source software and libraries, utilizing Java as the programming language. Augmented reality, on the other hand, was implemented by leveraging the two main frameworks supported by major mobile platforms: ARKit for iOS and ARCore for Android.

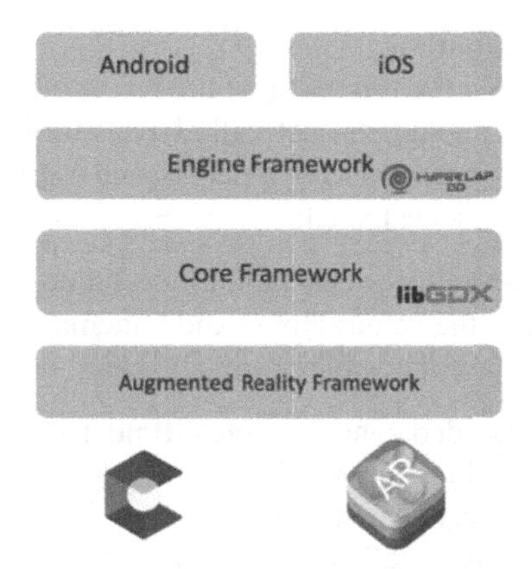

Fig. 2. Main Technological Components of the Prototype Architecture

The architecture was developed based on three levels of abstraction (Fig. 2). An augmented reality framework was developed specifically for the use case. The two backends related to ARCore and ARKit were concurrently developed.

4.2 Gameplay

The treasure hunt is presented to the user in the form of an interactive map, and through the mobile device's location tracking systems, the user is guided to reach the point of interest (Fig. 3).

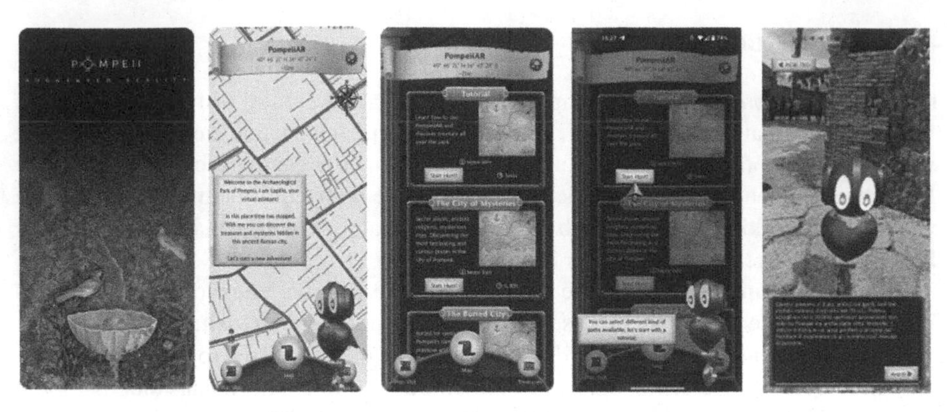

Fig. 3. Some Snapshots of the app at work

5 Results

The system was evaluated on two main aspects. Firstly, the Recommender System module for generating personalized paths was evaluated, and secondly, the game was assessed in an experimental phase and evaluated by target users through a questionnaire. Both results demonstrate that the system is indeed effective and able to enhance users' visiting experience.

Recommender System Validation
In the first experimental phase, the validation of the recommendation system is targeted. For this purpose, a state-of-the-art dataset called DePaulMovie [15] is utilized. The dataset consists of 5043 known ratings from ninety-seven users and seventy-nine movies. Additionally, three contexts are evaluated:

- Time, which can have the values weekend and weekday.
- Location, which can have the values Home and Cinema.
- Companion, which can have the values Alone, Partner, and Family.

For considering the embedded context, tuples (time, location, companion) are taken into account, while contextual recommendations are generated using Eq. (1).

To evaluate the obtained results, two accuracy techniques are employed: Mean Absolute Error (MAE) and Root Mean Squared Error (RMSE) [11]. Table 1 summarizes the results achieved by the proposed contextual recommendation approach compared to some state-of-the-art matrix factorization-based techniques. From the results obtained, it can be observed that the proposed contextual recommendation approach outperforms the comparison methods.

5.1 Users Valuation

The system was evaluated using a questionnaire consisting of nine questions, in which users were asked to express their preferences regarding various aspects of the visit and the game. The sample of thirty-two users was divided into age groups, and the results are summarized in Figs. 4 and 5.

Table 1. Recommender System validation results.

	MEA	RMSE
CAMF-C	0.7029	0.9571
CAMF-CI	0.7122	0.9660
Proposed RS	0.6591	0.9186

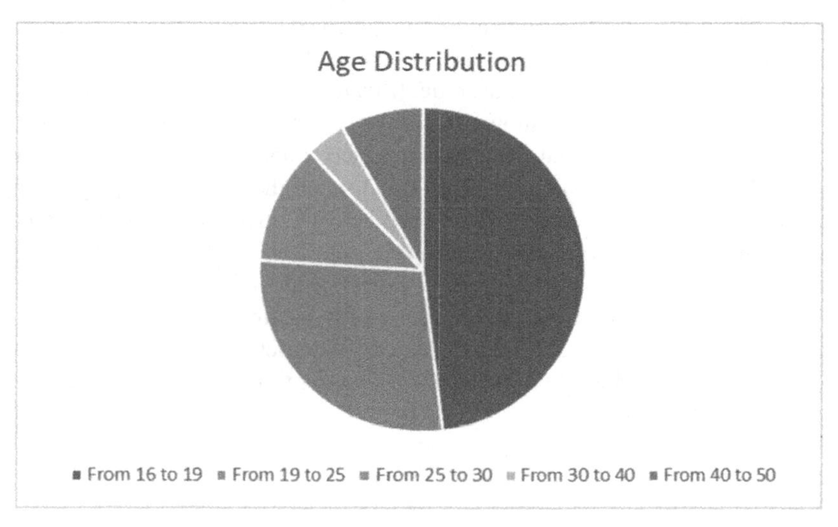

Fig. 4. Test users divided by age.

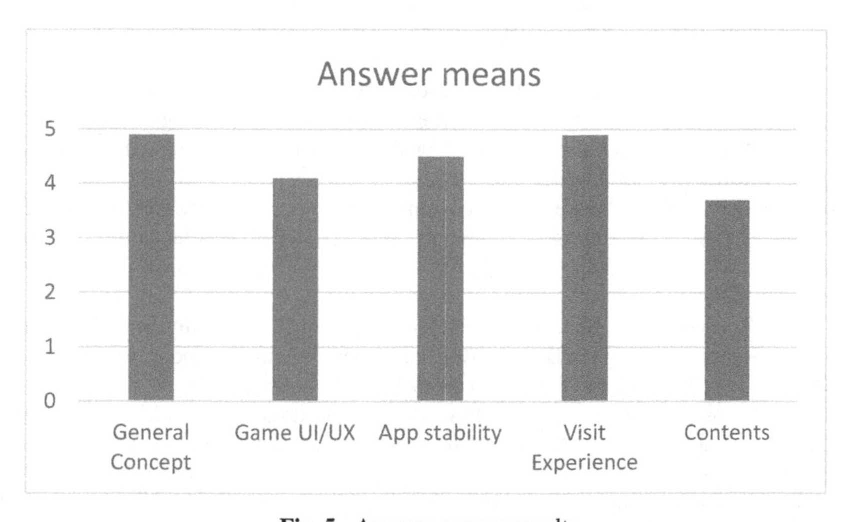

Fig. 5. Average answer results

The results highlighted some criticisms regarding usability and content; however, the overall user experience was extremely positive, demonstrating the effectiveness of this approach in enhancing experiences within archaeological parks.

6 Conclusions

This work aims to improve the visiting experience in large archaeological parks and museums using gamification approaches and the integration of innovative technologies. To achieve this goal, a methodological and technical framework has been developed, enabling the implementation of these techniques combined with the use of contextual recommendation systems for precise user identification and personalized visiting experiences. By applying gamification strategies and leveraging innovative technologies such as augmented reality and virtual reality, an interactive and engaging environment has been created for visitors. Gamification has transformed the visit into a playful experience where visitors are motivated to achieve objectives, solve puzzles, or participate in challenges, making the visiting experience more enjoyable and engaging for people of all ages. Additionally, the use of contextual recommendation systems has allowed for precise user identification and the adaptation of the visiting experience to their preferences and interests. This enables personalized itineraries, suggesting specific points of interest and activities based on the user's profile, optimizing their engagement and satisfaction during the visit.

References

1. Abowd, G.D., Dey, A.K., Brown, P.J., Davies, N., Smith, M., Steggles, P.: Towards a better understanding of context and context-awareness. In: Gellersen, H.-W. (ed.) HUC 1999. LNCS, vol. 1707, pp. 304–307. Springer, Heidelberg (1999). https://doi.org/10.1007/3-540-48157-5_29
2. Clarizia, F., Colace, F., Santo, M. De, Lorusso, A., Marongiu, F., Santaniello, D.: Augmented Reality and Gamification technics for visit enhancement in archaeological parks. In: 2022 IEEE 2nd IoT Vertical and Topical Summit for Tourism (IoTT), pp. 1–4. IEEE (2022). https://doi.org/10.1109/IoTT56174.2022.9925901
3. Tondello, G.F., Orji, R., Nacke, L.E.: Recommender systems for personalized gamification. In: Adjunct Publication of the 25th Conference on User Modeling, Adaptation and Personalization, pp. 425–430. ACM, New York, NY, USA (2017). https://doi.org/10.1145/3099023.3099114
4. Ricci, F., Rokach, L., Shapira, B.: Recommender systems: introduction and challenges. In: Ricci, F., Rokach, L., Shapira, B. (eds.) Recommender Systems Handbook. Springer, Boston, MA, pp. 1–34 (2015). https://doi.org/10.1007/978-1-4899-7637-6_1
5. Aldana-Burgos, L.M., Gaona-García, P.A., Restrepo-Rodriguez, A.O., Montenegro-Marín, C.E.: Gamification: preserving cultural heritage through immersive scenarios and second language strengthening. In: Abreu, A., Liberato, D., Garcia Ojeda, J.C. (eds.) Advances in Tourism, Technology and Systems. Smart Innovation, Systems and Technologies, vol. 293, pp. 287–297, Springer, Singapore (2022). https://doi.org/10.1007/978-981-19-1040-1_25
6. Yáñez de Aldecoa, C., Gómez-Trigueros, I.M.: Challenges with complex situations in the teaching and learning of social sciences in initial teacher education. Soc Sci. **11**, 295 (2022). https://doi.org/10.3390/socsci11070295

7. Donadio, M.G., Principi, F., Ferracani, A., Bertini, M., Del Bimbo, A.: Engaging museum visitors with gamification of body and facial expressions. In: Proceedings of the 30th ACM International Conference on Multimedia, pp. 7000–7002. ACM, New York, NY, USA (2022). https://doi.org/10.1145/3503161.3547744
8. Adomavicius, G., Mobasher, B., Ricci, F., Tuzhilin, A.: Context-aware recommender systems. AI Mag. **32**, 67–80 (2011). https://doi.org/10.1609/aimag.v32i3.2364
9. Ruotsalo, T., et al.: SMARTMUSEUM: a mobile recommender system for the web of data. J. Web Semant. **20**, 50–67 (2013). https://doi.org/10.1016/j.websem.2013.03.001
10. Alexandridis, G., Chrysanthi, A., Tsekouras, G.E., Caridakis, G.: Personalized and content adaptive cultural heritage path recommendation: an application to the Gournia and Çatalhöyük archaeological sites. User Model User-adapt Interact. **29**, 201–238 (2019). https://doi.org/10.1007/s11257-019-09227-6
11. Bartolini, I., et al.: Recommending multimedia visiting paths in cultural heritage applications. Multimed Tools Appl. **75**, 3813–3842 (2016). https://doi.org/10.1007/s11042-014-2062-7
12. Chen, W., Hsu, W., Lee, M.L.: Making recommendations from multiple domains. In: Proceedings of the 19th ACM SIGKDD International Conference on Knowledge Discovery and Data Mining, pp. 892–900. ACM, New York, NY, USA (2013). https://doi.org/10.1145/2487575.2487638
13. Baltrunas, L., Ludwig, B., Ricci, F.: Matrix factorization techniques for context aware recommendation. In: Proceedings of the Fifth ACM Conference on Recommender Systems, pp. 301–304. ACM, New York, NY, USA (2011). https://doi.org/10.1145/2043932.2043988
14. Bokde, D., Girase, S., Mukhopadhyay, D.: Matrix factorization model in collaborative filtering algorithms: a survey. Procedia Comput. Sci. **49**, 136–146 (2015). https://doi.org/10.1016/j.procs.2015.04.237
15. Ilarri, S., Trillo-Lado, R., Hermoso, R.: Datasets for context-aware recommender systems: current context and possible directions. In: 2018 IEEE 34th International Conference on Data Engineering Workshops (ICDEW), pp. 25–28. IEEE (2018). https://doi.org/10.1109/ICDEW.2018.00011

Classification of Turkish and Balkan House Architectures Using Transfer Learning and Deep Learning

Veli Mustafa Yönder[1,2(✉)], Emre İpek[1], Tarık Çetin[1], Hasan Burak Çavka[1], Mehmet Serkan Apaydın[1,3], and Fehmi Doğan[1]

[1] İzmir Institute of Technology, İzmir 35433, Turkey
velimustafa.yonder@dpu.edu.tr

[2] Department of Architecture, Faculty of Architecture, Kütahya Dumlupınar University, Kütahya 43100, Turkey

[3] Department of Computer Science, Faculty of Engineering and Natural Sciences, Acibadem Mehmet Ali Aydinlar University, İstanbul 34752, Turkey

Abstract. Classifying architectural structures is an important and challenging task that requires expertise. Convolutional Neural Networks (CNN), which are a type of deep learning (DL) approach, have shown successful results in computer vision applications when combined with transfer learning. In this study, we utilized CNN based models to classify regional houses from Anatolia and Balkans based on their architectural styles with various pretrained models using transfer learning. We prepared a dataset using various sources and employed data augmentation and mixup techniques to solve the limited data availability problem for certain regional houses to improve the classification performance. Our study resulted in a classifier that successfully distinguishes 15 architectural classes from Anatolia and Balkans. We explain our predictions using grad-cam methodology.

Keywords: architectural classification · resnet · inception · convnext · transfer learning · cnn · grad-cam

1 Introduction

In human history, people have transitioned from the hunter-gatherer phase to settled societies, during which they needed permanent structures for shelter. When constructing these buildings, factors such as the geographical and climatic conditions of the regions, available materials, tools and equipment at hand, and cultural beliefs have been influential in determining the shapes and styles of their constructions. Exploring the similarities and differences between structures in architecture, and explaining them, is an important area of study. Architectural factors have developed differently in each geography, influencing and diverging from one another. Each region has had its own distinctive homes and structural features. The study of these distinguishing features has traditionally been conducted through manual methods, but with advancements in artificial intelligence techniques, it is possible to study architectural theories using automated methods.

G. L. Foresti et al. (Eds.): ICIAP 2023 Workshops, LNCS 14366, pp. 398–408, 2024.
https://doi.org/10.1007/978-3-031-51026-7_34

Convolutional Neural Networks (CNNs) in computer vision demonstrated the most successful performance in the ImageNet classification competition for the first time in 2012, using deep learning. This approach, known as AlexNet [1], has since evolved with models such as ResNet [2] and VGG [3] and achieved even higher levels of success. As an alternative to CNNs, methods based on Vision Transformers have been proposed [5] for image classification. In these methods, the transformer architecture, which was initially devised for natural language processing, has been utilized in the context of computer vision applications. Although it has achieved higher success rates due to larger datasets and larger number of parameters, it has not taken advantage of the convolution operator which makes CNN's so much more efficient when applied to images. Facebook AI Research group introduced a model called ConvNeXt [4] in 2022, constructed solely from standard ConvNet (convolutional network aka CNN) modules. ConvNeXts demonstrated competitive performance compared to vision transformer models in terms of accuracy but with much better efficiency. ConvNeXt model is a CNN-based approach that achieves higher performance compared to vision transformers. It accomplishes this by modernizing the CNN architectures to the state-of-the-art methods in computer vision applications. Transfer learning is a method that allows to reuse the parameters of a trained model on a different dataset; by changing these parameters in a process called fine tuning, minimizing training time and resource requirements compared to training from scratch. Both CNN and Vision Transformer pretrained models are available for fine tuning.

Fast.ai is a software library that enables easy implementation of deep learning applications [6]. This library is built on PyTorch. It allows for easy loading and finetuning of pre-trained models, includes the implementation of many practical techniques from the literature collected from various papers for achieving state of the art results with few lines of code, such as learning rate finder and data augmentation methods, and is easy to get started with.

Ottoman architecture has left a distinct mark on Anatolia and the Balkans since the 19th century. Studying this architecture using artificial intelligence can help validate existing hypotheses and also integrate these insights into future architectural projects. In our study, we classified architectural structures obtained from 15 different regions in the Balkans and Anatolia based on building photographs. We utilized Fast.ai, PyTorch software libraries, as well as the Inception [7], ResNet and ConvNext architectures in our approach. We collected images of selected Ottoman architectural buildings through the Google Street View tool. We applied data augmentation techniques such as Mixup to improve class imbalance problems. We developed a classifier that can distinguishes 15 distinct architectural styles, achieving a validation accuracy of 94.6% and a test accuracy of 90.2%.

2 Previous Work

Many studies have been conducted on the classification of architectural structures. For instance, Bo Wang et al. [8] proposed a method for Architectural Style Classification using CNN and Convolutional Block Attention Module (CBAM) [9] with the Architectural Style Dataset and architectural heritage element dataset (AHE_Dataset). In the first step, they [8, p.100] applied "a method called Candidate Area Size Comparison

(CASC), which can sort the size of candidate areas and select the main building region", thus removing objects outside the building from the image. After this process, they [8] used CNN Feature Extractor to obtain the features of the images. Then, the obtained features were passed through the CBAM. It consists of two main components: Channel Attention Module (CAM) for extracting channel information from the feature map obtained from the CNN Feature Extractor, and Spatial Attention Module (SAM) for extracting spatial information. In other words, CAM focuses on what to pay attention to, while SAM focuses on where to pay attention. Finally, the feature maps obtained from the CNN Feature Extractor, CAM, and SAM are combined and sent to the classifier.

According to Obeso et al. [10, p.1], the utilization of "Deep Convolutional Neural Networks and Sparse Features" is proposed for the classification of architectural styles of historic structures in Mexico. In this study [10], a deep convolutional neural network (DCNN) is proposed to classify Mexican buildings based on their architectural styles using sparse features (SF). Additionally, a trained neural model is obtained by incorporating primary color pixel values, enabling the classification of Mexican buildings into three classes [10, p.1]: "pre-hispanic, colonial, and modern, with an accuracy of 88.01%". The utilization of data augmentation and oversampling techniques is proposed as a potential remedy for the problem of inadequate information within the training dataset, which arises due to the uneven distribution of cultural content [10].

3 Dataset

This study focuses on the classification of house images from different regions. During the data collection process, we used Google Street View technology to gather house images from a total of 15 regions, including 4 cities in Turkey and 11 cities in the Balkans. Due to the limited number of houses in our dataset, we included multiple facades of the same house in our dataset. Additionally, as can be seen in Fig. 2, we applied cropping to exclude elements other than buildings (such as cars or people) that can be encountered in Google Street View images from the house images.

The dataset (Table 1) consists of 528 collected images. The dataset is divided into approximately 70% training (365 images), 20% validation (112 images) and 10% test set (51 images). To prevent data leakage, we collectively placed images of the same house taken from different viewpoints into only one of the training, validation, or test sets. The number of images per region varies between 9 and 62 in the training set, between 2 and 20 in the validation set and between 1 and 10 in the test set (Figs. 1 and 3).

Table 1. Number of Images of Each Region in Train, Validation and Test Set

Regions	Train	Validation	Test
Kula (Turkey)	20	8	3
Kuzguncuk (Turkey)	11	4	1
Mardin (Turkey)	11	5	2
Safranbolu (Turkey)	27	9	4
Arbanasi (Balkans)	27	6	3
Plovdiv (Balkans)	18	7	2
Gjirokaster (Balkans)	21	5	3
Shkodër (Balkans)	13	5	2
Ampelakia (Balkans)	9	2	1
Ohrid (Balkans)	12	6	3
Pristina (Balkans)	23	6	3
Thessaloniki (Balkans)	34	7	3
Sozopol (Balkans)	47	14	7
Tirana (Balkans)	62	20	10
Varna (Balkans)	30	8	4

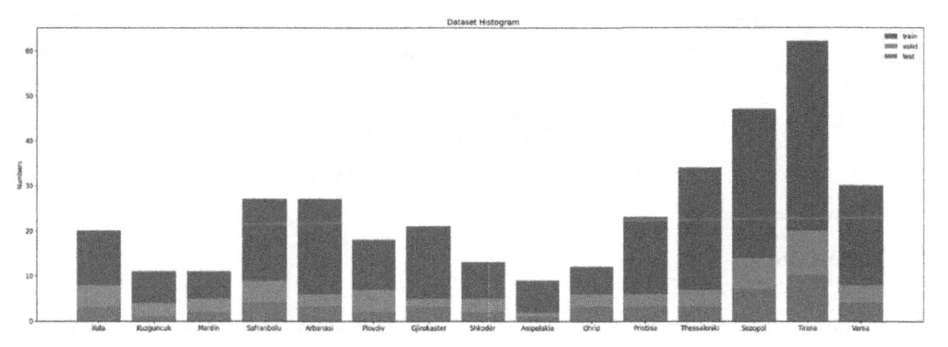

Fig. 1. Dataset Histogram

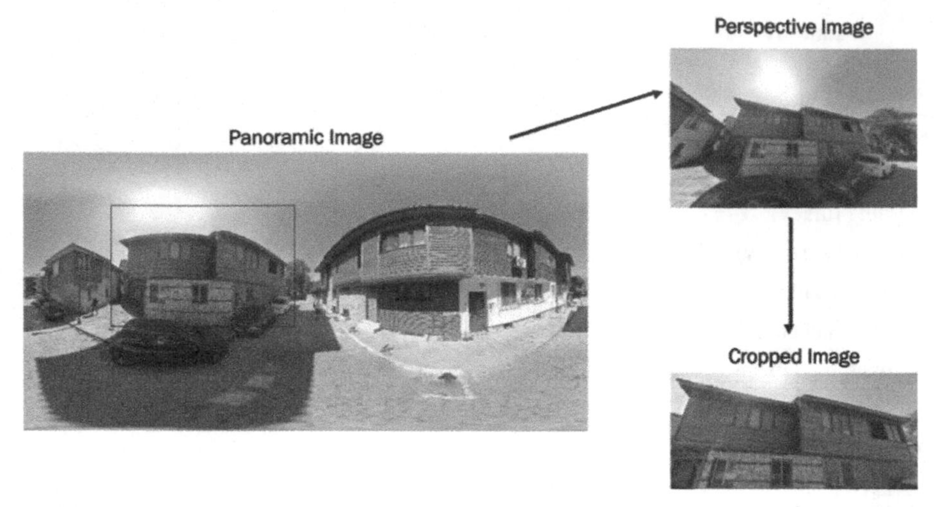

Fig. 2. Obtaining a House Image from Panoramic Image Acquired by Google Street View

Fig. 3. Some Images from Dataset

4 Methods

During the training stage, we trained the provided dataset using the Fast.ai library with pre-trained models Inception-V3, Inception-V4, ResNet-34, ResNet-101, ResNet-152, and ConvNeXt-Tiny. Among these models, ConvNext has the highest accuracy while also being very fast [11]. For all models, we utilized data augmentation and mixup techniques provided by FastAI to augment the limited amount of data in our dataset. We performed the training with transfer learning, which resulted in high performance training.

Data augmentation is a technique used to increase the diversity of a dataset by introducing variations to the existing data, thereby enhancing the generalization ability of the model. The Data Augmentation functions provided by FastAI allow manipulations of the training dataset through random cropping, random rotation, scaling, horizontal or vertical flipping, contrast or brightness adjustment, and other techniques to generate more diverse examples. In this process, we applied FastAI's default data augmentation parameters. We did not apply vertical flipping in this study since upside-down house images were not considered.

Mixup method is an effective technique used to diversify the dataset. Mixup creates a new example by blending two different image samples with certain ratios while also blending their labels [12]. This method helps to reduce overfitting issues and assists the model in better generalization. In this study, we applied the Mixup technique with a ratio of 0.5/0.5.

Transfer learning refers to the practice of using knowledge acquired from a specific deep learning task and applying it to another task. Inception, ResNet and ConvNeXt models have been trained on the ImageNet dataset to perform classification tasks, and their model parameters are available. By fine tuning these models on a novel dataset, one can achieve significant accuracies with little compute. Fine tuning involves starting from the pretrained parameters rather than from scratch and updating them in order to adapt them to the new dataset.

To obtain better training results during transfer learning, we normalized our data based on the ImageNet dataset statistics. This normalization process aligns the mean and standard deviation values of both datasets. These mean and standard deviation values are statistical measures derived from millions of images provided by the ImageNet dataset. Normalizing the data in hand during transfer learning is recommended to achieve better performance during training. This is one of the features already implemented within fast.ai library.

In conclusion, we increased the diversity of the small dataset by using data augmentation and mixup techniques. We performed training using Transfer Learning, which allowed us to leverage the pre-trained models. By applying these techniques, we enhanced the performance of the models despite the limited amount of data.

5 Evaluation

5.1 Comparison of Training Results

We conducted the training in the Google Colab environment. The classification success rates for the Inception-V3, Inception-V4, ResNet34, ResNet-101, ResNet-152 and ConvNeXt-Tiny models are provided in Table 2. Among these models, ResNet-50 model, which is the second best in terms of validation accuracy, provides 84.8% validation accuracy, while the ConvNeXt-Tiny model provides 94.6% for the same metric. Additionally, ResNet-101 model, which is the second best in terms of test accuracy, gives 80.4% accuracy, whereas ConvNeXt-Tiny gives 90.2% accuracy. Furthermore, there is a considerable difference between validation accuracy and test accuracy in Inception-V4, ResNet-50 and ResNet-152 models, which is referred to as overfitting. We attribute this overfitting problem to the relatively small and unbalanced dataset and the complexity of the classification task. However, ConvNeXt-Tiny model performs well in this regard.

TTA allows the model to make predictions on transformed images by applying data augmentation techniques during the testing phase. By providing the same image to the model multiple times with different transformations such as cropping, flipping, rotating, scaling, etc.., the average of these different predictions can be calculated, resulting in a more reliable and consistent evaluation outcome. This is also implemented within Fast.AI.

Table 2 includes TTA accuracy percentages for each model. ResNet-152 model, which is the second best for this metric, provides 88.4% TTA accuracy, while ConvNeXt-Tiny model provides 92.9% accuracy.

In Table 3, Matthew's Correlation Coefficient (MCC) score of validation predictions is presented to evaluate the performance of the ConvNeXt-Tiny model. MCC is known as a reliable performance metric, especially for imbalanced classification problems. It is computed by combining true positive and true negative predictions with false positive and false negative predictions, resulting in a score between -1 and + 1. As the score approaches + 1, it indicates high classification performance, while closer to -1 indicates poor performance. A score around 0 suggests performance close to random classification. The model's high MCC score serves as evidence of its robust capacity to effectively forecast classes and achieve excellent performance in the categorization of real-world data.

Table 2. Training Results

Model	Validation Accuracy (%)	TTA (%)	Test Accuracy (%)	Training Duration (min.)
Inception-V3	76.9	80.4	76.5	10
Inception-V4	81.2	80.4	64.7	12
ResNet-50	84.8	86.6	76.5	10
ResNet-101	82.1	87.5	80.4	13
ResNet-152	83.0	88.4	72.5	20
ConvNeXt-Tiny	**94.6**	**92.9**	**90.2**	13

Table 3. MCC Scores of the Models

Model	MCC Score
Inception-V3	0.7460
Inception-V4	0.7946
ResNet-50	0.8361
ResNet-101	0.8078
ResNet-152	0.8169
ConvNeXt-Tiny	**0.9420**

5.2 Gradient-Weighted Class Activation Mapping (Grad-CAM)

Grad-CAM [13] is a visualization tool for understanding how a deep learning model reached its classification decisions. The Grad-CAM technique demonstrates exceptional performance in models such as Convolutional Neural Networks (CNNs), which possess the ability to extract features at the pixel level. The backpropagation gradient is utilized between the final convolutional layer and the classification layer. The model's output class can be inferred from these gradients. GradCAM adds up these gradients per pixel by multiplying them by a weighted map based on the activation map of the convolutional layer. The outcome is a class activation map, which shows where the model placed the most emphasis when making its categorization. This enables the model's decision to be more transparent and interpretable. In this way, we understand which regions of the house are influential in the model's prediction.

Figure 4. Shows the Grad-CAM outputs of some images taken from the dataset. The first column contains the images given to the model for prediction. The second column shows the class activation map (Heatmap) obtained from the model, and in the third column, the input image is overlaid with the activation map.

Our Grad-Cam results suggest that the model's attention is concentrated towards the center of the house facade (rather than unrelated parts of the image such as the sky) and gives us confidence in the prediction of our classifier.

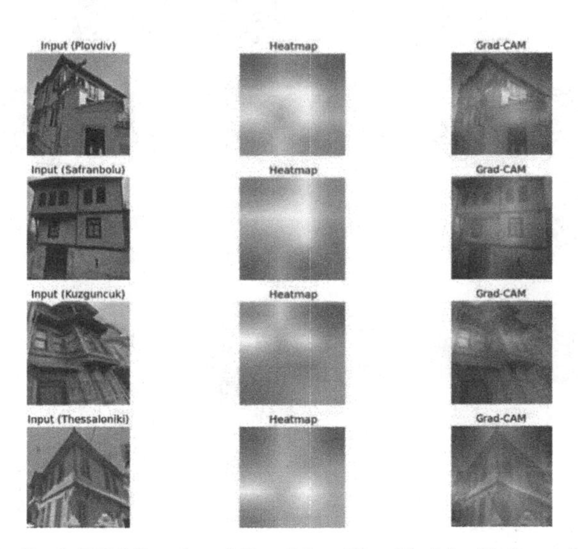

Fig. 4. Grad-CAM Results of ConvNext-Tiny Model with Some Images

5.3 Image Retrieval

In this study, we employed the ConvNeXt-Tiny model to extract embeddings (feature vectors) for each image in the dataset. These embeddings contain high-level visual features of the images. By calculating the Euclidean distances between these embeddings, we conducted an Image Retrieval task. In Fig. 5, given an input image, the model was utilized to retrieve the five most similar images from the dataset. The retrieved images were then visually presented in a graph, illustrating their closeness to the test (query) image. This Image Retrieval approach allowed us to demonstrate the capability of the ConvNeXt-Tiny model in capturing and matching visual patterns, facilitating efficient content-based image retrieval in real-world applications.

Fig. 5. Similar images inferred from ConvNext-Tiny model (In each row, there is a test image and the five closest images to it in terms of similarity, ranging from nearest to farthest).

6 Future Work

We are in the process of collecting more house photos from these regions. In future work, we will also obtain house images from more regions to increase the number of labels. We will also work with deeper ConvNeXt model architectures than ConvNext-Tiny such as ConvNeXt-Base, ConvNeXt-Large and ConvNeXt-XLarge. While working on these models, we will also work about the regularization of the overfitted Inception-V4, ResNet-50 and ResNet-152 models to prevent overfitting. Additionally, we will also work with the DINO model [14] to perform unsupervised segmentation and feature extraction and ProtoPNet [15] to interpret our classification on finer level details of the architectural elements.

Acknowledgements. The research work presented in this workshop is a component of the research project with the identification number 2022IYTE-2-0029 and is aligned with the doctoral studies of Veli Mustafa Yönder.

References

1. Krizhevsky, A., Sutskever, I., Hinton, G.E.: ImageNet classification with deep convolutional neural networks. In: Advances in Neural Information Processing Systems, pp. 1097–1105 (2012)
2. He, K., Zhang, X., Ren, S., Sun, J.: Deep residual learning for image recognition. arXiv:1512.03385 [cs.CV]
3. Simonyan, K., Zisserman, A.: Very deep convolutional networks for largescale image recognition. arXiv preprint arXiv:1409.1556, 2014[cs.CV]
4. Liu, Z., Mao, H., Wu, C.-Y., Feichtenhofer, C., Darrell, T., Xie, S.: A ConvNet for the 2020s. arXiv:2201.03545 [cs.CV]
5. Dosovitskiy, A., et al.: An image is worth 16 × 16 words: transformers for image Recognition at Scale. arXiv:2010.11929 [cs.CV]
6. Howard, J., Gugger, S.: Fastai: a layered API for deep learning. arXiv:2002.04688 [cs.LG]
7. Szegedy, C., et al.: Going deeper with convolutions. arXiv:1409.4842 [cs.CV]
8. Wang, B., Zhang, S., Zhang, J., et al.: Architectural style classification based on CNN and channel–spatial attention. SIViP **17**, 99–107 (2023). https://doi.org/10.1007/s11760-022-02208-0
9. Woo, S., Park, J., Lee, J.-Y.: In So Kweon. CBAM: convolutional block attention Module. arXiv:1807.06521 [cs.CV]
10. Obeso, A.M., Benois-Pineau, J., Acosta, A.Á.R. and Vázquez, M.S.G.: Architectural style classification of Mexican historical buildings using deep convolutional neural networks and sparse features. J. Electron. Imaging **26**, 011016 (2016). https://doi.org/10.1117/1.JEI.26.1.011016
11. Howard, J.: Which image models are best? (2023). Accessed 12 Jul 2023. https://www.kaggle.com/code/jhoward/which-image-models-are-best/
12. Zhang, H., Cisse, M., Dauphin, Y.N., Lopez-Paz, D.: Mixup: beyond empirical risk minimization. arXiv:1710.09412 [cs.LG]
13. Selvaraju, R.R., Cogswell, M., Das, A., Vedantam, R., Parikh, D., Batra, D.: Grad-CAM: visual explanations from deep networks via gradient-based localization. arXiv:1610.02391 [cs.CV]

14. Caron, M., Touvron, H., Misra, I., Jégou, H., Mairal, J., Bojanowski, P., Joulin, A.: Emerging properties in self-supervised vision transformers. In: Proceedings of the International Conference on Computer Vision (ICCV) (2021)
15. Chen, C., Li, O., Tao, C., Barnett, A.J., Su, J., Rudin, C.: This looks like that: deep learning for interpretable image recognition. arXiv:1806.10574 [cs.LG]

Method for Ontology Learning from an RDB: Application to the Domain of Cultural Heritage

Fabio Clarizia, Massimo De Santo, Rosario Gaeta, and Rosalba Mosca(✉)

DIIN University of Salerno, Fisciano, SA, Italy
rmosca@unisa.it

Abstract. For decades, ontologies have defined valuable terminology for describing and representing a knowledge domain, capturing relationships between concepts, and improving knowledge management.

Ontologies enable the exchange and sharing of information, extending syntactic and semantic interoperability: such advantages are also very useful in the cultural heritage (CH) field.

Nowadays, ontologies are often made manually, although various attempts have been made in the literature for their automatic generation (Ontology Learning).

This paper proposes a new way for the semi-automatic building of an ontology from a Relational Database (RDB).

Following an accurate review of existing methods, we propose the implementation of a Python library capable of converting an RDB into an OWL ontology and importing its data inside the ontology as concepts and properties instances. We present a case study on actual data from the cultural heritage world coming from the REMIAM project of the High Technology District for Cultural Heritage (Distretto ad Alta Tecnologia per i Beni Culturali - DATABENC).

Through interviews with experts in the field, a series of valid questions were identified for the experts' research work and the interrogation of the knowledge base.

The questions were then converted into SQL and SPARQL queries to assess the correctness of the method. The ability of the generated ontology to infer new knowledge on accurate data in the RDB will also be highlighted.

Keywords: Ontology Learning · Cultural Heritage · Relational Database

1 Introduction

Many projects were born to give free access to information relating to cultural heritage to promote it. However, the mere accessibility of data and information on various cultural assets is not sufficient to satisfy the need for experts to carry out cross-sectional research, data migrations and comparative studies. Therefore, high-level ontologies were born to integrate Cultural Heritage data (CIDOC CRM [1], Europeana [2]). However, much information in a specific domain, perhaps related to archaeological or historical-artistic sites that are not known (but any less attractive), risks likely to remain an island of data.

G. L. Foresti et al. (Eds.): ICIAP 2023 Workshops, LNCS 14366, pp. 409–421, 2024.
https://doi.org/10.1007/978-3-031-51026-7_35

To allow professionals to conduct research and studies on such data, interoperability at the semantic level is needed. Indeed, in semantics reside the characteristics of a domain that allows for broader searches, comparative studies, and data transfer.

Ontologies enable the exchange and sharing of information, extending syntactic and semantic interoperability. The role of ontologies, in fact, is to provide a common and shared representation of a domain that communicates that domain's information between people and systems [3].

Despite the many well-known advantages that the application of ontologies can bring, modeling an ontology manually is a slow process that requires the involvement of domain-specific experts, and the result obtained may need to be completed or approximate [4].

To cope with this limitation, semi-automatic or automatic approaches for ontology construction have appeared. The automation of ontology construction, known as "ontology learning", may reduce modeling time and creates ontologies that are fitter for purpose.

Ontology learning is thus the set of techniques that allow an ontology to be built from scratch or an existing ontology to be enriched semi-automatically [5].

Different ontology learning techniques depend on the types of data from which the ontology is generated. Data can be of three types: unstructured (text), semi-structured (HTML and XML documents), and structured (databases and dictionaries).

This article focuses on structured data, beginning with the authors' own review of methods for learning ontologies from relational databases (RDBs).

In this paper, we propose the implementation of a Python library capable of converting an RDB into an OWL ontology and importing its data. We propose a case study on actual data from the cultural heritage world coming from the REMIAM [30] project of the Distretto ad Alta Tecnologia per i Beni Culturali (Databenc) [23] which aims to protect, enhance and promote cultural heritage. The REMIAM project is designed to create a prototype demonstrator of a Museum Network and lay the consequent foundations for an integrated action aimed at the vast heritage of Campania's Museums.

The ***POR****tale per i* ***BE****ni* ***C****ulturali* (PORBEC) [22] is the gateway to all cataloging, fruition, and enhancement services implemented in the REMIAM project. The portal has been designed to connect works of art belonging to different museums by creating a physical or "virtual" path between works and places of art, even geographically distinct from each other. In particular, the proposed method has been applied to a relational database containing the data accessible through the PORBEC portal.

This work stems from the specific need of experts in the field to use the data stored in the PORBEC RDB for research purposes. To this end, a collection of requirements for research in the field of CH was carried out through interviews with experts in the field (archaeologists, art historians, among others). A list of questions useful for experts in their study and research work was constructed from these requirements. Such queries were then converted into SQL and SPARQL queries to evaluate the correctness of the method. The possibility of inferring new knowledge about the actual data present in the RDB through the use of the generated ontology will also be highlighted.

The rest of the document is organized as follows. Section 2 outlines an overview of studies on ontology learning from structured data. Section 2.1 describes the ontology learning algorithm. Section 3 illustrates the Python implementation of the algorithm. Section 4 describe the use case, with the presentation of the database, the resulting ontology, and the evaluation of the results. Finally, Sect. 5 provides concluding remarks.

2 Related Works

This paper was born after the drafting of the review by the same authors on ontology learning methods that take into input the schema and data of an RDB.

OntoBase [17] and **DataMaster** [18] are two Protégé plugins that are usable in its 3.x versions and that are not supported anymore. Analysis has shown that OntoBase outperforms DataMaster in conversion quality [16]. However, the mapping principles of both plugins are inconsistent with various mapping principles.

Ghawi, Raji, and Nadine Cullot [7] propose the DB2OWL tool, which uses 6 mapping rules to generate the ontology: it leaves the rules for constructing axioms from RDB constraints undefined.

In the work of **Bakkas et al.** [10], two algorithms create the terminological (T-Box) and assertional (A-Box) part of the ontology. In this, the mappings of not null and unique constraints are missing.

The system of **Yiqing et al.** [8] proposes 11 mapping rules. Still, it does not specify if it considers the porting of data into the ontology, as the EVis system presented by **Zhang, Hui, et al.** [9], which also proposes editing and visualization functionalities. In addition, Yiqing et al. convert an RDB into an OWL-Full ontology, as the work proposed by **Hazber, Mohamed AG, et al.** [12]. Unfortunately, an OWL-Full ontology is often not very useful because does not allow the use of reasoners.

The method presented by **Mona Dadjoo et al.** [11] also only imports concepts and not instances, but on the other hand proposes rules for the translation of triggers, differently from other work.

Liu and Gao [13] implement the WN_Graph method which integrates the model proposed by Mona Dadjoo et al. with WordNet to better extract hierarchical relationships from the RDB.

The approach proposed by **Batool Lakzaei and Mehrnoush Shamsfard** [15] is rich in implementing many conversion rules regarding tables, attributes, constraints hierarchy relationships, and queries. This method, like that of **Li, Man, Xiao-Yong Du, and Shan Wang** [6], translates the n-ary table in a non-optimal way.

The one by **Mahria et al.** [14] is one of the methods which implements more rules. It creates also a SymmetricProperty when a Foreign Key (FK) that refers to the same table is encountered (like *Hazber, Mohamed AG, et al.* [12]) but this translation is not always correct because there are cases of non-symmetric properties even with the same domain and range.The Python library we propose implements a method capable of applying various rules for converting SQL patterns into OWL patterns. it has been obtained from a synthesis of rules from the literature analyzed in our review, but introducing some variation to eliminate the problems previously described and allowing the user to intervene to resolve ambiguities and situations that cannot be automated.

SQL2OWL Algorithm

The one proposed in this paper is a semi-automatic method that extracts and explicates TBOXs and ABOXs knowledge implicitly present in a relational database (RDB) to generate an ontology.

In Fig. 1, the architecture with the main component of the method is shown. In particular, the method acquires RDB schema characteristics, mapping also the data inside the database to this schema. Once the data are recovered, the conversion method is applied: in particular, the TBOX Algorithm analyzes all the tables taken from the RDB schema and applies various conversion rules to obtain a partial ontology with a series of concepts, data properties, object properties, and axioms finally. Then the data inside the database tables are used to feed the ABOX Conversion algorithm, which populates the ontology using the mapping established in the previous TBOX algorithm. Then the method gives the OWL File as output.

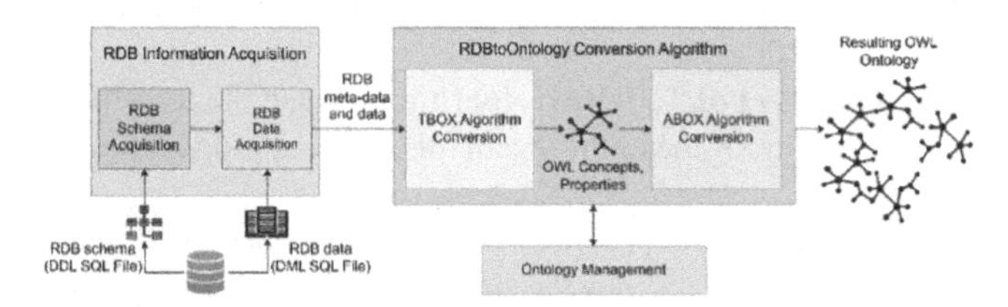

Fig. 1. Proposed Method Architecture

2.1 DDL Conversion into TBOX Algorithm

The designed method converts a relational database into an OWL ontology using a set of mapping rules. The method is a semi-automatic conversion approach because, although it allows the creation of an ontology in an almost entirely automated manner, it requires human intervention in the case of the detection of a possible class-subclass relationship, Symmetric property or Transitive property to decide whether or not such elements exist. Algorithm 1 firstly processes all the tables creating the corresponding classes, and the object properties (if there are FKs): the tables without FK columns or with 1 or more than 2 FKs are all converted into classes. For the tables with at least an FK, the algorithm must also control if the created class is a candidate to be a subclass of the referred class and if it is so, the user is allowed to choose whether or not such class is a subclass. Similar decisions must also be taken when a possible Symmetric or Transitive Property is detected: in Fig. 2 such a process is shown more clearly.

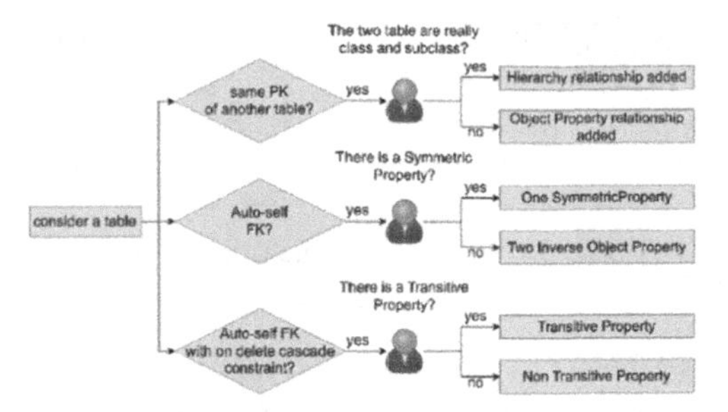

Fig. 2. Human Intervention

The tables which have precisely 2 FKs can be converted in two ways:

If the table has at least one column, which is not an FK, the table is translated into a class. Otherwise, the table is translated into two object properties.

Then, the method creates the data properties for each created class, converting the type of the relational database into OWL type, and finally, all the restrictions are added to the ontology.

The following is a sketch of the proposed algorithm:

Algorithm1: TBOX Conversion Algorithm

```
     Input: ddl_sql_file (DDL Database Schema)
     Output: owl_file (Ontology Schema)
1    rdb ← parse ddl_sql_file
2    Initialize the ontology
3    for all table ∈ rdb.tables():
4        if the number of foreign key columns is 0:
5            add to the ontology a class "table.name" and that is subclass of "Thing"
6        if the number of foreign key columns is 1:
7            If a possible hierarchy relationship is detected:
8              user decide or not to create a class-subclass relationship
9            Else:
10             add to the ontology a class "table.name" and that is subclass of "Thing"
11           convert all the foreign keys to object properties
12       if the number of foreign key columns is 2:
13           if the table is not a bridge table:
14               If a possible hierarchy relationship is detected:
15                 user decide or not to create a class-subclass relationship
16               Else:
17                 add to the ontology a class "table.name" and that is subclass of "Thing"
18               convert all the foreign keys to object properties
19           Else:
20               convert the table into two object properties
21       If the number of foreign key columns is greater than 2:
22           If a possible hierarchy relationship is detected:
23             user decide or not to create a class-subclass relationship
24           Else:
25             add to the ontology a class "table.name" and that is subclass of "Thing"
26           convert all the foreign keys to object properties
27   convert the no FK columns of all the tables became classes into data properties
28   add all the restrictions to object and data properties
```

Algorithm 1 applies a series of rules for the translation from SQL to OWL:

All the tables that are not bridge tables, i.e., those that are realized to convert a many-to-many relationship between two entity an Entity Relationship (ER) diagram into the relational model, are converted into classes. This means that all tables with at least one column that is not a foreign key, or that have 0, 1, or more than 2 FKs must be translated in class. In Fig. 3 an example of conversion is reported.

SQL Pattern	OWL Pattern
CREATE TABLE media.mav_internal_autore (id bigint NOT NULL PRIMARY KEY, nome character varying(10000), dati_anagrafici character varying(10000));	<owl:Class rdf:about="#mav_internal_autore"> <rdfs:subClassOf rdf:resource="http://www.w3.org/2002/07/owl#Thing"/> </owl:Class>

Fig. 3. Not Bridge Table Conversion

If a table is a bridge table, i.e., have only two foreign keys and no other columns, it is translated into 2 inverse object properties with domain and range given by the classes resulting from the 2 tables referenced by the bridge table. In such cases, the object properties are not characterized by any axiom or restriction (Figs. 4 and 5).

SQL Pattern	OWL Pattern
CREATE TABLE media.mav_internal_card_materia_tecnica (id_card bigint NOT NULL, id_internal_materia_tecnica bigint NOT NULL, PRIMARY KEY (id_card, id_internal_materia_tecnica), FOREIGN KEY (id_card) REFERENCES media.mav_internal_card(id), FOREIGN KEY (id_internal_materia_tecnica) REFERENCES media.mav_materia_tecnica(id));	<owl:ObjectProperty rdf:about="#mav_internal_card_materia_tecnica"> <rdfs:domain rdf:resource="#mav_internal_card"/> <rdfs:range rdf:resource="#mav_materia_tecnica"/> <owl:inverseOf rdf:resource="#INV_mav_internal_card_materia_tecnica"/> </owl:ObjectProperty> <owl:ObjectProperty rdf:about="#INV_mav_internal_card_materia_tecnica"> <rdfs:domain rdf:resource="#mav_materia_tecnica"/> <rdfs:range rdf:resource="#mav_internal_card"/> <owl:inverseOf rdf:resource="#mav_internal_card_materia_tecnica"/> </owl:ObjectProperty>

Fig. 4. Bridge Table Conversion

The method also considers the possibility to have a hierarchy relationship between tables:

SQL Pattern	OWL Pattern
ALTER TABLE ONLY media.mav_spes ADD CONSTRAINT mav_spes_pkey PRIMARY KEY (id); CREATE TABLE media.mav_contatti (id bigint NOT NULL, address character varying(10000), . . link_visita_virtuale character varying(10000), PRIMARY KEY (id), FOREIGN KEY (id) REFERENCES media.mav_spes(id));	<owl:Class rdf:about="#mav_spes"> <rdfs:subClassOf rdf:resource="http://www.w3.org/2002/07/owl#Thing"/> </owl:Class> <owl:Class rdf:about="#mav_contatti"> <rdfs:subClassOf rdf:resource="http://www.w3.org/2002/07/owl#Thing"/> </owl:Class>

Fig. 5. Hierarchy relationship

When the two tables are correlated for other reasons (for example, a person and his driver's license). For such reason, when the possibility of a hierarchy relationship is detected, the method asks the user to confirm or not the class-subclass relationship (Fig. 2).

Each column with no foreign key constraint is converted into a data property (except for any identifiers used by the database), converting its datatype into an OWL. Each data property name has been obtained as *"name of the column"* + _*of*_ + *"name of the* *table*". Such properties must always be defined as functional because in a database, each row can have only a value for a column (Fig. 6).

SQL Pattern

```sql
CREATE TABLE media.mav_collection (
    id bigint NOT NULL PRIMARY KEY,
    title character varying(1000),
    description character varying(1000),
    .
    .
    .
    click_app bigint DEFAULT 0
);
```

OWL Pattern

```xml
<owl:DatatypeProperty
rdf:about="#image_of_mav_collection">
  <rdf:type
  rdf:resource="http://www.w3.org/2002/07/owl#FunctionalProperty"/>
  <rdfs:domain rdf:resource="#mav_collection"/>
  <rdfs:range
  rdf:resource="http://www.w3.org/2001/XMLSchema#string"/>
</owl:DatatypeProperty>
```

Fig. 6. Conversion of a not foreign key column into data property

In addition, if the original column is defined as nullable, the restriction defined is max_cardinality = 1; if the original column is not nullable, a restriction on cardinality = 1 must be defined on the property. A direct translation of unique and primary key columns is impossible in OWL-DL because the inverse functional properties are not definable on data properties.

Each column with an FK constraint is converted into two object properties, one the inverse of the other. The first object property has the class corresponding to the table with the FK as the domain, while the range is the class of the table referenced by the FK (Fig. 7). Each of the object property names is in the form "*name of domain class*" + "*_has_a_*" + "*name of the range class*" in the case of direct property, while "*name of domain class*" + "*_belongs_to_*" + "*name of the range class*" is used for the inverse properties. This consideration is valid also when a foreign key is referred to the table itself, in such case domain and range of the property will be the same.

SQL Pattern

```sql
CREATE TABLE media.mav_internal_racconto (
    id bigint NOT NULL PRIMARY KEY,
    id_card bigint,
    descrizione character varying(10000),
    profilo character varying(100),
    lingua character varying(5),
    FOREIGN KEY (id_card)
    REFERENCES media.mav_internal_card (id)
);
```

OWL Pattern

```xml
<owl:ObjectProperty
rdf:about="#mav_internal_racconto_has_a_mav_internal_card">
  <rdf:type rdf:resource="http://www.w3.org/2002/07/owl#FunctionalProperty"/>
  <rdfs:domain rdf:resource="#mav_internal_racconto"/>
  <rdfs:range rdf:resource="#mav_internal_card"/>
  <owl:inverseOf
  rdf:resource="#mav_internal_card_belongs_to_mav_internal_racconto"/>
</owl:ObjectProperty>
<owl:ObjectProperty
rdf:about="#mav_internal_card_belongs_to_mav_internal_racconto">
  <rdfs:domain rdf:resource="#mav_internal_card"/>
  <rdfs:range rdf:resource="#mav_internal_racconto"/>
  <owl:inverseOf
  rdf:resource="#mav_internal_racconto_has_a_mav_internal_card"/>
</owl:ObjectProperty>
```

Fig. 7. Conversion of column with FK into Object Properties

Such object properties can have different characteristics depending on the constraints defined on the original columns:

- If the original column is nullable and not unique, the object property is functional, while If the original column is not nullable and not unique, the object property is functional, and there is a restriction on cardinality = 1

- Suppose the original column is nullable and unique. In that case, the object property is both functional and inverse functional. In contrast, if the original column is not nullable and unique (PK included), the object property is functional, inverse functional, and there is a restriction on cardinality = 1.

However, when a column that has an FK constraint which is referred to the table itself, is detected, the method asks to the user (in real-time) if it must be converted into a SymmetricProperty or not (for example an *hasFather* property is not symmetrical while *hasSibling* is symmetrical).

When a column that has an FK constraint which is referred to in the table itself, has also an **on-delete cascade** constraint, a possibility of TransitiveProperty is detected, also in this case the user must take the final decision. When a TransitiveProperty is created, the property is not declared functional (because of OWL-DL incompatibility).

Unlike other automatic methods like [12] and [14], our method introduces human interaction to avoid creating incorrect properties because of the ambiguities of such situations. Since this interaction takes place in real-time as a simple choice of alternatives that is done in real-time, such a decision has little influence on the automaticity of the method.

2.2 DML Conversion into ABOX Algorithm

The ABOX algorithm takes in input all the rows for each table and uses the information about the mapping between the tables and the concepts and properties previously created by the TBOX conversion algorithm to correctly populate the resulting ontology.

Algorithm2: ABOX Conversion Algorithm

```
     Input: dml_sql_file (DML Database Insert), rdb, ontology
     Output: owl_file (Ontology Schema)
1    statements ← parse a dml sql file
2    for all table ∈ rdb.tables():
3        for all row ∈ table.get_rows():
5            if the table has been previously converted into a class:
6                create an instance with the name obtained from the primary key of the row
7                for all column ∈ columns:
8                    Create a data property instance with the value of the column as a range
11   for all table ∈ rdb.tables():
12       if the number of fk columns in table is different from 0:
14           if table.get_table_name() in ontology.get_created_classes():
15               for all row ∈ table.get_rows():
16                   Instantiate an object property with the involved instances
17           else:
18               for all row ∈ table.get_rows():
19                   Instantiate an object property considering the instances identified by the 2 FKs
```

The approach for each row creates an instance of the corresponding class generated by the TBOX algorithm. Each instance has been named starting from the membership class name of the instance and from the value of its primary key.

Looking at the database schema, concepts, and properties, the method associates the values coming from each row with a data property or an object property created during the execution of the TBOX algorithm.

3 Implementation

The method previously described has been implemented as a Python Library that can be used for the conversion of SQL files into an OWL file.

Indeed, the proposed method takes in input two SQL files:

The first must contain an SQL file containing Data Definition Language (DDL) statements for the definition of the schema of the database (tables, columns, foreign keys, constraints, and others) while the second is an SQL File containing Data Manipulation Language (DML) insert statements for the definition of the content of the tables defined by the previous SQL file.

The library contains modules for the management of the RDB schema and data, for the creation of the ontology starting from such RDB information, and for the execution of the TBOX and ABOX algorithms.

In the implementation, much attention has been made to the design of a method that is as much as possible automated, avoiding forcing the user to create configuration files that would make the method much less usable.

The developed method is available in a GitHub repository [20].

4 Use Case

4.1 Database Porbec

The database taken into consideration is a PostgreSQL relational database that collects the data coming from the REMIAM and PORBEC projects. In such a database, the data coming from various physical and virtual exhibitions and museums are stored. The database is composed by 16 tables (2 of which are bridge table), 94 not FK columns, 17 FK columns and 20322 total rows (4657 of which are of bridge tables).

4.2 Resulting Porbec Ontology

Once the method is applied, the OWL Ontology is generated. In the resulting ontology, all tables have been translated into classes, except for the bridge tables that implement the many-to-many relation (in the conceptual schema). For each of the other FK in the tables, two object properties are created for connecting the classes resulting from the tables connected by such foreign keys. *Mav_internal_autore* presents two object properties with the same domain and range. The first of these is a transitive property. Each of the not foreign key columns in the original database is converted into a data-property and each of the rows is in instances. Finally, the ontology is composed of 14 classes, 82 data properties, 30 object properties, and 15665 instances: 14 classes are obtained by the 16 tables minus 2 bridge tables, 82 data properties from the 94 not FK columns minus 12 IDs, 30 object properties are obtained from the 17 FKs, 4 of these are inside bridge tables which become 4 object properties: the remaining 13 Fks become 26 object properties (it must be considered that each bridge table and FK of not bridge table has been translated into 2 object property). The 20322 rows in the database become 15665 instances because 4657 of the rows are inside bridge tables.

4.3 Evaluation

In this paper, we decide to test if our method can generate an OWL ontology without loss of data testing it against a series of queries that are obtained from experts of the domain that are capable of understanding what are the question to which a CH system must be able to answer.

For our evaluation, the experts we involved are university professors, researchers, freelancers, managers, and officials of the Italian state working in the field of reference (particularly, Ministry of Culture - MIC). The choice of experts also considered the different disciplinary specificities and time frames (ancient, medieval, modern, and contemporary) that affect cultural heritage.

A questionnaire was constructed with the experts to ask them about data useful for their research work. This phase allowed the formulation of natural language queries useful for verifying the correct generation of the ontology with the proposed method.

Such queries must be translated into SPARQL queries to interrogate an ontology. In order to perform an evaluation of the capability of our method in generating an OWL Ontology without loss of data, we decided to create for each of the query found by the experts the corresponding SPARQL query, together with the SQL query. In such a way, we can compare the response of such queries executed on both the database and the ontology so that is possible to verify that the porting of the data has been successfully executed and that both systems can respond to the identified questions.

In Table 1 it is possible to see some examples of the transposition of the Natural Language Queries into SQL and SPARQL queries. For reasons of space, we have reported only some queries.

Table 1. Examples of Query in various languages

NL Query	SQL Query	Sparql Query

What are the work in a SPES X?	SELECT distinct card.titolo collate "C"
	FROM media.mav_spes as spes
	left join media.mav_internal_card
	as card on (spes.id = card.id_spes)
	WHERE spes.description = "{}"
	order by card.titolo collate "C"
		PREFIX porbec:
		<http://www.semanticweb.org/rosar/ontologies/2023/0/porbec#>
		SELECT DISTINCT ?titolo
		WHERE {?spes porbec:description_of_mav_spes "{}"^^xsd:string.
		?card porbec:mav_internal_card_has_a_mav_spes ?spes.
		?description porbec:mav_description_has_a_mav_contatti ?contatti.
		OPTIONAL{
		?card porbec:titolo_of_mav_internal_card ?titolo.
		}
		} order by (?titolo)

What are the authors of the works in a collection X?	SELECT distinct autore.nome collate "C"
	FROM media.mav_collection as collection
	left join media.mav_artworks_collection
	as art_coll on
	(collection.id = art_coll.id_collection)
	left join media.mav_internal_card
	as card on (card.id = art_coll.id_card)
	left join media.mav_card_internal_autore
	as card_autore on
	(card_autore.id_card = art_coll.id_card)
	left join media.mav_internal_autore
	as autore on
	(autore.id = card_autore.id_internal_autore)
	WHERE collection.title = '{}'
	and autore.nome != 'null'
	order by autore.nome collate "C"
		PREFIX porbec:
		<http://www.semanticweb.org/rosar/ontologies/2023/0/porbec#>
		SELECT DISTINCT ?nome
		WHERE {?collection porbec:title_of_mav_collection
		"{}"^^xsd:string.
		?art_coll porbec:mav_artworks_collection_has_a_mav_collection
		?collection.
		?art_coll porbec:mav_artworks_collection_has_a_mav_internal_card
		?card.
		?card porbec:mav_card_internal_autore ?autore.
		OPTIONAL{
		?autore porbec:nome_of_mav_internal_autore ?nome.
		}
		} order by ?nome

When was realized the work X?	SELECT card.titolo collate "C", card.cronologia collate "C" FROM media.mav_internal_card as card WHERE card.titolo = '{}' order by card.titolo collate "C", card.cronologia collate "C"	PREFIX porbec: <http://www.semanticweb.org/rosar/ontologies/2023/0/porbec#> SELECT ?titolo ?cronologia WHERE {?card porbec:titolo_of_mav_internal_card "{}"^^xsd:string. OPTIONAL{?card porbec:cronologia_of_mav_internal_card ?cronologia.} OPTIONAL{?card porbec:titolo_of_mav_internal_card ?titolo.} } order by ?titolo ?cronologia

To validate our method, we translated natural language queries formulated with the help of experts in SQL and SPARQL queries. To have a method that correctly performs the conversion from a SQL database to an OWL ontology, all rows and records of the original database must have been translated into instances, data properties and object properties without loss of information. Therefore, every query executed in SQL or SPARQL must return the same result.

In total, 32800 pairs of SQL and SPARQL queries were executed, with the results always being the same.

An example of the result of executing these queries in both systems (RDB and generated Ontology) can be seen in Table 2.

Table 2. Example of results of such queries

NL Query	RDB Result	Ontology Result
What are the works made of "cera"?	titolo character varying (10000) 1 Armonia 2 Ombra sul rosso	titolo "Armonia"^^<http://www.w3.org/2001/XMLSchema#string> "Ombra sul rosso"^^<http://www.w3.org/2001/XMLSchema#string>
When was realized the work "Armonia"?	cronologia character varying (10000) ante 1940	cronologia "ante 1940"^^<http://www.w3.org/2001/XMLSchema#string>

Obviously, for reasons of space, we cannot report all the queries performed for validation.

However, all queries and results are repeatable using the material stored in this GitHub repository [16].

Once the ontology is correctly generated, the semantics made explicit by the method can be used to make inferences about the data in the database, generating new knowledge. An example in our database is the transitive property derived from the foreign key of the descendant column in the author table. In the database, this property can link an author with a direct ancestor in the database. This attribute was converted into a transitive property with the class mav_author_interior (Author) as domain and range in our generated ontology. There are three authors in the database: Andrea Malinconico with id 811, Nicola Malinconico (son of Andrea) with id 812 and Carlo Malinconico (son of Nicola). In the OWL ontology, after the work of a reasoner, a descent of the relationship between Carlo and Andrea can be deduced (Fig. 8).

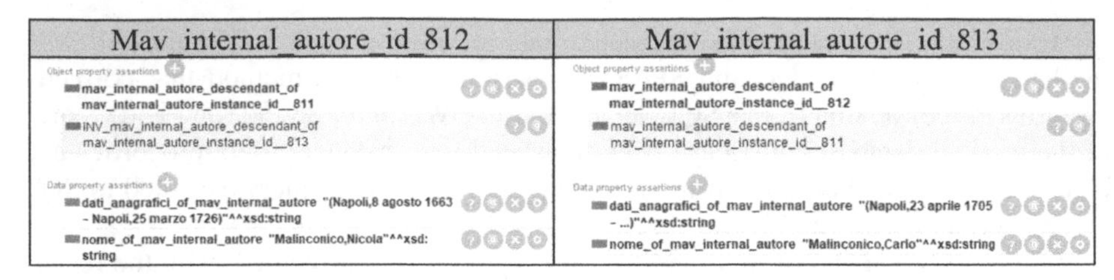

Fig. 8. Example of inference connected to the transitive property

5 Conclusions

The work proposes a Python Library implementing a semi-automatic method for converting an RDB database into an OWL-DL compliant Ontology File.

Such a Python library has been tested on a Relational Database containing data of the cultural heritage domain made available using the PORBEC platform, developed in the context of the REMIAM project. Through interviews with various experts in the cultural heritage field, a series of questions of interest have been identified: such questions have been translated as SQL and SPARQL queries to verify that the data porting has been successfully executed. The paper has also shown how the generated ontology can be used to infer new knowledge starting from the data inside the database: a reasoner can make inferences on the properties and data explicating the semantic which was implicit in the database, exploiting the support of the user which use the library.

Future works will focus on the generation of ontology starting from the semi-structured and not structured sources of data and on the application of the generated ontologies (in the field of cultural heritage but also in others) in the context of big data filters and to improve the performance of dialogue management systems.

References

1. Doerr, M.: The CIDOC conceptual reference module: an ontological approach to semantic interoperability of metadata. AI Mag. **24**(3), 75 (2003)
2. Doerr, M., et al.: The Europeana data model (EDM): object representations, context and semantics. In: 76th IFLA General Conference and Assembly, Gothenburg (2010)
3. Karoui, L., Aufaure, M.A., Bennacer, N.: Ontology discovery from web pages: Application to tourism. In: the Workshop of Knowledge Discovery and Ontologies (2004)
4. Shamsfard, M., Barforoush, A.A.: The state of the art in ontology learning: a framework for comparison. Knowl. Eng. Rev. **18**(4), 293–316 (2003)
5. Gomez-Perez, A., Manzano-Macho, D.: OntoWeb Deliverable 1.5: A Survey of Ontology Learning Methods and Techniques. Universidad Politecnica de Madrid (2003)
6. Li, M., Du, X.-Y., Wang, S.: Learning ontology from relational database. In: 2005 International Conference on Machine Learning and Cybernetics, vol. 6, IEEE (2005)
7. Ghawi, R., Cullot, N.: Database-to-ontology mapping generation for semantic interoperability. In: Third International Workshop on Database Interoperability (InterDB 2007), vol. 91 (2007)
8. Yiqing, L., Lu, L., Chen, L.: Automatic learning ontology from relational schema. In: 2012 IEEE Symposiumon Robotics and Applications (ISRA), IEEE (2012)

9. Zhang, H., et al.: EVis: a system for extracting and visualizing ontologies from databases with webinterfaces. In: 2012 Fourth International Symposium on Information Science and Engineering, IEEE (2012)
10. Bakkas, J., Bahaj, M., Marzouk, A.: Direct migration method of RDB to ontologywhile keeping semantics. In: International Journal of Computer Applications, vol. 65, no. 3 (2013)
11. Dadjoo, M., Kheirkhah, E.: An approach for transforming of relational databases to OWLontology. arXiv preprint arXiv:1502.05844 (2015)
12. Hazber, M.A.G., et al.: Integration mapping rules: transforming relational database to semantic web ontology. Appl. Math. **10**(3), 1–21 (2016)
13. Liu, X., Gao, F.: An approach for learning ontology from relational database. In: Proceedings of the 2018 International Conference on Algorithms, Computing and Artificial Intelligence (2018)
14. Ben Mahria, B., Chaker, I., Zahi, A.: A novel approach for learning ontology from relational database: from the construction to the evaluation. J. Big Data **8**(1), 1–22 (2021)
15. Lakzaei, B., Shamsfard, M.: Ontology learning from relational databases. Inf. Sci. **577**, 280–297 (2021)
16. Mogotlane, K.D., Fonou-Dombeu, J.V.: Automatic conversion of relational databases into ontologies: a comparative analysis of Prot\'eg\'e plug-ins performances. arXiv preprint arXiv:1611.02816 (2016)
17. https://protegewiki.stanford.edu/wiki/OntoBase
18. https://protegewiki.stanford.edu/wiki/DataMaster
19. Sequeda, J.F., et al.: Survey of directly mapping SQL databases to the semantic web. Knowl. Eng. Rev. **26**(4), 445–486 (2011)
20. https://github.com/RosarioG99/rdb_to_owl_hand.git
21. https://databenc.it/project/remiam/
22. https://www.porbec.it/
23. https://databenc.it/

A Novel Writer Identification Approach for Greek Papyri Images

Nicole Dalia Cilia[1], Tiziana D'Alessandro[2(✉)], Claudio De Stefano[2], Francesco Fontanella[2], Isabelle Marthot-Santaniello[3], Mario Molinara[2], and Alessandra Scotto Di Freca[2]

[1] Department of Computer Engineering, University Kore Enna, Enna, Italy
nicoledalia.cilia@unikore.it
[2] Department of Electrical and Information Engineering (DIEI), University of Cassino and Southern Lazio, Lazio, Italy
{tiziana.dalessandro,destefano,fontanella,m.molinara,a.scotto}@unicas.it
[3] Universität Basel, Basel, Switzerland
i.marthot-santaniello@unibas.ch

Abstract. Papyrology is the field of study dedicated to ancient texts written on papyri. One significant challenge faced by papyrologists and paleographers is the identification of writers, also referred to as scribes, who penned the texts preserved on papyri. Traditionally, paleographers relied on qualitative assessments to differentiate between writers. However, in recent years, these manual techniques have been complemented by computer-based tools that enable the automated measurement of various quantities such as letter height and width, character spacing, inclination angles, abbreviations, and more. Digital palaeography has emerged as a new approach combining advanced Machine Learning (ML) algorithms with high-quality digital images. This fusion allows for extracting distinctive features from the manuscripts, which can be utilized for writer classification using ML algorithms or Deep Learning (DL) systems. Integrating powerful computational methods and digital imagery has opened up new avenues in palaeography, enabling more accurate and efficient analysis of ancient manuscripts. After applying image processing and segmentation techniques, we exploited the power of Convolutional Neural Networks to characterize a scribe's handwriting.

Keywords: Greek Papyri · Writer Identification · Deep Learning

1 Introduction

Palaeography and papyrology are closely intertwined and have significantly enhanced our understanding of ancient societies [8]. Palaeography involves examining and interpreting ancient handwriting, encompassing various aspects such as letter forms, writing styles, punctuation, abbreviations, and other characteristics of written texts. On the other hand, papyrology studies ancient documents

G. L. Foresti et al. (Eds.): ICIAP 2023 Workshops, LNCS 14366, pp. 422–436, 2024.
https://doi.org/10.1007/978-3-031-51026-7_36

written on papyrus, a commonly used writing material in the ancient Mediterranean region, mostly preserved in Egypt. Scholars have gained valuable insights into the customs, cultures, and historical contexts of past civilizations through these interconnected disciplines. Papyri records are a treasure of valuable information, offering profound glimpses into bygone eras. Within these documents, various elements such as illustrations, decorative motifs, shapes, letters, and signatures convey explicit content and reflect various cultural and social characteristics through the evolution of writing styles. With their expertise in this field, paleographers are particularly interested in tasks that involve identifying a specific scribe, discerning a manuscript's origin and date, and spotting fragments that used to belong to the same document. These endeavours naturally demand substantial experience and a deep understanding of the domain.

Technological advancements have profoundly transformed palaeography and papyrology, revolutionizing how ancient manuscripts and documents are preserved and accessed. Through digital imaging technology, generating high-quality images of ancient texts has become easier, enabling their preservation and widespread dissemination through online platforms. This accessibility has opened up fresh avenues for research and collaboration, allowing scholars from around the globe to engage in large-scale projects and share their discoveries with a broader audience. Moreover, the collaborative efforts among paleographers, papyrologists, historians, linguists, and archaeologists have significantly enriched our comprehension of the ancient world. By leveraging the collective expertise of professionals from diverse fields, interdisciplinary research initiatives have shed new light on various aspects of ancient texts, contexts, and cultures. This collaborative approach has deepened our understanding of the intricate interactions and relationships that existed within the ancient world. Software solutions have emerged as valuable tools in studying papyri, offering researchers powerful means to analyze and interpret these ancient documents. These software solutions encompass a range of functionalities and features designed specifically for the unique challenges of studying papyri. Often they include advanced image processing and analysis capabilities. These features aid in tasks such as text segmentation, character recognition, and restoration of damaged or faded portions of the manuscript. By leveraging algorithms and machine learning techniques, these software solutions can assist in deciphering complex scripts, identifying handwriting patterns, and extracting meaningful information from texts.

This research focuses on the problem of identifying writers based on their handwriting on papyrus. We considered the Papyrow dataset [5], a dataset of row images from ancient Greek Papyri. All the papyri images had to face a preprocessing step to enhance the handwriting and uniform the deteriorated background. Moreover, images were segmented into rows and patches, extending the data availability. We employed Deep Learning techniques to extract features from each row using several pre-trained Convolutional Neural Networks (CNNs). The results are presented at both the row and document levels, with the latter achieved by aggregating row-level decisions through a majority voting mechanism. The remainder of the paper is organized as follows: Sect. 2 illustrates the related work, Sect. 3 presents the data used and the elaboration steps, Sect. 4 describes the experimental setting while Sect. 5 the results obtained. Conclusions and future work are eventually left to Sect. 6.

2 Related Work

The digitization of ancient manuscripts involves converting physical, often fragile, and historically significant handwritten documents into digital formats. This transformation aims to preserve the content and make it accessible to a broader audience while enabling advanced analysis through computational methods. The article [13] presents advancements in enhancing and digitizing ancient manuscripts and inscriptions. In particular, it thoroughly analyses various methods to enhance degraded ancient images, such as dealing with low resolution, minimal intensity differences between text and background, show-through effects, and uneven backgrounds. The digitization of historical documents brings many advantages related to preservation, accessibility, and collaboration. Besides this, one of the greatest benefits comes from the fact that once obtained the digital version of a document, it is possible to analyze it with different software solutions which represent important supporting tools for experts. With the digital version of historical documents, many researchers focused on enhancing their readability and interpretability, totally changing the traditional approaches used by paleographers to analyze sources. An example is given by [10], which proposes techniques to improve the quality of scanned images of old manuscripts by addressing the challenge of degraded text portions that have become unreadable over time. To tackle this issue, the authors introduced an automated identification process for detecting the degraded text, which utilizes a matched wavelet-based text extraction algorithm. Regarding the enhancement of ancient documents, in [3], the authors explore theoretical, algorithmic, perceptual, and interactive dimensions of improving script legibility in the visible light spectrum. Its primary focus is to enhance the scholarly editing of papyrus texts. In [2] is described the System for Paleographic Inspections (SPI) developed at the University of Pisa, which is a tool to assist scholars, historians, and paleographers in analyzing and studying ancient scripts and manuscripts. A more recent open-source software is READ (Research Environment for Ancient Documents) [1], designed to facilitate the scholarly examination of ancient texts on their physical mediums. Its purpose is to provide various functionalities such as linking images of inscribed objects with transcriptions, managing multiple transcriptions of the same object simultaneously, connecting original-language texts with translations, and generating glossaries and paleographic charts.

Writer identification is determining the writer of a given text or document by analyzing unique writing characteristics that form his writing style. Thanks to digitising historical manuscripts, experts can now use AI techniques to support them in the writer identification task. An important study is given by [19], which thoroughly examines various writer identification techniques and aims to create a clear overview of datasets, feature extraction methods, and classification approaches (conventional and deep learning-based). Regarding Greek papyri, recent interesting results were obtained in [4,16,17] and [18], where different image processing and ML or DL techniques are employed to reach the common purpose of writer identification. In [16], the authors propose a writer identification method on the GRK-Papyri dataset [15]. First, they processed the digitized papyrus images and divided them into 512×512 patches using dense sampling. The processing involved several binarization techniques, but the selected one was DeepOtsu [12]. Then they considered different models of CNN (VGG16, ResNet 50 and InceptionV3) for the classification and the leave-one-out approach, well known in DL or ML systems, to deal with small datasets. In detail, they used two different fine-tuning (FT) schemes for training these networks. The first FT scheme (FT1) consisted in using the networks pre-trained on ImageNet, and then finetuned on the task dataset. For the second scheme (FT2), they pre-trained the network on ImageNet and fine-tuned it on the IAM dataset [14] before giving papyrus patches as input. It should be underlined that they used the 10 writers from GRK-Papyri, meaning the same writers as us, except for Dios, who was not in that dataset. Furthermore, they discarded some papyrus images to have a more balanced dataset. The other interesting research is [18], in which the authors propose a novel DL system for writer identification and test it on two datasets, one of them is the first variant of the PapyRow dataset, but they resized its rows to a size of 512×128. They developed a modified version of ResNet34 and trained it in a supervised way with several versions of the dataset based on the chosen preprocessing technique. PapyRow samples were used, in fact, in three different versions, first as they are, then binarized with Sauvola's algorithm and finally binarized through U-Net. In their experimental setting, they considered 21 writers and applied a 4-fold cross-validation technique. Section 5.3 will compare our approach and the two previously described.

3 Data and Preprocessing

This section overviews the dataset and outlines the various preprocessing techniques to enhance its images. Due to significant image degradation, employing image enhancement methods to improve the effectiveness of any computational approach involved in handwriting analysis becomes essential.

3.1 Reference Dataset: PapyRow

The reference dataset considered for this work is PapyRow [5], which contains images of Greek papyri from the 6th century CE. The writer of each papyrus of this collection is ensured by the presence of notary subscriptions, analyzed in [7]. Experts examined and selected every image to retain only the portion written by the notary, excluding any content from potential parties or witnesses. The PapyRow dataset comprises 120 papyrus images, representing a collection of 23 writers. In particular, 50 of these images came from the GRK-Papyri dataset [15] and were unevenly distributed over 10 writers. Most of the documents come from the extensive archive of Dioscorus of Aphrodito, regarded as one of the most comprehensive collections of the Byzantine period [9,20], to which were added two geographical outsiders, Menas and Dios, notaries from the big cities of Hermopolis and Antinoopolis several hundred kilometres north. For this work, we selected from the PapyRow dataset the data concerning the 10 writers already in GRK-Papyri dataset. We included an eleventh one, Dios, because besides being a geographical outsider, as said above, only one very long document is preserved from his hand, located (and thus digitized) in a single collection. Dios offers, therefore, the ideal test case for large homogeneous data. The first column of Table 1 shows the name of the writers, while the second column contains the number of papyrus images acquired for every writer, the third column reports the number of papyrus documents, and columns fourth and fifth the overall number of rows and patches extracted for each writer.

Table 1. Subpart of the PapyRow dataset considered and its organization. The first column contains the names of the eleven writers, while the second column contains the number of digital images, not to be confused with the third column, containing the number of papyrus documents for each writer. The fourth and fifth columns show the number of rows/patches extracted from the first (rows of 1232px in width) and second (patches of 500px in width) segmentation methods, respectively.

Writer	N. Images	N. Papyri	N. Rows	N. Patches
Abraamios	21	17	585	1608
Andreas	4	1	342	907
Dios	15	1	745	2083
Dioscorus	5	4	491	1320
Hermauos	5	5	429	1141
Isak	8	5	428	1154
Kyros1	8	8	324	870
Kyros3	5	5	466	1246
Menas	5	4	242	654
Pilatos	8	8	425	1216
Victor1	10	7	361	1004

3.2 Data Processing

In [5] is explained all the processing applied to every papyrus image to constitute a uniform dataset. A heavy phase of image elaboration was necessary due to the extensive degradation of most papyrus fragments. The initial step was background smoothing; it consisted of standardizing the background and filling in any hole with the average colour of the papyrus background. Subsequently, all the images were converted to greyscale, resulting in a dataset with a uniform appearance. The ultimate objective was to extract all the lines from each papyrus, requiring a resizing phase to ensure uniform line dimensions, which is essential for properly utilising convolutional neural networks. With this uniform dataset in place, the next step involved labelling all the text lines within the images. Since the images had varying widths, an additional step of row segmentation was executed to ensure consistent image sizes. Two distinct segmentation methods were developed to generate images with similar widths: Rows segmentation and Patches segmentation.

The rows segmentation method identified the line image with the narrowest width among all available images. This width value was determined to be 1232 pixels. Subsequently, all other images were divided based on this identified width.

Alternatively, patches segmentation method involved cutting all the lines using a predetermined threshold value for the width, which was set at 500 pixels. This approach allowed to collect a larger quantity of material per line, potentially reducing the loss of helpful writing.

Figure 1 shows the sequence of steps developed to enhance the handwritten traits and extract the rows and patches, while Fig. 2 shows the example of an image involved in the elaboration. In definitive, we used in this research both variants of the Papyrow dataset.

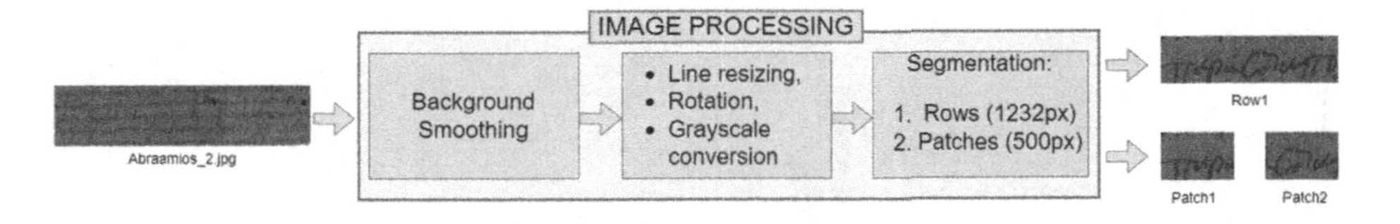

Fig. 1. Representation of the image processing approach.

4 Experimental Method

This study focuses on creating a writer identification system specifically designed for Greek papyri in the dataset discussed earlier. The research methodology comprises two distinct stages: row classification and papyrus classification. In the first stage, three widely recognized CNN models are employed to classify the rows. For the second stage, the majority vote rule (MV) is applied to the predictions obtained from the rows belonging to the same papyrus. The subsequent subsections provide detailed explanations of the procedures employed in each stage, also illustrated in Fig. 3. The entire experimental setting is iterated over the two variants of the dataset coming from the different segmentation methods considered.

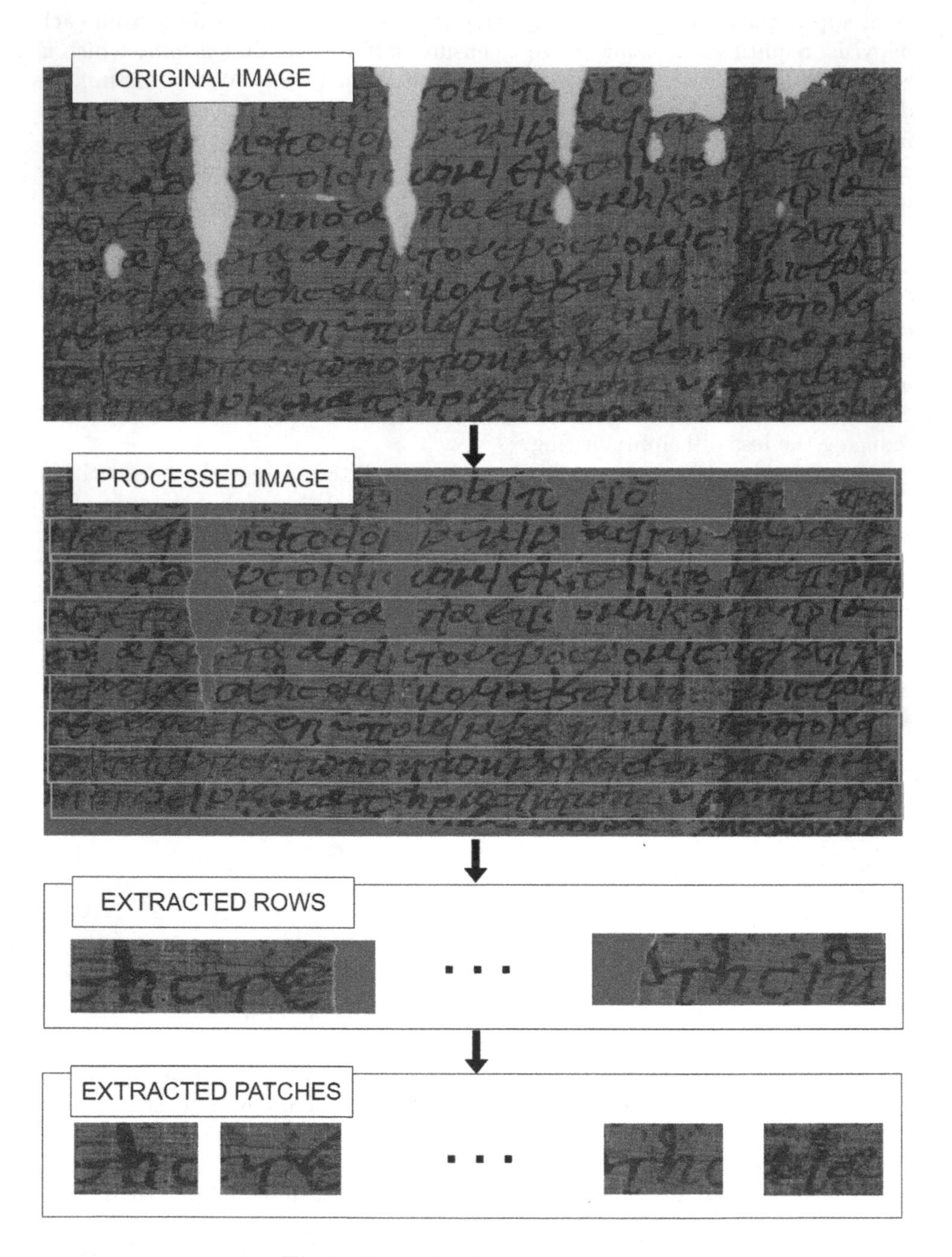

Fig. 2. Example of image processing.

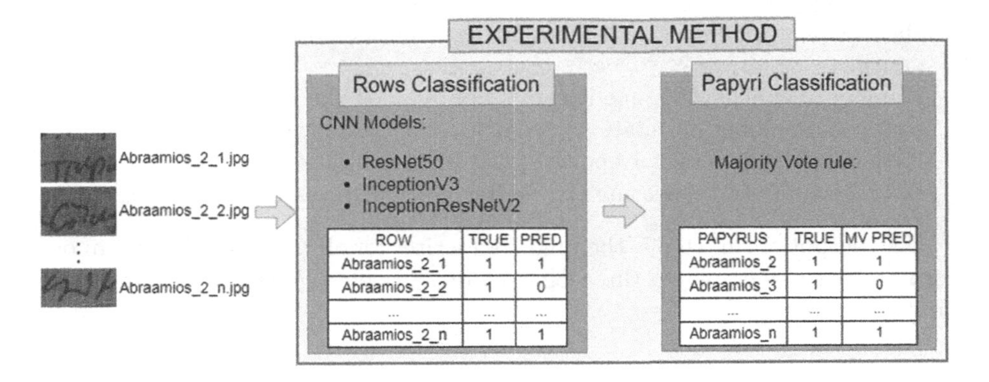

Fig. 3. Scheme of the experimental method developed.

4.1 Row/Patches Classification

The first stage of our experimental setting consists of the classification of rows (or patches), where the ground truth for each row is the scribe who wrote the papyrus from which it was extracted. We only considered 11 authors, shown in Table 1 with their corresponding number of papyri and rows. This part of the experimental approach involves the use of three CNN models: ResNet50 [11], InceptionV3 [22], and InceptionResNetV2 [21]. They are convolutional neural network architectures widely used for various computer vision tasks, including image classification and object recognition. While they share the goal of extracting meaningful features from input images, there are notable differences in their design and performance. Besides the differences, every CNN comprises two parts: a feature extractor (FE) and a classifier (C). A fully connected architecture substituted the original classifier of each model with two input layers and as many output neurons as the authors in our problem. We employed transfer learning by pre-training the feature extraction (FE) component on the well-known ImageNet dataset [6] to extract meaningful features. This approach is widely used to leverage the knowledge gained from training on a large dataset and apply it to a specific task. We implemented a k-fold cross-validation strategy during the experiment, a popular machine learning technique used to assess the performance of a model and mitigate overfitting. The process involves dividing the dataset into k subsets (folds) of approximately equal size. In our case, the number of folds (k) is 4. One fold is kept as a test set, while the remaining k-1 folds are used for training the model. The model is then evaluated on the test set, and this process is repeated four times. By averaging the performance across all four iterations, the K-fold cross-validation provides a more robust estimate of the model's generalization ability than a single train-test split. We tuned the CNN hyperparameters after preliminary experiments. We selected the following values:

- Optimization function: Stochastic Gradient Descent (SGD) with learning rate 0.001 and momentum 0.9.
- Loss function: Categorical cross-entropy.

- Batch size: 16, number of training images considered in each iteration of an epoch.
- Number of epochs: 200, one epoch is one pass on the entire training set and contains a number of iterations equal to $(trainingsetsize)/batch$.
- Patience: 20, the limit for epochs if the validation loss does not improve.
- Accuracy: metric to measure the performance, expressed in percentage.

The output of this step is the predicted writer for all the rows/patches in our dataset. Since we employed three CNN models, we obtained three sets of output predictions.

4.2 Papyrus Classification

In the previous section, we described the classification process that gave as output a set of predictions, one for each row/patch, independently from its belonging papyrus. After that, we considered the majority vote rule to get a classification on the papyrus level. The majority vote combination rule is a simple and effective ensemble technique, very popular in classification tasks. It involves combining the predictions of rows/patches from the same papyrus and selecting the class that receives the most votes as the final prediction. This method is particularly useful in reducing bias and variance, improving overall predictive accuracy, and enhancing the robustness of the ensemble by leveraging the collective knowledge of the individual classifiers.

5 Experimental Results

This section describes the performance obtained from the experimental setting explained earlier. To evaluate the performance of our system, we computed Top1, Top2, and Top3 accuracy, which are commonly used evaluation metrics in multi-class classification tasks. They represent the accuracy of a model in predicting the correct class label for a given input sample. Top1 accuracy measures the percentage of samples for which the model's top predicted class matches the true class label. In other words, it checks if the model's most confident prediction is correct. A higher accuracy score indicates that the model makes more correct predictions, while a lower score suggests that the model is struggling to classify the samples correctly. TopN accuracy, instead, considers the top N predictions made by the model. It calculates the percentage of samples for which the true class label is among the model's N highest-confidence predictions. In our case, we computed from Top1 to Top3 accuracy. We considered the same metrics on the row and papyrus classification sides.

5.1 PapyRow Rows Results

This section contains a discussion about results obtained on the variant of the dataset containing rows of 1232px of width. Table 2 shows in the first column the

CNN models used to classify the rows, and the following columns represent the Top1, Top2 and Top3 accuracies obtained, first for the row level classification and then for the papyrus one. All the values are expressed in percentages, and the best result is in bold for each column. Looking at the row-level results in the first multicolumn of the table, the model that reached the best performance was InceptionV3, with a value of 39.49% for the Top1 accuracy, 54.79% for the Top2 and 63.93% for the Top3. As expected, from Top1 to Top3, the accuracies scores increased. Top-1 accuracy measures the percentage of predictions where the model's top-ranked prediction matches the ground truth label exactly. This evaluation metric is strict and requires the model to make the most accurate prediction. On the other hand, Top-n accuracy allows for a slightly more lenient evaluation.

The results on the papyrus level are shown in the second multicolumn of the Table. For every CNN model, all the metrics decrease. In the worst case, the Top1 accuracy decreases by more than 10%. This trend is due to the unbalance of the dataset in terms of number of rows for each papyrus and writer. Also, concerning the papyrus level classification, the best results were achieved using the InceptionV3 CNN.

Table 2. Row and papyrus level results achieved by the first variant of the dataset (rows of 1232px in width) for every CNN model tested

Variant 1	**Row level**			**Papyrus level**		
Model	**Top1**	**Top2**	**Top3**	**Top1**	**Top2**	**Top3**
ResNet50	38.21	52.98	61.44	26.86	43.28	50.74
InceptionV3	**39.49**	**54.79**	**63.93**	**34.32**	44.77	**56.71**
Inc.Res.V2	37.37	54.68	62.64	31.34	**46.26**	49.25

5.2 PapyRow Patches Results

Here are discussed results obtained on the variant of the dataset containing patches of 500px of width. As in the previous table, Table 3 contains in the first column the CNN models considered, and the following columns represent the Top1, Top2 and Top3 accuracies obtained, first for the patch level classification and then for the papyrus one. Again, all the values are expressed in percentages; the best result is in bold for each column. Considering the patch level results, the model that reached the best performance was InceptionV3, with a value of 40.81% for the Top1 accuracy. The best Top2 and Top3 accuracies are obtained with InceptionResNetV2 with the following values, respectively: 56.39% and 65.01%. ResNet50 achieves the worst performance. Like what happened for the first variant of the dataset, the results increased from Top1 to Top3, as expected.

Regarding the papyrus level performance, for every CNN model, all the metrics decrease compared to the patch level. This trend is due to the unbalance of the dataset in terms of number of patches for each papyrus and writer.

Figure 4 shows the confusion matrix given by InceptionV3 during the classification of patches. A confusion matrix is a table that provides a summary of the performance of a classification model. It compares the predicted labels (columns) against the actual labels (rows) and shows the number of correct and incorrect predictions for each class. This confusion matrix shows that the classification performance is extremely high regarding Dios, our test case for geographical outsiders with large homogeneous data (2083 patches from one document). The second best classified is also the second most homogeneous case (and the chronological outsider writing almost forty years after the others, see graph): Andreas from whom the 907 patches come from only one document but preserved in two institutions. Abraamios, who is the second most largely documented writer with 1608 patches but coming from 17 papyri spanning over 20 years, is not so clearly discriminated. Neither is Menas, the second geographical outsider, whose 654 patches come from four documents covering more than 25 years. What is, however, encouraging is that for seven out of eleven writers, the most frequently predicted label is the correct one: only Isak, Pilatos and Victor1 were most frequently confused with other precise writers than correctly classified. The last one, Victor1 has especially bad predictions, less than one percent and all the other writers, but one being more frequently attributed to his patches than himself. This would deserve further investigation, Victor1 has indeed heterogeneous data with 1004 patches coming from 7 papyri, but he is not the only one in this case (see, for instance, Kyros1 with 870 patches from 8 papyri); besides, Victor1 is the most ancient writer of the dataset, which could have helped to discriminate him. To better understand the discussion above, please refer to Fig. 5 showing the chronological distribution of writers.

Table 3. Row and papyrus level results achieved by the second variant of the dataset (patches of 500px in width) for every CNN model tested

Variant 2	Patch level			Papyrus level		
Model	**Top1**	**Top2**	**Top3**	**Top1**	**Top2**	**Top3**
ResNet50	38.59	55.22	64.26	32.83	**52.23**	**61.19**
InceptionV3	**40.81**	53.98	62.64	29.85	40.29	49.25
Inc.Res.V2	39.79	**56.39**	**65.01**	**34.32**	47.76	58.20

5.3 Comparison

In [16] and [18], the authors proposed writer identification techniques for Greek papyri based on handwriting, a description of their research is in Sect. 2. This section compares the results obtained in these articles and our study. Table 4 is

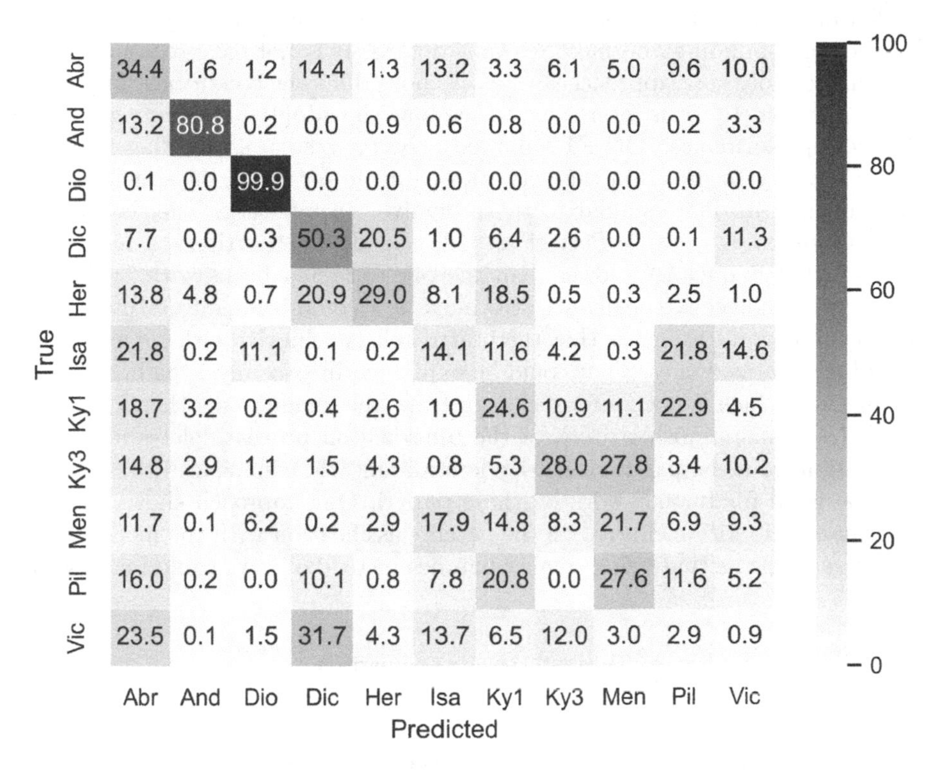

Fig. 4. Confusion Matrix given by InceptionV3 for the patch level classification.

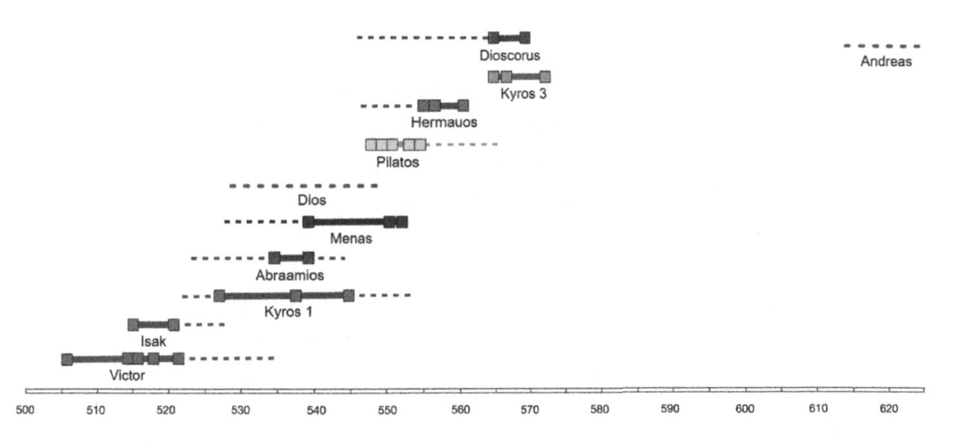

Fig. 5. Chronological distribution of the GRK-Papyri writers to which Dios has been added. Squares represent precisely dated texts in GRK-Papyri, while broken lines mark the activity span of these writers, known otherwise. Note that Dios and Andreas are only known by one text whose date can only be assessed in a time range marked by a dotted line.

divided into three sub-tables, showing the performance achieved by each study and the corresponding and most relevant details. It is not easy to compare the performances of these approaches as each chose different conditions. From Siddiqi's approach [16], it is interesting to see how the adopted fine-tuning method influences performance. The FT2 allowed a better result, illustrating that it is a good practice to pre-train the network on ImageNet and a dataset like IAM, where images are more similar to those involved in our task. The second sub-table is interesting because Peer et al. [18] considered 21 writers, achieving an accuracy that is only 10% lower than the one obtained by us with 11 writers. Their approach needs to be investigated more as it could bring interesting results. Another important aspect is that the best result is achieved with colour images instead of binarized ones. That could be explained in two ways: the background information, which is not uniform enough from one image to another, is used by the CNNs to distinguish writers, or the binarization process deletes important information from the handwriting. Understanding what influenced these results could be good for further work on these papyri. Our approach shows the best performance, 40.81%, achieved on the patch classification with the model InceptionV3 and tells us that using smaller patches instead of rows could increase the results.

Table 4. Results comparison

Siddiqi et al.			
N. writers	**Model**	**Acc FT1**	**Acc FT2**
10	VGG16	14.00	16.00
10	ResNet50	**30.00**	**33.00**
10	Inc.V3	24.00	27.00
Peer et al.			
N. writers	**Model**	**Processing**	**Acc**
21	Mod. ResNet34	Color	**28.70**
21	Mod. ResNet34	Sauvola	28.20
21	Mod. ResNet34	U-Net	27.80
Our Approach			
N. writers	**Model**	**Row Acc**	**Patch Acc**
11	ResNet50	38.21	38.59
11	Inc.V3	**39.49**	**40.81**
11	IncRes.V2	37.37	39.79

6 Conclusions and Future Work

The proposed research aimed to implement a classification system based on DL for the writer identification task of Greek papyri. This is challenging due to several factors. First, the availability and quality of data is a real issue, as the training of AI models requires a large dataset of known writers and their corresponding handwriting samples. However, when it comes to ancient papyri, the available data is often limited and may not cover a wide range of writers or styles. Also, the degradation and preservation of ancient papyri makes it harder to extract clear and accurate handwriting samples. Fading ink, smudging, tears, or other damages of the papyrus material condition can obscure the original writing, making it challenging for AI to analyze.

Despite these challenges, AI can still assist in identifying writers from papyri. Researchers can employ techniques such as image processing, pattern recognition, and machine learning algorithms to analyze the available data and make informed assessments. The preliminary results obtained with our research are encouraging and in line with those obtained in other studies. Further work will involve several changes. First of all, we will use a balanced dataset, even if this means deleting a lot of samples. We will also consider applying a fine-tuning procedure with a dataset of handwritten texts. Finally, we will consider binarization approaches, trying to delete the background information so that the network can only focus on writers' handwriting.

References

1. https://github.com/readsoftware/read
2. Aiolli, F., Ciula, A.: A case study on the system for paleographic inspections (SPI): challenges and new developments. In: Proceedings of the 2009 Conference on Computational Intelligence and Bioengineering: Essays in Memory of Antonina Starita, pp. 53–66. IOS Press (2009)
3. Atanasiu, V., Marthot-Santaniello, I.: Personalizing image enhancement for critical visual tasks: improved legibility of papyri using color processing and visual illusions. Int. J. Doc. Anal. Recogn. (IJDAR) **25**(2), 129–160 (2021). https://doi.org/10.1007/s10032-021-00386-0
4. Christlein, V., Marthot-Santaniello, I., Mayr, M., Nicolaou, A., Seuret, M.: Writer retrieval and writer identification in Greek papyri. In: Carmona-Duarte, C., Diaz, M., Ferrer, M.A., Morales, A. (eds.) Intertwining Graphonomics with Human Movements, pp. 76–89. Springer International Publishing, Cham (2022). https://doi.org/10.1007/978-3-031-19745-1_6
5. Cilia, N.D., De Stefano, C., Fontanella, F., Marthot-Santaniello, I., Scotto di Freca, A.: PapyRow: a dataset of row images from ancient Greek papyri for writers identification. In: Del Bimbo, A., et al. (eds.) ICPR 2021. LNCS, vol. 12667, pp. 223–234. Springer, Cham (2021). https://doi.org/10.1007/978-3-030-68787-8_16
6. Deng, J., Dong, W., Socher, R., Li, L.J., Li, K., Fei-Fei, L.: ImageNet: a large-scale hierarchical image database. In: CVPR, pp. 248–255. IEEE Computer Society (2009)
7. Diethart, J., Worp, K.: Notarsunterschriften im byzantinischen Ägypten (Byz. Not.). No. v. 1 in Mitteilungen aus der Papyrussammlung der Österreichischen Nationalbibliothek, Hollinek (1986). https://books.google.it/books?id=HqfezQEACAAJ

8. Fontanella, F., Colace, F., Molinara, M., Scotto Di Freca, A., Stanco, F.: Pattern recognition and artificial intelligence techniques for cultural heritage. Pattern Recogn. Lett. **138**, 23–29 (2020)
9. Fournet, J. (ed.): Les archives de Dioscore d'Aphrodité cent ans après leur découverte, histoire et culture dans l'Égypte byzantine. Actes du Colloque de Strasbourg. Études d'archéologie et d'histoire ancienne, Paris (2008)
10. Gupta, A., Kumar, S., Gupta, R., Chaudhury, S., Joshi, S.: Enhancement of old manuscript images. In: Ninth International Conference on Document Analysis and Recognition (ICDAR 2007), vol. 2, pp. 744–748. IEEE (2007)
11. He, K., Zhang, X., Ren, S., Sun, J.: Deep residual learning for image recognition. In: 2016 IEEE Conference on Computer Vision and Pattern Recognition (CVPR), pp. 770–778 (2016)
12. He, S., Schomaker, L.: DeepOtsu: document enhancement and binarization using iterative deep learning. Pattern Recogn. **91**, 379–390 (2019). https://doi.org/10.1016/j.patcog.2019.01.025
13. Jayanthi, N., Indu, S., Hasija, S., Tripathi, P.: Digitization of ancient manuscripts and inscriptions - a review. In: Singh, M., Gupta, P.K., Tyagi, V., Sharma, A., Ören, T., Grosky, W. (eds.) ICACDS 2016. CCIS, vol. 721, pp. 605–612. Springer, Singapore (2017). https://doi.org/10.1007/978-981-10-5427-3_62
14. Marti, U.V., Bunke, H.: The IAM-database: an English sentence database for offline handwriting recognition. Int. J. Doc. Anal. Recogn. **5**, 39–46 (2002)
15. Mohammed, H., Marthot-Santaniello, I., Margner, V.: GRK-Papyri: a dataset of Greek handwriting on papyri for the task of writer identification. In: Proceedings of the 2019 International Conference on Document Analysis and Recognition (ICDAR), pp. 726–731 (2019)
16. Nasir, S., Siddiqi, I.: Learning features for writer identification from handwriting on papyri. In: Djeddi, C., Kessentini, Y., Siddiqi, I., Jmaiel, M. (eds.) Pattern Recognition and Artificial Intelligence, pp. 229–241. Springer International Publishing, Cham (2021). https://doi.org/10.1007/978-3-030-71804-6_17
17. Nasir, S., Siddiqi, I., Moetesum, M.: Writer characterization from handwriting on papyri using multi-step feature learning. In: Barney Smith, E.H., Pal, U. (eds.) ICDAR 2021. LNCS, vol. 12916, pp. 451–465. Springer, Cham (2021). https://doi.org/10.1007/978-3-030-86198-8_32
18. Peer, M., Sablatnig, R.: Feature mixing for writer retrieval and identification on papyri fragments (2023)
19. Rehman, A., Naz, S., Razzak, M.I.: Writer identification using machine learning approaches: a comprehensive review. Multimedia Tools Appl. **78**, 10889–10931 (2019)
20. Ruffini, G. (ed.): Life in an Egyptian Village in Late Antiquity: Aphrodito Before and After the Islamic Conquest. Cambridge University Press, Cambridge (2018)
21. Szegedy, C., Ioffe, S., Vanhoucke, V.: Inception-v4, Inception-ResNet and the impact of residual connections on learning. In: AAAI (2016)
22. Szegedy, C., Vanhoucke, V., Ioffe, S., Shlens, J., Wojna, Z.: Rethinking the inception architecture for computer vision. In: 2016 IEEE Conference on Computer Vision and Pattern Recognition (CVPR), pp. 2818–2826 (2016)

Convolutional Generative Model for Pixel–Wise Colour Specification for Cultural Heritage

Furnari Giuseppe[1(✉)], Anna Maria Gueli[2,3], Stanco Filippo[1], and Dario Allegra[1]

[1] Department of Mathematics and Computer Science (DMI), University of Catania, Viale A. Doria 6, Catania, Italy
giuseppe.furnari@phd.unict.it

[2] Department of Physics and Astronomy Ettore Majorana, University of Catania, via S. Sofia 64, Catania, Italy
https://www.dfa.unict.it/

[3] INFN-CHNet Sez CT, via S. Sofia 64, Catania, Italy
http://web.dmi.unict.it/

Abstract. Colour specification can be carried out using different instruments or tools. The biggest limitation of these existing instruments consists of the region in which they can be applied. Indeed, they can only work locally in small regions on the surface of the object under examination. This implicates a slow process, errors while repeating the procedure and sometimes the impossibility of measuring the colour depending on the object's surface. We present a new way to perform colour specification in the CIELab colour space from RGB images by using Convolutional Generative Model that performs the transformation needed to remove all the shading effect on the image, producing an albedo image which is used to estimate the CIELab value for each pixel. In this work, we examine two different models one based on autoencoder and another based on GANs. In order to train and validate our models we present also a dataset of synthetic images which have been acquired using a Blender–based tool. The results obtained using our model on the generated dataset prove the performance of this method, which led to a low average colour error ($\Delta E00$) for both the validation and test sets. Finally, a real-scenario test is conducted on the head of the god Hades and a half-bust depicting the goddess Persephone, both are from the archaeological Museum of Aidone (Italy).

Keywords: Color measurement · Color specification · Autoencoder · GANs

1 Introduction

Several instruments can be used to perform colour specification, such as colourimeters and spectrophotometers, in the past years also software–based tools have

G. L. Foresti et al. (Eds.): ICIAP 2023 Workshops, LNCS 14366, pp. 437–448, 2024.
https://doi.org/10.1007/978-3-031-51026-7_37

been proposed to perform the colour measurement. In general, the colour specification can be useful for many domains such as dentistry [7,16], medicine [20] or branding and marketing [23,30]. Cultural heritage also benefits from the colour specification. Indeed, it is a qualitative instrument to guarantee the exact copy of colours in digital images, for instance, in online collections and virtual reality experiences. The predominant role of the colour specification is in the conservation, documenting, and understanding of cultural heritage [18,21,29,32], principally, by supplying useful data for conservation and analysis purposes and confirming the correct replica of tints in authentic objects. Distinct approaches have been investigated to perform the colour specification for different colour systems; in [24] authors explore five different models (direct, gamma, linear, quadratic, and neural) to achieve the RGB to Lab transformation. Neural networks and their variants have been widely used for RGB to CIELab and RGB to XYZ transformations in different contexts [9,25]. In [34] authors investigated different approaches based on Support Vector Regression (SVR), Artificial Neural Networks (ANN), Deep Neural Networks (DNN), and Radial Basis Function (RBF) to perform the CMYK to CIELab transformation. Even, Markov Chain Monte Carlo (MCMC) has been used to perform the colour transformation [6]. In a previous work [26], we approached it as a classification problem using the Support Vector Classifier (SVC). The choice was motivated by the fact that the set of all possible HVC coordinates in the Munsell soil colour charts (MSCCs), commonly used by archaeologists, is discrete. While, the first time in [1], we have investigated the RGB to CIELab conversion in the 3D setting, operating a 3D scanner and training a deep learning model to predict the colour given an RGB-D image (RGB image with an extra depth channel). One of the biggest challenges in this field is that colour perception is impacted by different factors [10], and one of the limitations of instruments and solutions which perform the colour measurement is that they work locally, in a single region which often is very small. This makes the colour measurement a slow process, and sometimes infeasible depending on the surface of the item. In this work, we explore whether it is possible to perform a pixel–by–pixel colour transformation using different generative models. We present a new approach to achieve the colour specification in the CIELab colour space. The key idea is to train a generative model to remove all the shading effects on the objects, generating an albedo image and then simply transforming the resulting image in the CIELab colour space. We investigate two different architectures to achieve our goal, an Autoencoder based and a GANs (Generative Adversarial Networks) based. Furthermore, we apply our methodology to two authentic artefacts to provide a real use case. The artefacts consist of a head of the god Hades and a half-bust depicting the goddess Persephone both are from the archaeological Museum of Aidone (Italy). To the best of our knowledge, this approach has never been used to perform the colour transformation, while albedo images are widely used in image decomposition [5,8,11]. In a nutshell, image decomposition aims to split an image into two parts: an albedo image, which consists of the reflectance properties (known as intrinsic properties) of the object and a shade image, which conserves all the shading effects (also referred to as extrinsic properties). The applications of such a technique include editing the scene's lighting and integrating new objects into photographs [19]. In our work, we focus on the albedo image because it is a useful component to perform the colour specification. Setting up a simulated environment, we obtain both the rendered and the albedo version of images containing objects with different colours that will be used to train our model to perform the transformation. Although several datasets have been proposed to accomplish the image decomposition task [4,22,27], in our study case it's not directly possible to use them, since we need to estimate the colour error after removing all the shading effects and they don't provide the real CIELab map. Nevertheless, we think that future investigations into such approaches may also be useful for the purpose of colour measurement.

To summarize the main contributions of our work are:

- Presenting a new method to perform pixel–wise colour specification based on generating an albedo (or shadeless) image through a generative model;
- Investigating two different approaches based on Convolutional Autoencoder and GANs;
- Realising a Blender-based tool to generate datasets for colour specification;
- Realising a synthetic dataset consisting of $128K$ images to perform colour specification;
- Showing a real use case on two different artefacts.

The implementation of the paper is available at github.com/giuseppefrn/PixelwiseColourSpecification [13].

2 Materials and Methods

The goal of the model is to produce images without all the shading effects created by the rendering process, giving as output the albedo image of the object. In our experiment setup, the albedo image or shadeless version can be used to perform the colour transformation from RGB to CIELab by simply applying the *rgb2lab* function implemented in the Python library scikit-image to perform this colour transformation.

2.1 Dataset

Using a slightly different version of the Blender-based tool proposed in previous work [12], we generated a new dataset that consists of images of 5 objects (also referred to as shapes), which have been acquired using 1600 colours grouped as 20 primary hues and 80 sub-colours for each hue. For each object, 8 different images are obtained by moving the camera partially around the item, in this dataset we avoided the views of the dark side of the objects which don't represent a real use case for this kind of application (poor illumination will let all the colours appear almost black and it is impossible to discriminate them using a single view approach). For our experiments, we produced images using illumination that reproduce the *D65* illuminant, and a *shadeless* version of the image which is used as ground truth (GT). The final dataset counts a total of $128K$ images: $64K$ for each rendered and shadeless version. The shadeless version (or albedo) consists of the image without any shading effect, which means the colour will be perceived as uniform around the shape of the objects.

2.2 Autoencoder Based Network

The proposed model consists of a Convolutional Neural Network based on an Autoencoder model with skip connections between the encoder and the decoder blocks. An autoencoder is a neural network that learns to compress the input data into a latent space, and then uncompress that into something that closely matches the original data. Autoencoders are commonly used for their ability to learn compressed and meaningful representations of data [3]. These representations can be used for a variety of tasks, such as data compression [33], denoising [2], and anomaly detection [14]. In our experiments, we want to exploit this ability to generate an albedo image from the latent representation of the rendered version. As the diagram shows in Fig. 1, we used three blocks for encoding and three for decoding, each encoder block consists of a 2D Convolution layer, a Batch Normalization, a Leaky Relu activation function and a 2D Max Pool. While the decoder's block consists of an Upsample Layer, which uses a bilinear function, a 2D Convolution Layer, Batch Normalization and a Leaky Relu activation function. The head of the model is made by a 2D Convolutional Layer and a Sigmoid activation function.

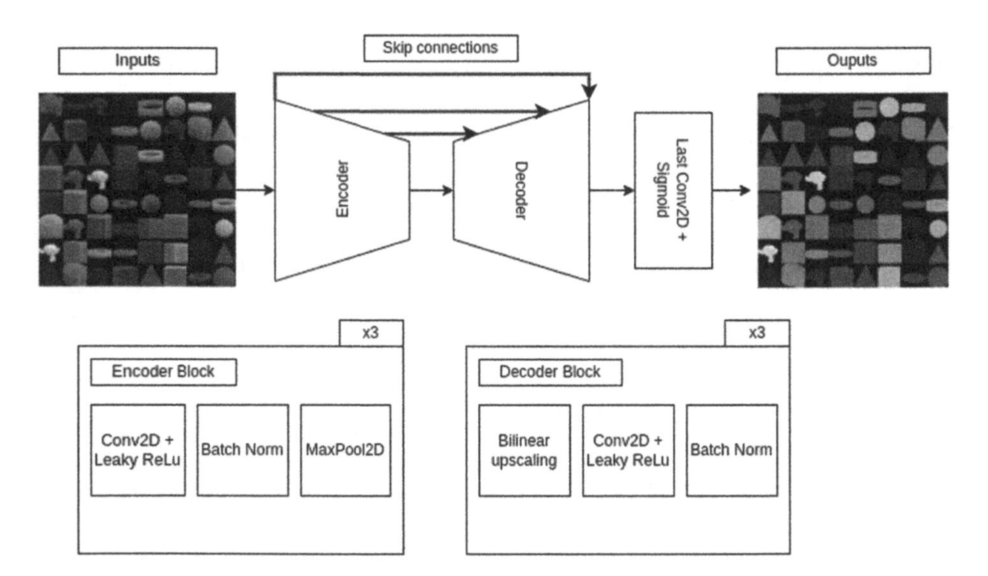

Fig. 1. Diagram representing the network architecture. The model consists of an encoder and decoder with skip connections for each convolutional block.

2.3 GANs

In our experiments, we have tested different architectures based on Generative Adversarial Networks (GANs). GANs are a type of neural network architecture that is used to generate new data that is similar to the training data. GANs consist of two neural networks: a generator and a discriminator. The generator creates new data samples that are similar to the training data, while the discriminator tries to distinguish between real and fake data. The two networks are trained together in an adversarial process where the generator attempts to create better fake data and the discriminator tries to get better at distinguishing between real and fake data [15]. In our work, the Generator is trained to generate an albedo image of a given rendered picture. Basically, it acts as the autoencoder discussed earlier, while the discriminator is trained to distinguish between real albedo and the generated ones. We have tested different versions of this framework by varying the architectures and the training mode. We can distinguish the following two main experimental setups: classical GANs and Siamese Discriminator. In both experimental setups, the Generator is the same as the Autoencoder discussed earlier, while the discriminator is a Convolutional Neural Network which performs binary classification, it is made by three convolutional blocks, which consist of a 2D Convolutional Layer, Batch Normalization and Leaky ReLu activation function, followed by a MaxPool Layer, a Fully Connected (FC) layer and a Sigmoid activation function. In the Siamese Discriminator experimental setup, the Discriminator still performs binary classification but it acts as a Siamese network, it receives two images as input, an albedo image and the rendered version and tries to learn if the albedo image is proper for the given rendered version. The backbone of the architecture is the same, the only difference is that the two feature vectors extracted by the Max Pool layer are merged together before the last FC layer. The idea is to force the generator to create albedo images that are more coherent with the given render version.

3 Experiments and Results

3.1 Autoencoder Results

In our experiments, we randomly split the dataset into training and validation sets using the 80% of the images for the training and 20% for the validation. In addition, for each colour, a sub-colour is selected to be used as a test. The *Autoencoder* model is trained using the *Mean Squared Error* (MSE) as a loss function and *ADAM* as an optimizer with an initial learning rate of 10^{-3}, the batch size is set to 64 and the model is trained for 150 epochs. In Fig. 2a the average loss value over epochs is shown for both the training and validation set. In order to evaluate the model performance over each epoch, we used the CIEDE00, also known as $\Delta E00$, which is a Euclidean distance-based formula to estimate the distance between two colours in the CIELAB colour space. In Fig. 2b the average $\Delta E00$ on the validation over epochs is shown. This value is also used to detect the best model during the training that in our experiment achieved a $\Delta E00$ of 1.19. After the end of the training, we used the best model to perform the colour transformation for the test set. The obtained average $\Delta E00$ for the test set is 1.64, while the average $\Delta E00$ for each colour and shape is shown in Table 1 and 2. A visual result can be seen in Fig. 3 where the rendered version of the image, the ground truth and the predicted images are shown respectively from left to right.

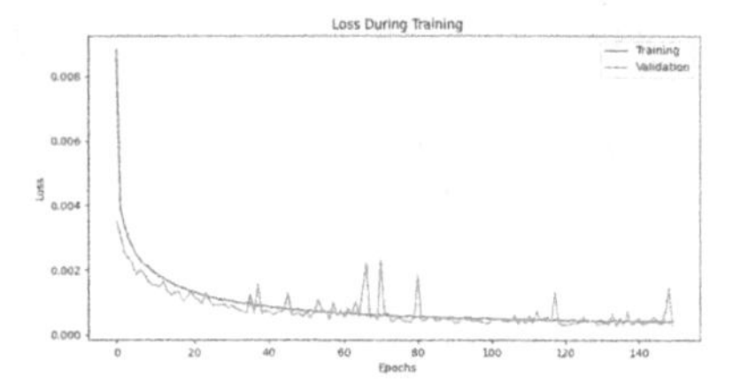

(a) Loss function(MSE) over epochs for training and validation dataset.

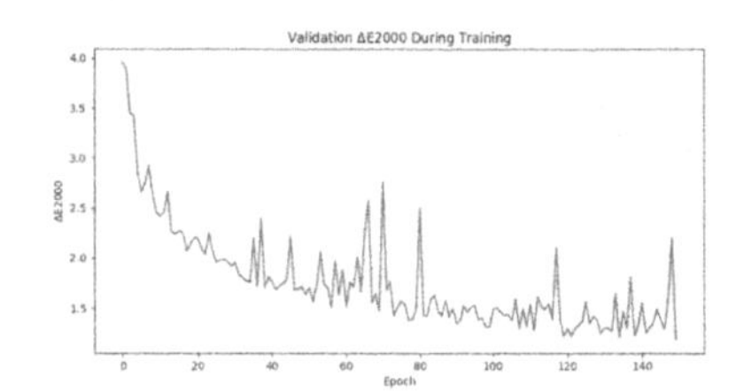

(b) $\Delta E00$ over epochs for the validation dataset.

Fig. 2. MSE and $\Delta E00$ over epochs.

Table 1. Average $\Delta E00$ on test-set grouped by main hue.

Hue	0	1	2	3	4	5	6	7	8	9	10	11	12	13	14	15	16	17	18	19
Average $\Delta E00$	1.21	1.28	1.26	1.17	1.11	1.06	1.04	1.04	1.14	1.22	1.3	1.29	1.22	1.13	1.1	1.04	0.99	**0.95**	1.06	1.17

Table 2. Average $\Delta E00$ on test-set grouped by shape (all colours).

Shape	Cone	Cube	Icosphere	Suzanne	Torus
Average $\Delta E00$	0.88	2.44	1.4	1.29	**0.78**

3.2 GANs Results

As mentioned previously, we tested two different GANs based on architectures. In the first setup, the *classical GANs*, the Generator takes as input the rendered version of the scene and generates an albedo version, while the discriminator takes in input an albedo image and classifies it depending on if it is a true or a fake one. In the *Siamese Discriminator*, the only difference is that the Discriminator takes as inputs an albedo (generated or true) image and its rendered

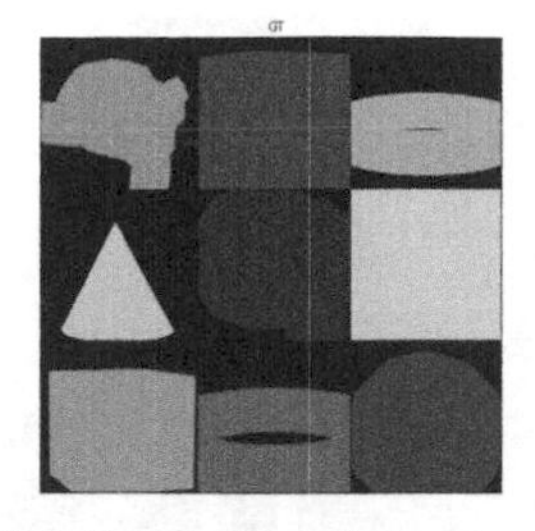

(a) Rendered images. (b) Ground truth images. (c) Generated images.

Fig. 3. From left to right: rendered version of a batch (used as input for the model), ground truth images and generated images from the model.

version. The reason why we used the Siamese Discriminator is to force the Generator to produce albedo images that are more coherent with the rendered versions. Indeed, we noticed that the generator used to create new albedo images without considering the original colours (basically it generates new images maintaining the shape the same but re-colouring them). By giving the original rendered version of the image to the Discriminator it will easily classify as fake not coherent albedo images. Although it let the visual results a bit better, the limitation of this setup is the Discriminator easily wins the "game" against the Generator by predicting 0 or 1 with very little uncertainty, also after applying the *instance noise* proposed in [31]. On the other hand, the Generator is not able to generate better albedo images and still generates images with completely wrong colouration (Fig. 4).

3.3 Test on Real Data

The Autoencoder-based model was applied to authentic artefacts and, in particular, to original artefacts from the archaeological museum of Aidone (Italy). The museum houses lithic and terracotta artefacts from the ancient archaeological site of Morgantina. Among these, a head of the god Hades and a half-bust depicting the goddess Persephone were selected for research. The artefacts were acquired in a non–controlled environment and using an unknown illuminant, generally before testing a model on real data it requires fine–tuning or domain adaptation steps. In our case, this is not possible due to the lack of real data, and the complexity (or inability) of obtaining the CIELab values for each "point" in a real scenario. Nevertheless, we decided to test our model in such a challenging scenario to examine its performance. Although these tests don't provide us with any metric of the model's performance, we can still examine them to investigate the general behaviour of the model in a real scenario, focusing on analysing the colouration preservation of details. In Fig. 5 and 6, images representing details of the artefacts and the albedo generated by the model are shown respectively. In Fig. 5a and 5d it is possible to note how the model is able to keep the brown stain in the bottom–left of the face, on the other hand, the contour of the eyes

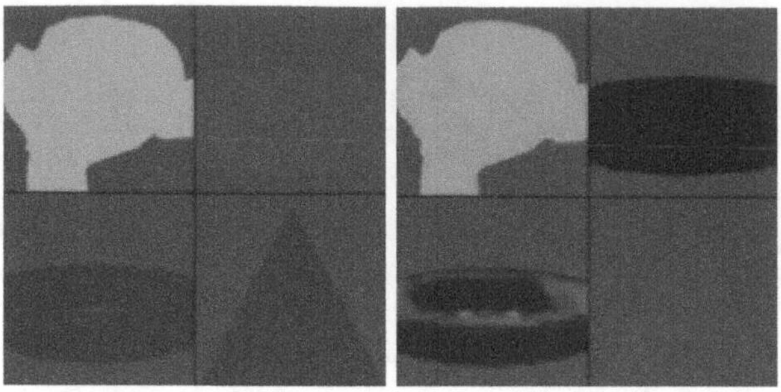

(a) Real albedo. (b) Generated image.

Fig. 4. Real albedo image and generated version from the Generator using the GANs approach.

has been treated as shade and then removed. The inputs provided in Figs. 5b and 5c are very challenging because the shape of the beard and hairs is very irregular and curly, however, we can note how the model is able to maintain colours detail such as the grey/blue colour of the beard and the red colouration of the hairs. Similar results are shown on the Persephone goddess, in Fig. 6c it is possible to note the lips' details are maintained, and, in Figs. 6b and 6d almost all the colouration is preserved.

4 Discussion

In this section, we discuss the obtained results and we give more information about the implementation choices we have made. We favoured the *bilinear function* to perform the upscaling instead of the common *Transpose Convolution* layer, the reason is Transpose Convolution can lead to artefacts in the generated images, like the chessboard pattern [28], while performing the upscaling using a bilinear function and then applying a convolution avoids this kind of artefacts. Our models use the *Leaky ReLU* function because it has an advantage over the standard *ReLU*, indeed it can help to improve the performance of deep neural networks by solving the dying ReLU problem. We have chosen the $\Delta E00$ as the evaluation metric, because the Euclidean Distance (known as $\Delta E76$ in the colour measurement terminology) demonstrated to affect heavily our results, especially for the saturated regions of the CIELab colour space, giving a high $\Delta E76$ although where the colours were the same. On the other hand, the $\Delta E00$ demonstrated to be more suitable and more consistent for our case study. This can be explained since CIELAB space is not as perceptually uniform as it was intended, especially in the saturated regions.

Looking at the loss function curve in Fig. 2a for the Autoencoder model, we can see how the model rapidly decreases this value within a few epochs. Although, the validation $\Delta E00$ shown in Fig. 2b, still decreases for almost all the epochs, letting to the best model in the last epoch. Both the validation and test

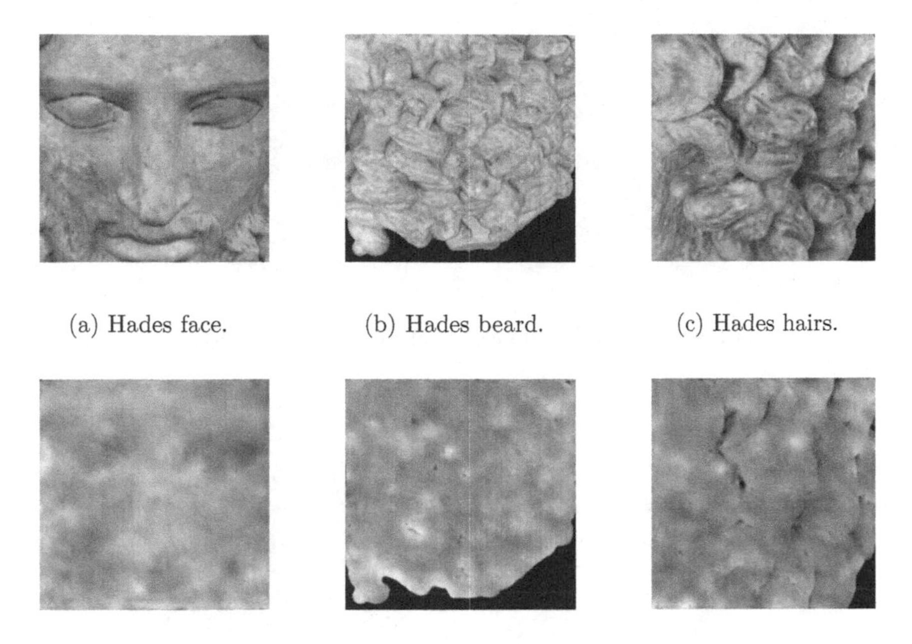

(a) Hades face. (b) Hades beard. (c) Hades hairs.

(d) Generated albedo of a. (e) Generated albedo of b. (f) Generated albedo of c.

Fig. 5. Results on real samples of a head of the god Hades from the archaeological museum of Aidone.

$\Delta E00$ are lower than the common threshold of 2 which represents the smallest colour difference the human eye can catch. Of course, in our experiment, that value denotes the average $\Delta E00$ between the Ground Truth and the generated image, which means the colour difference could be higher locally. While the average $\Delta E00$ is almost the same for all the hues 1, that is not true for the different shapes 2, we can note the average error for the cubes is quite higher compared to the others shapes (although it is still close to the threshold of 2). These results are not totally unexpected, indeed the other shapes are more smooth while the cube is affected more by a heavy and drastic colour change depending on the point of view, making it harder to generate a correct albedo image for all of them.

As it turned out from our experiments, the GANs–based models seem to be weak in dealing with this task, indeed, it is hard to force the generator to produce the correct colouration without generating new ones. Moreover, as is well known it is hard to stabilize the training of this kind of model and avoid model collapse. We think future investigations are possible, for instance, using the WGAN and applying solutions to improve their training [17].

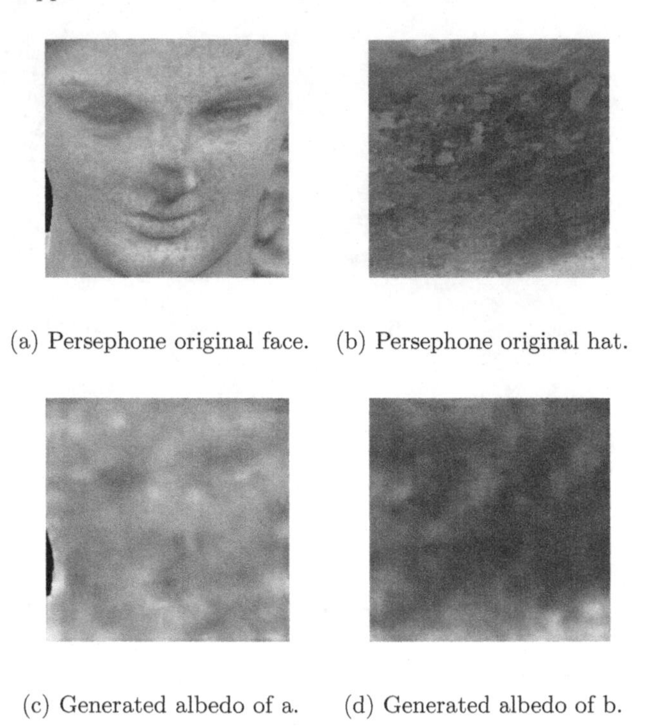

(a) Persephone original face. (b) Persephone original hat.

(c) Generated albedo of a. (d) Generated albedo of b.

Fig. 6. Results on real samples of Persephone from the archaeological museum of Aidone.

5 Conclusion

In this work, we presented a method to perform pixel–wise colour specification by generating an albedo image through generative models. We investigated a Convolutional Autoencoder with skip-connection and different approaches GANs–based to perform the generation of the albedo. The Autoencoder model seems to be able to remove almost all the shading effects and therefore the output images can be used to conduct the colour measurement in the CIELab colour space. The results obtained using our synthetic dataset have proved the performance of this method, which led to a low average $\Delta E00$ for both the validation and test sets. We have also tested our methodology on two real artefacts from the archaeological museum of Aidone. Although at the moment it is not possible to provide an evaluation metric for the real case tests, we have shown the general behaviour of the model in a real context, like the ability to keep some details but also its pitfalls. A strategy to fine–tune and apply the method in real scenarios should be drawn in future works. Moreover, our experiments have shown the limitations of the GANs approaches, although we think future investigations should be conducted to provide an exhaustive investigation of these approaches. Several challenges still exist in the field of colour measurement such as: performing the colour transformation in a multi-colour environment; making the albedo piecewise constant, and exploring the solutions in bigger and different datasets.

Acknowledgement. The research activity was funded by the University of Catania (Italy) through the PIAno di inCEntivi per la RIcerca di Ateneo (PIACERI) linea 2 project CLEAR - CoLor rEndering Accuracy in cultuRal heritage.

References

1. Allegra, D., et al.: A method to improve the color rendering accuracy in cultural heritage: preliminary results. In: Journal of Physics: Conference Series, vol. 2204, p. 012057. IOP Publishing (2022)
2. Bajaj, K., Singh, D.K., Ansari, M.A.: Autoencoders based deep learner for image denoising. Procedia Comput. Sci. **171**, 1535–1541 (2020)
3. Bank, D., Koenigstein, N., Giryes, R.: Autoencoders. arXiv preprint: arXiv:2003.05991 (2020)
4. Bell, S., Bala, K., Snavely, N.: Intrinsic images in the wild. ACM Trans. Graph. (TOG) **33**(4), 1–12 (2014)
5. Chen, Q., Koltun, V.: A simple model for intrinsic image decomposition with depth cues. In: Proceedings of the IEEE International Conference on Computer Vision, pp. 241–248 (2013)
6. Chen, Y., Liu, D., Liang, J.: A new method for RGB to CIELAB color space transformation based on Markov chain monte Carlo. In: MIPPR 2013: Parallel Processing of Images and Optimization and Medical Imaging Processing, vol. 8920, pp. 102–108. SPIE (2013)
7. Chu, S.J., Trushkowsky, R.D., Paravina, R.D.: Dental color matching instruments and systems. Review of clinical and research aspects. J. Dentistry **38**, e2–e16 (2010)
8. Fan, Q., Yang, J., Hua, G., Chen, B., Wipf, D.: Revisiting deep intrinsic image decompositions. In: Proceedings of the IEEE Conference on Computer Vision and Pattern Recognition, pp. 8944–8952 (2018)
9. Fdhal, N., Kyan, M., Androutsos, D., Sharma, A.: Color space transformation from RGB to CIELAB using neural networks. In: Muneesawang, P., Wu, F., Kumazawa, I., Roeksabutr, A., Liao, M., Tang, X. (eds.) PCM 2009. LNCS, vol. 5879, pp. 1011–1017. Springer, Heidelberg (2009). https://doi.org/10.1007/978-3-642-10467-1_97
10. Finlayson, G., Hordley, S., Schaefer, G., Tian, G.Y.: Illuminant and device invariant colour using histogram equalisation. Pattern Recogn. **38**(2), 179–190 (2005)
11. Forsyth, D., Rock, J.J.: Intrinsic image decomposition using paradigms. IEEE Trans. Pattern Anal. Mach. Intell. **44**(11), 7624–7637 (2021)
12. Giuseppe, F., Dario, A., Anna, G., Filippo, S.: CIELab color measurement through RGB-D images. In: Rousseau, J.J., Kapralos, B. (eds.) Pattern Recognition, Computer Vision, and Image Processing. Lecture Notes in Computer Science, vol. 13645, pp. 15–20. Springer, Cham (2023). https://doi.org/10.1007/978-3-031-37731-0_2
13. Giuseppe, F., Gueli, A.M., Stanco, F., Allegra, D.: PixelwiseColourSpecification. https://github.com/giuseppefrn/PixelwiseColourSpecification/
14. Gong, D., et al.: Memorizing normality to detect anomaly: memory-augmented deep autoencoder for unsupervised anomaly detection. In: Proceedings of the IEEE/CVF International Conference on Computer Vision, pp. 1705–1714 (2019)

15. Goodfellow, I., et al.: Generative adversarial networks. Commun. ACM **63**(11), 139–144 (2020)
16. Gueli, A.M., Pedullà, E., Pasquale, S., La Rosa, G.R., Rapisarda, E.: Color specification of two new resin composites and influence of stratification on their chromatic perception. Color. Res. Appl. **42**(5), 684–692 (2017)
17. Gulrajani, I., Ahmed, F., Arjovsky, M., Dumoulin, V., Courville, A.C.: Improved training of Wasserstein GANs. In: Advances in Neural Information Processing Systems, vol. 30 (2017)
18. Iturbe, A., Cachero, R., Canal, D., Martos, A.: Virtual digitization of caves with parietal paleolithic art from Bizkaia. Scientific analysis and dissemination through new visualization techniques. Virtual Archaeol. Rev. **9**(18), 57–65 (2018)
19. Karsch, K., Hedau, V., Forsyth, D., Hoiem, D.: Rendering synthetic objects into legacy photographs. ACM Trans. Graph. (TOG) **30**(6), 1–12 (2011)
20. Kawanabe, T., et al.: Quantification of tongue Colour using machine learning in Kampo medicine. Eur. J. Integr. Med. **8**(6), 932–941 (2016)
21. Korytkowski, P., Olejnik-Krugly, A.: Precise capture of colors in cultural heritage digitization. Color. Res. Appl. **42**(3), 333–336 (2017)
22. Kovacs, B., Bell, S., Snavely, N., Bala, K.: Shading annotations in the wild. Comput. Vis. Pattern Recogn. (CVPR) (2017)
23. Labrecque, L.I., Milne, G.R.: Exciting red and competent blue: the importance of color in marketing. J. Acad. Mark. Sci. **40**(5), 711–727 (2012)
24. Leon, K., Mery, D., Pedreschi, F., Leon, J.: Color measurement in lab units from RGB digital images. Food Res. Int. **39**(10), 1084–1091 (2006)
25. MacDonald, L.: Color space transformation using neural networks. In: Color and Imaging Conference, vol. 2019, pp. 153–158. Society for Imaging Science and Technology (2019)
26. Milotta, F.L.M., et al.: Challenges in automatic Munsell color profiling for cultural heritage. Pattern Recogn. Lett. **131**, 135–141 (2020)
27. Murmann, L., Gharbi, M., Aittala, M., Durand, F.: A multi-illumination dataset of indoor object appearance. In: 2019 IEEE International Conference on Computer Vision (ICCV) (2019)
28. Odena, A., Dumoulin, V., Olah, C.: Deconvolution and checkerboard artifacts. Distill (2016). https://doi.org/10.23915/distill.00003, http://distill.pub/2016/deconv-checkerboard
29. Ruiz, J.F., Pereira, J.: The colours of rock art. Analysis of colour recording and communication systems in rock art research. J. Archaeol. Sci. **50**, 338–349 (2014)
30. Singh, S.: Impact of color on marketing. Manage. Decis. **44**(6), 783–789 (2006)
31. Sønderby, C.K., Caballero, J., Theis, L., Shi, W., Huszár, F.: Amortised map inference for image super-resolution. arXiv preprint: arXiv:1610.04490 (2016)
32. Stanco, F., Battiato, S., Gallo, G.: Digital Imaging for Cultural Heritage Preservation. Analysis, Restoration, and Reconstruction of Ancient Artworks (2011)
33. Theis, L., Shi, W., Cunningham, A., Huszár, F.: Lossy image compression with compressive autoencoders. arXiv preprint: arXiv:1703.00395 (2017)
34. Velastegui, R., Pedersen, M.: CMYK-CIELAB color space transformation using machine learning techniques. In: London Imaging Meeting, vol. 2021, pp. 73–77. Society for Imaging Science and Technology (2021)

3D Modeling and Augmented Reality in Education: An Effective Application for the Museo dei Saperi e delle Mirabilia of the University of Catania

Germana Barone[1], Raissa Garozzo[2], Gloria Russo[2], Cettina Santagati[2], Diego Sinitò[3], Marilisa Yolanda Spironello[4(✉)], and Filippo Stanco[3]

[1] Department of Biological, Geological and Environmental Sciences, University of Catania, Corso Italia, 57, 95129 Catania, Italy
germana.barone@unict.it

[2] Department of Civil Engineering and Architecture, University of Catania, via Santa Sofia 64, 95123 Catania, Italy
{raissa.garozzo,cettina.santagati}@unict.it, gloria.russo@phd.unict.it

[3] Department of Mathematics and Computer Science, University of Catania, via Santa Sofia 64, 95123 Catania, Italy
dsinito@dmi.unict.it, filippo.stanco@unict.it

[4] Department of Human Science, University of Catania, Piazza Dante 32, 95124 Catania, Italy
marilisa.spironello@phd.unict.it

Abstract. Augmented reality is the process of using technology to superimpose video, images, text or sound onto what a person can already see with their own eyes in the reality around them.

It is enough to have a smartphone or tablet to change the reality in front of you. In fact, by means of an app, the user can display on his or her device content related to what is in front of him or her, accessing an 'altered' version of reality. Augmented Reality content can add value to any museum or art gallery and offer interactive solutions to entertain people and create rewarding engagement opportunities.

This technology, applied to the field of Cultural Heritage and museum enjoyment, brings a significant enhancement to experiential feedback, attracting an ever-widening audience. In 2021, the National Gallery in London sought to take the collections of the National Gallery, the National Portrait Gallery and the Royal Academy of Arts beyond the museum walls with an Augmented Reality experience that the public could access via smartphones. Users used an app to activate artworks marked with QR codes and the initiative was very successful.

The Museo dei Saperi e delle Mirabilia Siciliane at the University of Catania has set itself the goal of using this technology to display 'digital versions' of selected artefacts, bringing them to life in contexts outside the museum space. There are already many institutions using Augmented Reality and many more are being added all the time.

Keywords: Digitisation · augmented reality · university museums

G. L. Foresti et al. (Eds.): ICIAP 2023 Workshops, LNCS 14366, pp. 449–461, 2024.
https://doi.org/10.1007/978-3-031-51026-7_38

1 Introduction

The latest reports published by the National Institute of Statistics[1] highlight an inflection of public expenditure in the protection and enhancement of cultural heritage, which is accompanied by a decrease in cultural participation, especially among the young generations. A further consequence is also the increased risk of educational poverty and school drop-out, especially in young people belonging to families with limited financial resources[2].

For this reason, the role played on the territory by schools, universities and cultural institutions is pivotal for the development of innovative, educational and welfare programmes, capable of responding to the specific needs of their audience, including disadvantaged young people[3].

For years, the University of Catania has been implementing policies of openness and collaboration in line with the objectives of the Third Mission and with the objective to be part of the cultural system of the territory, promoting the dissemination of culture also using its museums. To do this, the SiMuA (Sistema Museale d'Ateneo – network of university museums) has promoted initiatives aimed at achieving the sustainable development goals launched by the UNESCO, especially regarding issues such as technological implementation, accessibility and inclusivity, and at paying more attention to audience development strategies, broadening and diversifying the audience.

The paper shows the results of an educational project entitled "Manufatti 3D al Museo dei Saperi e delle Mirabilie Siciliane' (3D artifacts at the Museums of Knowledge and Sicilian Mirabilia) which has been conceived as an experimental workshop for the improvement of high school students' digital skills in the 3D digitalization processes. The project involved an interdisciplinary team of experts - museologists, art historians, geologists, engineers, architects, archaeologists and computer scientists - and has been focused on the creation of 3D content through photogrammetric methodologies which would constitute the basis for Augmented Reality (AR) applications.

The paper is structured as follows: after a brief introduction on the national Italian plane for digital school, the project and its underpinning methodology are illustrated; then, the operational steps regarding 3D acquisition and processing are shown as well as the Augmented Reality process; then conclusions are given.

2 Digital Skills in Education

The discussion about the skills that individuals must acquire to ensure their full development has been carried out for years at European level. This is a central issue, with far-reaching implications that affect the issues of training, education, and job orientation. The result of this process was the elaboration of the eight European key competences,

[1] ISTAT, *La situazione del Paese. Rapporto annuale,* 2022.; ISTAT, *Il Benessere Equo e Sostenibile in Italia. Rapporto BES,* 2023.

[2] SAVE THE CHILDREN, *Povertà educativa: necessario un cambio di passo nelle politiche di contrasto,* 2022.

[3] UNESCO, *Recommendation on Participation by the people at large in Cultural Life and their contribution to it,* 1976.

which the EU Member States are called upon to transpose, facilitating their acquisition by all citizens. The reference text that sets and defines them is the Recommendation on key competences for lifelong learning (with its annexed European Reference Framework), approved by the European Parliament on 22 May 2018[4]. The activity refers to and emphasizes the strengthening of 'digital competence', understood as the ability to use new technologies with familiarity, through processes of computer literacy and the creation of digital content. This objective is also the basis of the National Digital School Plan (PNSD), the main tool for planning the digital transformation of Italian schools, introduced by Article 1, paragraphs 56–59 of Law no. 107 of 13 July 2015[5].

The plan is based on a vision of education in the digital age that starts from a renewed idea of the school, conceived as an open space for learning and not only as a physical place. In this new configuration, technologies become formidable tools at the service of school activity, with reference to all those activities oriented towards training and learning with direct repercussions extended to the territory, as in the case of the PCTO experience (Percorsi per le Competenze Trasversali e l' Orientamento, 2018)[6], previously called 'Alternanza Scuola-Lavoro'and recounted in this contribution.

The importance of digital education is also confirmed by the High-Level Conference of the European Commission in December 2014[7], by various publications of the Centre for Educational Research and Innovation of the OECD[8], by the New Vision for Education Report of the World Economic Forum[9] and by research such as "Ripensare l'Educazione del XXi secolo"[10] of the Ambrosetti think tank.

Currently, the PNSD has been enhanced with a multilevel strategy for the adoption, in all schools, of digital curricula (DigComp 2.2 - DigCompEdu), of the European reference frameworks on digital competences mentioned above, of up to date of didactic methodologies, of innovative learning environments, and it interfaces with the new scenarios designed by the 'Piano Nazionale di Ripresa e Resilienza' (PNRR) and by the European structural funds (ESF-ESF). The proposed experience aimed, therefore, at enhancing and improving the use of specific devices, making a stimulating journey of knowledge, which made the participants skilled and aware in their use. Of course, it is a well-known fact that the best way to understand a procedure, a task or an activity is to have direct experience, and the PCTO set itself this objective, allowing young people to approach new and different digital methodologies, appreciating all the possibilities of use.

[4] Council Recommendation of 22 May 2018 on key competences for lifelong learning (2018/C 189/01) C 189/1: https://eur-lex.europa.eu

[5] *Piano Nazionale Scuola Digitale* (PNSD) (Article 1, paragraphs 56-59 of Law No 107 of 13 July 2015): https://scuoladigitale.istruzione.it/pnsd

[6] *Percorsi per le Competenze Trasversali e per l'Orientamento* - Guidelines (Pursuant to Article 1, Section 785, Law No. 145 of 30 December 2018): https://www.miur.gov.it/documents/2018

[7] High-Level Conference - The Future of Work: Today. Tomorrow. For All: https://ec.europa.eu/social

[8] Centre for Educational Research and Innovation - Ceri: https://www.oecd.org/education/ceri/

[9] New Vision for Education. Unlocking the Potential of Technology: https://widgets.weforum.org

[10] Ripensare l'Educazione nel XXI secolo: incontri per riflettere, proporre, agire: https://www.raicultura.it/raicultura/eventi

3 The Project "Manufatti 3D al Museo dei Saperi e delle Mirabilie Siciliane'

The SiMuA is structured in an integrated network of as many as twenty-one museums, archives and collections, to become the central node of a network that stimulates visitors to further their interests in the existing museum structures (peripheral nodes). Among these, the Museo dei Saperi e delle Mirabilia Siciliane was established in 2018 with the aim of making available in the form of a miscellany the numerous items (16th-20th century) of historical, cultural and scientific relevance of the SiMuA.

The museum is located on the ground floor of University's Central Building in Catania city centre. This location encourages its role as trait d'union with the other city's museum structures, sharing cultural initiatives and co-planning activities, including recreational ones[11]. The scientific head of the Museo dei Saperi e delle Mirabilie Siciliane is Professor Germana Barone, the Rector's delegate for the University Museum System. She has launched a programme of initiatives that foresees the SiMuA at the centre of a wealth of new and targeted projects aimed at implementing digital media for better use of the museum spaces, also with a view to improving accessibility, among these, DREAMIN Digital Hub[12].

The didactic project "Manufatti 3D al Museo dei Saperi e delle Mirabilie Siciliane" (3D artefacts at the Museum of Sicilian Knowledge and Mirabilies) was conceived as an experimental workshop that allowed a class of students from the Liceo Artistico Statale "Emilio Greco" of Catania (4^A - Architecture and Design) to improve their learning of 3D digitisation and AR technologies for museum collections.

Indeed, the digitisation of museum collections helps to preserve cultural and artistic heritage in case of physical damage or disasters. Digital resources can also enrich the educational aspect of the museum by providing detailed information, explanatory texts, instructional videos and interactive materials, while facilitating academic research and collaboration with other cultural and academic institutions.

The project took place within the rooms of the Museum and was organised with the aim of enhancing and making public the heritage of knowledge and assets that represent the fruit of centuries of research, teaching and dissemination activities of the University.

It was decided to focus on the creation of 3D content through photogrammetric methodologies to be used in Augmented Reality (AR) apps. Furthermore, the 3D products realized by the students have found their way into the DREAMIN Digital Hub. The digital hub is one of the outputs of the project DREAMIN, created in response to the spread of the Covid-19 infection, and provides easy access to all digitisation activities conducted by the University of Catania museums and its research centres. Hence the hub allows people from all over the world to explore museum collections without having to physically go there. This expands the museum's audience, reaching a wider and more diverse target audience, including those who may have difficulty visiting the museum in person due to disabilities, handicaps or other limitations.

---

[11] Regulations of Sistema Museale d'Ateneo (SiMuA): https://www.unict.it/sites/Regolamento/SiMuA

[12] Digital HUB website: https://dreamin.unict.it

The educational project "Manufatti 3D al Museo dei Saperi e delle Mirabilie Siciliane" took place between January and June 2023 and was divided into several operational phases, involving the collaboration of a highly qualified team. It was carried out in different locations: at the Rector's Office (lectures, visits and 3D acquisitions at the Museum of Sicilian Knowledge and Mirabilies) and at the Museum of Representation (creation and processing of 3D models, ot-timing for inclusion on the Sketchfab platform).

3.1 Methodological Approach

The methodology employed involved the exploration of 'bottom-up' strategies with the aim of actively and collaboratively engaging high school students in the creation of digital content.

Fundamental was the integration of multidisciplinary knowledge and skills pertaining to diverse scientific fields, which enabled students to understand the importance of collaboration in study and research activities.

Tailored training workflows were developed for high school students who actively participated in the digitization of specific objects within the Mirabilia museum.

The methodological approach foresaw several steps:

- Introductory lectures on technology with practical examples in the field at the University's Central Building (3 h)
- In-depth lecture on photogrammetry and museum collections with specific reference to the Museo dei Saperi e delle Mirabilie siciliane at the University Central Building (3 h)
- Scanning of previously selected artifacts at the Museo dei Saperi e delle Mirabilie siciliane (9 h)
- Model processing at MuRa (Museo della Rappresentazione) (9 h)
- Input of photogrammetric models on Dreamin platform, creation of augmented reality model and related qr code at MuRa (Museo della Rappresentazione) (3 h)
- Completion Inserting photogrammetric models and launching the augmented reality (AR) initiative at MuRa (Museum of Representation) (3 h)

The project lasted a total of thirty hours.

4 SfM Photogrammetry for the Digitalization and Fruition of Museum Collections

In recent years, digital survey and 3D modelling has gained significant attention within the museum sector, revolutionizing the way cultural heritage is preserved and shared. By leveraging these techniques, museums can offer virtual access to their collections, allowing to explore cultural artifacts remotely or creating immersive digital experiences for museum visitors. It provides a gateway to virtual content consumption, enhancing their overall experience and enabling them to access collections beyond the limitations of physical space [5, 6].

In recent time, multiview photogrammetry has emerged as a highly valuable tool within the museum sector, particularly due to its democratic nature. Indeed, this advanced technique has revolutionized the preservation and accessibility of cultural heritage, enabling museums to reach wider audiences and foster inclusivity.

By capturing multiple images of an object or artifact from various angles, photogrammetry reconstructs a three-dimensional digital representation. This virtual model can be explored through interactive platforms, offering visitors a close-up examination, detailed visual inspection, and even virtual manipulation of objects that may otherwise be inaccessible due to preservation concerns or restricted physical access. Furthermore, multiview photogrammetry opens up new possibilities for remote access, enabling individuals from all corners of the globe to engage with cultural heritage, fostering inclusivity and democratization of knowledge.

Also, it facilitates the learning of digitalization processes and provides students with the chance to delve into museum collections, even from the confines of their classrooms [1, 2]. For young students, multiview photogrammetry serves as a valuable educational tool, introducing them to the processes and technologies involved in digital preservation and reconstruction. By actively participating in the creation of digital models, students gain a hands-on understanding of the complexities and nuances of this cutting-edge methodology. Furthermore, multiview photogrammetry offers students an avenue for studying and analysing museum collections. This expands their educational horizons and enables them to explore various art forms, historical artifacts, and cultural objects, thereby deepening their appreciation and understanding of diverse cultural heritages.

The integration of photogrammetry in museum settings presents a transformative approach to heritage preservation, exhibition, and education. Its application allows visitors to virtually experience collections and provides remote access to cultural artifacts. Additionally, it empowers students by imparting practical knowledge of digitalization processes and facilitating in-depth exploration of museum collections. Through the seamless fusion of technology and cultural heritage, photogrammetry plays a pivotal role in enhancing visitor engagement, expanding educational opportunities, and preserving our shared heritage for generations to come.

4.1 Methodology

To introduce the students to the theory and practice of photogrammetry, the following operational methodology was implemented:

- **Theoretical lecture on digital surveying techniques, with a particular focus on photogrammetry and its applications in the context of cultural heritage**. In this context, operational guidelines for image acquisition and subsequent processing were provided. Students learned about acquiring high-resolution images, aligning and stitching photographs, and generating three-dimensional models.
- **Creation of working groups**. Three students per group were chosen to ensure active participation, tutored by qualified experts in acquisition procedures and knowledgeable about the museum's heritage.
- **Selection of the items to be digitized**. The selection process considered the feasibility of acquisition, and the variety of objects present in the museum, as it housed artifacts from various university museum collections. Given the advantages of multi-image photogrammetry, objects that were compatible with this technique were preferred.

- **Image acquisition process** for the photogrammetric survey using smartphones or professional cameras, employing Structure from Motion (SfM) methodology or smartphone applications such as Scaniverse, Polycam, and MagiScan, which leverage both the camera and LiDAR sensor in the devices to obtain three-dimensional models.
- **Acquired photographic datasets processing** for the creation of 3D digital models.
- **Editing of the 3D models** including scaling, cleaning, decimation, and refinement, to prepare the models for subsequent stages of virtual utilization.

4.2 3D Acquisition and Processing

In the acquisition process, conducted at the premises of the Museo dei Saperi e delle Mirabilie Siciliane, the students implemented the methodology as instructed. Specifically, in the selection of objects to be scanned, they prioritized artifacts with textured surfaces that displayed distinguishable chromatic contrasts. Conversely, smooth, reflective, and transparent objects were avoided due to the potential challenges they posed in capturing precise and detailed 3D models using photogrammetry.

The combination of SfM technology and specialized smartphone applications allowed the students to efficiently capture the required high-resolution images of the selected artifacts. This approach provided a cost-effective and accessible means for the creation of detailed 3D models, contributing to the preservation and documentation of the diverse cultural heritage objects host in the museum.

In the model processing stage, they were given the option to choose between two applications: the cloud-connected Autodesk Recap Photo app and the desktop app Agisoft Metashape. This allowed them to explore different processing methods and gain practical experience in model creation.

After the processing, the students were asked to perform scaling, cleaning, decimation, and refinement of the models. These editing steps aimed to enhance the quality and optimize the 3D models, making them suitable for further applications, such as virtual presentations or interactive experiences. Finally, they were requested to compile the results of the acquisitions in a table, including a significant image of the 3D model, the name of the object and the collection it belongs to, the device used for the acquisitions, the corresponding number of images, and the number of faces in the produced mesh (Table 1).

As an example, in this paper we show the workflow for the 3D acquisition of the Tanagrina, a terracotta figurine depicting a woman (Fig. 1). The clay figurine wears chiton and himation, which also covers the head and lower part of the face. The hands of the female figure are wrapped in himation, the folds of which form a characteristic triangular motif between the left arm and the right hand.

The drapery retains traces of pink, as do the foot and hair, which are red and orange in colour. The detail of the cloak covering the face, rendered with a clay handkerchief applied separately from the garment, is the clue that would indicate its modern make. The specimen from Centuripe is in fact a fake imitating the original Tanagrino type, a name derived from the Greek city of Tanagra, where from the 4th century BC there is evidence of a large production of draped and polychrome figurines. The same matrix used and produced by the forger Antonino Biondi is at the origin of at least two other specimens

Table 1 .

3D view	Object name	Camera	Photos n.	Faces n.
	Stucchi frammentari – motivi vegetali Collezione Museo della Fabbrica	Nikon 5100	46	604.072
	Terracotta figurata – tipo Tanagrina Collezione Museo di Archeologia	Nikon D5600	33	44.133
	Olpe etrusco-corinzia Collezione Museo di Archeologia	Panasonic DMC-G80/85	95	246.642
	Quarzo con inclusione fluida e argilla Collezione Museo di Mineralogia, Petrografia e Vulcanologia	Nikon D5300	64	1.460.949
	Clypeaster Collezione Museo di Paleontologia	Nikon D5600	30	377.246
	Superficie di strato a potamidi Collezione Museo di Paleontologia	Panasonic DMC-G80/85	68	375.472
	Phoenicopterus roseus pallas Collezione Museo di Zoologia	Panasonic DMC-G80/85	65	177.291

purchased by the National Museum in Leiden and another belonging to the Libertini Collection, which, however, has the head uncovered, a variation that was supposed to make the 20th century specimens on sale more 'original'.

The dataset of the Tanagrina consists of thirty-three photographic images captured using a Nikon D5600 camera. Utilizing Agisoft Metashape software, it facilitated the creation of a three-dimensional model comprising 44.133 triangles.

Fig. 1. 3D model of Tanagrina created through digital photogrammetry technique.

5 AR for the Enhancement and Fruition of Museum Collections

In the past years, many applications designed for museums, and more in general for cultural heritage, were developed. In [3] authors show a smart guide for mineralogical-petrographic exhibitions where visitors who use this application can get information about the artefacts from micro to macro scale. In [4] authors describe the use of serious games, to create a simulator where users can assume the role of researchers in mineralogy and petrography, study the minerals constituting the rocks and, thus, conduct analysis such as X-ray Fluorescence (XRF), Fourier-Transform Infrared-Spectroscopy (FTIR) and Raman Spectroscopy (Raman).

Augmented reality (AR) offers several interesting possibilities to enrich the visitor experience in a museum. The AR can show 3D models of artefacts or structures that might not be physically displayed in the museum. This gives visitors the opportunity to explore complex details or parts of objects that would otherwise be inaccessible.

Nowadays, thanks to the continued development of mobile devices such as smartphones, this technology has become commonplace and does not require expensive devices such as viewers or glasses. Smartphones, tablets and other devices used for the augmented reality experience are equipped with sensors such as the camera, gyroscope, accelerometer and proximity sensor. These sensors enable the device to understand the user's position, orientation and movement in the surrounding space. The device's cameras are crucial for capturing the real environment so that digital elements can be superimposed appropriately.

AR technology can be used without the use of dedicated software thanks to modern web browsers. In fact, the two largest mobile operating system manufacturers have implemented two frameworks that allow the power of augmented reality to be exploited natively in Android and iOS.

ARCore and ARKit are respectively the augmented reality frameworks developed by Google and Apple. Both provide developers with the necessary tools to create augmented reality applications for Android and iOS mobile devices. These frameworks allow users to interact with digital content superimposed on the real world through the cameras of their devices [7, 8].

ARCore is an augmented reality framework developed by Google for Android devices. Launched in 2017, ARCore uses the Android device's camera and sensors to detect and understand its surroundings in real time, allowing developers to create immersive and interactive AR experiences.

The main features of ARCore are:

- Motion tracking: ARCore uses the device's gyroscopes and accelerometers to detect the user's movement and position in space. This allows virtual objects to be precisely positioned within the real environment;
- Flat Surface Detection: ARCore detects flat surfaces such as floors and tables, allowing virtual objects to be realistically positioned on these surfaces;
- Light Estimation: ARCore measures ambient lighting to make virtual objects more realistic and consistent with the light in the real environment.
- Point of Interest Recognition: ARCore can identify and recognize specific points of interest, such as objects, buildings or monuments, enabling location-based AR experiences.

ARKit is the augmented reality framework developed by Apple for iOS devices. Launched in 2017, ARKit leverages the advanced capabilities of the iPhone and iPad, including the TrueDepth camera, face detection and motion detection, to provide immersive augmented reality experiences. The main features of ARKit are:

- Flat Surface Detection: ARKit can detect and track flat surfaces, allowing virtual objects to be positioned realistically and stably on the real environment.
- Facial Tracking: Using the TrueDepth camera on some iOS devices, ARKit can detect and track the user's face to create immersive and interactive AR experiences based on facial expressions.
- Light Estimation: ARKit measures the lighting of the surrounding environment to make virtual objects more realistic and consistent with the lighting of the real environment.
- Integration with SceneKit and SpriteKit: ARKit is integrated with Apple's SceneKit and SpriteKit frameworks, simplifying the development of 3D content and animations for augmented reality.

Both ARCore and ARKit have contributed significantly to the expansion and popularity of augmented reality applications on mobile devices, opening up new possibilities for games, educational applications, shopping experiences and more.

5.1 The Sketchfab Platform

The Sketchfab platform was used to enable museum visitors to interact with the 3D models developed during the project. Sketchfab is an online platform specialising in sharing, discovering and visualising interactive 3D models. Founded in 2012, Sketchfab's main goal is to provide a space where artists, designers, developers, companies and 3D modelling enthusiasts can share their creations with the world and enjoy an interactive visualisation experience. It offers an AR visualisation service thanks to the ARCore and ARKit frameworks (Fig. 2).

The 3D models have been uploaded to the platform including the metadata and descriptions associated with the items (Fig. 3).

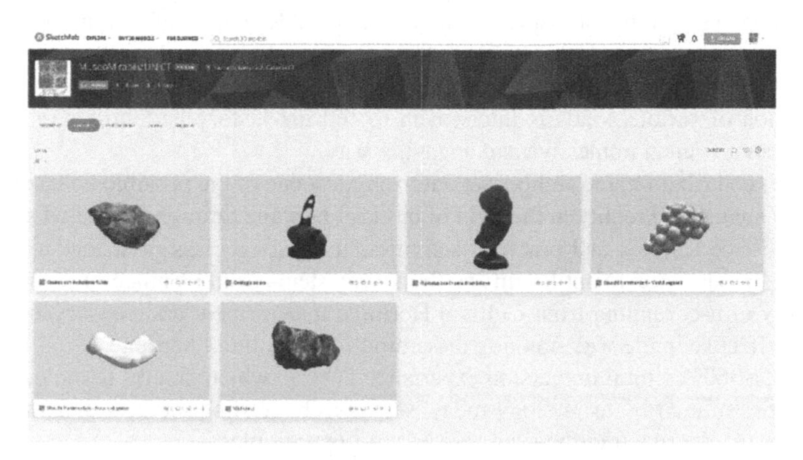

Fig. 2. Sketchfab page of Museum of Knowledge and Mirabilia.

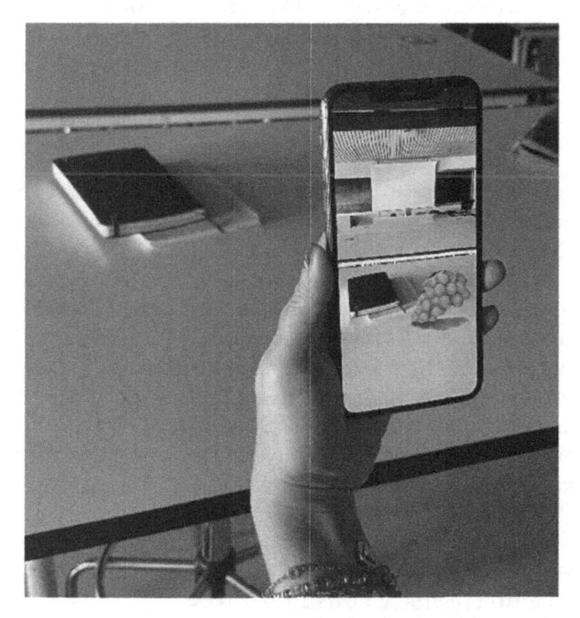

Fig. 3. Interaction with the 3D model via Sketchfab AR visualization.

6 Conclusions

AR is a technology whose main feature is the possibility of seeing virtual objects and interacting with them with a device, adding layers of knowledge to the real world.

In recent years and after the Covid -19 infection, augmented reality applications have been increasingly used in museums with the aim of fostering an effective connection and deeper engagement between museum spaces and visitors, resorting, mainly, to applications on smartphones. Mobile technology remains, in fact, one of the most accessible and promising solutions for displaying user-oriented AR content of Cultural Heritage.

In this context, the combination of technology and cultural heritage becomes of fundamental importance to provide the user with the best possible experience, where information of various kinds is interwoven to tell the history and stories of a given cultural context in an immersive and engaging way.

In this contribution, the authors set out to suggest one of the possible fields of application of augmented reality in the field of cultural heritage through a detailed schedule of face-to-face lectures and practical activities, aimed at a class of secondary school students. The decision to opt for this methodology stems from its inherent potential. Its versatility of use, ranging from Cultural Heritage management and conservation, also appears effective in the wide-ranging dissemination of cultural heritage.

Thus, a 360° cultural immersion experience is born, which, thanks to the numerous possibilities offered by technology today, seems to be leading the new generations of students towards an extraordinary future full of opportunities.

Acknowledgement. The work was carried out thanks to funding from the Ministry of Culture, Measure M1-C3 - 1.2 Removal of physical and cognitive barriers in museums, libraries and archives to enable wider access to and participation in culture CUP E67B22001040006.

References

1. Galizia, M., D'Agostino, G., Garozzo, R., La Russa, F.M., Seminara, G., Santagati, C.: Novel cultural experiences for the communication of museum collections: the Francesco Fichera projects fund at Museo della Rappresentazione in Catania. DISEGNARECON **12**(23), 8–1 (2019)
2. Garozzo, R., Pasqualino, G., Allegra, D., Santagati, C., Stanco F.: Augmented reality for the valorization and communication of ruined architecture. In: Cristani, M., Prati, A., Lanz, O., Messelodi, S., Sebe, N. (eds.) New Trends in Image Analysis and Processing – ICIAP 2019. LNCS, vol. 11808, pp. 170–178. Springer, Cham (2019). https://doi.org/10.1007/978-3-030-30754-7_17, ISBN: 978-3-030-30753-0, ISSN: 0302-9743
3. Linowes, J., Babilinski, K.: Augmented Reality for Developers: Build Practical Augmented Reality Applications with unity, ARCore, ARKit, and Vuforia. Packt Publishing Ltd., Birmingham (2017)
4. Oufqir, Z., El Abderrahmani, A., Satori, K.: Arkit and arcore in serve to augmented reality. In: 2020 International Conference on Intelligent Systems and Computer Vision (ISCV), pp. 1–7 (2020).https://doi.org/10.1109/ISCV49265.2020.9204243
5. Santagati, C., Noto, V., La Russa, F.: A project for Museo Civico Castello Ursino in Catania: breaking through museum walls and unlocking collections to everyone. Journal **2**(5), 99–110 (2018)

6. Santagati, C., et al.: Sperimentazione di tecnologie low cost 3D per la divulgazione delle collezioni museali. In: 3D Modeling&BIM - Nuove frontiere. 19 April 2018, Roma. ISBN 978 88496194 1 6 (2018)
7. Sinitò, D., et al.: I-peter (interactive platform to experience tours and education on the rocks): a virtual system for the understanding and dissemination of mineralogical-petrographic science. Pattern Recogn. Lett. **131**, 85–90 (2020). https://doi.org/10.1016/j.patrec.2019.12.002
8. Sinitò, D., et al.: A simulator for minero-petrographic and chemical research instruments. In: Rousseau, J.J., Kapralos, B. (eds.) Pattern Recognition, Computer Vision, and Image Processing. LNCS, vol. 13645, pp. 36–49. Springer, Cham (2022). https://doi.org/10.1007/978-3-031-37731-0_4

Visual Processing of Digital Manuscripts: Workflows, Pipelines, Best Practices (ViDiScript)

Writer Identification in Historical Handwritten Documents: A Latin Dataset and a Benchmark

Alessio Fagioli[1(✉)], Danilo Avola[1], Luigi Cinque[1], Emanuela Colombi[2], and Gian Luca Foresti[3]

[1] Department of Computer Science, Sapienza University, Via Salaria 113, 00198 Rome, Italy
{fagioli,avola,cinque}@di.uniroma1.it

[2] Department of Humanist Studies and Cultural Heritage, University of Udine, Vicolo Florio, 2/b, 33100 Udine, Italy
emanuela.colombi@uniud.it

[3] Department of Mathematics, Computer Science and Physics, University of Udine, Via delle Scienze 206, 33100 Udine, Italy
gianluca.foresti@uniud.it

Abstract. Writer identification refers to the process of determining or attributing the authorship of a document to a specific individual through the analysis of various elements such as writing style, linguistic characteristics, and other textual features. This is a relevant task in heterogeneous fields such as cybersecurity, forensics, or linguistics and becomes particularly challenging when considering historical documents. In fact, the latter might present deterioration due to time, often lack signatures, and could be authored by multiple people. Complicating matters further, scribes were trained to mimic handwriting meticulously when copying manuscripts, making author identification of such documents even more difficult. In this context, this paper introduces a curated collection of Latin documents from the Genesis and Gospel of Matthew specifically gathered for the purpose of exploring the writer identification task. In particular, the dataset comprises over 400 pages, written by nine distinct persons. The primary objective is to explore the efficacy of state-of-the-art deep learning architectures in accurately ascribing historical texts to their rightful authors. To this end, this paper conducts extensive experiments, utilizing varying training set sizes and employing diverse preprocessing techniques to assess the performance and capabilities of these renowned models on the writer identification task while also providing the community with a baseline on the introduced collection.

Keywords: Writer Identification · Historical Handwritten Documents · Deep Learning · Benchmark

1 Introduction

Writer identification (WI) is a complex task that focuses on the authorship attribution of documents. This task is of particular interest when historical

G. L. Foresti et al. (Eds.): ICIAP 2023 Workshops, LNCS 14366, pp. 465–476, 2024.
https://doi.org/10.1007/978-3-031-51026-7_39

documents come into play, as writer identification holds a pivotal role in unraveling the mysteries and stories encapsulated within the pages of time-worn manuscripts [11]. Furthermore, historical documents often lack explicit attribution to their authors, making it crucial to determine the writers behind these texts for accurate historical analysis, understanding cultural shifts, and tracing the evolution of language and writing styles [2]. In fact, historians can gain valuable insights into the lives, perspectives, and influences of individuals who shaped the past through the identification of writers. This knowledge enhances the comprehension of historical events, literary movements, and societal contexts, enabling a deeper connection to the shared human heritage [18]. WI also extends beyond historical documents into a contemporary context, finding applications in various domains such as forensic analysis, plagiarism detection, and security authentication. For instance, in the field of forensics, the ability to attribute anonymous or disputed documents to specific authors can aid criminal investigations by shedding light on potential suspects or perpetrators, as well as verifying the authenticity of a signature [5]. Similarly, in the academic and publishing world, writer identification can be used to detect cases of plagiarism, thereby upholding the integrity of scholarly work [20]. Furthermore, in the realm of digital security, writer identification can contribute to the development of more robust authentication methods, utilizing an individual's unique writing style as an additional layer of verification [13]. Deep learning approaches have revolutionized heterogeneous fields, such as medical image analysis [4,7], emotion recognition [6,10], and object classification [8,9], including the contemporary task of WI where they harness the power of neural networks to automatically learn and extract intricate patterns from text data. For instance, convolutional neural networks (CNNs) and recurrent neural networks (RNNs) are commonly employed to capture the spatial and temporal dependencies in writing styles. The former architecture excels in extracting local features, such as stroke patterns and pixel arrangements, from scanned images of handwritten text [21]. Conversely, the latter class of models is designed to handle sequential data, which is critical for capturing the temporal aspects of writing styles [23]. However, when applied to historical documents, writer identification faces unique challenges stemming from the scarcity and condition of the available data. Historical texts, often aged and deteriorated, can present difficulties during the extraction and recognition of handwriting features. Furthermore, the small number of labeled datasets for historical writer identification [32] poses limitations on the application of deep learning techniques, which generally thrive on large labeled collections. This scarcity necessitates the development of creative solutions, such as transfer learning, to adapt modern deep learning methods to the intricacies of historical writing styles. Despite these challenges, the potential for uncovering hidden authorship and shedding light on the past continues to drive advancements in writer identification technology across historical documents, along with a growing need for new labeled datasets.

To contribute to the advancement of WI research, this paper introduces a novel dataset specifically curated for addressing the task of writer identification.

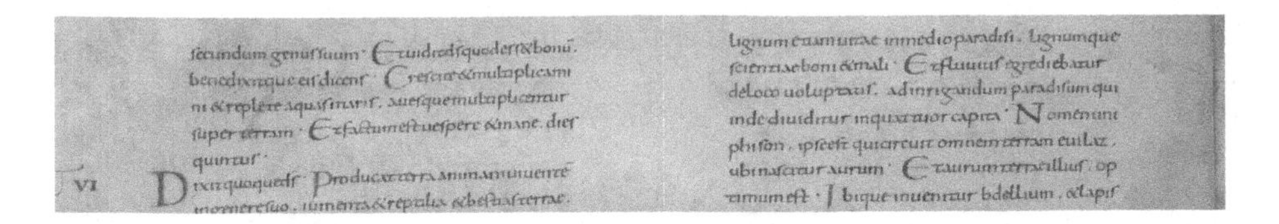

Fig. 1. Portion of a scanned page

In-depth, the dataset comprises over 400 pages from the books of Genesis and the Gospel of Matthew, transcribed by nine scribes from distinct historical eras. An example of a scanned page is shown in Fig. 1. Moreover, this paper presents a benchmark involving five well-known deep neural networks: AlexNet, DenseNet, EfficientNet B3, VGG16, and VGG19. This benchmark utilizes various configurations applied to the aforementioned dataset. Specifically, this study emphasizes the effects of class balance, over-sampling, and under-sampling on model performance. Various training set sizes are utilized throughout the experiments to showcase how an inadequate or excessive number of samples can amplify the inherent complexity of the writer identification task. Additionally, supplementary experiments are carried out to evaluate the efficacy of data augmentation, involving the exploration of different strategies during model testing.

The subsequent sections are organized as follows. Section 2 introduces related works addressing WI on historical documents. Section 3 provides a brief description of the renowned models used in creating the benchmark. Section 4 reports the extensive experiments performed on the novel WI collection. Finally, Sect. 5 presents the conclusion of this paper.

2 Related Work

Over the years, various approaches to WI have been developed using different datasets. For instance, the method proposed in [16] extracts computer vision features from the page layout, including margins, the number of rows and columns, and relationships among rows, e.g., interline distance. These features are then used to apply five univariate measures for recognizing the authors of medieval manuscripts. The latter are also analyzed by [15], where an object detector is first used to find and select rows in the documents, while a CNN classifier is implemented to classify them. Finally, majority voting is exploited to perform writer identification by accounting for all rows of a given page. The study presented in [12] utilizes a feature descriptor, such as SIFT, to detect features in Latin documents. Following this, the authors train a CNN on patches extracted by the descriptor. They encode the CNN output using a vector of locally aggregated descriptors (VLAD) and ultimately identify the author through an exemplar support vector machine (E-SVM). A comparable strategy is employed by the approach discussed in [31], which utilizes a descriptor to detect keypoints and construct patches around text. Subsequently, a CNN, specifically a ResNet, is employed to extract and classify features from these patches. Ultimately, a majority voting scheme is applied to achieve writer identification of papyri. In contrast to the previous approaches, the authors of [22] utilize a dual-branch multi-task CNN. This CNN is designed to focus, on the one hand, on single words extracted from the input documents. On the other hand, it concurrently addresses an additional task involving lexical context, word length, and character attributes to identify the writers of English and German texts. An alternative strategy is outlined in [27], where a self-supervised learning procedure involving teacher/student encoders is employed. These encoders generate bag of words (BoW) representations from input documents, which are subsequently classified using a linear model coupled with triplet margin loss. This approach enables writer identification for documents from a custom collection of the Vatican Apostolic Library. Lastly, the approach presented in [3] introduces an RGB algorithmic analysis of the ink embedded on the page in conjunction with the background paper. Specifically, the authors employ the fractal Minkowski dimension to decipher both the gall ink utilized and writing traits of the scribe that was responsible for the handwritten text.

The available public datasets have been instrumental in driving progress within the field of WI for historical manuscripts. These datasets serve as benchmarks for researchers, enabling them to assess and refine their methods across a wide range of historical contexts and languages. For instance, the IAM Historical Documents [29] dataset provides a comprehensive collection of medieval and early modern manuscripts written in multiple languages, allowing researchers to explore writer identification across diverse linguistic backgrounds. The KERTAS [1] dataset focuses on historical Malay documents, offering insight into writer identification within a specific language and cultural context. The GRK-Papyri [30] dataset presents ancient Greek inscriptions and papyri, while ScribbleLens [19] offers a varied selection of historical documents, spanning different time periods and linguistic traditions. In the realm of Arabic handwriting, both the ICDAR19-HDRC-IR and ICFHR20 HisFragIR20 [14] datasets contribute valuable resources. The Hugin-Munin [28] dataset provides Norse and Old Icelandic manuscripts, further enriching the landscape. The Avila Bible [17] dataset showcases manuscripts written in Latin, while the CVL [25] dataset encompasses various historical documents in multiple languages. Lastly, the Vatican Apostolic Library [27] dataset contains diverse manuscripts, contributing to the understanding of historical writing across languages and cultures. These publicly accessible datasets have not only facilitated the evaluation and refinement of writer identification techniques but have also fostered collaboration and knowledge-sharing within the expansive domain of historical manuscript analysis, thereby also inspiring the collection of the dataset presented in this work.

3 Proposed Method

This section briefly describes the well-known architectures used to create the benchmark on the novel dataset presented in this work.

3.1 AlexNet

The AlexNet architecture [26] has had a significant impact on the field of computer vision over the last decade. Its success is closely tied to its remarkable performance in the ImageNet LSVRC-2012 competition, where it outperformed the competition by a substantial margin, achieving an accuracy increase of 10.9%. The architecture is composed of five convolutional layers, with kernel sizes of 11×11, 5×5, and 3×3 for the last three layers. Additionally, max pooling, dropout, and ReLU activation functions are applied after the first, second, and fifth convolutional layers, respectively. These operations collectively extract a vector of relevant features with a dimensionality of 9216. Ultimately, these extracted features are fed into three fully connected layers with 4096, 4096, and 1000 neurons in each layer, respectively. This final classification step effectively categorizes the input data.

3.2 DenseNet121

A more recent model is the DenseNet architecture [24], developed to address the vanishing/exploding gradient problem that is commonly encountered in deep neural networks with numerous layers. To tackle this challenge, the authors structured their network into dense blocks interleaved with transition layers. These transition layers can act as bottlenecks, effectively reducing the number of architectural parameters and input size. Furthermore, within a dense block, multiple convolutions (e.g., 6, 12, or 24) can be incorporated. Notably, each of these operations takes all of the feature maps produced by the preceding layers as input. This strategy facilitates the propagation of more information throughout a given block, resulting in a more compact configuration due to the subdivision of dense blocks and improved overall performance. In this model, the classifier consists of a single dense layer taking as input a feature vector of size 1024.

3.3 EfficientNet B3

Another effective architecture is the EfficientNet B3 model [34]. This model introduces a compound scaling approach that harmonizes depth, width, and resolution, resulting in models that offer impressive performance and computational efficiency. The central building block of this architecture is the mobile inverted bottleneck (MBConv) with squeeze-and-excitation optimization. The base architecture, denoted as EfficientNet B0, is composed of a convolutional layer, seven MBConv layers with varying kernel sizes, and a final convolutional layer serving as a feature extractor. This is followed by a single dense layer employed for classification purposes. The B3 variant, much like the other models in the EfficientNet series, is a scaled version of the base architecture achieved through a compound scaling coefficient. It maintains the same underlying structure while benefiting from the scaling factor to enhance its capabilities.

3.4 VGG16

One of the pioneering networks to incorporate a deeper architecture is the OxfordNet model, commonly referred to as VGG16 [33]. The Oxford Visual Geometry Group achieved remarkable performance by stacking up to 13 convolutional layers, interspersed with five max pooling operations applied after the second, fourth, seventh, tenth, and thirteenth convolutions, serving to reduce the input size. This structured approach results in the extraction of a vector containing meaningful features with a size of 25088 from the input image. This vector is then fed into a classifier composed of three dense layers, effectively discriminating between the 1000 classes in the ImageNet dataset. It is noteworthy that VGG16 is one of the first models to employ smaller kernels in its convolutions, i.e., 3×3, to capture intricate details within extremely small receptive fields.

3.5 VGG19

The last architecture under consideration is a variation of VGG16, developed by the same group, i.e., the VGG19 architecture [33]. The main difference with the previous model lies in its 19 convolutional layers using 3×3 kernels. Max-pooling operations are applied after several of these convolutions, facilitating multi-scale feature extraction, while ReLU activation functions maintain non-linearity throughout the model. Finally, fully connected layers complete the architecture to perform object classification. While resource-intensive due to its depth and parameter count, VGG19 offers potent representation learning capabilities and is considered an extremely effective model.

4 Experiments

This section presents the benchmark on the proposed dataset. In detail, Sect. 4.1 describes the acquired collection. Section 4.2 lists the implementation details to reproduce the obtained results. Finally, Sect. 4.3 discusses the performance of the chosen models on the WI task.

4.1 Dataset

The novel dataset, which focuses on Latin documents from the Genesis and Gospel of Matthew, was collected by the Department of Humanist Studies and Cultural Heritage at the University of Udine. Specifically, the acquired images encompass the work of nine scribes from various eras, spanning from around the year 820 to approximately 1320, resulting in a total of 414 pages across the two books. Due to the varying conditions of the manuscripts, the scanned pages, denoted as $\mathcal{D}_0$, have different resolutions. To tackle this variation, the pages underwent pre-processing by extracting patches with dimensions of 300×500, as depicted in Fig. 2. Moreover, datasets with differing numbers of patches were curated to explore the significance of class balance, over-sampling, and under-sampling for the WI task. To elaborate, the first dataset, denoted as $\mathcal{D}_1$, was

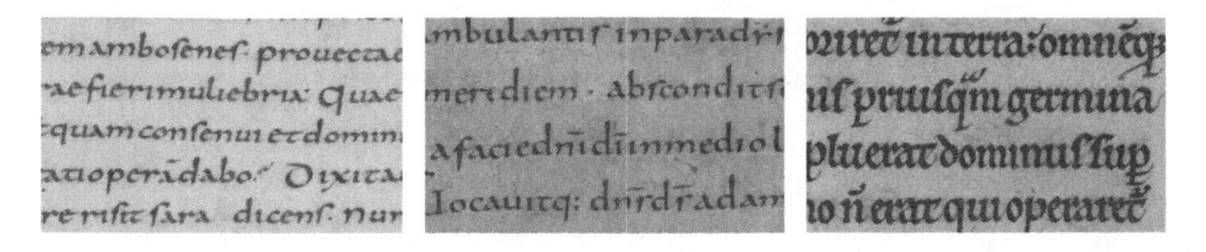

Fig. 2. Patch samples from pages associated with three different scribes.

Table 1. Datasets built to create the benchmark.

	S1	S2	S3	S4	S5	S6	S7	S8	S9	Total	Augm.
$\mathcal{D}_0$	48	56	54	45	14	39	49	58	51	414	No
$\mathcal{D}_1$	9369	7179	518	7900	635	1719	9295	7336	43937	87888	Yes
$\mathcal{D}_2$	8832	8848	8856	9000	8400	9009	8820	9048	9027	79840	Yes
$\mathcal{D}_3$	144	112	108	135	98	117	98	116	102	1030	Yes
$\mathcal{D}_4$	144	112	108	135	98	117	98	116	102	1030	No

constructed by extracting partially overlapping patches-employing offsets of half the patch size in both the x and y directions. However, due to the disparity in page resolutions, this dataset exhibited a pronounced class imbalance, where a single category covered roughly half of the dataset samples. To remedy this, $\mathcal{D}_2$ was generated by uniformly sampling each class up to a specified patch count of approximately 9000 patches. Given the likelihood of significant overlap in patches per page due to the high number of generated images, the third dataset, $\mathcal{D}_3$, was curated by uniformly sampling up to approximately 100 patches per class, resulting in non-overlapping samples. The fourth collection, $\mathcal{D}_4$, followed a similar procedure to $\mathcal{D}_3$, although, unlike the previous sets, it did not incorporate a data augmentation strategy during training, as elaborated in Sect. 4.2. This decision was taken so that the effect of data augmentation strategies could also be analyzed at test time. A summary reporting the total number of samples of each dataset is shown in Table 1.

4.2 Implementation Details

All experiments were conducted using the same hyperparameters across all models. Specifically, the datasets were divided into 70%/30% training and test sets. The models were trained for 50 epochs using the Adam optimizer with a learning rate of 1e−3 and a batch size of 32, facilitating a balanced mix of samples. Additionally, for datasets $\mathcal{D}_1$, $\mathcal{D}_2$, and $\mathcal{D}_3$, a data augmentation strategy was employed during training by applying random cropping to dimensions of 224×224, random rotations up to 25°, and color jitter. For all test sets, a center crop of 224×224 was applied during evaluation for consistency reasons.

Regarding the chosen architectures, all models were implemented using the PyTorch framework. All experiments considered standard classification metrics and were executed on a Quadro RTX 6000 GPU with 24 GB of RAM.

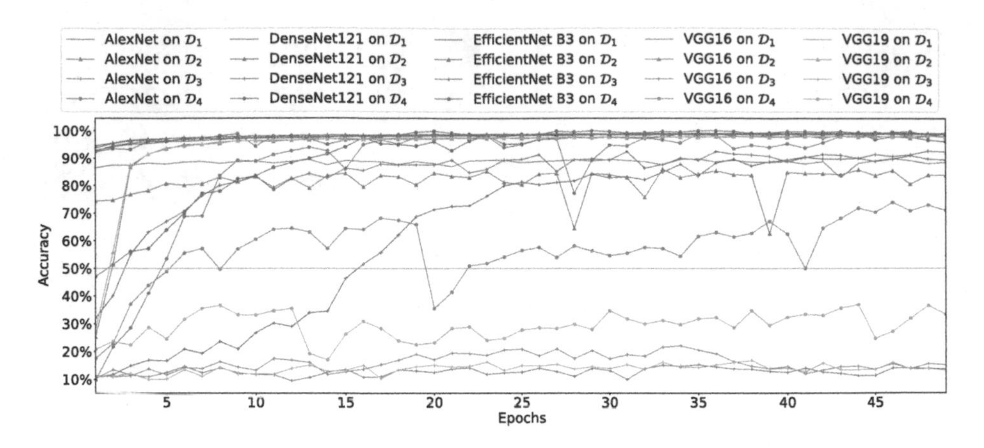

Fig. 3. Training accuracy comparison using datasets $\mathcal{D}_1$, $\mathcal{D}_2$, $\mathcal{D}_3$, and $\mathcal{D}_4$.

4.3 Performance Evaluation

To establish an effective benchmark, extensive experiments were conducted using the four collections: $\mathcal{D}_1$, $\mathcal{D}_2$, $\mathcal{D}_3$, and $\mathcal{D}_4$. The corresponding results on training and test sets are presented in Fig. 3 and Fig. 4, respectively. From Fig. 3, it is evident that DenseNet121 and EfficientNet B3 models achieve remarkable performance across all datasets. Conversely, other models struggle to grasp the WI task, as seen with VGG19 on $\mathcal{D}_3$, displaying convergence to an exceedingly low accuracy rate. This divergence in performance can be attributed, on the one hand, to the internal structure of the chosen models, which can extract more or less discriminative features. On the other hand, it could be related to the inherent complexity of the dataset, potentially challenging the capability of a given architecture. These observations are substantiated by the confusion matrices depicted in Fig. 4. Notably, VGG16 and VGG19 demonstrate a tendency to overfit on dataset $\mathcal{D}_1$, which is considerably affected by class imbalance. In contrast, the more diverse architectures adeptly manage this scenario. Furthermore, when considering over- or under-sampling (i.e., $\mathcal{D}_2$ and $\mathcal{D}_3$), the models exhibit significant shifts in behavior. For instance, in the over-sampling setting, all architectures manage to achieve near-perfect accuracy. Conversely, when considering the under-sampling case, also the AlexNet architecture suffers from clear signs of overfitting, which is confirmed by the training accuracy of AlexNet on $\mathcal{D}_3$ shown in Fig. 3. As a matter of fact, this outcome is closely tied to the aggressiveness of the data augmentation strategy chosen for the models. Indeed, when considering the last dataset, $\mathcal{D}_4$, that comprises the same sample count as $\mathcal{D}_3$, as outlined in Table 1, but does not implement the data augmentation strategy described in Sect. 4.2, all models-with the exception of VGG19-effectively learn the WI task. On a different note, when evaluating other standard classification metrics such as precision, recall, and F1-score, as reported in Table 2 for datasets $\mathcal{D}_2$ and $\mathcal{D}_4$, all models achieve consistent results across all measures, indicating that the models can accurately discern the various scribes.

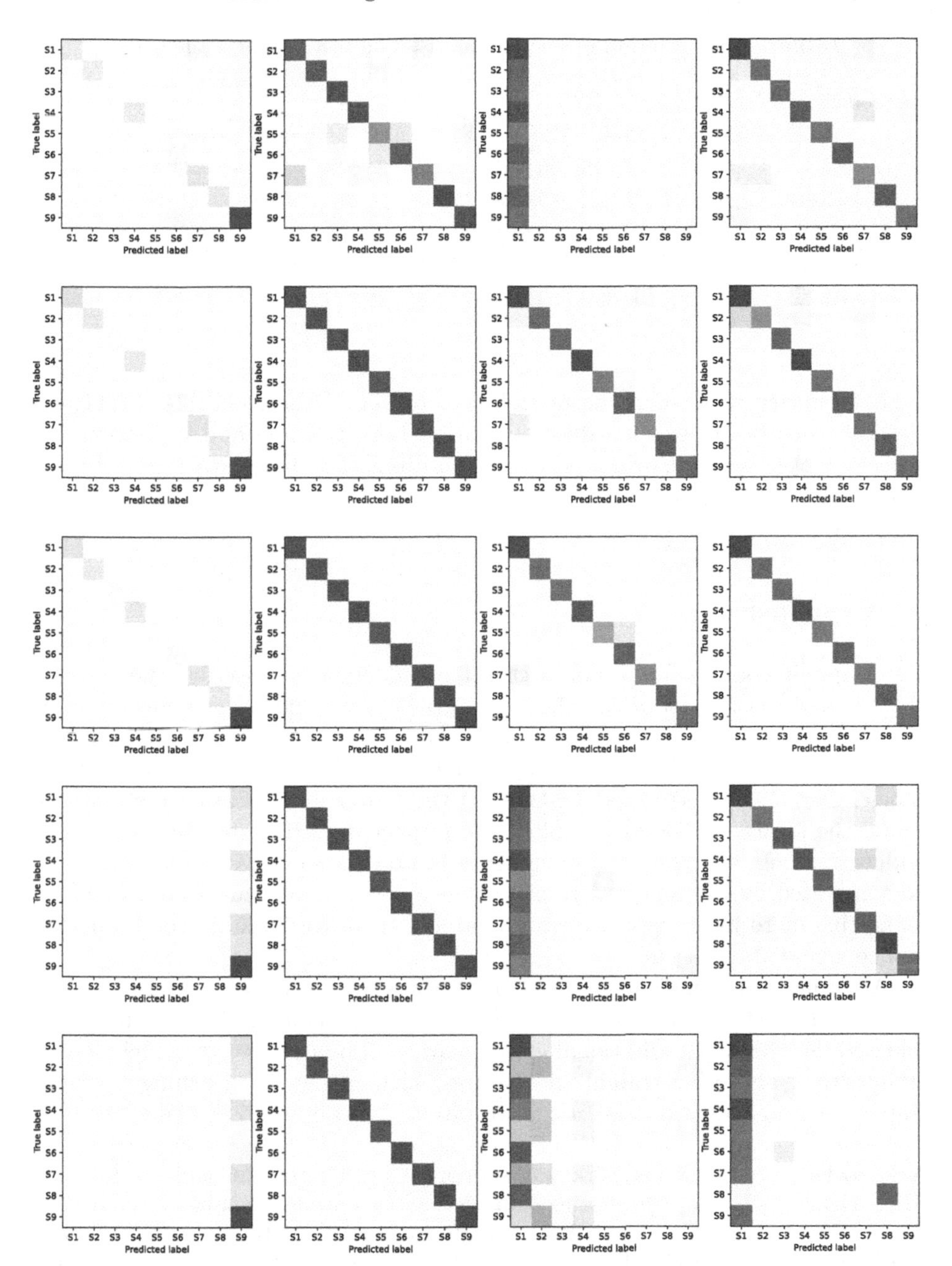

Fig. 4. From left to right: confusion matrices for the test sets of datasets $\mathcal{D}_1$, $\mathcal{D}_2$, $\mathcal{D}_3$, and $\mathcal{D}_4$. From top to bottom rows: matrices associated with AlexNet, DenseNet121, EfficientNet B3, VGG16, and VGG19 models.

Table 2. Classification metrics comparison on test sets of $\mathcal{D}_2$ and $\mathcal{D}_4$.

Model	$\mathcal{D}_2$				$\mathcal{D}_4$			
	Acc.	Prec.	Rec.	F1	Acc.	Prec.	Rec.	F1
AlexNet	87.13%	86.93%	86.92%	86.61%	94.82%	95.05%	94.86%	94.85%
DenseNe121	99.74%	99.74%	99.75%	99.74%	95.46%	96.64%	95.53%	95.65%
EfficientNet B3	99.61%	99.62%	99.61%	99.61%	99.67%	99.74%	99.67%	99.70%
VGG16	99.19%	99.21%	99.18%	99.19%	88.02%	90.05%	88.69%	88.40%
VGG19	99.29%	99.30%	99.29%	99.29%	25.56%	16.06%	22.59%	15.16%

In summary, while more recent architectures, i.e., DenseNet121 and EfficientNet B3, adeptly navigate all datasets during both training and validation, the aspects of class balance, over- and under-sampling, as well as data augmentation, significantly influence model performance. This leads to substantial performance disparities, even when evaluating a single architecture.

5 Conclusion

This paper introduces a novel dataset of Latin documents based on the books of Genesis and the Gospel of Matthew, comprising over 400 pages transcribed by nine scribes spanning various years. From this dataset, four collections were generated to assess the effectiveness of five renowned models: AlexNet, DenseNet121, EfficientNet B3, VGG16, and VGG19, on the writer identification task. Specifically, the created collections explore the impact of data imbalance, over- and under-sampling, as well as data augmentation on classification performance. As demonstrated by the reported results, more recent architectures are capable of extracting more meaningful features, enabling them to maintain robust performance across all datasets.

As future work, additional pages and scribes will be incorporated to further increase the dataset complexity so that a more challenging collection can be released. Furthermore, additional experiments will be performed to include new architectures and other training procedures, such as few-shot learning, a crucial step since writer identification scenarios often involve limited available samples.

Acknowledgements. This work was supported by "A Brain Computer Interface (BCI) based System for Transferring Human Emotions inside Unmanned Aerial Vehicles (UAVs)" Sapienza Research Projects (Protocol number: RM1221816C1CF63B); and Departmental Strategic Plan (DSP) of the University of Udine - Interdepartmental Project on Artificial Intelligence (2020–25); and "A proactive counter-UAV system to protect army tanks and patrols in urban areas (PROACTIVE COUNTER-UAV)" project of the Italian Ministry of Defence (Number 2066/16.12.2019); and the MICS (Made in Italy - Circular and Sustainable) Extended Partnership and received funding from Next-Generation EU (Italian PNRR - M4 C2, Invest 1.3 - D.D. 1551.11-10-2022, PE00000004). CUP MICS B53C22004130001.

References

1. Adam, K., Baig, A., Al-Maadeed, S., Bouridane, A., El-Menshawy, S.: KERTAS: dataset for automatic dating of ancient Arabic manuscripts. Int. J. Doc. Anal. Recogn. **21**, 283–290 (2018)
2. Amelin, K., Granichin, O., Kizhaeva, N., Volkovich, Z.: Patterning of writing style evolution by means of dynamic similarity. Pattern Recogn. **77**, 45–64 (2018)
3. Andronache, I., Liritzis, I., Jelinek, H.F.: Fractal algorithms and RGB image processing in scribal and ink identification on an 1819 secret initiation manuscript to the "Philike Hetaereia". Sci. Rep. **13**(1), 1735 (2023)
4. Avola, D., Bacciu, A., Cinque, L., Fagioli, A., Marini, M.R., Taiello, R.: Study on transfer learning capabilities for pneumonia classification in chest-x-rays images. Comput. Methods Programs Biomed. **221**, 106833 (2022)
5. Avola, D., Bigdello, M.J., Cinque, L., Fagioli, A., Marini, M.R.: R-signet: reduced space writer-independent feature learning for offline writer-dependent signature verification. Pattern Recogn. Lett. **150**, 189–196 (2021)
6. Avola, D., Cascio, M., Cinque, L., Fagioli, A., Foresti, G.L.: Affective action and interaction recognition by multi-view representation learning from handcrafted low-level skeleton features. Int. J. Neural Syst. 2250040 (2022)
7. Avola, D., Cinque, L., Fagioli, A., Filetti, S., Grani, G., Rodolà, E.: Multimodal feature fusion and knowledge-driven learning via experts consult for thyroid nodule classification. IEEE Trans. Circuits Syst. Video Technol. **32**(5), 2527–2534 (2021)
8. Avola, D., Cinque, L., Fagioli, A., Foresti, G.L.: Sire-networks: convolutional neural networks architectural extension for information preservation via skip/residual connections and interlaced auto-encoders. Neural Netw. **153**, 386–398 (2022)
9. Avola, D., et al.: Medicinal boxes recognition on a deep transfer learning augmented reality mobile application. In: Proceedings of the International Conference on Image Analysis and Processing, pp. 489–499 (2022)
10. Avola, D., Cinque, L., Fagioli, A., Foresti, G.L., Massaroni, C.: Deep temporal analysis for non-acted body affect recognition. IEEE Trans. Affect. Comput. **13**(3), 1366–1377 (2020)
11. Bulacu, M., Schomaker, L.: Text-independent writer identification and verification using textural and allographic features. IEEE Trans. Pattern Anal. Mach. Intell. **29**(4), 701–717 (2007)
12. Chammas, M., Makhoul, A., Demerjian, J.: Writer identification for historical handwritten documents using a single feature extraction method. In: International Conference on Machine Learning and Applications, pp. 1–6 (2020)
13. Chen, Z., Yu, H.X., Wu, A., Zheng, W.S.: Level online writer identification. Int. J. Comput. Vis. **129**(5), 1394–1409 (2021)
14. Christlein, V., Nicolaou, A., Seuret, M., Stutzmann, D., Maier, A.: ICDAR 2019 competition on image retrieval for historical handwritten documents. In: International Conference on Document Analysis and Recognition, pp. 1505–1509 (2019)
15. Cilia, N.D., De Stefano, C., Fontanella, F., Marrocco, C., Molinara, M., Di Freca, A.S.: An end-to-end deep learning system for medieval writer identification. Pattern Recogn. Lett. **129**, 137–143 (2020)
16. De Stefano, C., Fontanella, F., Maniaci, M., Scotto di Freca, A.: A method for scribe distinction in medieval manuscripts using page layout features. In: International Conference on Image Analysis and Processing, pp. 393–402 (2011)
17. De Stefano, C., Maniaci, M., Fontanella, F., di Freca, A.S.: Reliable writer identification in medieval manuscripts through page layout features: the "Avila" bible case. Eng. Appl. Artif. Intell. **72**, 99–110 (2018)

18. Decker, S., Hassard, J., Rowlinson, M.: Rethinking history and memory in organization studies: the case for historiographical reflexivity. Hum. Relat. **74**(8), 1123–1155 (2021)
19. Dolfing, H.J., Bellegarda, J., Chorowski, J., Marxer, R., Laurent, A.: The "Scribble-Lens" Dutch historical handwriting corpus. In: International Conference on Frontiers in Handwriting Recognition, pp. 67–72 (2020)
20. Foltýnek, T., Meuschke, N., Gipp, B.: Academic plagiarism detection: a systematic literature review. ACM Comput. Surv. (CSUR) **52**(6), 1–42 (2019)
21. Gan, J., Wang, W., Lu, K.: Compressing the CNN architecture for in-air handwritten Chinese character recognition. Pattern Recogn. Lett. **129**, 190–197 (2020)
22. He, S., Schomaker, L.: Deep adaptive learning for writer identification based on single handwritten word images. Pattern Recogn. **88**, 64–74 (2019)
23. He, S., Schomaker, L.: GR-RNN: global-context residual recurrent neural networks for writer identification. Pattern Recogn. **117**, 107975 (2021)
24. Huang, G., Liu, Z., Van Der Maaten, L., Weinberger, K.Q.: Densely connected convolutional networks. In: IEEE Conference on Computer Vision and Pattern Recognition, pp. 4700–4708 (2017)
25. Kleber, F., Fiel, S., Diem, M., Sablatnig, R.: CVL-database: an off-line database for writer retrieval, writer identification and word spotting. In: International Conference on Document Analysis and Recognition, pp. 560–564 (2013)
26. Krizhevsky, A., Sutskever, I., Hinton, G.E.: ImageNet classification with deep convolutional neural networks. In: Advances in Neural Information Processing Systems, vol. 25, pp. 1097–1105 (2012)
27. Lastilla, L., Ammirati, S., Firmani, D., Komodakis, N., Merialdo, P., Scardapane, S.: Self-supervised learning for medieval handwriting identification: a case study from the Vatican apostolic library. Inf. Process. Manag. **59**(3), 102875 (2022)
28. Maarand, M., Beyer, Y., Kåsen, A., Fosseide, K.T., Kermorvant, C.: A comprehensive comparison of open-source libraries for handwritten text recognition in Norwegian. In: International Workshop on Document Analysis Systems, pp. 399–413 (2022)
29. Marti, U.V., Bunke, H.: The IAM-database: an English sentence database for offline handwriting recognition. Int. J. Doc. Anal. Recogn. **5**, 39–46 (2002)
30. Mohammed, H., Marthot-Santaniello, I., Märgner, V.: GRK-Papyri: a dataset of Greek handwriting on papyri for the task of writer identification. In: International Conference on Document Analysis and Recognition, pp. 726–731 (2019)
31. Nasir, S., Siddiqi, I., Moetesum, M.: Writer characterization from handwriting on papyri using multi-step feature learning. In: International Conference on Document Analysis and Recognition Workshop, pp. 451–465 (2021)
32. Nikolaidou, K., Seuret, M., Mokayed, H., Liwicki, M.: A survey of historical document image datasets. Int. J. Doc. Anal. Recogn. **25**(4), 305–338 (2022)
33. Simonyan, K., Zisserman, A.: Very deep convolutional networks for large-scale image recognition. arXiv preprint arXiv:1409.1556 preprint, pp. 1–14 (2014)
34. Tan, M., Le, Q.: EfficientNet: rethinking model scaling for convolutional neural networks. In: International Conference on Machine Learning, pp. 6105–6114 (2019)

Synthetic Lines from Historical Manuscripts: An Experiment Using GAN and Style Transfer

Chahan Vidal-Gorène[1,2](✉), Jean-Baptiste Camps[1], and Thibault Clérice[3]

[1] Centre Jean Mabillon, PSL – École nationale des Chartes, Paris, France
chahan.vidalgorene@chartes.psl.eu
[2] Calfa, Paris, France
[3] ALMAnaCH, Inria Paris, Paris, France

Abstract. Given enough data of sufficient quality, HTR systems can achieve high accuracy, regardless of language, script or medium. Despite growing pooling of datasets, the question of the required quantity of training material still remains crucial for the transfer of models to out-of-domain documents, or the recognition of new scripts and under-resourced character classes. We propose a new data augmentation strategy, using generative adversarial networks (GAN). Inspired by synthetic lines generation for printed documents, our objective is to generate handwritten lines in order to massively produce data for a given style or under-resourced character class. Our approach, based on a variant of Scrabble-GAN, demonstrates the feasibility for various scripts, either in the presence of a high number and variety of abbreviations (Latin) and spellings or letter forms (Medieval French), in a situation of data scarcity (Armenian), or in the instance of a very cursive script (Arabic Maghribi). We then study the impact of synthetic line generation on HTR, by evaluating the gain for out-of-domain documents and under-resourced classes.

Keywords: gan · handwritten text recognition · data augmentation

1 Introduction

Generative Adversarial Networks (GANs) have shown promising results in generating realistic images, audio, and video. In this paper, we explore the use of GANs to generate synthetic text lines for historical documents, particularly ancient and medieval manuscript in Latin scripts (French and Latin) and non-Latin scripts (Arabic and Armenian). Given a text input, our goal is to obtain the generation of one or several realistic lines. The ability to generate synthetic text lines can have multiple applications, such as data augmentation for handwritten text recognition systems, or generating fake lines under specific constraints such as style, script, or date. This can help in filling the gaps in ancient manuscripts, which are often incomplete or have missing sections, and can also serve as a tool for researchers to study and analyse different writing styles and scripts.

G. L. Foresti et al. (Eds.): ICIAP 2023 Workshops, LNCS 14366, pp. 477–488, 2024.
https://doi.org/10.1007/978-3-031-51026-7_40

Here, we focus on the goal of data augmentation, to try to measure the improvements that can be obtained on some frequent issues in handwritten text recognition. Data augmentation could potentially help alleviating the central problem in HTR for ancient documents that is the availability of training data and the cost (in terms of human time and expertise) needed to create them, which is often a bottleneck in workflows. In particular, we wish to investigate the success of data augmentation regarding:

1. the prediction of rare character classes, such as abbreviation signs, diacritics or variant letter forms;
2. the situation of extreme data scarcity due to the rarity of transcribed material;
3. very complex scripts, due to a high level of cursiveness, ligatures or variation.

More generally, we also wish to address the problem of generalisation and alleviate the issue of overfitting, since the generated lines can also be used to augment the training data with variations in writing style, script, and other factors that can help to improve the robustness of the recognition system.

In this work, we propose an approach based on a variant of ScrabbleGAN [10], trained on datasets containing Medieval manuscripts and incunabula in Medieval French and Latin [7,15,17,21,23], as well as Arabic manuscripts [20] and finally Armenian manuscripts [19], which we compare with a style transfer approach based on cycleGAN [24]. We use very different datasets to check the feasibility and validity of the method on a wide variety of cases: Latin and non-Latin spellings, right-left (e.g. Arabic) and left-right (e.g. Latin, Medieval French and Armenian) reading directions, complex abbreviation system (e.g. Latin and Medieval French) or minimal abbreviation system (e.g. Armenian and Arabic), large variety of hands (e.g. Latin, Medieval French and Armenian) or more restricted dataset (e.g. Arabic).

2 Background and Related Work

2.1 Background

Compared to printed documents, historical manuscripts present several challenges for training recognition models. The variation in individual letter shapes, even within the same hand, coupled with larger variations between scribes, script types, regions and periods, complicate the task of training more general models, as well as creating synthetic data. Moreover, historical manuscripts often present a large spectrum of non standard signs, that include abbreviative markers, diacritics, correction marks, allographs, or ligatures and letter fusions [2,4,5]. In addition, the creation of training data is made more complicated by the required level of expertise for transcribing such documents. Even when transcription exists, the pooling of datasets is made very difficult because there often does not exist a single unambiguous way to encode non standard characters and abbreviations, necessitating many normalisation and mapping efforts between datasets before being able to handle the diversity of the data [3]. Finally,

Fig. 1. Example of lines in Medieval French manuscripts.

physical damage to documents as well as different digitisation methods add a supplementary layer of variation and complexity for recognition tasks (Fig. 1).

To experiment with challenges of varying natures, we choose to work on three different datasets. The Latin and Medieval French dataset offers a large chronological (8th to 15th century) and linguistic span, contains different scripts (Caroline, Gothic scripts) and even early types (incunabula in Gothic types). Moreover, it contains important allographetic and graphematic variation: Latin manuscripts can be heavily abbreviated with a diversity of abbreviations (contractions, suspensions, tachograph signs, etc.), while Medieval French presents important linguistic variation in space and time. Both contain allographs for many letters. The Armenian case is characterised by a strong scarcity of available data, while Maghrebi Arabic dataset presents very cursive scripts, with many ligatures.

2.2 Related Work

Creating synthetic data to improve training of recognition systems has already been done with success for printed materials [9]. GANs appear as a relevant answer to this challenge, as they reach very convincing results in different scenarii. State-of-the-art results rely on a style-transfer approach [14], in which a generator tries to map on a targeted dataset unsupervised learned features from another. Such an approach has already been successfully applied to historical manuscripts to create realistic Latin manuscripts [22] with very constrained styles in training (page-level), or for contemporary cursive hand-writings [11] (line-level with a semi-supervised approach). For contemporary scripts, the IAM dataset is mainly used to evaluate and to compare line or word generation, using ScrabbleGAN at character level [10], or similar approaches at the word level [13], even with transformers [1]. This dataset nevertheless suffers from a low variation in the handwritten styles, and from the low level of noise in the images, in comparison with historical documents. Scrabble's latest enhancements also allow control over the style produced by the generator, which increases the benefits of generating artificial handwritten lines for HTR in order to target a specific hand [12]. GANs have also been successfully applied to other tasks on historical documents and manuscripts, such as artificial colourisation [6].

3 Proposed Method

We evaluate two approaches for line synthesis, ScrabbleGAN and CycleGAN.

3.1 ScrabbleGAN

The first one is based on ScrabbleGAN, whose relevance has already been demonstrated for non-cursive modern scripts [10] and for contemporary Arabic lines [8], on the condition that the script doesn't contain too much ligatures. ScrabbleGAN is based on the simultaneous training of a generator/discriminator, and a CNN used as a recogniser. Basically, this GAN is trained to generate each character independently, and the line is then recomposed by putting each character image end to end. This splitting, which forces the GAN to use the different possible input and output contexts of a character (upper and lower), certainly only makes sense in the context of a very unequal distribution of character classes. The GAN therefore has two contextual informations: the character and its neighborhood, and within the character the possible inputs and outputs. In addition to the traditional discrimination of a GAN between a real and false image, the network must also confirm that the generated image is well readable, beyond its plausible aspect. Our dataset (see *below*) includes a large number of cursive cases and superscript/subscript signs (particularly Latin abbreviations), leading us to introduce a small change in the behaviour of Scrabble: after a first classic training without modification allowing the network to identify the space occupied by a character in a line of text, each character is split into four sub-images (see Fig. 2) in order to force the network to focus on 'paleographic' features shared by letters or specific to some ligatures/abbreviations. In addition, in order to prevent overfitting of the recogniser on the dataset (leading to its ability to recognise a character despite a bad generation), we reset the weights of the recogniser every 100 epochs. Main ScrabbleGAN hyperparameters are kept. We favour the decomposition into strokes because of the large number of ambiguities and ligatures that our datasets present, in particular in Arabic and contemporary Armenian scripts (see below).

One of the foreseable limits, that we wish to investigate, is that this strategy will perhaps find itself also limited regarding the production of under-resourced classes: it is indeed possible that the GAN will perform badly on those classes, similarly to HTR systems (Fig. 3).

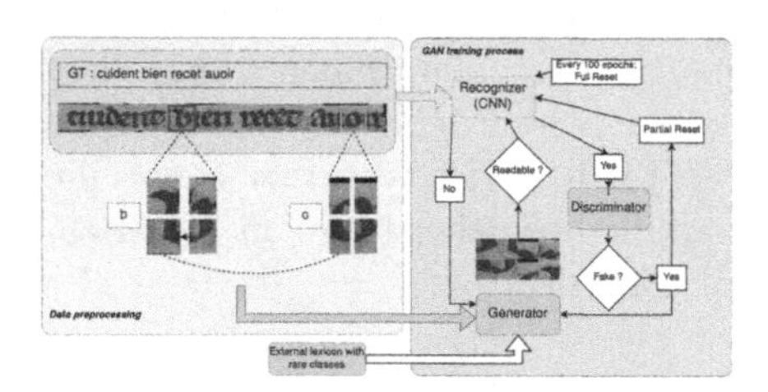

Fig. 2. Training strategy of ScrabbleGAN using a recogniser and a constrained discriminator

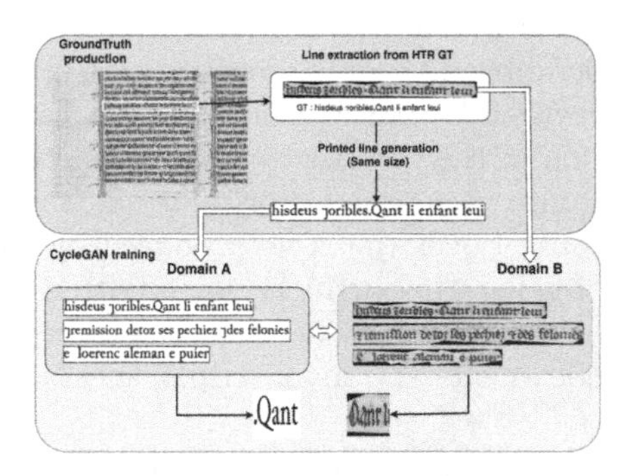

Fig. 3. Training strategy of CycleGAN using non-aligned pair images

3.2 CycleGAN

To try to overcome this problem of under-resourced classes, our second approach is based on CycleGAN [24], which adopts a "style transfer" strategy, in order to apply to target images (domain B) the characteristics of a training dataset (domain A). It is an image-to-image translation strategy, usually applied to unpaired data on generation tasks of a particular style, for example artistic. The use of CycleGAN has already been proven on historical documents, at the page level to generate a look-like Latin manuscript page, with real text [22]. In particular, the authors trained CycleGAN by providing images of manuscripts (domain B) and natively printed PDF texts (domain A), to train GAN to convert from one to the other by giving the PDF a handwritten look. The results produced are quite convincing. The authors use in particular a recogniser in addition to ensure, as for ScrabbleGAN, that the result produced has real relevance and is not simply a decorum added to the PDF. In our case, we limit ourselves to the level of the line of text, by creating a dataset of paired printed and handwritten lines (without correspondence at the pixel level). Our assertion is that it may not be necessary to use a recogniser as backup, and that the task of generating lines could be a task of transferring a handwritten style to a line of printed text. We want to check at this stage if this strategy would make it possible to easily generate lines with under-resourced classes in order to compensate for the lack of data and improve their recognition.

4 Experimental Setup

Training and evaluation of models are performed with Kraken (4.1.2). Training used architecture with CTC Decoding (Fig. 4) with a learning rate of 0.001 halved every 5 epochs of plateau.

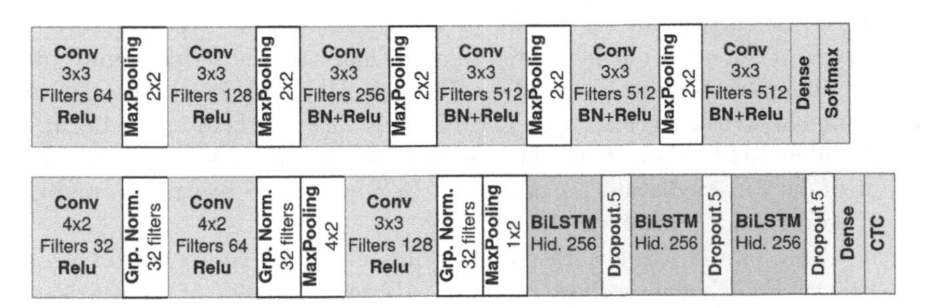

Fig. 4. Differences of architectures used for HTR. Top: Architecture used during GAN training to help discriminator. Bottom: Architecture used for HTR experiments

Latin and Medieval French Experiment. We consider two training subsets for Latin and Medieval French, based on type (print vs. manuscript) and periods (Table 1). Both subsets follow the same "graphetic" transcription guidelines [16]. A graphetic transcription implies the non resolution of abbreviation

Table 1. Composition of Medieval Latin and French (left) and Classical Armenian (right) datasets

Subset	Ref.	Chars*	Lines	Period	Lang
A	[15]	**593k**	22269	12-15th	fro
A	[7]	236k	6040	12-15th	lat
A	[17]	169k	5937	15th	fro
B	[23]	17k	439	9-12th	lat
B	[21]	87k	2826	9th	lat
B	[18]	245k	7607	15th	fro
Total		1149K	45118		

* Space characters excluded.

Manuscript	Period	Script	Lines
M1	1633	*bolorgir*	512
M6	16th	*bolorgir*	819
M982	1460	*bolorgir*	3035
MAF52	15-16th	*bolorgir*	954
MAF54	16th	*bolorgir*	1253
MAF62	18th	*bolorgir*	522
TBI122	17th	*bolorgir*	1209
V. Melkonian Arch.	19th	*Shghagir*	500
V. Trianz Arch.	19th	*Shghagir*	500
Total			9304

and the use of common characters for different shape of signs, *e.g.* long s (ſ) are transcribed as s, or more commonly for abbreviated manuscripts, combining horizontal abbreviating marks are transcribed using a tilde ($\sim$).

Subset A combines three different dataset which have a large amount of data around the end of the Middle Ages in both Latin and French (1100–1500). Latin represents around 23.4% of the subset's characters count but has a far higher abbreviation rate than Medieval French, which implies more rare classes with higher frequency. Subset A is used for generating synthetic data, as it is far more cohesive in terms of written shapes than subset B and more diverse in terms of characters and manuscripts.

Subset B serves as a secondary dataset to augment Subset A and analyze if data augmentation of out-of-domain data improves results. Subset B consists mostly of early medieval Latin data written in Caroline Minuscule as well as one dataset of early print (*incunabula*) written in Gothic types close to the manuscripts of Subset A but far more uniform than its manuscript counterparts.

Four manuscripts (not included in Table 1) are used for evaluation: they present similar writing style as seen in Subset A, three of them are in Latin, one in Medieval French[1]. They are used to test the models capacities (see Fig. 5). The amount of data in Medieval French and Latin allows us to test more conditions than in Arabic and Armenian.

The sigma dataset was generated using out-of-domain transcription of Latin and Medieval French manuscripts, randomly abbreviating 30% of the words using ad-hoc rules. We generated 8,000 lines with character length of 5 to 40 characters.

Arabic. For Arabic, we used the RASAM dataset [20], which contains 7,540 lines and 483,725 characters. The dataset is built from three very different BULAC manuscripts, offering either very dense scripts, or on the contrary with

[1] Folders `Ms_KBR_9232_Examen_Moraux` in CREMMA Medieval; folders `SBB_PK_Hdschr25`, `UBL758` and `BGO-511` in CREMMA Medii Aevi.

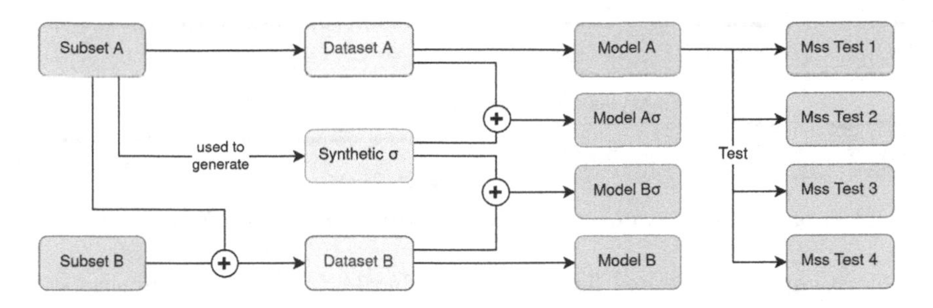

Fig. 5. Roles of subsets and models for the Medieval French and Latin experiment.

very large loops. The dataset nevertheless presents a limited variability, each manuscript being represented equally, guaranteeing for the training of GANs a fair distribution of classes for each targeted style. The number of classes is also limited to 54 (including 7 vocalisation signs). For the training of the GANs, we carried out an equal distribution of the data, ensuring the same number of lines for each manuscript (1725 lines for each).

Armenian. The Armenian dataset (Table 1) is composed of 8 Armenian manuscripts in a very regular *bolorgir* script [19], from the 15th to the 18th century, to which we add two manuscripts in cursive *Shghagir* script from the 19th century.

5 Experiments

5.1 Task 1: Line Generation

Medieval Latin and French. Line generation using ScrabbleGAN for Medieval Latin and French provided encouraging results, as seen in Fig. 6. The model reproduced in part the fact that training data contained black mask and has a tendency to leave very little horizontal padding at the end or beginning of lines. Extremely rare classes of characters have a tendency to be less well represented, such as the sign ÷ (Fig. 6b), used as an abbreviation for *est* in Latin. The model learnt some form of medieval decoration habits, such as colour-shading of *litterae notabiliores* transcribed as uppercase (Fig. 6c).

Line generation using CycleGAN gave however rather bad results. The model focused more on the manuscript colours (back- and foreground) and on the mask than on the shape of the input (See Fig. 7). Lines can sometimes be read but they are mostly indecipherable.

Armenian and Arabic. The GAN generates completely believable lines in Armenian and Arabic (Fig. 8). The lines generated show little variety in the styles produced but appear to have been written by a human. A blurred effect

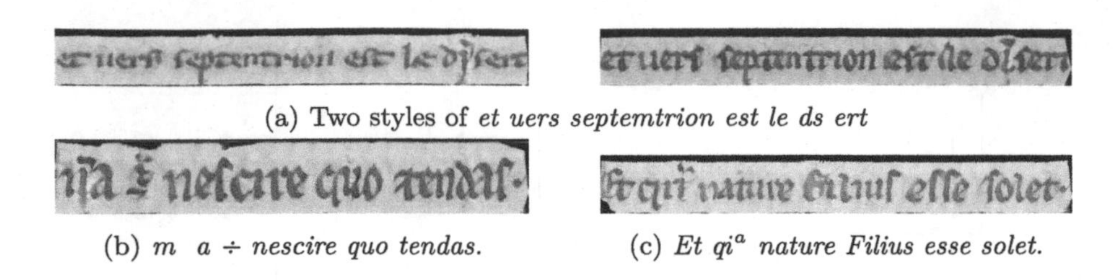

(a) Two styles of *et uers septemtrion est le ds ert*

(b) *m a* ÷ *nescire quo tendas.*

(c) *Et qi*^{a} *nature Filius esse solet.*

Fig. 6. Example of generated lines using ScrabbleGAN in Medieval French (a) and Latin (b–c).

Fig. 7. Example of generated lines using CycleGan.

on the Arabic lines nevertheless betrays the GAN. For the Armenian, the abusive and erroneous use of the abbreviation sign (a superscript horizontal line) also indicates that the GAN randomly generates the latter, which a human would not do. As a general rule, the Arabic GAN has largely specialized on the MS.ARA.417 manuscript.

5.2 Task 2: HTR Usage

Medieval Latin and French. The impact of the HTR fake lines can be considered as non-existant, or a result of local minimums (see Table 2). On the small dataset A & Aσ, the model shows a decrease of 0.1% of the score on the Medieval French dataset, which might be due to an already high score (97.80%). Testing on the Latin dataset, which is the least represented language in the training set A and the most represented one in the σ subset, shows an increase of 2.2% points of the accuracy using fake lines. However, this result is contrasted in the context of the B* models, where the addition of the σ dataset always results in a decrease of the score, including for the Latin testing set.

The analysis of the training curves (Fig. 9a) do not provide us with any indication of a real impact of the fake lines on the model training, specifically given the fact that the *sigma* subset was only used in the training split of the

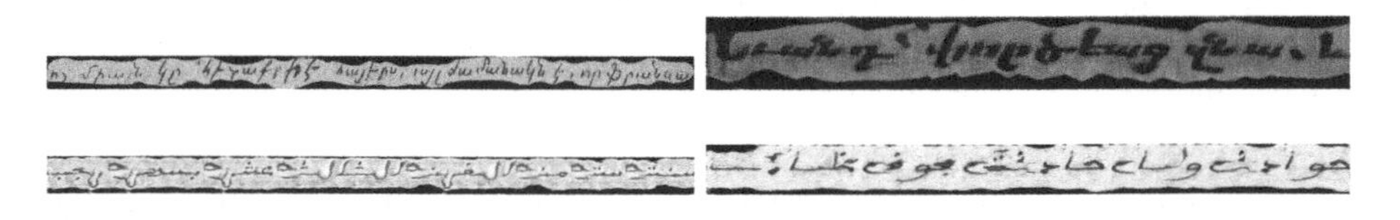

Fig. 8. Example of generated lines using ScrabbleGAN in Armenian (top 2) and Arabic (bottom 2).

Table 2. Accuracy on out of domain test set depending and the language and the training set.

Test set	A	Aσ	B	Bσ
Med. French	**97.78**	97.66	**97.87**	97.70
Latin	87.80	**89.00**	**89.84**	88.61

model. Analysis of the rolling average of the difference between non-sigma and sigma accuracy (Fig. 9b) shows a very limited impact at training time (less than 0.5% of accuracy) on the first 5 to 10 epochs depending on the size of the training set, and a null or negative impact afterwards.

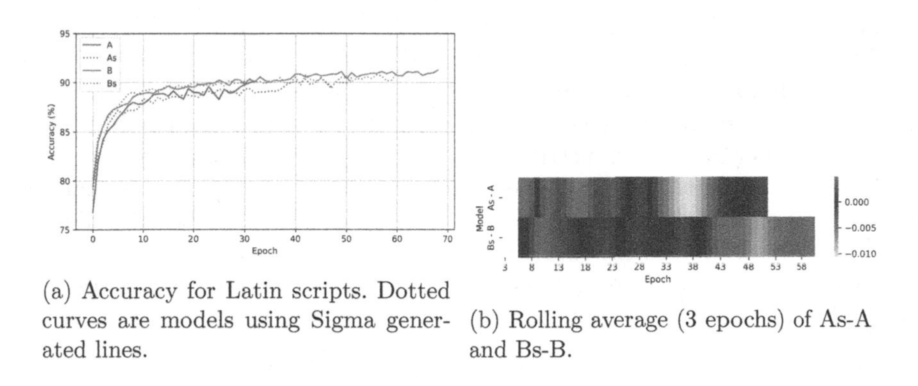

(a) Accuracy for Latin scripts. Dotted curves are models using Sigma generated lines.

(b) Rolling average (3 epochs) of As-A and Bs-B.

Fig. 9. Validation accuracy for Latin

Armenian. At this stage, generated images with their corresponding transcription can already be used as a relevant data augmentation. We have led a very short experiment with the Armenian dataset. We have divided our dataset in 4 subsets: train (827 real lines), val (1.703 real lines), test in-domain (3.772 real lines) and test out-of-domain (2.002 lines). In-domain data are composed of *bolorgir* manuscripts and out-of-domain data composed of *shghagir* scripts. We have performed 4 training:

1. default: train set only
2. default+gan250: train set + 250 generated lines
3. default+gan500: train set + 500 generated lines
4. default+gan1000: train set + 1.000 generated lines

Training curves (Fig. 10) show that we significantly speed-up training with fake images, as we could have with a bigger training dataset. CER achieved by default model and default+gan500 model are respectively 6.13% and 5.63% on in-domain test, and 22.72% and 10.79% on out-of-domain test.

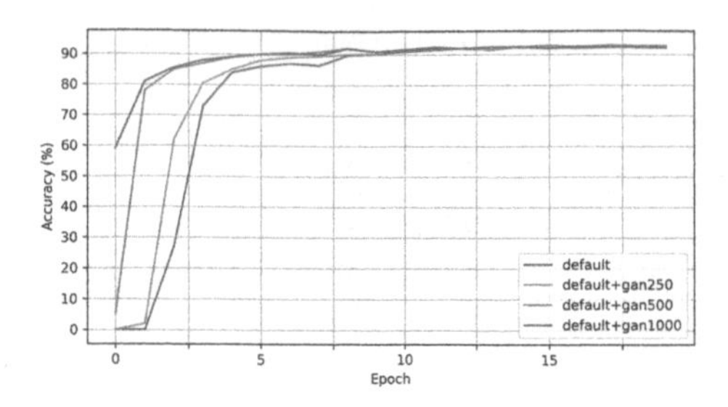

Fig. 10. Validation accuracy on the Armenian dataset.

6 Conclusion

In our experiments, the ScrabbleGan approach proved a lot more effective than the CycleGan approach. The lines generated by the former produce synthetic lines in Medieval French, Latin, Arabic and Armenian that are both usually very legible and similar to lines found in authentic documents. Though we lack, for now, a proper evaluation metric for verisimilitude, a qualitative expert assessment shows that some synthetic lines are hard to differentiate from actual lines encountered in historical documents (N.B.: this type of expert based evaluation could be systematised in the future with a stricter protocol).

Concerning the improvements of our generation as data augmentation techniques, the results tend to show that:

1. in the case of rare character classes in large datasets, the GAN seems as much in trouble to handle them as the HTR engines, and somehow fails to generate plausible outputs that can help in learning said classes. In consequence, the improvements in HTR seem for now insignificant, if any.
2. On the other hand, improvements in case of small datasets (e.g., Armenian dataset) is more substantial. In particular, models tend to learn in less epochs and, more importantly, generalise better, if a small training set (e.g. around 1000 lines) is up to doubled with artificial lines.

Further research could focus on defining metrics for the assessment of generated lines, based on expert consensus and in improvements in automated tasks. In addition, we plan to work on controlling the different 'styles' of the output [12], in order to be able to generate several outputs for the same line, while specifying a given script (*e.g.* Caroline, Gothic *Textualis*,...), a specific hand, etc. The use of diffusion models to generate a precise style is a possible field of experimentation [25].

Beyond mere increase in the scores of automated tasks, artificial lines generation could be useful to paleographers and philologists for goals such as creating virtual restoration of damaged manuscripts or studying the evolution of scripts and their variation.

Acknowledgement. This research was funded by the project "Artificial Pasts: lost texts and manuscripts that never were" (Paris Artificial Intelligence Research Institute – PRAIRIE, SHS call).

References

1. Bhunia, A.K., Khan, S., Cholakkal, H., Anwer, R.M., Khan, F.S., Shah, M.: Handwriting transformers. In: Proceedings of the IEEE/CVF International Conference on Computer Vision, pp. 1086–1094 (2021)
2. Camps, J.B.: La 'Chanson d'Otinel': édition complète du corpus manuscrit et prolégomènes à l'édition critique. thèse de doctorat, dir. dominique boutet, Paris-Sorbonne (2016). https://doi.org/10.5281/zenodo.1116735. https://halshs.archives-ouvertes.fr/tel-01664932
3. Camps, J.B., Vidal-Gorène, C., Stutzmann, D., Vernet, M., Pinche, A.: Data diversity in handwritten text recognition. Challenge or opportunity? In: Digital Humanities 2022, pp. 160–165 (2022)
4. Cappelli, A.: Dizionario di abbreviature latine ed italiane: usate nelle carte e codici specialmente nel medio-evo ripprodotte con oltre 13000 segni incisi. Hoepli (1899)
5. Careri, M., et al.: Album de manuscrits français du XIIIe siècle. Viella (2001)
6. Clérice, T., Pinche, A.: Artificial colorization of digitized microfilms: a preliminary study (2022). https://hal-lirmm.ccsd.cnrs.fr/MOM/hal-03335326v2, preprint
7. Clérice, T., Vlachou-Efstathiou, M., Chagué, A.: CREMMA Medii Aevi: Literary manuscript text recognition in Latin (2023). https://hal-enc.archives-ouvertes.fr/hal-03828353, preprint
8. Eltay, M., Zidouri, A., Ahmad, I., Elarian, Y.: Generative adversarial network based adaptive data augmentation for handwritten Arabic text recognition. PeerJ Comput. Sci. **8**, e861 (2022)
9. Etter, D., Rawls, S., Carpenter, C., Sell, G.: A synthetic recipe for OCR. In: 2019 International Conference on Document Analysis and Recognition (ICDAR), pp. 864–869. IEEE (2019)
10. Fogel, S., Averbuch-Elor, H., Cohen, S., Mazor, S., Litman, R.: ScrabbleGAN: semi-supervised varying length handwritten text generation. In: CVPR 2020, WiDS TLV 2020 (2020). https://www.amazon.science/publications/scrabblegan-semi-supervised-varying-length-handwritten-text-generation
11. Fogel, S., Averbuch-Elor, H., Cohen, S., Mazor, S., Litman, R.: ScrabbleGAN: semi-supervised varying length handwritten text generation (2020). https://doi.org/10.48550/ARXIV.2003.10557
12. Gan, J., Wang, W., Leng, J., Gao, X.: HiGAN+: handwriting imitation GAN with disentangled representations. ACM Trans. Graph. **42**(1) (2022). https://doi.org/10.1145/3550070
13. Kang, L., Riba, P., Wang, Y., Rusiñol, M., Fornés, A., Villegas, M.: GANwriting: content-conditioned generation of styled handwritten word images. In: Vedaldi, A., Bischof, H., Brox, T., Frahm, J.-M. (eds.) ECCV 2020. LNCS, vol. 12368, pp. 273–289. Springer, Cham (2020). https://doi.org/10.1007/978-3-030-58592-1_17
14. Karras, T., Laine, S., Aila, T.: A style-based generator architecture for generative adversarial networks (2018). https://doi.org/10.48550/ARXIV.1812.04948
15. Pinche, A.: Generic HTR Models for Medieval Manuscripts the CREMMALab Project. https://hal.science/hal-03837519, preprint
16. Pinche, A.: Guide de transcription pour les manuscrits du Xe au XVe siècle (2022). https://hal.archives-ouvertes.fr/hal-03697382, working paper

17. Pinche, A., Gabay, S., Leroy, N., Christensen, K.: Données HTR manuscrits du 15e siècle. https://github.com/Gallicorpora/HTR-MSS-15e-Siecle, dataset
18. Pinche, A., Gabay, S., Leroy, N., Christensen, K.: Données HTR incunables du 15e siècle (2022). https://github.com/Gallicorpora/HTR-incunable-15e-siecle, dataset
19. Vidal-Gorène, C., Decours-Perez, A.: A computational approach of Armenian paleography. In: Barney Smith, E.H., Pal, U. (eds.) ICDAR 2021. LNCS, vol. 12917, pp. 295–305. Springer, Cham (2021). https://doi.org/10.1007/978-3-030-86159-9_20
20. Vidal-Gorène, C., Lucas, N., Salah, C., Decours-Perez, A., Dupin, B.: RASAM – a dataset for the recognition and analysis of scripts in Arabic Maghrebi. In: Barney Smith, E.H., Pal, U. (eds.) ICDAR 2021. LNCS, vol. 12916, pp. 265–281. Springer, Cham (2021). https://doi.org/10.1007/978-3-030-86198-8_19
21. Vlachou-Efstathiou, M.: Voss.lat.o.41 - eutyches "de uerbo" glossed (2022). https://github.com/malamatenia/Eutyches, dataset
22. Vögtlin, L., Drazyk, M., Pondenkandath, V., Alberti, M., Ingold, R.: Generating synthetic handwritten historical documents with OCR constrained GANs. In: Lladós, J., Lopresti, D., Uchida, S. (eds.) ICDAR 2021. LNCS, vol. 12823, pp. 610–625. Springer, Cham (2021). https://doi.org/10.1007/978-3-030-86334-0_40
23. White, N., Karaisl, A., Clérice, T.: Caroline minuscule by rescribe (2022). https://github.com/rescribe/carolineminuscule-groundtruth, dataset
24. Zhu, J.Y., Park, T., Isola, P., Efros, A.A.: Unpaired image-to-image translation using cycle-consistent adversarial networks. In: 2017 IEEE International Conference on Computer Vision (ICCV) (2017)
25. Zhu, Y., Li, Z., Wang, T., He, M., Yao, C.: Conditional text image generation with diffusion models. In: Proceedings of the IEEE/CVF Conference on Computer Vision and Pattern Recognition, pp. 14235–14245 (2023)

Is ImageNet Always the Best Option? An Overview on Transfer Learning Strategies for Document Layout Analysis

Axel De Nardin[1(✉)], Silvia Zottin[1], Emanuela Colombi[2], Claudio Piciarelli[1], and Gian Luca Foresti[1]

[1] Department of Mathematics, Informatics and Physics (DMIF), University of Udine, Udine, Italy
{denardin.axel,zottin.silvia}@spes.uniud.it,
{claudio.piciarelli,gianluca.foresti}@uniud.it

[2] Department of Humanities and Cultural Heritage (DIUM), University of Udine, 33100 Udine, Italy
emanuela.colombi@uniud.it
https://www.dmif.uniud.it/

Abstract. Semantic segmentation models have shown impressive performance in the context of historical document layout analysis, but their effectiveness is reliant on having access to a large number of high-quality annotated images for training. A popular approach to address the lack of training data in other domains is to rely on transfer learning to transfer the knowledge learned from a large-scale, general-purpose dataset (e.g. ImageNet) to a domain-specific task. However, this approach has been shown to lead to unsatisfactory results when the target task is completely unrelated to the data employed for the pre-training process, which is the case when working on document layout analysis. For this reason, in the present paper, we provide an overview of domain-specific transfer learning for document layout segmentation. In particular, we show how relying on document-related images for the pre-training process leads to consistently improved performance and faster convergence compared to training from scratch or even relying on a large, general purpose, dataset such as ImageNet.

Keywords: Document Layout Analysis · Fine Tuning Approach · Page Segmentation

1 Introduction

Historical document layout analysis refers to the process of analyzing the structure and organization of historical documents. It involves the extraction and understanding of various layout elements such as texts, headings, paragraphs, images, paratexts and other graphical components present in the document.

A. De Nardin and S. Zottin—Equally contributed.

G. L. Foresti et al. (Eds.): ICIAP 2023 Workshops, LNCS 14366, pp. 489–499, 2024.
https://doi.org/10.1007/978-3-031-51026-7_41

The goal of historical document layout analysis is to interpret the document's layout, which may have been disrupted due to many factors like aging, degradation, or scanning artifacts. By analyzing the layout, humanists and historians can gain valuable insights into the document's content, context, and historical significance [1]. Furthermore, by segmenting the document images into semantic classes, computer scientists can analyze and understand the layout structure of historical documents, as well as develop algorithms and models for tasks such as document understanding, optical character recognition, or text extraction.

In recent years the requests for new methods to analyze historical documents automatically, or as a support for scholars, are growing and to this end machine learning and deep learning techniques are increasingly used. Although the use of deep learning techniques for these tasks are promising, they require a large amount of data to be trained effectively. In the literature, the availability of a sufficient amount of high-quality, annotated data from historical documents is limited. This is mainly due to the specialist nature of the content. Indeed, annotating data requires domain expertise or specialist knowledge, and obtaining Ground Truth (GT) from domain experts can be time-consuming and costly. To address this problem, several alternative approaches have been proposed in the literature to perform the aforementioned task while relying on a limited amount of data. These approaches involve the adoption of unsupervised learning, few-shot learning and transfer learning techniques. Transfer learning, which is the focus of the present work, involves the preliminary training of a model on a large annotated dataset in order to learn a set of baseline parameters that will serve as an initialization for the fine-tuning on the dataset connected to the target task. Transfer learning has become a widely adopted technique in the field of computer vision [2,11,17], especially following the release of large-scale, general-purpose, datasets such as ImageNet [7], CIFAR-10 [12], PASCAL [9] and COCO [13]. While these datasets provide a valuable resource for the pre-training of deep learning models on a wide array of tasks, they fall short when the downstream task is related to a very specific domain (e.g. medical imaging, handwritten document analysis, etc.). For this reason in the present paper we provide a study on the effects of pre-training on domain-specific datasets as opposed to general-purpose ones in the context of document layout analysis of ancient manuscripts. In particular, we show how performing the pre-training step on a dataset that is related to the downstream tasks substantially improves the performance of the model and allows for a faster convergence compared to performing the pre-training on general-purpose dataset or training the model from scratch.

The rest of the paper is organized as follows: Sect. 2 provides a brief review of historical document layout analysis approaches that consider the lack of annotated data; Sect. 3 focuses on presenting a detailed description and setup of the proposed experiments; in Sect. 4 the results of the experiments are presented; finally, in Sect. 5, the conclusions and future works are drawn.

2 Related Works

The lack of large amounts of labeled data in the domain of ancient manuscript analysis is due to the fact that its creation requires the expertise of domain specialists and a significant time and cost investment, especially when the documents are characterized by a complex layout. Therefore, a natural step is to aim at building systems that can achieve good performance while relying on small amounts of annotated data. So far, however, very few examples of such systems are available in the literature.

The work by Tarride et al. [16] introduces a few-shot learning approach called Deep&Syntax for segmenting historical handwritten registers. Few-shot learning is a learning paradigm that enables models to generalize from a limited number of examples, making it suitable for scenarios with limited annotated data. The proposed method employs a hybrid system that leverages recurring patterns to delineate individual records. This hybrid system combines U-shaped networks, commonly used in image segmentation tasks, with logical rules including filtering and text alignment.

A more recent example of a few-shot learning approach for document layout segmentation has been proposed by De Nardin et al. [6], where only two GT images per manuscript are used to train the segmentation model and obtain results comparable to supervised models representing the current state-of-the-art for the task at hand. The proposed framework combines a novel data augmentation approach, used to train a semantic segmentation backbone, with a segmentation refinement module being represented by a traditional computer vision approach for local thresholding, to fully leverage the limited amount of data available while still achieving competitive performance on the Diva-HisDB dataset.

Droby et al. [8] address the issue of limited GT availability by introducing an unsupervised deep learning method for page segmentation. They utilize a Siamese neural network to distinguish between patches based on their measurable properties, specifically focusing on the number of foreground pixels. The objective is to ensure that spatially adjacent patches exhibit similarities in their properties. Once the network is trained, these learned features are employed for page segmentation.

Finally, Studer et al. [15] proposed a comprehensive study focusing on transfer learning and in particular the application of ImageNet pre-trained models for a variety of tasks connected to historical document image analysis, such as layout segmentation, writer identification and style classification. Regarding the layout segmentation of ancient documents, this work reports mixed results. While pre-training proves beneficial for some of the document classes, it results in lower performance compared to training the models from scratch in other cases.

In the present paper, we expand the analysis of the effectiveness and potential benefits of using transfer learning techniques for the layout analysis of ancient manuscripts compared to training the model from scratch. In particular, we highlight the benefits of using pre-training on specialized domain datasets, compared to using pre-training on ImageNet.

3 Methods

This section provides a comprehensive overview of the chosen architecture and training setup, as well as the detailed datasets used for the experiments and evaluation metrics.

3.1 Model Architecture

For our experiments, we chose a recent and popular model for semantic segmentation called DeepLabv3+ [5]. This architecture improves upon its predecessor DeepLabv3 [4], by incorporating an encoder-decoder architecture that combines the benefits of both high-resolution and low-resolution features. It utilizes a powerful backbone network, such as ResNet or Xception, as the encoder to extract high-level semantic features from the input image. These features are then fed into an Atrous Spatial Pyramid Pooling module, which captures multi-scale contextual information. In addition, DeepLabv3+ employs a decoder network that refines the segmentation results by upsampling the low-resolution features and combining them with the high-resolution features from the encoder. This decoder network helps to recover spatial details and produce more accurate segmentation maps. This model and DeepLabv3 have already been used in other works on the layout segmentation task of ancient manuscript datasets with excellent results [6,15]. For the experiments presented in this paper, the ResNet50 version was chosen as the backbone of the DeepLabv3+ architecture.

3.2 Datasets

For the pre-training of the selected model, we relied on two datasets. The first one is the popular ImageNet dataset [7], which serves as our general-purpose pre-training baseline. The second one, on the other hand, is a private domain-specific dataset consisting of 150 manually annotated images belonging to three different Latin manuscripts published between the 6th and 12th centuries A.D. Each instance of this dataset has been manually annotated by a humanities expert, which provided a segmentation mask consisting of 6 different classes, namely: main text, decorations, titles, chapter headings and paratext. Out of the 150 images, 120 have been used to train the model and the remaining 30 for the validation step. A sample for each of the manuscript classes and its respective GT is provided in Fig. 1.

To test our approach we chose to use two popular datasets for document layout analysis: Diva-HisDB dataset [14] and the dataset of Bukhari et al. [3]. The layout of these manuscripts is particularly challenging and very different from each other, as visible in Fig. 2.

Diva-HisDB dataset [14], comprises three distinct medieval manuscripts: CB55, CSG18, and CSG863. Each manuscript consists of 40 pages that have been annotated at the pixel level of which 20 form the training set, 10 for the validation set and the remaining 10 for the test set. The semantic segmentation classes are main text, decoration, comment and background.

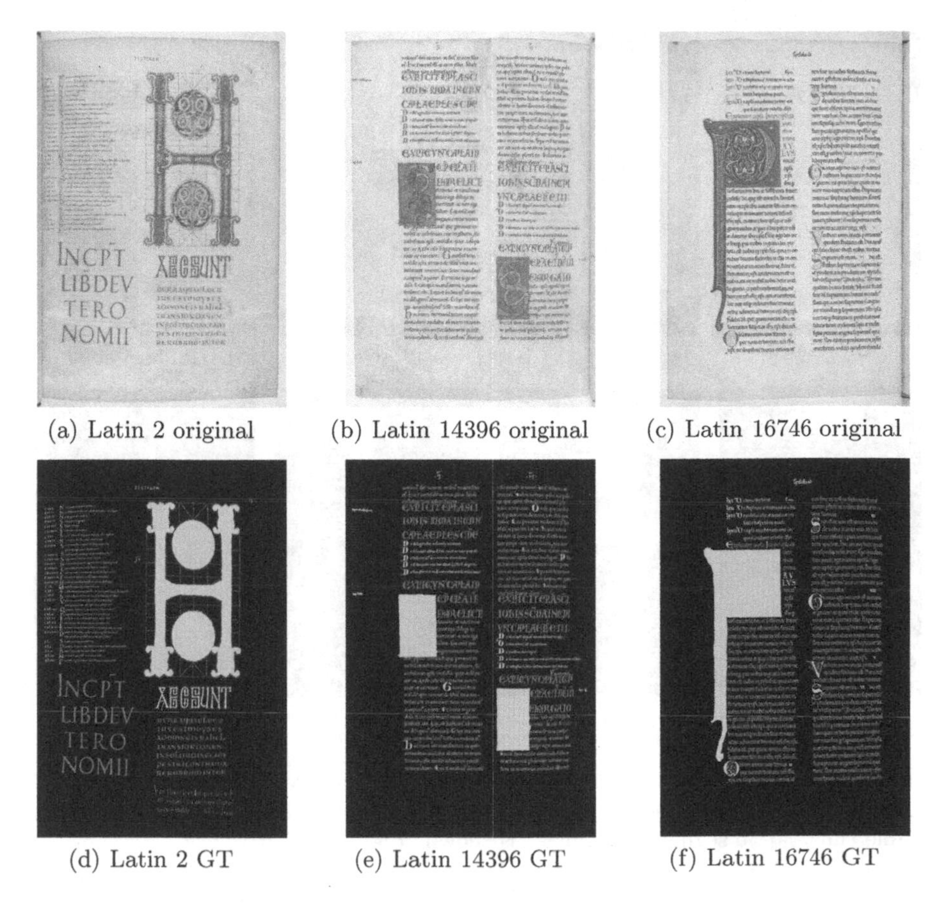

(a) Latin 2 original (b) Latin 14396 original (c) Latin 16746 original

(d) Latin 2 GT (e) Latin 14396 GT (f) Latin 16746 GT

Fig. 1. Images showing a sample instance of each manuscript (Fig. 1(a)– 1(c)) characterizing the adopted pre-training dataset together with its corresponding ground truth mask (Fig. 1(d)– 1(f))

The dataset introduced by Bukhari et al. [3] consists of 32 images, with each image representing a page from one of three distinct Arabic historical manuscripts. For the training process, 24 out of the available samples are used, while the remaining 8 samples are reserved for testing and evaluation purposes. The semantic segmentation classes in this dataset are: main text, side text and background.

3.3 Evaluation

To evaluate the performance of our proposed approach, we employ four metrics, namely Precision, Recall, Intersection over Union (IoU), and F1-Score. In particular, for the present study, we considered the class-wise macro average of the selected metrics. These metrics are computed separately for each manuscript in the Diva-HisDB dataset and are defined as follows:

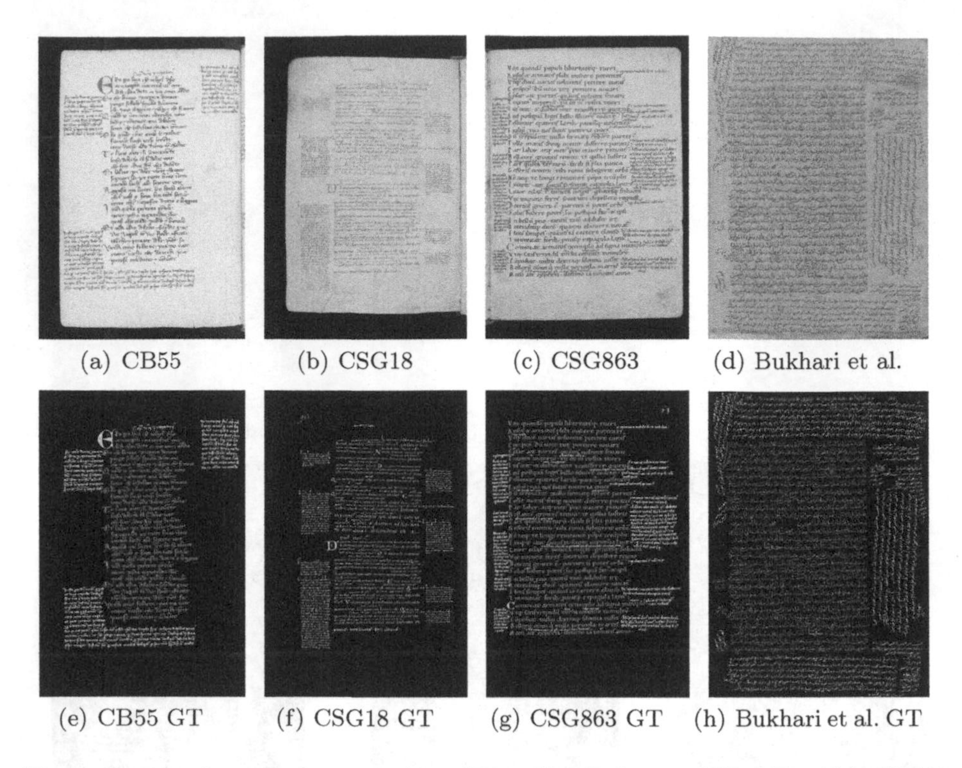

(a) CB55 (b) CSG18 (c) CSG863 (d) Bukhari et al.

(e) CB55 GT (f) CSG18 GT (g) CSG863 GT (h) Bukhari et al. GT

Fig. 2. Samples from the 3 manuscripts of Diva-HisDB dataset (CB55 Fig. 2(a), CSG18 Fig. 2(b), and CSG863 Fig. 2(c)) and from Bukhari et al. dataset (Fig. 2(d)). For each sample the relative segmentation mask is shown (Fig. 2(e)– 2(h))

$$\text{Precision} = \frac{\text{True Positive}}{\text{True Positive} + \text{False Positive}} \tag{1}$$

$$\text{Recall} = \frac{\text{True Positive}}{\text{True Positive} + \text{False Negative}} \tag{2}$$

$$\text{IoU} = \frac{\text{True Positive}}{\text{True Positive} + \text{False Positive} + \text{False Negative}} \tag{3}$$

$$\text{F1-Score} = \frac{2 \times \text{Precision} \times \text{Recall}}{\text{Precision} + \text{Recall}} \tag{4}$$

3.4 Training and Fine-Tuning Setup

The pre-training of the DeepLabv3+ architecture on the private, domain-specific, dataset was carried out by relying on the popular Adam optimizer with a learning rate of 10^{-4}, a weight decay of 10^{-5} and a batch size of 3. The images were only slightly resized to 1344×2016 px, roughly one-third of the original size, to retain as much detail as possible. The same hyper-parameters were also used for the

Table 1. Classes distribution (%) for each manuscript of the private dataset used for domain-specific pre-training.

	Latin 2	Latin 14396	Latin 16746
Background	92.8	89.2	88
Paratext	0.1	0.1	0.3
Decoration	1.5	2	3
Main Text	4.7	7.6	7.8
Title	0.4	0.5	0.1
Chapter Headings	0.5	0.6	0.8

Table 2. Classes distribution (%) for each manuscripts of Diva-HisDB and for Bukhari et al. dataset.

	CB55	CSG18	CSG863	Bukhari et al.
Background	82.41	85.16	77.82	86.07
Comment	8.36	6.78	6.35	4.71
Decoration	0.55	1.47	1.83	—
Text	8.68	6.59	14.00	9.22

fine-tuning process and for training the model from scratch on the dataset of the downstream task.

To address the high-class imbalance in each manuscript of the dataset, we selected a weighted Cross Entropy Loss as the loss function for training the model. Specifically, the weight assigned to each class, denoted as W_i, is calculated as the square root of the reciprocal of the occurrence frequency of the corresponding class in the dataset (as shown in Eq. 5). The frequencies F_i of each class in the private dataset are shown in Table 1.

$$W_i = \sqrt{\frac{1}{F_i}} \tag{5}$$

The same loss function was used for training the model from scratch and for the fine-tuning process, adjusting the weights to the respective segmentation classes present in the Bukhari et al. and Diva-HisDB datasets (Table 2).

The pre-trained ImageNet model relied on the one made available in the Segmentation Models Pytorch repository [10].

On the other hand, when pre-training on the domain-specific dataset we allowed a maximum of 500 epochs while introducing an early stop mechanism that was triggered in case the performance of the model on the validation set didn't improve in the previous 20 epochs. Finally, when training the models from scratch and for the fine-tuning process, the architectures were trained for a total of 100 epochs, with a checkpoint saved every 10. The fine-tuning process, in particular, was carried out by freezing the weights of the encoder, which performs

Table 3. Results on the test set of the different manuscripts of Diva-HisDB and Bukhari et al. datasets. Each metric is calculated for training from scratch and for fine-tuning of ImageNet pre-training and domain specific pre-training. The results regarding the effect of fine-tuning from ImageNet and from domain-specific is reported in Δ.

Manuscript	Metric	Scratch	Pre-Trained (ImageNet)	Pre-Trained (domain specific)	Δ (ImageNet)	Δ (domain specific)
Bukhari	Precision	0.730	0.738	**0.780**	+0.008	+0.050
	Recall	**0.910**	0.906	0.902	−0.004	−0.008
	IoU	0.678	0.685	**0.706**	+0.007	+0.028
	F1-Score	0.798	0.802	**0.818**	+0.004	+0.020
CB55	Precision	0.722	0.703	**0.737**	−0.019	+0.015
	Recall	0.914	**0.942**	0.920	+0.028	+0.006
	IoU	0.681	0.673	**0.698**	−0.008	+0.017
	F1-Score	0.799	0.791	**0.812**	−0.008	+0.013
CSG18	Precision	0.730	0.735	**0.740**	+0.005	+0.010
	Recall	0.917	**0.936**	0.924	+0.019	+0.007
	IoU	0.698	0.700	**0.702**	+0.002	+0.004
	F1-Score	0.813	**0.815**	**0.815**	+0.002	+0.002
CSG863	Precision	0.778	0.786	**0.792**	+0.008	+0.014
	Recall	0.891	0.896	**0.898**	+0.005	+0.007
	IoU	0.719	0.735	**0.736**	+0.016	+0.017
	F1-Score	0.828	0.834	**0.839**	+0.006	+0.011

the feature extraction, while updating the weights of the decoder module of the network together with the new segmentation head, introduced to match the number of classes present in the downstream dataset.

4 Results

In Table 3 is reported the performance of the DeepLabv3+ model when trained from scratch on the target datasets and when pre-trained on both the ImageNet and the domain-specific datasets. As we can see the model pre-trained on the domain-specific dataset consistently outperforms the other 2 across all the selected metrics on all the manuscript classes, with the exception of the recall on the Bukhari et al. dataset. In particular, domain-specific pre-training leads to a 1% average improvement on the classes belonging to the Diva-HisDB dataset. Surprisingly, this strategy proved to be even more effective for the Bukhari et al. dataset, which is the only one containing documents written in an alphabet compared to the one present in the pre-training dataset, with improvements going from a 2% on the F1-score metric to a 5% for the Precision.

Furthermore, in Fig. 3 we report the performance, in terms of IoU, at different stages during the training process on the downstream task. As we can see, pre-training on a domain-specific dataset (blue line), leads to improved performance in the early stages of training, allowing for a faster convergence on all the classes with the exception of the Arabic manuscripts contained in the Bukhari dataset, for which the curves described by the 3 models appear very similar to each other. Conversely, pre-training the model on a general-purpose dataset such as ImageNet doesn't seem to have the same positive effects, on the contrary, it seems to lead to a decreased performance in the initial phase of

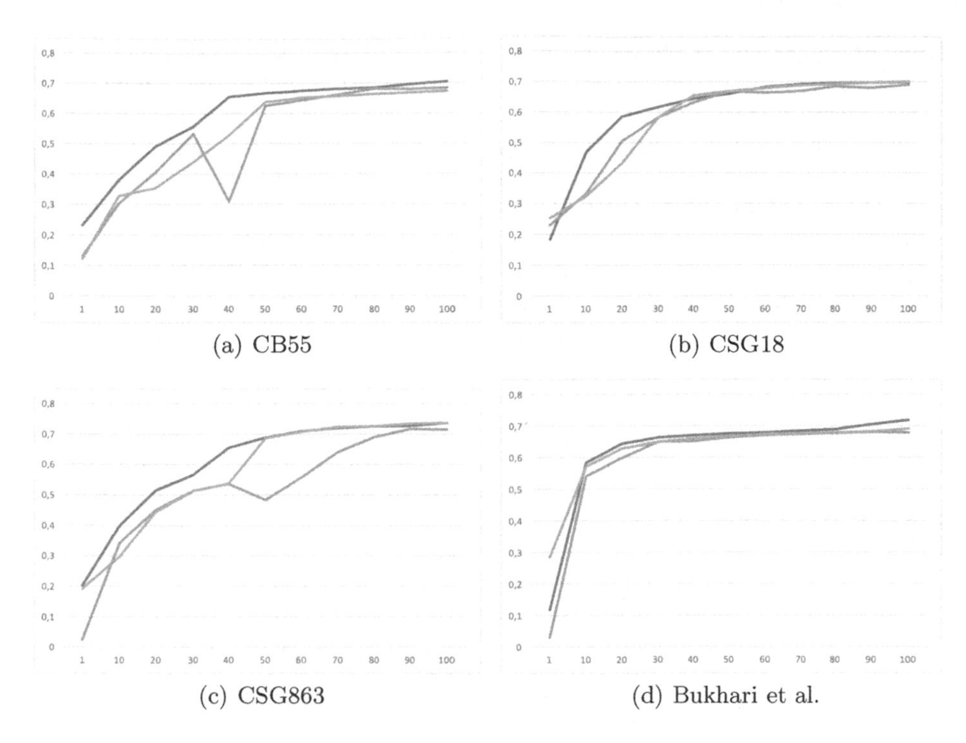

(a) CB55

(b) CSG18

(c) CSG863

(d) Bukhari et al.

Fig. 3. Overview of the performance of the DeepLabv3+ architecture at different stages of the training process for the 4 document classes. The orange line represents the model trained from scratch on the target datasets, the gray line the model pre-trained on ImageNet and the blue line the model pre-trained on the domain-specific dataset (Color figure online)

the training for two of the classes, namely the CB55 and CSG18 ones. Finally, pre-training on a domain-specific dataset consistently leads to a more stable performance curve during training. This effect is particularly visible for the CB55 and CSG863 document classes where the model trained from scratch presents some very noticeable downward spikes in performance around halfway during the training process.

5 Conclusion and Future Work

In the present work, we have shown the advantages of relying on a domain-specific dataset for the pre-training of models in the context of handwritten document layout segmentation. Even when relying on a small amount of data for this purpose the performance of the final model, fine-tuned on the target dataset, is consistently improved compared with the one of a model that has been trained from scratch on the latter or even compared with a model pre-trained on a large scale, general purpose, dataset such as ImageNet. While the obtained results are already promising we believe that an important limiting factor of the present study has been represented by the relatively small amount of data present in the private dataset we used for the pre-training, which is not even comparable to the large datasets commonly used for this purpose. A further limitation is represented by the homogeneity of the said dataset, mainly in terms of the alphabet in it contained. For these reasons in future works, we plan to expand and make publicly available the dataset employed in this work so as to understand the effects it would have on the performance achieved on the target task as well as to make the results obtained reproducible. Finally, we would like to gain a better understanding of the effectiveness of employing a transfer learning approach in a cross-task scenario where the model is pre-trained on a set document analysis task and fine-tuned on a different one.

Acknowledgements. Partial financial support was received from Piano Nazionale di Ripresa e Resilienza (PNRR) DD 3277 del 30 dicembre 2021 (PNRR Missione 4, Componente 2, Investimento 1.5) - Interconnected Nord-Est Innovation Ecosystem (iNEST).

References

1. Andrist, P.: Toward a definition of paratexts and paratextuality: the case of ancient Greek manuscripts, pp. 130–150. De Gruyter, Berlin, Boston (2018). https://doi.org/10.1515/9783110603477-010
2. Brodzicki, A., Piekarski, M., Kucharski, D., Jaworek-Korjakowska, J., Gorgon, M.: Transfer learning methods as a new approach in computer vision tasks with small datasets. Found. Comput. Decision Sci. **45**(3), 179–193 (2020). https://doi.org/10.2478/fcds-2020-0010
3. Bukhari, S.S., Breuel, T.M., Asi, A., El-Sana, J.: Layout analysis for Arabic historical document images using machine learning. In: 2012 International Conference on Frontiers in Handwriting Recognition, pp. 639–644 (2012). https://doi.org/10.1109/ICFHR.2012.227
4. Chen, L., Papandreou, G., Schroff, F., Adam, H.: Rethinking atrous convolution for semantic image segmentation. CoRR abs/1706.05587 (2017)
5. Chen, L.C., Zhu, Y., Papandreou, G., Schroff, F., Adam, H.: Encoder-decoder with atrous separable convolution for semantic image segmentation. In: Ferrari, V., Hebert, M., Sminchisescu, C., Weiss, Y. (eds.) Computer Vision - ECCV 2018, pp. 833–851. Springer International Publishing, Cham (2018)
6. De Nardin, A., Zottin, S., Piciarelli, C., Colombi, E., Foresti, G.L.: Few-shot pixel-precise document layout segmentation via dynamic instance generation and local thresholding. Int. J. Neural Syst. **33**(10), 2350052 (2023). https://doi.org/10.1142/S0129065723500521, PMID: 37567858
7. Deng, J., Dong, W., Socher, R., Li, L.J., Li, K., Fei-Fei, L.: Imagenet: A large-scale hierarchical image database. In: 2009 IEEE Conference on Computer Vision and Pattern Recognition, pp. 248–255 (2009). https://doi.org/10.1109/CVPR.2009.5206848
8. Droby, A., Barakat, B.K., Madi, B., Alaasam, R., El-Sana, J.: Unsupervised deep learning for handwritten page segmentation. In: 2020 17th International Conference on Frontiers in Handwriting Recognition (ICFHR), pp. 240–245. Dortmund, Germany (2020). https://doi.org/10.1109/ICFHR2020.2020.00052

9. Everingham, M., Gool, L.V., Williams, C.K.I., Winn, J.M., Zisserman, A.: The pascal visual object classes (voc) challenge. Int. J. Comput. Vision **88**, 303–338 (2010)
10. Iakubovskii, P.: Segmentation models pytorch (2019). https://github.com/qubvel/segmentation_models.pytorch
11. Kornblith, S., Shlens, J., Le, Q.V.: Do better imagenet models transfer better? In: Proceedings of the IEEE/CVF Conference on Computer Vision and Pattern Recognition (CVPR) (June 2019)
12. Krizhevsky, A., Hinton, G., et al.: Learning multiple layers of features from tiny images (2009)
13. Lin, T.Y., et al.: Microsoft coco: Common objects in context. In: Fleet, D., Pajdla, T., Schiele, B., Tuytelaars, T. (eds.) Computer Vision - ECCV 2014, pp. 740–755. Springer International Publishing, Cham (2014)
14. Simistira, F., Seuret, M., Eichenberger, N., Garz, A., Liwicki, M., Ingold, R.: Diva-hisdb: a precisely annotated large dataset of challenging medieval manuscripts. In: 2016 15th International Conference on Frontiers in Handwriting Recognition (ICFHR), pp. 471–476. Shenzen, China (2016). https://doi.org/10.1109/ICFHR.2016.0093
15. Studer, L., et al.: A comprehensive study of imagenet pre-training for historical document image analysis. In: 2019 International Conference on Document Analysis and Recognition (ICDAR), pp. 720–725 (2019). https://doi.org/10.1109/ICDAR.2019.00120
16. Tarride, S., Lemaitre, A., Coüasnon, B., Tardivel, S.: Combination of deep neural networks and logical rules for record segmentation in historical handwritten registers using few examples. Int. J. Doc. Anal. Recogn. (IJDAR) **24**(1), 77–96 (2021). https://doi.org/10.1007/s10032-021-00362-8
17. Zhuang, F., et al.: A comprehensive survey on transfer learning. Proc. IEEE **109**(1), 43–76 (2021). https://doi.org/10.1109/JPROC.2020.3004555

Correction to: Vision Transformers for Breast Cancer Histology Image Classification

Giulia L. Baroni, Laura Rasotto, Kevin Roitero, Ameer Hamza Siraj, and Vincenzo Della Mea(✉)

Correction to:
Chapter 2 in: G. L. Foresti et al. (Eds.): *Image Analysis and Processing - ICIAP 2023 Workshops*, LNCS 14366, https://doi.org/10.1007/978-3-031-51026-7_2

In the originally published version of chapter 2, the author name V. Della Mea has been updated to Vincenzo Della Mea.

The updated version of this chapter can be found at
https://doi.org/10.1007/978-3-031-51026-7_2

G. L. Foresti et al. (Eds.): ICIAP 2023 Workshops, LNCS 14366, p. C1, 2024.
https://doi.org/10.1007/978-3-031-51026-7_42

Author Index

G. L. Foresti et al. (Eds.): ICIAP 2023 Workshops, LNCS 14366, pp. 501–505, 2024.
https://doi.org/10.1007/978-3-031-51026-7

Printed in the United States
by Baker & Taylor Publisher Services

Printed in the United States
by Baker & Taylor Publisher Services